Disaster Risk Reduction

Methods, Approaches and Practices

Series Editor

Rajib Shaw, Keio University, Shonan Fujisawa Campus, Fujisawa, Japan

Disaster risk reduction is a process that leads to the safety of communities and nations. After the 2005 World Conference on Disaster Reduction, held in Kobe, Japan, the Hyogo Framework for Action (HFA) was adopted as a framework for risk reduction. The academic research and higher education in disaster risk reduction has made, and continues to make, a gradual shift from pure basic research to applied, implementation-oriented research. More emphasis is being given to multi-stakeholder collaboration and multi-disciplinary research. Emerging university networks in Asia, Europe, Africa, and the Americas have urged process-oriented research in the disaster risk reduction field. With this in mind, this new series will promote the output of action research on disaster risk reduction, which will be useful for a wide range of stakeholders including academicians, professionals, practitioners, and students and researchers in related fields. The series will focus on emerging needs in the risk reduction field, starting from climate change adaptation, urban ecosystem, coastal risk reduction, education for sustainable development, community-based practices, risk communication, and human security, among other areas. Through academic review, this series will encourage young researchers and practitioners to analyze field practices and link them to theory and policies with logic, data, and evidence. In this way, the series will emphasize evidence-based risk reduction methods, approaches, and practices.

All submitted proposals are carefully reviewed by the series editor.

Publication Ethics and Publication Malpractice Statement: https://www.springernature.com/gp/policies/book-publishing-policies

Ethical, Conflict-of-Interest, and Editorial Policy Statement: https://www.springer.com/us/authors-editors/journal-author/journal-author-helpdesk/before-you-start/before-you-start/1330#c14214

Pritam Ghosh · Soumita Banerjee ·
Ranit Chatterjee · Rajib Shaw
Editors

Climate Crises and Public Health Challenges

With a Focus on South and Southeast Asia

Springer

Editors
Pritam Ghosh
Department of Geography
Hijli College
Kharagpur, West Bengal, India

Soumita Banerjee
Amity University
Kolkata, West Bengal, India

Ranit Chatterjee
RIKA Institute
Noida, Uttar Pradesh, India

Rajib Shaw
Graduate School of Media and Governance
Keio University
Fujisawa, Kanagawa, Japan

ISSN 2196-4106 ISSN 2196-4114 (electronic)
Disaster Risk Reduction
ISBN 978-981-95-9708-6 ISBN 978-981-95-9709-3 (eBook)
https://doi.org/10.1007/978-981-95-9709-3

This Springer imprint is published by the registered company Springer Nature Singapore Pte Ltd.
The registered company address is: 152 Beach Road, #21-01/04 Gateway East, Singapore 189721, Singapore

Preface

This century has been marked by compounding crises, of which climate change alone stands as both a catalyst and an effect of risks. These changing aspects usually get more visible in South and Southeast Asia than in any other region of the world, where most of the population lives in the coastal megacities and by default resides at the forefront of climatic extremes while concurrently trying to cope with developmental changeovers and health disparities. In this context, the idea for *Climate Crises and Public Health Challenges—With a Focus on South and Southeast Asia* was born out of a collective acknowledgement among scholars and practitioners that climate change is essentially a public health encounter as much as it is an ecological or developing one. This volume begins from an interdisciplinary discussion that carries together contributors from diverse theoretical and professional experiences, each of whom delivers standpoints that are at once local and yet globally pertinent. The chapters in this book validate how the climate–health link is facilitated by geography, governance, and social vulnerability. They also prompt us that climate change influences are erratically dispersed, unreasonably encumbering sidelined groups, including women, children, migrant workers, and indigenous members. The encounter, therefore, is not merely a technical version but also a structural alteration toward equality, justice, and resilience.

A noteworthy feature of this volume is its focal point being on the empirically significant case studies from across South and Southeast Asia. Ranging from the urban heat islands in megacities and coastal flooding in low-lying deltas to the revival of vector-borne diseases in tropical settings, the book is able to hold a wide spectrum of trials. It also permits the grassroots techniques, indigenous adaptation practices, and institutional feedback that propose lessons beyond provincial boundaries. These understandings are very crucial at a time when global policy frameworks—such as the Sendai Framework for Disaster Risk Reduction, the Paris Agreement, and the Sustainable Development Goals—emphasize the call for cohesive and inclusive approaches.

As editors, we interpret this work as both an academic involvement and an implementational resource. It is envisioned to extend theoretical discussions in climate and health scholarship, to assist as a reference for students and researchers, and to deliver evidence-based understandings for policymakers. It is also our optimism that the work will kindle serious interactions that would be cross-disciplinary and beyond sectoral boundaries, and that it will cheer more comprehensive arrangements of knowledge co-production that esteem local insight while appealing with global science.

We sincerely extend our gratitude to every scholarly author who has brought their knowledge, passion, and devotion to this project. Their work aptly underlines the importance of collective scholarship in tackling challenges of such extent. We are also obliged to Springer Nature, under whose *Disaster Risk Reduction* series this book is published, for delivering a stand that advances both theoretical rigor and practical significance.

In the end, we would like to convey that this book summons to reconsider the behaviors we comprehend the links between climate and health, to distinguish the insistence of building adaptive abilities, and to act inclusively to confirm a future that is both recovered and more resilient.

Kharagpur, India Pritam Ghosh
Kolkata, India Soumita Banerjee
Noida, India Ranit Chatterjee
Fujisawa, Japan Rajib Shaw

About the Book

Climate Crises and Public Health Challenges—With a Focus on South and Southeast Asia is an inclusive scholarly book that dives into one of the most alarming issues today: the context among climate change, disaster risk, and public health. South and Southeast Asia, a home to over two billion people, falls under the category of the most globally exposed regions to climate adversities. The region often faces a sequence of cyclones, resulting in floods, erratic monsoons, droughts, heatwaves, and landslides, while also battling rapid and haphazard urban growth, dense populations, widening socio-economic inequalities, and inadequate, fragile healthcare systems. This multipart interface makes the region a vital location for understanding how climate crises lead to health problems, and how health vulnerabilities can increase disaster risks.

This edited collection usually draws attention from multiple sources along with multiple facets of the climate–health interface. It tries to explore in a new way it addresses the direct impacts of climate excesses on physical and mental health; the growth of vector-borne and waterborne diseases under ever-changing ecological conditions; challenges of food security and nutrition; and the stress placed on public health systems during and after disasters. Equally, the book explores the indirect and systemic effects of climate change—such as forced migration, gendered vulnerabilities, and mental health stresses—that are often underrepresented in mainstream policy discourses.

The enriching contributions to the volume are rooted in several interdisciplinary perspectives, encompassing geography, disaster studies, epidemiology, urban planning, and environmental governance. Pragmatic case studies from India, Bangladesh, Sri Lanka, Indonesia, the Philippines, and other countries have provided acute and comparative understandings into region-specific backdrops. The book underlines not only the setbacks but also pioneering coping strategies, extending from community-based disaster preparedness and nature-based solutions to unified policy frameworks. By placing the indigenous experiences within a much wider regional and global debate on climate change and health, this work aims to inform academic research, policymaking, and practice. It will be of immense value to scholars and students in geography, disaster risk reduction, climate studies, public health, and sustainable development, as well as to policymakers, NGOs, and international bodies looking to foster resilience and equity in climate-vulnerable regions.

Reviewers

Satyabrata Mondal Department of Geography, Banwarilal Bhalotia College, Asansol, India
Binay Krishna Pal Department of Geography, Sitalkuchi College, Coochbehar, India
Bikash Das Department of Geography, School of Education, Central University of Kashmir, Ganderbal, Jammu & Kashmir, India
Shahnaz Basheer Loyola College of Social Sciences, Trivandrum, Kerala, India
Sourav Chowdhury Department of Geography, Shishuram Das College, South 24 Parganas, India
Moslem Hossain Department of Geography, School of Earth Sciences, Central University of Karnataka, Kalaburagi, India
Mriganka Dolui Department of Geography, School of Earth Sciences, Central University of Karnataka, Kalaburagi, India
Arpita Trivedy Department of Geography, Aliah University, Kolkata, West Bengal, India
Mahashweta Chakrabarty Department of Geography, Banaras Hindu University, Varanasi, India
Sweta Chatterjee School of Water Resources Engineering, Jadavpur University, Kolkata, India
Naseem Nawaz Institute of Geography, University of the Punjab Lahore, Pakistan
Amiya Gayen Centre for Disaster Management and Research at IITG, Guwahati, Assam, India
Sakhi Roy Amity School of Economics, Amity University, Kolkata, India
Pamela Deb Amity Institute of Social Science, Amity University, Kolkata, India
Tithee Dutta Amity Institute of Social Science, Amity University, Kolkata
Samik Chakraborty Manbhum Mahavidyalaya, Purulia
Kunaljeet Roy Department of Humanities and Social Sciences, BITS-Pilani (Hyderabad)
Rameshwar Mukherjee Department of Geography, Samsi College, Malda

Contents

Editors and Contributors

About the Editors

Dr. Pritam Ghosh is an Assistant Professor and HOD at the Department of Geography, Hijli College, Kharagpur, Paschim Medinipur, India, and a Human Geographer with a Doctoral Degree from the University of Calcutta, Kolkata, India. His area of interest covers Medical and Health Geography, Epidemiology, and Public Health. As a dedicated researcher specializing in Health Geography and Social Sciences, his work is centred around investigating the geospatial, sociocultural, and environmental dimensions of public health. His research critically examines issues like childhood malnutrition, infant feeding practices, access to water, sanitation, and hygiene, and the role of allied public policies in India. He also contributes to emerging discourses on climate change and health, particularly in South Asia. With over 35 peer-reviewed international journal articles and several book chapters to his name, Dr. Ghosh is a prolific scholar and thought leader in his field. He currently serves as an Academic Editor for *PLOS ONE* and *Discover Social Science and Health Journal*, and reviews more than 35 international journals spanning geography, public health, and the social sciences. He is also serving as an Editor for several books on health geography under contract with a leading international publisher.

Dr. Soumita Banerjee is an Assistant Professor in the Department of Geography at the Amity School of Social Sciences, Amity University Kolkata. She earned her doctoral degree in Disaster Management from the Department of Geography, Jadavpur University, Kolkata. Her research expertise lies in Disaster Management and Urban Environmental Geography, focusing on the causes and socio-cultural impacts of urban hazards and disasters. Her work delves into key aspects of urban infrastructure, such as drainage systems, waste management, and community engagement. She adopts a multidisciplinary approach to advance urban environmental research, aiming to develop effective disaster management strategies and climate change adaptation policies. She received a Gold Medal for her Master's degree in Geography from the

University of Calcutta and has been honored with prestigious international fellowships from DAAD and ERASMUS MUNDAS. Her scholarly output includes 11 publications spanning journals, esteemed science magazines, and an edited book. She has also presented eight papers at both international and national conferences. She has been invited to speak on various topics, including floods, the fundamentals of geography, and Sustainable Development Goals (SDGs). Additionally, she serves as a reviewer for several journals and books with prominent publishing houses.

Dr. Ranit Chatterjee is an architect-turned-disaster management professional with a Ph.D. in Environmental Management from Kyoto University, Japan. With over a decade of experience, he has contributed to a range of projects from architectural planning and conservation to disaster risk reduction across Asia, including India, Japan, Nepal, and Laos. His diverse career includes collaborations with UN organizations, national and local governments, the private sector, and academic institutions. He is a co-founder of RIKA Institute, a research-driven startup focused on generating evidence for disaster risk reduction and climate change adaptation. He is a senior policy researcher at Keio University, Japan, and a visiting professor at Rashtriya Raksha University, India. Ranit is an Integrated Research on Disaster Risk (IRDR) Young Scientist Fellow, an IAS residential fellow at Loughborough University, a CEM member of the IUCN business and biodiversity group, and an Advisory member of UNDRR's Stakeholder Engagement Mechanism (SEM). He has authored more than 35 academic journals and book chapters and has contributed to various UN technical reports.

Prof. (Dr.) Rajib Shaw is a professor at Graduate School of Media and Governance of Keio University, Japan. He is the co-founder of a Delhi (India)-based social entrepreneur startup Resilience Innovation Knowledge Academy (RIKA), and RIKA Institute. He is the Co-chair of the United Nations Asia-Pacific Science Technology Advisory Group (AP-STAG) and was CLA for IPCC's 6th Assessment Report. Professor Shaw is the recipient of prestigious "Pravasi Bharatiya Samman Award (PBSA)" in 2021 for his contribution to the education sector. He also received the United Nations Sasakawa Award in 2022 for his lifetime contributions in the field of disaster risk reduction. He published 86 books and over 450 research papers. More about his work can be found in www.rajibshaw.org.

Contributors

Mohamed A. E. AbdelRahman Division of Environmental Studies and Land Use, National Authority for Remote Sensing and Space Sciences (NARSS), Cairo, Egypt

Maisarah Abdul Mutalib School of Graduate Studies, Postgraduate Centre, Management and Science University, University Drive, Off Persiaran Olahraga,

Shah Alam, Selangor, Malaysia;
International Medical School, Management and Science University, University Drive, Off Persiaran Olahraga, Shah Alam, Selangor, Malaysia

Sarah Afsari North South University, Dhaka, Bangladesh

Tanvir Ahmad International Centre for Diarrhoeal Disease Research, Mohakhali, Dhaka, Bangladesh

Waqas Ahmad Department of Clinical Sciences, University of Veterinary and Animal Sciences, Lahore, Narowal Campus, Pakistan

Mumtaz Alam Department of Social Sciences, College of Humanities and Education, Fiji National University, Nasinu, Fiji

Mahin Al Nahian Initiative for Climate Change and Environment, International Centre for Diarrhoeal Disease Research, (icddr,b), Mohakhali, Dhaka, Bangladesh; International Centre for Diarrhoeal Disease Research, Mohakhali, Dhaka, Bangladesh

K. H. Anagha Indian Institute of Technology, Roorkee, Uttarakhand, India

S. Elika Assumi National Law University Meghalaya, Shillong, Meghalaya, India

Azliyana Azhari Monash Climate Communication Hub (MCCH), School of Arts and Social Sciences, Monash University Malaysia, Subang Jaya, Malaysia

Puteri Amirah Nabilah Azman Institute for Environment and Development (LESTARI), Universiti Kebangsaan Malaysia (UKM), Bangi, Selangor, Malaysia

Soumita Banerjee Amity School of Social Science, Amity University, Kolkata, India

Prianka Barman International Centre for Diarrhoeal Disease Research, Mohakhali, Dhaka, Bangladesh

Soham Basu Department of Forest Ecology, Mendel University, Brno, Czech Republic

Muhammad Mubashar Beig Institute of Microbiology, University of Veterinary and Animal Sciences, Lahore, Pakistan

Bhupal Bhattacharya Department of Law, Raiganj University, Raiganj, West Bengal, India

Sakshi Bhutani Chandigarh, India

Tanisha Biswas Department of Dental Surgery, Sankalchand Patel University (SPU), Visnagar, Gujarat, India

V. Chaitra School of Law, Christ University, Bengaluru, Karnataka, India

Akash Chakraborty School of Environment and Disaster Management, Ramakrishna Mission Vivekananda Educational and Research Institute, Kolkata, West Bengal, India

Ayeshik Chakraborty Department of Public Health, Institute of Public Health Kalyani, The West Bengal University of Health Sciences, Kolkata, West Bengal, India

Ranit Chatterjee Co-Founder and Director, RIKA Institute, New Delhi, India

Pamela Deb Department of Geography, Amity Institute of Social Sciences, Amity University Kolkata, Kolkata, India

Arunabha Dey Department of Geography, McGill University, Montreal, Canada

Basil N. D. Diengdoh National Law University Meghalaya, Shillong, Meghalaya, India

Arjun Dubey Institute for Human Development, Delhi, India

Vibhor Dudhraj School of Health and Allied Sciences, Sohna, Sohna Rural, Haryana, India

Mithun Dutta Save the Children, Gulshan—2, Dhaka, Bangladesh

Asmita Ghosh Department of Surgical and Interventional Sciences, McGill University, Montreal, Canada

Pritam Ghosh Department of Geography, Hijli College, Kharagpur, West Bengal, India

Kunal Avishek Gounder College of Humanities and Education, Fiji National University, Nasinu, Fiji

Ajay Goyal National Institute of Mental Health and Neuroscience (NIMHANS), Bangalore, Karnataka, India

K. Hema School of Law, Christ University, Bengaluru, Karnataka, India

Kritika Jain Department of Public Health, Torrens University, Surry Hills, NSW, Australia

Hamimatunnisa Johar Jeffrey Cheah School of Medicine and Health Sciences, Monash University Malaysia, Subang Jaya, Malaysia;
Heidelberg Institute of Global Health (HIGH), Faculty of Medicine and University Hospital, Heidelberg University, Heidelberg, Germany

Renuka Joseph School of Law, Christ University, Bengaluru, Karnataka, India

Nitish Karn School of Agriculture, Uttaranchal University, Dehradun, Uttarakhand, India

Atin Kumar School of Agriculture, Uttaranchal University, Dehradun, Uttarakhand, India

Hari Kumar GeoHazards Society, New Delhi, India

Khai Ern Lee Institute for Environment and Development (LESTARI), Universiti Kebangsaan Malaysia (UKM), Bangi, Selangor, Malaysia

Aditi Madan Institute for Human Development, Delhi, India

Sujan Mandal School of Environment and Disaster Management, Ramakrishna Mission Vivekananda Educational and Research Institute, Kolkata, West Bengal, India

Rumbidzai Matindike Department of Geography Environmental Management Energy Studies, University of Johannesburg, Johannesburg, Gauteng, South Africa

Saqib Mehmood Department of Pathology, University of Veterinary and Animal Sciences, Lahore, Pakistan

Shreya Mitra School of Environment and Disaster Management, Ramakrishna Mission Vivekananda Educational and Research Institute, Kolkata, West Bengal, India

Muhammad Amirulamri Mohd Akhairi Institute for Environment and Development (LESTARI), Universiti Kebangsaan Malaysia (UKM), Bangi, Selangor, Malaysia

Mohd Shahrul Mohd Nadzir Department of Earth Sciences and Environment, Universiti Kebangsaan Malaysia, Bangi, Selangor, Malaysia

Mohd Mudassir United Nations Development Programme, New Delhi, India

Nurfashareena Muhamad Institute for Environment and Development (LESTARI), Universiti Kebangsaan Malaysia (UKM), Bangi, Selangor, Malaysia

Rameswar Mukherjee Department of Geography, Samsi College, Kandaran, Malda, India

Debraj Mukhopadhyay Department of Medical Sciences and Public Health, University of Cagliari (UniCa), Monserrato, Cagliari, Sardinia, Italy

Bushra Noor North South University, Dhaka, Bangladesh

Johanna R. Ñaupari Climate Change and Sustainable Development, San Antonio Abad del Cusco University, Cusco, Peru

Aniket Patil Department of Biostatistics and Demography, International Institute of Population Sciences (IIPS), Deonar, Mumbai, Maharashtra, India

Muhammad Naim Rosli School of Graduate Studies, Postgraduate Centre, Management and Science University, University Drive, Off Persiaran Olahraga, Shah Alam, Selangor, Malaysia

Suchismita Roy School of Environment and Disaster Management, Ramakrishna Mission Vivekananda Educational and Research Institute, Kolkata, West Bengal, India

Sharad Sachan Department of Agricultural Economics and Extension, School of Agriculture, Lovely Professional University, Phagwara, Punjab, India

MD Shiyan Sadik North South University, Dhaka, Bangladesh

Soheli Saha School of Environment and Disaster Management, Ramakrishna Mission Vivekananda Educational and Research Institute, Kolkata, West Bengal, India

Priyanka Sarkar Department of Medical Biotechnology, University of Siena (UniSi), Siena, Tuscany, Italy

Aida Soraya Shamsuddin Institute for Environment and Development (LESTARI), Universiti Kebangsaan Malaysia (UKM), Bangi, Selangor, Malaysia

Rajib Shaw Keio University, Graduate School of Media and Governance, Fujisawa, Kanagawa, Japan

Roopam Shukla Indian Institute of Technology, Roorkee, Uttarakhand, India

Ashraful Alam Siddique International Centre for Diarrhoeal Disease Research, Mohakhali, Dhaka, Bangladesh

K. Sree Lakshmi National Institute of Mental Health and Neuroscience (NIMHANS), Bangalore, Karnataka, India

Peter Kim Streatfield International Centre for Diarrhoeal Disease Research, Mohakhali, Dhaka, Bangladesh

Tin Tin Su Regional Hub for Asia Climate Change and Health (REACH), Jeffrey Cheah School of Medicine and Health Sciences, Monash University Malaysia, Subang Jaya, Malaysia;
Heidelberg Institute of Global Health (HIGH), Faculty of Medicine and University Hospital, Heidelberg University, Heidelberg, Germany

J. Swaminathan School of Allied Health Sciences and Management (SAHSM), Delhi Pharmaceutical Sciences and Research University (DPSRU), Sainik Farm, New Delhi, India

Sharifah Norkhadijah Syed Ismail Department of Environmental and Occupational Health, Faculty of Medicine and Health Sciences, Universiti Putra Malaysia (UPM), Serdang, Selangor, Malaysia

Shubham Tandon United Nations Development Programme, New Delhi, India

Pin Cheng Wong Monash Medical Centre, Clayton, VIC, Australia

Sarmad Zia University of Veterinary and Animal Sciences, Lahore, Pakistan

Aziemah Zulkifli Institute for Environment and Development (LESTARI), Universiti Kebangsaan Malaysia (UKM), Bangi, Selangor, Malaysia

Muhammad**Hassnain** University of Veterinary and Animal Sciences, Lahore, Narowal Campus, Narowal, Pakistan

Chapter 1
Climate Crises and Public Health Challenges—With a Focus on South and Southeast Asia

Pritam Ghosh, Soumita Banerjee, Ranit Chatterjee, and Rajib Shaw

Abstract South and Southeast Asia are widely recognised as global hotspots facing the escalating consequences of climate change, characterised by rising climate-induced disease burdens and systemic vulnerabilities in public health systems. As the introductory chapter of this edited volume, this chapter outlines the key pathways linking climate crises and public health challenges, examines the multifaceted impacts of climate change on population health, and discusses potential pathways for building resilience based on existing scholarly literature. It also provides a brief overview of the chapters included in this volume. The chapter highlights the direct association of various extreme climatic events with increased human mortality, injuries, the spread of infectious diseases, rising morbidity, and disruptions to healthcare infrastructure and the continuity of health services across South and Southeast Asian countries. In addition, various climate change–related processes, mediated through both physical and social environments, indirectly exacerbate food and nutritional insecurity, water and sanitation challenges, air pollution, the urban heat island effect, and human displacement, all of which adversely affect regional health outcomes. The complex interaction of direct, indirect, and intersectional pathways between climate crises and public health also contributes to heightened psychological and social stress, post-disaster mental health disorders, and the amplification of

P. Ghosh (✉)
Department of Geography, Hijli College, Kharagpur 701 306, West Bengal, India
e-mail: mr.ghosh.pritam@gmail.com

S. Banerjee
Amity School of Social Science, Amity University, Kolkata, India
e-mail: soumita905@gmail.com

R. Chatterjee
Co-Founder and Director, RIKA Institute, New Delhi, India
e-mail: ranit@rikaindia.com

R. Shaw
Keio University, Graduate School of Media and Governance, Fujisawa, Kanagawa, Japan
e-mail: shaw@sfc.keio.ac.jp

P. Ghosh et al. (eds.), *Climate Crises and Public Health Challenges*, Disaster Risk Reduction, https://doi.org/10.1007/978-981-95-9709-3_1

existing disease burdens. The convergence of high climate vulnerability, a substantial pre-existing disease burden, and fragile health systems creates a "triple burden" that significantly intensifies public health risks in this region. Strengthening adaptive governance and community-based resilience—through early warning systems, climate-informed health surveillance, and coordinated cross-sectoral policies—is therefore essential for effectively managing climate-induced public health risks in South and Southeast Asia.

Keywords Climate change · Climate crises · Public health · Healthcare · South and Southeast Asia

A rapidly changing climate is not a new pathogen, but the defining public health challenge of the twenty-first century. Regarding climate change, the scientifically accepted opinion as publicly expressed by the Intergovernmental Panel on Climate Change (IPCC) holds that human activity has caused global warming, with broad and accelerating shifts in weather and climate extremes taking place in all corners of the world (IPCC 2022). The World Health Organization (WHO) defines this not only as an ecological problem but as the greatest health threat to humanity, approximating tens of thousands of thousands of more deaths each year because of global warming, undernourishment, malaria, and heat stress by the central of the century (WHO, 2021). Yet, this international story fails to highlight one important fact: the effect is disproportionately unequal. This book hypothesizes that the front lines of this struggle are the colorful, overcrowded, and ecologically rich South and Southeast Asian areas where the triple burden of climate vulnerability, high disease burden, and systemic vulnerability forms a perfect storm of health risks. The pressing challenge now is to understand this convergence, and the ways to be resilient.

1.1 Climate Crises and Public Health: Global and Regional Hotspots

Kitchell's community is one of billions expected to face a wider world of possibilities about who survives and who doesn't once the Intergovernmental Panel on Climate Change's (IPCC) Sixth Assessment Report (AR6) drops (IPCC 2022). These are the trends that echo in a kind of gut-wrenchingly specific local way. It means salinization of drinking water and loss of ancestral homes to sea-level rise for millions who live in the low-lying deltas of Bangladesh and the Mekong. For the urban poor in megacities like Mumbai, Manila and Jakarta, it means lethal urban heat islands and cataclysmic flooding. For the farmers of India's heartland and Nepal's hills, it spells failed crops from erratic monsoons and melting glaciers (World Bank 2021).

The book breaks down the global scale to the climatic hotspots regions of South and Southeast Asia—a region that houses more than a quarter of the global population. The regional profiles do vary, but they are also similar in terms of vulnerability as follows:

First, South Asia (Bangladesh, India, Nepal): extremely populated, highly agricultural, highly vulnerable to risk experience (such as cyclones in the Bay of Bengal, riverine flooding, and scorchers throughout the Indo-Gangetic Plain) (Mani et al. 2018).

Second, due to its coastline length and area, this is a global area where sea-level rise and intrusion of sea water and intense tropical storms are commonly experienced (Southeast Asia e.g. Thailand, Indonesia, Philippines, Vietnam). The local economies also depend on coastal fisheries, tourism, and low-lying agriculture (Yusuf and Francisco 2009).

1.2 The Social Determinants of Health: Poverty, Inequality, and Governance

Viewing climate change merely as an environmental issue is a misunderstanding of its intrinsic nature. It is a "threat multiplier," amplifying existing social inequalities and worsening health inequities through the social determinants of health (WHO 2021).

Poverty is the primary risk multiplier. A poor family living in an informal settlement does not have air conditioning during a heatwave, savings to recover from a flood, or secured food supply when a drought diminishes harvest. They are at elevated risk of exposure to climate hazardous events, face greater susceptibility to the diseases those events produce, and have the least response capacity to adapt (Hallegatte et al. 2020).

Inequality form, in all its many forms, determines how risk is distributed. Gender inequality that constrains women's mobility can exacerbate mortality in a flooding event. Economic inequality that allows individuals to build walls and purchase clean water means the burdens of pollution, disease, and displacement fall primarily to the poor. The best protection against climate-related adverse health effects is a robust system of governance which functions with flexibility and fairness. Climate-sensitive illnesses can be managed through an effective healthcare system.

1.3 Pathways of Climate Crises and Public Health Challenges

The impact of climate change—or the broader climate crisis—on health outcomes and health infrastructure is well documented. Today, climate change is recognized as one of the most critical global challenges affecting human health (Romanello et al. 2021; Perera et al. 2020). The pathways of its influence can be both direct and indirect. These pathways are multifaceted, often overlapping and reinforcing each other, leading to a wide range of adverse health outcomes and negative impacts on health systems and infrastructure (WHO 2023).

Direct pathways involve the immediate effects of various climate change phenomena or climatic extreme events—such as heatwaves, floods, and storms—on human mortality, injury, morbidity, and other physical and mental health outcomes and social well-being. Indirect pathways, on the other hand, operate through changes in the physical, social, and economic environments, influencing nutrition, mental health, disease transmission, and overall well-being. Understanding these complex pathways between climate crises and public health challenges is essential for identifying vulnerable regions and populations, building resilience, developing adaptive strategies, and strengthening health systems—particularly across South and Southeast Asia.

1.3.1 Direct Pathways: Immediate Effect on Health and Healthcare

Direct physical exposure to various climatic extreme events, such as heatwaves, floods, droughts, cyclones, lightning, landslides, and storms, directly affects different physical and mental health outcomes.

1.3.1.1 Mortality and Injury from Extreme Climatic Events

The Asia–Pacific region is widely regarded as the most disaster-prone area in the world. Various extreme climatic events have directly caused human deaths and injuries, threatening lives through physical trauma and morbidity. Particularly outdoor labourer, women, elderly population children are the vulnerable population affected by various extreme events. For instance, during last decade the frequency of different extreme events has found to be increased in India (Tolani et al. 2024; Ray et al. 2021; Mohapatra and Sharma 2019; Mukherjee et al. 2018) and has been significantly associated with rising human mortality (Ray et al. 2021; Yadava et al. 2020; Mahapatra et al. 2018). Several researchers have highlighted that fatalities caused by lightning and flood events have increased considerably during this period (Ray et al. 2021, 2019). Singh and Patwardhan (2012) further emphasized that floods and tropical cyclones are the major contributors to disaster-related mortality in India. Similarly, the past decade has witnessed a surge in cyclonic and storm events across the coastal regions of India, Bangladesh, and the Philippines, leading to numerous deaths and leaving many survivors with long-term injuries and disabilities (Ray et al. 2021).

1.3.1.2 Heat Stress and Morbidity

Rising temperatures have led to an increasing incidence of heatwaves across India, Bangladesh, and other tropical South Asian countries during the summer months. Heatwave-related illnesses such as dehydration, heat exhaustion, and heatstroke have become increasingly common in both urban areas and arid rural regions of South and Southeast Asia. The health risks associated with heat stress are particularly high among outdoor laborers, the elderly, and children. Over the past decade, several South Asian countries—including India, Bangladesh, Nepal, and Pakistan—have reported a significant rise in heatwave-induced mortality and heat stress–related morbidity (Nissan et al. 2017; Saleem et al. 2018; Shrestha et al. 2016; Rohini et al. 2016).

1.3.1.3 Respiratory Illness

Rising incidents of air pollution and forest fires—often intensified by prolonged droughts and increasing temperatures—have been closely linked to a range of respiratory illnesses, including asthma, chronic obstructive pulmonary disease (COPD), and lung cancer. Numerous studies have established a strong association between elevated levels of fine particulate matter ($PM_{2.5}$) and respiratory infections across several Indian cities (Bhuyan et al. 2025; Kakkad et al. 2022).

1.3.1.4 Vector and Water Borne Diseases

Changing patterns of temperature, rainfall, and the increasing frequency of floods due to shifting climatic conditions have been linked to alterations in the seasonality and distribution of disease vectors such as flies, mosquitoes, and ticks across various geographical regions. Warmer and wetter local environments accelerate vector breeding and pathogen replication, contributing to the spread of vector-borne diseases such as dengue, encephalitis, malaria, and chikungunya in South and Southeast Asia (Liu-Helmersson et al. 2016, 2014; Morin et al. 2013; Lambrechts et al. 2011). The incidence of dengue and chikungunya has risen sharply across South and Southeast Asia over the past two decades (WHO, 2014). Several scholars have also highlighted strong correlations between climate change phenomena and the increasing prevalence of vector-borne diseases like dengue and malaria in Bangladesh and other South Asian countries (Jibon 2024; Banu 2014). Moreover, the growing frequency of floods, droughts, and storms has degraded water quality in countries such as Pakistan, contributing to a surge in communicable diseases, including dengue, malaria, cholera, and typhoid (Ahmed 2016).

1.3.1.5 Maternal and Child Health

Changing climatic conditions and the increasing occurrence of extreme events—such as air pollution, floods, and heatwaves—are directly associated with a range of health risks, including inadequate pregnancy care, preterm births, low birth weight, and even maternal mortality. Climate-induced food insecurity further undermines the nutritional status of mothers and children, exacerbating health vulnerabilities. Globally, nearly 80 million children are exposed to six or more overlapping climatic and environmental hazards, with the Child Climate Risk Index (CCRI) ranking particularly high in South Asian countries (UNICEF 2021). In India, numerous studies have established a strong link between exposure to fine particulate matter ($PM_{2.5}$) and premature deaths (Saini and Sharma 2020; Chowdhury and Dey 2016; Ghude et al. 2016).

1.3.1.6 Immediate Effect on Healthcare Infrastructure, Service

The occurrence of extreme climatic events directly disrupts healthcare infrastructure and service delivery in disaster-prone regions (Tsakonas et al., 2024; Mosadeghrad et al., 2023). Such disruptions often lead to significant variations in health outcomes during and after disaster events, as healthcare systems struggle to respond to increased demand amid damaged facilities and limited resources. This situation underscores the urgent need for **climate-resilient healthcare infrastructure** and strengthened health systems capable of withstanding and adapting to the growing impacts of climate change.

1.3.2 Indirect Pathways: Mediated Through Physical and Social Environment

1.3.2.1 Food and Nutritional Insecurity

Changing patterns of rainfall, along with increasing frequency, intensity, and duration of droughts and other extreme events—such as floods and saltwater intrusion into coastal agricultural fields—have severely affected agricultural productivity. The decline in agricultural output is closely linked to food insecurity and reduced food availability, which in turn contribute to undernutrition and micronutrient deficiencies, particularly among children in South Asian countries such as India, Bangladesh, Indonesia, and Nepal (Hanifi et al. 2022; Thiede and Gray 2020; Tiwari et al. 2017;

Rodriguez et al. 2011). As a large proportion of the population in South and Southeast Asia depends on agriculture for livelihood and sustenance, the ongoing climate crisis over the past decade has significantly undermined agricultural productivity in the region, deepening food and nutritional insecurity (Das et al. 2025; Rabbi et al. 2021; Mahapatra et al. 2021; McMahon and Gray 2021; Barnett 2011).

1.3.2.2 Water Scarcity and Sanitation Challenges

Changing climatic conditions, along with the increasing frequency and intensity of extreme events, have significant impacts on water security, which in turn exacerbate sanitation and hygiene challenges. Sea-level rise, cyclones, storm surges, and seawater intrusion into coastal areas and water bodies reduce access to safe drinking and agricultural water across South Asian countries (Nongbri et al. 2025; WaterAid 2019; Anik et al. 2023; Singh et al. 2021; Cai et al. 2015; Piao et al. 2011). Similarly, prolonged droughts in arid and semi-arid regions of India, Pakistan, and other tropical countries in South and Southeast Asia create severe water shortages during the summer months. These shortages compromise safe sanitation and hygiene practices, further increasing vulnerability (Ashrafuzzaman et al. 2023; Ahmed et al. 2021). Conversely, during floods and seawater intrusions, water from local wetlands, water bodies, and other sources of drinking water becomes contaminated, significantly raising the risk of waterborne diseases in affected regions of South and Southeast Asia (Jibon 2024; Banu 2014; Ahmed 2016).

1.3.2.3 Air Pollution and Urban Heat Island

Urbanization is rapidly increasing over south and southeast Asian countries during last few years. Due to unplanned urbanization different cities of this region specially the metropolitan areas are currently facing dual burden of rising temperature and deterioration of air quality. The interplay between PM 2.5 and PM 10 pollution along with extreme heat during summer intensifying the risk of different respiratory and cardiovascular morbidities in these cities (Irfan 2024; Kaur and Pandey 2021; Shaw and Gorai 2020). Indian cities also, the increase of PM 2.5 and PM 10 pollution linked with premature mortality and neo natal death (Mahapatra et al. 2020; Roy et al., 2026).

1.3.2.4 Climate Induced Displacement and Health Consequences

Extreme climatic events are driving large-scale human displacement, forcing people to migrate from their homes to new locations. Displaced populations often face limited or unsafe access to basic amenities, including healthcare services, which contributes to adverse health outcomes. For example, sea-level rise, cyclones, storm surges, and coastline changes have forced large numbers of people to leave their homes in coastal areas of South and Southeast Asia. These displaced individuals frequently experience physical and mental health challenges, along with reduced access to healthcare, which in turn undermines their overall social well-being (Berry et al. 2010; Adger et al. 2021; McMichael and Powell 2021; Haque et al. 2020; Pardhi et al. 2020; Patel and Giri 2019; Rakib et al. 2019).

1.3.3 Direct, Indirect, and Intersectional Pathways: Climate Crises and Phycological Stress and Anxiety

1.3.3.1 Phycological Stress and Anxiety

Frequent and repeated exposure to extreme climatic events can induce feelings of helplessness, anxiety, and chronic stress among populations living in disaster-prone areas of the region. For instance, farmers who repeatedly experience crop failures due to floods, cyclones, or erratic rainfall patterns often suffer from climate-related distress, which in some cases has been linked to increased suicidal tendencies in various parts of South Asia (Barbani 2025; Tripathi et al. 2025; Basistha et al. 2024; Dasgupta and Basu 2023; Gibson et al. 2020; Viswanathan et al. 2019).

1.3.3.2 Livelihood Loss and Social Stress

Socioeconomically marginalized populations—particularly those dependent on crop production, fishing, forestry, or other agriculture-based livelihoods—are especially vulnerable to instability caused by changing climatic conditions and repeated extreme events. These disruptions are closely linked to social and economic insecurity, as well as displacement within affected communities. The resulting stress often manifests as heightened physiological strain, increased domestic violence, and weakening of community bonds, further compounding the burden of mental stress and depression (Burrows et al. 2021; Ayeb-Karlsson et al. 2020).

1.3.3.3 Trauma and Post Disaster Mental Health Disorder

Communities living in hazard-prone areas often experience stress and depression both before and during extreme climatic events. Additionally, populations affected by floods, droughts, cyclones, and other disasters frequently exhibit high levels of post-traumatic stress disorder (PTSD), depression, and insomnia. These psychological impacts can persist long after the immediate disaster, continuing even after physical reconstruction and the restoration of affected areas. Such outcomes highlight the urgent need to integrate mental health and stress management strategies into climate disaster management plans (Asim et al. 2022; Patwary et al. 2024).

1.3.4 Amplification of Existing Health Burden

Socioeconomically marginalized communities across South and Southeast Asia already face multiple public health challenges, including malnutrition, limited access to nutritional supplements, unsafe drinking water, inadequate sanitation and hygiene facilities, insufficient housing, and restricted access to healthcare services. The occurrence of extreme climatic events further exacerbates their livelihood vulnerabilities and health inequities. For example, heatwaves can worsen cardiovascular and renal diseases, floods intensify the spread of diarrheal diseases, and prolonged droughts reduce crop productivity, indirectly contributing to food insecurity and undernutrition among these marginalized populations.

1.4 The Threefold Impact: A Combination of the Southeast and South Asian Risks

The combination of three difficulties that together produce a higher risk profile than each of their own would produce is what gives this region an elevated risk profile. The combination of three complementary burdens, which results in a risk characteristic larger than every one of its individual components, this is what leaves this region particularly vulnerable.

1.4.1 High Weather Vulnerability

This area's geography makes it a target for a number of growing physical threats, such as cyclones and sea-level rise (Eckstein et al. 2021).

1.4.2 High Existing Health Burden

The region currently contends with a lot of illnesses that are both communicable and not. Several ailments that are affected by the weather, such as malaria, dengue, and gastrointestinal illnesses. Such illnesses are propagating and last deeper because of rising temperatures and changing patterns of rainfall (Watts et al. 2021). The increase in non-transmissible illnesses, like heart and lung problems, makes people even more vulnerable to heat tension and pollutants in the air, which are often made worse by blazes caused by climate change.

1.4.3 *Infrastructural and Systemic Fragility*

Many health systems in the region are already stretched thin, lacking the universal coverage, workforce, and financial resilience to handle routine demands, let alone the surge capacity required for climate-amplified disasters (Global Fund 2020). This fragility extends to weak water and sanitation infrastructure, unstable food supply chains, and inadequate housing.

A vicious loop is created by this "triple burden": an environmental shock (such as a flood) overwhelms vulnerable facilities, ruining clinics and polluting water, which in turn causes a spike in the disease load, further impairing response and recuperation systems.

1.5 Pathways to Resilience: Adaptive Governance and Community Action

The story is not one of being a passive victim in the face of this difficult triple burden. The main difference between a crisis and a managed challenge is how well adaptive governance works and how strong the community's resilience is. According to Pahl-Wostl (2009), adaptive governance is about systems that can change and learn in order to deal with uncertainty.

1.5.1 *Early Alert Systems*

The nation of Bangladesh's impressive success in lowering cyclone deaths through precise forecasting as well as grassroots evacuation procedures has become an outstanding instance (Paul 2009).

1.5.2 *Climate-Based Health Detection*

Combining healthcare monitoring as well as weather information to forecast and get ready for the spread of dengue and various other epidemics (Lowe et al. 2017).

1.5.3 Multidisciplinary Policy

Dismantling silos to allow the ministries of water, health, ecology, along with urban design to work together on coordinated approaches like creating cities which promote sustainability and elegance.

The area's biggest advantage is also its most significant weakness: local knowledge, interpersonal relationships, and traditional methods of coping are the initial line of defence. For instance, in India, farmers lead harvesting of water in areas that are affected by drought, as well as in Nepal, communities manage forests that stabilise slopes (Berkes 2007). This kind of neighbourhood agency is not going away because of effective governance; however, it will assist in helping it grow, learn from it, and support it.

1.6 Overview of the Edited Book

The escalating climate crises in South and Southeast Asia pose unprecedented challenges to public health, demanding urgent and multifaceted responses. This edited volume aims to provide a comprehensive exploration of how climate change is impacting the physical and mental health of people, healthcare infrastructure, and practices in this critically vulnerable region, and to propose adaptive strategies for resilience and future preparedness.

With a Focus on South and Southeast Asia offers a robust and interdisciplinary examination of how the climate emergency is reshaping health landscapes in some of the world's most vulnerable areas. The volume brings together different lenses, pulling from geography, public health, environmental science, and social policy to examine the complex network of relationships and interactions climate change, human well-being, and socioecological resilience. The volume is divided into three thematic sections, moving from identifying health impacts to interventions to governance, policy innovation, and sustainable solutions.

The *first segment* reviews how climate changes affect human health and well-being in many multi-faceted ways; direct and indirect. The section also provides a mapping of spatial and social vulnerabilities for South and Southeast Asia and demonstrates how geographical vulnerability and urbanization, together with socio-economic factors, contextually overlap to suffer increased vulnerability. The chapters rely on the ensuing health reflections of rising temperatures, extreme weather events and declining air quality and frame health impacts from respiratory illness, infectious disease to heat stress and malnourishment. The section also discusses an equally important consideration of recognizing mental health as a new field of research in climate. The mental tolls from displacement, loss of livelihoods or climate trauma are reflected alongside community-based coping strategies and community resilience strategies. Through these analyses collectively, the section demonstrates how health vulnerabilities are contextualized and calls for localized approaches to addressing health vulnerabilities and recognizing through studies multi-levels of science, social and cultural factors. This section encourages inter-sectoral collaboration and supports adopting frameworks such as One Health, which recognize the health interlinkages from environmental, animal and human health to sustainable wellbeing.

The *second section* addresses the intersection of climate change and health care delivery system. This introduces the importance of climate-related stressors in impacting occupational health, health system infrastructure and care delivery systems. In this section, beyond personal vulnerability, institutional preparedness and resilience is also examined. It strives to conceptualize climate-resilient health facilities that can provide sustainable care while facing disasters and are designed to be resilient against disasters. This section also captures the importance of gender-responsive adaptation and indigenous knowledge systems in promoting maternal health and community well-being. Also introduced is the innovative application of technology, specifically telemedicine, as a revolutionary way to promote access to health care in remote and climate-stressed settings. Discussion also highlights legal, ethical and infrastructure barriers that must be addressed to universalize such solutions. Collectively, this section conspires that climate-resilient health system will require system change and all levels of governance.

The *last section* expands the scope of analysis to study how people govern their health, their surveillance, and their resilience to climate change. It reviews the response of national and regional policies in South and Southeast Asia to the increasing number of health crises associated with environmental degradation. The ability of surveillance systems to identify and act on emerging climate-sensitive diseases is put under critical scrutiny. The debate extends beyond institutional arrangements to social resilience—to how communities can adapt to risks like sea-level rise, flooding and food insecurity, particularly along the coasts. Women and adolescents are given special consideration because their sexual and reproductive health tends to be undermined during climate induced emergencies. The section mentions the successes and failures of existing policy responses, citing a broken system of governance, a lack of data, and a lack of community integration as systemic obstacles. In its final considerations, the section highlights the urgency to incorporate climate adaptation in national health strategies, as well as their consistency with the concepts of environmental justice, equity, and sustainability. It demands a shared vision of the region that would increase cooperation, institutional capacity, and focus on the health of individuals and ecosystems in times of climate uncertainty.

1.7 Conclusion

South and Southeast Asia, home to over a third of the world's population, are on the frontlines of the climate crisis. Rising sea levels, changing coastlines, more frequent extreme weather events, rising temperatures, and changing disease patterns are all endangering public health, putting pressure on healthcare systems, and posing problems for policy frameworks. In the face of intensifying cyclones, floods, and droughts, healthcare systems are increasingly strained, particularly in vulnerable and underserved communities. As these regions grapple with the multifaceted impacts of climate change, the need for a comprehensive, interdisciplinary approach to public health has never been more urgent.

In the context of the United Nations' Sustainable Development Goals (SDGs), addressing climate change and its impact on public health is imperative for achieving sustainable development. South and Southeast Asia, regions critically vulnerable to climate change, are witnessing significant disruptions in public health due to rising temperatures, extreme weather events, and shifting disease patterns. These challenges threaten to derail progress toward SDGs, particularly Goal 3 (Good Health and Well-being) and Goal 13 (Climate Action). This edited volume brings together cutting-edge research and case studies from this geographical region to provide a comprehensive overview of the health risks, healthcare infrastructure, practice intervened by climate change. It also aims to inform and empower policymakers, healthcare professionals, and researchers, fostering a resilient and health-secure future in alignment with the SDGs for South and Southeast Asia.

References

Adger WN, de Campos RS, Siddiqui T, Gavonel MF, Szaboova L et al (2021) Human security of urban migrant populations affected by length of residence and environmental hazards. J Peace Res 58:50–66

Ahmed F, Shahid M, Cao Y, Qureshi MG, Zia S, Fatima S et al (2021) A qualitative exploration in causes of water insecurity experiences, and gender and nutritional consequences in South-Punjab, Pakistan. Int J Environ Res Public Health 18(23):12534. https://doi.org/10.3390/ijerph182312534

Ahmed T, Scholz M, Al-Faraj F, Niaz W (2016) Water-related impacts of climate change on agriculture and subsequently on public health: a review for generalists with particular reference to Pakistan. Int J Environ Res Public Health 13:1051

Anik AH, Sultan MB, Alam M, Parvin F, Ali MM, Tareq SM, Rahman MM (2023) The impact of climate change on water resources and associated health risks in Bangladesh: a review. Water Secur 18:100133. https://doi.org/10.1016/j.wasec.2023.100133

Ashrafuzzaman M, Gomes C, Guerra J (2023) The changing climate is changing safe drinking water, impacting health: a case in the southwestern coastal region of Bangladesh (SWCRB). Climate 11(7). https://doi.org/10.3390/cli11070146

Asim M, Sathian B, Van Teijlingen E, Mekkodathil AA, Babu MGR, Rajesh E, Kumar RN, Simkhada P, Banerjee I (2022) A survey of post-traumatic stress disorder, anxiety and depression among flood affected populations in Kerala, India. Nepal J Epidemiol 12(2):1203–1214. https://doi.org/10.3126/nje.v12i2.46334

Ayeb-Karlsson S, Kniveton D, Cannon T (2020) Trapped in the prison of the mind: notions of climate-induced (im)mobility decision-making and wellbeing from an urban informal settlement in Bangladesh. Palgrave Commun 6:62. https://doi.org/10.1057/s41599-020-0466-7

Banu S, Hu W, Guo Y, Hurst C, Tong S (2014) Projecting the impact of climate change on dengue transmission in Dhaka, Bangladesh. Environ Int 63:137–142. https://doi.org/10.1016/j.envint.2013.11.002

Barbani SM (2025) The indirect consequences of climate change and extreme environmental events on mental health: a literature review. J Public Health (Berl). https://doi.org/10.1007/s10389-025-02513-1

Barnett J (2011) Dangerous climate change in the Pacific Islands: food production and food security. Reg Environ Change 11(1):229–237

Basistha B, Paul FA, Ali A, Saikia SJ (2024) Environmental quality of life and psychological distress of emerging adults in flood-affected areas of Assam, India: a cross-sectional study. J Psychiatry Spectr 3(1):41–46. https://doi.org/10.4103/jopsys.jopsys_15_23

Berkes F (2007) Understanding uncertainty and reducing vulnerability: lessons from resilience thinking. Nat Hazards 41(2):283–295. https://doi.org/10.1007/s11069-006-9036-7

Berry HL, Bowen K, Kjellstrom T (2010) Climate change and mental health: a causal pathways framework. Int J Public Health 55:123–132

Bhuyan A, Bordoloi T, Debnath R et al (2025) Assessing AQI of air pollution crisis 2024 in Delhi: its health risks and nationwide impact. Discov Atmos 3:13. https://doi.org/10.1007/s44292-025-00041-x

Burrows K, Pelupessy DC, Khoshnood K, Bell ML (2021) Environmental displacement and mental wellbeing in Banjarnegara, Indonesia. Environ Health Perspect 129(11):117002. https://doi.org/10.1289/EHP7767

Cai X, Zhang X, Noël PH, Shafiee-Jood M (2015) Impacts of climate change on agricultural water management: a review. Wiley Interdiscip Rev Water 2(5):439–455. https://doi.org/10.1002/wat2.1089

Chowdhury S, Dey S (2016) Cause-specific premature death from ambient PM2.5 exposure in India: estimate adjusted for baseline mortality. Environ Int 91:183–190. https://doi.org/10.1016/j.envint.2016.03.004

Das A, Kumar S, Kasala K, Nedumaran S, Paithankar P, Kumar A, Jain A, Avinandan V (2025) Effects of climate change on food security and nutrition in India: a systematic review. Curr Res Environ Sustain 9:100286. https://doi.org/10.1016/j.crsust.2025.100286

Dasgupta R, Basu M (2023) Mental health and socio-psychological manifestations of cyclone-induced water insecurity in the Indian Sundarban delta. Int J Disaster Risk Reduct 98:104085. https://doi.org/10.1016/j.ijdrr.2023.104085

Eckstein D, Kunzel V, Schafer L, Winges M (2021) Global climate risk index 2021. Germanwatch. https://www.germanwatch.org/en/19777

Ghude SD, Chate DM, Jena C, Beig G, Kumar R, Barth MC, Pfister GG, Fadnavis S, Pithani P (2016) Premature mortality in India due to PM2.5 and ozone exposure. Geophys Res Lett 43:4650–4658

Gibson KE, Barnett J, Haslam N, Kaplan I (2020) The mental health impacts of climate change: findings from a Pacific Island atoll nation. J Anxiety Disord 73:102237. https://doi.org/10.1016/j.janxdis.2020.102237

Global Fund (2020) The impact of climate change on health and the fight against HIV, TB and malaria. The Global Fund to Fight AIDS, Tuberculosis and Malaria. https://www.theglobalfund.org/media/10776/publication_climatechange_impact_report_en.pdf

Hallegatte S, Fay M, Barbier EB (2020) Poverty and climate change: Introduction. In: Hallegatte S, Fay M, Barbier EB (eds) The economics of poverty and climate change. The World Bank. https://doi.org/10.1596/1813-9450-5066

Hanifi SM, Ahmed M, Menon N, Quisumbing A (2022) The impact of climate change on children's nutritional status in coastal Bangladesh. Soc Sci Med 294:114704. https://doi.org/10.1016/j.socscimed.2022.114704

Haque R, Parr N, Muhidin S (2020) Climate-related displacement, impoverishment and healthcare accessibility in mainland Bangladesh. Asian Popul Stud 16:220–239

Hsu HT, Wu CD, Chung MC, Shen TC, Lai TJ, Chen CY et al (2021) The effects of traffic-related air pollutants on chronic obstructive pulmonary disease in the community-based general population. Respir Res 22:1–12. https://doi.org/10.1186/s12931-021-01812-x

IPCC (2022) Climate change 2022: Impacts, adaptation and vulnerability. Contribution of Working Group II to the Sixth Assessment Report of the Intergovernmental Panel on Climate Change. Cambridge University Press. https://www.ipcc.ch/report/ar6/wg2/

Irfan H (2024) Air pollution and cardiovascular health in South Asia: a comprehensive review. Curr Probl Cardiol 49(2):102199. https://doi.org/10.1016/j.cpcardiol.2023.102199

Jibon MJN, Ruku SMRP, Islam ARMT, Khan MN, Mallick J, Bari ABMM, Senapathi V (2024) Impact of climate change on vector-borne diseases: exploring hotspots, recent trends and future outlooks in Bangladesh. Acta Trop 259:107373. https://doi.org/10.1016/j.actatropica.2024.107373

Kakkad KM, Oza C, Dutta P et al (2022) Linking PM pollution to the respiratory health of children: a cross-sectional study from Ahmedabad City in Western India. Aerosol Air Qual Res 22:220038. https://doi.org/10.4209/aaqr.220038

Kaur R, Pandey P (2021) Air pollution, climate change, and human health in Indian cities: a brief review. Front Sustain Cities 3:705131. https://doi.org/10.3389/frsc.2021.705131

Lambrechts L, Paaijmans KP, Fansiri T, Carrington LB, Kramer LD, Thomas MB, Scott TW (2011) Impact of daily temperature fluctuations on dengue virus transmission by Aedes aegypti. Proc Natl Acad Sci 108(18):7460–7465

Liu-Helmersson J, Quam M, Wilder-Smith A, Stenlund H, Ebi K, Massad E, Rocklöv J (2016) Climate change and Aedes vectors: 21st century projections for dengue transmission in Europe. EBioMedicine 7:267–277

Liu-Helmersson J, Stenlund H, Wilder-Smith A, Rocklöv J (2014) Vectorial capacity of Aedes aegypti: effects of temperature and implications for global dengue epidemic potential. PLoS ONE 9(3):e89783

Lowe R, Gasparrini A, Van Meerbeeck CJ, Lippi CA, Mahon R, Trotman AR, Rollock L (2017) Non-stationary influence of El Niño-southern oscillation on dengue in the Caribbean. PLoS Negl Trop Dis 11(9):e0005957. https://doi.org/10.1371/journal.pntd.0005957

Mahapatra B, Walia M, Rao CAR, Raju BMK, Saggurti N (2021) Vulnerability of agriculture to climate change increases the risk of child malnutrition: evidence from a large-scale observational study in India. PLoS ONE 16(6):e0253637

Mahapatra B, Walia M, Saggurti N (2018) EWEs induced mortalities in India 2001–2014: trends and differentials by region, sex, and age group. Weather Clim Extrem 21:110–116

Mahapatra, B, Walia M, Avis WR, Saggurti N (2020) Effect of exposure to PM10 on child health: evidence based on a large-scale survey from 184 cities in India. BMJ Glob Health 5:e002597. https://doi.org/10.1136/bmjgh-2020-002597

Mani M, Bandyopadhyay S, Chonabayashi S, Markandya A, Mosier T (2018) South Asia's hotspots: the impact of temperature and precipitation changes on living standards. World Bank. https://doi.org/10.1596/978-1-4648-1155-5

McMahon K, Gray C (2021) Climate change, social vulnerability and child nutrition in South Asia. Glob Environ Chang 71:102414. https://doi.org/10.1016/j.gloenvcha.2021.102414

McMichael C, Powell T (2021) Planned relocation and health: a case study from Fiji. Int J Environ Res Public Health 18(8):4355

Mohapatra M, Sharma M (2019) Cyclone warning services in India during recent years: a review. Mausam 70(4):635–666

Morin CW, Comrie AC, Ernst K (2013) Climate and dengue transmission: evidence and implications. Environ Health Perspect 121(11–12):1264–1272

Mosadeghrad AM, Isfahani P, Eslambolchi L, Zahmatkesh M, Afshari M (2023) Strategies to strengthen a climate-resilient health system: a scoping review. Global Health 19:62

Mukherjee S, Aadhar S, Stone D, Mishra V (2018) Increase in extreme precipitation events under anthropogenic warming in India. Weather Clim Extrem 20:45–53

Nissan H, Burkart K, Coughlan de Perez E, VanAalst M, Mason S (2017) Defining and predicting heat waves in Bangladesh. J Appl Meteorol Climatol 56:2653–2670

Nongbri RW, Sarkar A, Mandal SK (2025) Impact of temperature shocks on household water poverty in India. Sci Rep 15:16618. https://doi.org/10.1038/s41598-025-01732-6

Pahl-Wostl C (2009) A conceptual framework for analysing adaptive capacity and multi-level learning processes in resource governance regimes. Glob Environ Chang 19(3):354–365. https://doi.org/10.1016/j.gloenvcha.2009.06.001

Pardhi A, Jungari S, Kale P, Bomble P (2020) Migrant motherhood: maternal and child health care utilization of forced migrants in Mumbai, Maharashtra, India. Child Youth Serv Rev 110:104823

Patel A, Giri J (2019) Climate change, migration and women: analysing construction workers in Odisha. Social Change 49:97–113

Patwary MM, Bardhan M, Haque MA, Moniruzzaman S, Gustavsson J et al (2024) Impact of extreme weather events on mental health in South and Southeast Asia: two decades of systematic review of observational studies. Environ Res 250:118436. https://doi.org/10.1016/j.envres.2024.118436

Paul BK (2009) Why relatively fewer people died? The case of Bangladesh's cyclone Sidr. Nat Hazards 50(2):289–304. https://doi.org/10.1007/s11069-008-9340-5

Perera ATD, Nik VM, Chen D, Scartezzini JL, Hong T (2020) Quantifying the impacts of climate change and extreme climate events on energy systems. Nat Energy 5:150–159. https://doi.org/10.1038/s41560-020-0558-0

Piao S et al (2010) The impacts of climate change on water resources and agriculture in China. Nature 467(7311):43–51. https://doi.org/10.1038/nature09364

Piao S, Ciais P, Huang Y, Shen Z, Peng S, Li J, Zhou L, Liu H, Ma Y, Ding Y, Friedlingstein P, Liu C, Tan K, Yu Y, Zhang T, Fang J (2011) The impacts of climate change on water resources and agriculture in China. Nature 467(7311):43–51. https://doi.org/10.1038/nature09364

Rabbi SE, Ali M, Costa LC, Pradhan P, Rahman A, Yunus FM, Kropp JP (2021) Identifying climatic and non-climatic determinants of malnutrition prevalence in Bangladesh: a country-wide cross-sectional spatial analysis. Spat Spat-Temporal Epidemiol 37:100422. https://doi.org/10.1016/j.sste.2021.100422

Rakib MA, Sasaki J, Matsuda H, Fukunaga M (2019) Severe salinity contamination in drinking water and associated human health hazards increase migration risk in the southwestern coastal part of Bangladesh. J Environ Manag 240:238–248

Ray K, Giri RK, Ray SS, Dimri AP, Rajeevan M (2021) An assessment of long-term changes in mortalities due to extreme weather events in India: a study of 50 years' data, 1970–2019. Weather Clim Extrem 32:100315. https://doi.org/10.1016/j.wace.2021.100315

Ray K, Pandey P, Pandey C, Dimri AP, Kishore K (2019) On the recent floods in India. Curr Sci 117(2):204–218

Rodriguez-Llanes JM, Ranjan-Dash S, Degomme O, Mukhopadhyay A, Guha-Sapir D (2011) Child malnutrition and recurrent flooding in rural eastern India: a community-based survey. BMJ Open 1(2):e000109

Rohini P, Rajeevan M, Srivastava A (2016) On the variability and increasing trends of heat waves over India. Sci Rep 6:26153. https://doi.org/10.1038/srep26153

Romanello M, McGushin A, Napoli CD, Drummond P, Hughes N, Jamart L, Kennard H, Lampard P, Rodriguez BS, Arnell N et al (2021) The 2021 report of the Lancet Countdown on health and climate change: code red for a healthy future. Lancet 398:1619–1662. https://doi.org/10.1016/S0140-6736(21)01787-6

Roy C, Ghosh P, Sati VP, Marzo RR, Kumar S, Pandey K (2026) Association between PM2.5 exposure and child mortality from lower respiratory infections (LRIs) in India: a systematic analysis of the global burden of disease study (1990–2021). Arch Public Health 84(43):1–21. https://doi.org/10.1186/s13690-025-01824-9

Saini P, Sharma M (2020) Cause and age-specific premature mortality attributable to PM2.5 exposure: an analysis for million-plus Indian cities. Sci Total Environ 710:135230. https://doi.org/10.1016/j.scitotenv.2019.135230

Saleem M, Shah SZ, Azam S (2018) Heat wave killings in Pakistan and possible strategies to prevent future heat wave fatalities. Res Rev: Healthc Open Access J 1:1–5

Shaw N, Gorai AK (2020) Study of aerosol optical depth using satellite data (MODIS Aqua) over Indian territory and its relation to particulate matter concentration. Environ Dev Sustain 22:265–279. https://doi.org/10.1007/s10668-018-0198-8

Shrestha SL, Shrestha IL, Shrestha N (2016) Region-wise effects of climate-sensitive variables on some specific disease burdens in Nepal. Open Atmos Sci J 10:63–83

Shue H (2014) Climate justice: vulnerability and protection. Oxford University Press

Singh A, Patwardhan A (2012) Spatio-temporal distribution of extreme weather events in India. APCBEE Proc 1:258–262

Singh C, Jain G, Sukhwani V, Shaw R (2021) Losses and damages associated with slow-onset events: urban drought and water insecurity in Asia. Curr Opin Environ Sustain 50:72–86. https://doi.org/10.1016/j.cosust.2021.02.006

Thiede BC, Gray C (2020) Climate exposures and child undernutrition: evidence from Indonesia. Soc Sci Med 265:113298. https://doi.org/10.1016/j.socscimed.2020.113298

Tiwari S, Jacoby HG, Skoufias E (2017) Monsoon babies: Rainfall shocks and child nutrition in Nepal. Econ Dev Cult Change 65(2):167–188. https://doi.org/10.1086/689308

Tolani H, Neogi SB, Gupta SD et al (2024) Analyzing dynamics of extreme weather events (EWE) in India: unfolding trends through statistical assessment of 50 years' data (1970–2019). BMC Environ Sci 1:12. https://doi.org/10.1186/s44329-024-00012-4

Tripathi A, Jeswani P, Shukla R, Patel V (2025) Unveiling climate distress: examining the psychological impact of adverse weather events in India—a systematic review. Int J Soc Psychiatry. Advance online publication. https://doi.org/10.1177/00207640251362911

Tsakonas K, Badyal S, Takaro T, Buse CG (2024) Rapid review of the impacts of climate change on the health system workforce and implications for action. J Clim Chang Health 19:100337. https://doi.org/10.1016/j.joclim.2024.100337

UNICEF (2021) The climate crisis is a child rights crisis: introducing the children's climate risk index. United Nations Children's Fund.

United Nations Children's Fund (2021) The climate crisis is a child rights crisis: introducing the children's climate risk index. UNICEF, New York

Viswanathan DJ, Veerakumar AM, Kumarasamy H (2019) Depression, suicidal ideation, and resilience among rural farmers in a drought-affected area of Trichy district, Tamil Nadu. J Neurosci Rural Pract 10(2):238–244. https://doi.org/10.4103/jnrp.jnrp_257_18

WaterAid (2019) Climate change and resilient WASH in South Asia. https://washmatters.wateraid.org/sites/g/files/jkxoof256/files/climate-change-and-resilient-wash-in-south-asia.pdf

Watts N, Amann M, Arnell N, Ayeb-Karlsson S, Beagley J, Belesova K, Boykoff M, Costello A (2021) The 2020 report of The Lancet Countdown on health and climate change: responding to converging crises. Lancet 397(10269):129–170. https://doi.org/10.1016/S0140-6736(20)32290-X

World Bank. (2021). *Climate risk country profile: Bangladesh.* The World Bank Group. https://climateknowledgeportal.worldbank.org/sites/default/files/2021-08/15900-WB_Bangladesh%20Country%20Profile-WEB.pdf

World Health Organization (2014) A global brief on vector-borne diseases (No. WHO/DCO/WHD/2014.1). World Health Organization

World Health Organization (2021) *Climate change and health.* https://www.who.int/news-room/fact-sheets/detail/climate-change-and-health

World Health Organization (2023) Climate change and health. https://www.who.int/news-room/fact-sheets/detail/climate-change-and-health

World Health Organization (2024) COP29 special report on climate change and health: health is the argument for climate action. World Health Organization, Geneva

Yadava PK, Soni M, Verma S (2020) The major lightning regions and associated casualties over India. Nat Hazards 101:217–229

Yusuf AA, Francisco HA (2009) Climate change vulnerability mapping for Southeast Asia. Economy and environment program for Southeast Asia (EEPSEA). https://www.idrc.ca/sites/default/files/sp/Documents%20EN/Climate-Change-Vulnerability-Mapping-for-Southeast-Asia.pdf

Part I
Impact of Climate Crises on Physical Health and Mental Health

Chapter 2
Mapping Vulnerabilities: A Geographical Perspective on Climate Change and Health Risks in South and Southeast Asia

Aida Soraya Shamsuddin, Nurfashareena Muhamad, Aziemah Zulkifli, Maisarah Abdul Mutalib, Muhammad Amirulamri Mohd Akhairi, Puteri Amirah Nabilah Azman, Khai Ern Lee, Muhammad Naim Rosli, and Sharifah Norkhadijah Syed Ismail

Abstract Climate change poses growing threats to public health, with impacts that are unevenly distributed across geographies. This narrative review explores how geography shapes the health impacts of climate change, with a particular focus on South and Southeast Asia. The objective is to examine the spatial variability of climate-health risks, the role of geographic determinants in vulnerability, and the application of geospatial techniques in identifying and addressing these

---

A. S. Shamsuddin (✉) · N. Muhamad · A. Zulkifli · M. A. Mohd Akhairi · P. A. N. Azman · K. E. Lee
Institute for Environment and Development (LESTARI), Universiti Kebangsaan Malaysia (UKM), Bangi, Selangor, Malaysia
e-mail: aidasoraya@ukm.edu.my

N. Muhamad
e-mail: fasha@ukm.edu.my

A. Zulkifli
e-mail: aziemah_zul@ukm.edu.my

M. A. Mohd Akhairi
e-mail: amirulamri@ukm.edu.my

P. A. N. Azman
e-mail: puteriamirah.seadpri@gmail.com

K. E. Lee
e-mail: khaiernlee@ukm.edu.my

M. Abdul Mutalib · M. N. Rosli
School of Graduate Studies, Postgraduate Centre, Management and Science University, University Drive, Off Persiaran Olahraga, Shah Alam, Selangor, Malaysia
e-mail: maisarah_abdulmutalib@msu.edu.my

M. N. Rosli
e-mail: muhd_rosli@msu.edu.my

P. Ghosh et al. (eds.), *Climate Crises and Public Health Challenges*, Disaster Risk Reduction, https://doi.org/10.1007/978-981-95-9709-3_2

challenges. The review synthesises recent literature across disciplines to analyse how socioeconomic conditions, environmental exposure, population displacement, disease patterns, and institutional capacity interact geographically to influence health outcomes. It also highlights key geospatial approaches including geographical information system (GIS), remote sensing, and spatial modelling are used to map climate-induced health risks, identify vulnerability hotspots, and inform policy responses. The review concludes that geographically informed analysis is essential for effective climate-health adaptation. Strengthening spatial data integration and inclusive governance is the key to reducing vulnerability and advancing health equity in the context of a changing climate.

Keywords Climate change · Public health · Geography · Vulnerability · Adaptation strategies

2.1 Introduction

Climate change is increasingly recognized as a major threat to global public health. Rising temperatures, extreme weather events, sea-level rise, and altered precipitation patterns not only disrupt ecosystems but also compromise key social and environmental determinants of health, including clean air, safe drinking water, sufficient food, and adequate shelter (Palmeiro-Silva et al. 2023). Kamel Boulos and Wilson (2023) estimate that climate change could result in approximately 250,000 additional deaths annually between 2030 and 2050 due to heat stress, malaria, malnutrition, and diarrhoeal diseases. The Intergovernmental Panel on Climate Change (IPCC) has identified climate change as one of the defining challenges of our era, affecting public health globally through multiple direct and indirect pathways (Napoli et al. 2022; Tiwari et al. 2022; Li et al. 2023). Direct health impacts include cardiovascular, respiratory, and renal conditions, injuries and fatalities, and adverse mental health outcomes resulting from extreme temperatures, shifting weather patterns, and climate-related disasters (Li et al. 2023; Howlins et al. 2024). Indirect effects occur as climate change alters ecological and human systems, increasing the transmission of vector-borne and water-borne diseases, threatening crop yields and food security, straining inadequately adapted healthcare systems, and potentially escalating social conflicts, all of which amplify health losses (Berrang-Ford et al. 2021; Tiwari et al. 2022; Li et al. 2023; Parums et al. 2024). These impacts are unevenly

M. Abdul Mutalib
International Medical School, Management and Science University, University Drive, Off Persiaran Olahraga, Shah Alam, Selangor, Malaysia

S. N. Syed Ismail
Department of Environmental and Occupational Health, Faculty of Medicine and Health Sciences, Universiti Putra Malaysia (UPM), Serdang, Selangor, Malaysia
e-mail: norkhadijah@upm.edu.my

distributed, shaped by geographical, socioeconomic, and institutional factors, thereby highlighting critical issues of climate justice and health equity (Greeham et al. 2024).

Geography variation plays a pivotal role in shaping population health outcomes under a changing climate. The spatial distributions of exposure to climate hazards such as floods, droughts, and heatwaves varies considerably across regions due to differences in latitude, elevation, land use, and proximity to coasts or river basins. Similarly, the sensitivity and adaptive capacity of populations differ based on socioeconomic status, urbanisation, access to healthcare, infrastructure robustness, and governance (Sultana et al. 2024). For instance, urban population in low-lying coastal areas may be highly exposed to both flooding and vector-borne diseases, while rural communities in arid regions may face challenges related to water scarcity and food insecurity. These geographically determined disparities underscore the importance of adopting a spatial lens in analysing climate-health interactions and planning effective responses (Lv et al. 2024). These factors are not causative but can act as covariables in amplifying risk. For example, about half of the world's eight billion people live in regions that are highly exposed to the risk of climate change including flooding events, drought and rise in sea level (Makharia et al. 2023). In many parts of the world, particularly in low- and middle-income countries, climate change disproportionately affects populations already burdened by socioeconomic inequalities (Tiwari et al. 2022; Ngcamu 2023; Sarkar et al. 2024).

Mapping climate-related health vulnerabilities is essential for identifying populations and regions at greatest risk. Vulnerability is defined as increased exposure or heightened susceptibility to climate-related hazards (Makharia et al. 2023). While environmental changes affect all populations, some groups experience disproportionately greater adverse health outcomes due to individual and structural risk factors (Makharia et al. 2023). Vulnerable populations including children, older adults, and individuals with pre-existing health conditions are particularly susceptible to the health impacts of climate change (Tiwari et al. 2022; Ngcamu 2023; Sarkar et al. 2024). These vulnerability factors, which amplify health risks, encompass geographical, socioeconomic, sociopolitical, biological, and demographic dimensions.

Although existing studies have explored the general health implications of climate change, there remains a lack of comprehensive synthesis focusing specifically on how geography shapes these impacts particularly within the context of South and Southeast Asia. Current literature often addresses climate-health linkages in a fragmented manner, without systematically examining how spatial disparities in exposure, sensitivity, and adaptive capacity influence health outcomes. This chapter addresses this gap by offering a geographically grounded synthesis of how climate change affects health across different subnational contexts in the region. It highlights the interplay between climate hazards, socioeconomic factors, and institutional responses in shaping health vulnerabilities. In particular, this study seeks to examine how geographic factors shape climate-related health risks in South and Southeast Asia. It also explores the ways in which socioeconomic and health system disparities across different geographic locations influence the vulnerability of populations. Furthermore, the study investigates how geospatial techniques can be employed to identify and respond effectively to climate-health vulnerabilities.

The objective of this chapter is to synthesize current evidence on the intersection of geography and public health in the context of climate change, using a narrative review approach. By doing so, it contributes to the literature by emphasizing the need for location-specific strategies that integrate geospatial data, public health indicators, and socioeconomic variables. The insights offered in this chapter aim to support the development of more targeted and equitable public health interventions in the face of climate change.

2.2 Climate Change and Health Interactions in South and Southeast Asia

In the Asian region, there is compelling evidence that there has been an increase in the intensity and frequency of extreme heat events and decrease in the intensity and frequency of extreme cold events in recent decades due to the climate change (Sentian et al. 2022). Asia is defined as the land and territories of 51 countries or regions. It can be broadly divided into six sub-regions based on geographic position and coastal peripheries; (1) Central Asia (five countries); (2) East Asia (seven countries/ regions); (3) North Asia (two countries); (4) West Asia (17 countries); (5) South Asia (eight countries), and Southeast Asia (12 countries) (Fig. 2.1) (IPCC 2022). South and Southeast Asia encompass a diverse range of geographic features, including densely populated coastal areas, fertile plains, mountainous regions, and tropical forests. These geographic characteristics, combined with socioeconomic disparities, make the region particularly vulnerable to climate-induced health risks (Nilsson et al. 2021; Zain et al. 2024).

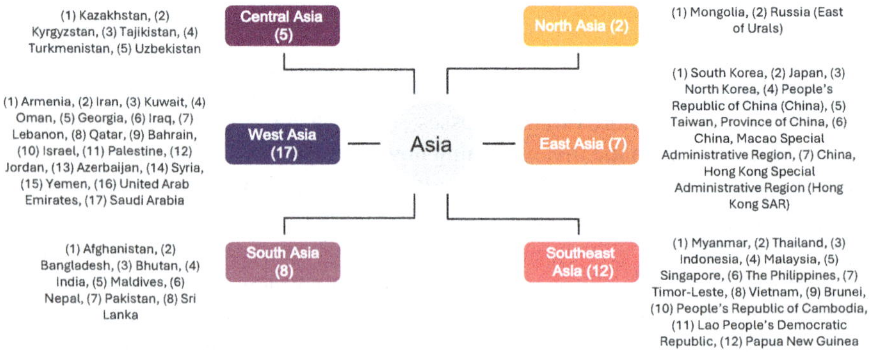

Fig. 2.1 Countries and regions in Asia

Climate change affects human health through complex and interrelated pathways. These interactions are mediated by a range of vulnerability factors, climate-related hazards, and patterns of exposure, which together influence health outcomes and the resilience of health systems. Vulnerability factors include demographic profiles, geographic location, baseline health status, and broader sociopolitical and economic conditions (Sultana et al. 2024). Gender and equity considerations also play a critical role, as marginalised populations often face disproportionate risks (Thomas et al. 2018; Tesfaye et al. 2025).

The primary hazards associated with climate change include extreme weather events such as floods, storms, and heatwaves, as well as slower-onset changes like sea level rise, air pollution, shifts in vector ecology, reduced food production, and water scarcity (Berrang-Ford et al. 2021). These hazards interact with existing environmental, health, and social systems, creating differential exposures across regions and populations. For instance, people living in informal settlements or low-lying coastal zones may be more directly exposed to coastal flooding, submergence, and erosion because of relative sea level rise, while urban populations often experience intensified heat due to the urban heat island effect (Chaudhry 2024; Hussainzad and Gou 2024; Bharadwaz et al. 2025). In addition, exposure encompasses not only the physical proximity of people to climate hazards but also the vulnerability of critical systems that support health. These include the health workforce, infrastructure, energy systems, water and food supply chains, and the health system itself (Thomas et al. 2018). The interplay between exposure and vulnerability determines the extent to which climate hazards translate into actual health risks.

The direct impacts of climate change on human health are significant and widespread. These effects include injuries and deaths caused by extreme weather events such as floods, storms, and heatwaves, as well as the worsening of existing health problems like respiratory conditions and chronic diseases (Ebi and Hess 2020; Issa et al. 2023). Climate change also increases the risk of mental health issues and affects the spread and pattern of infectious diseases, including those that are transmitted through water, animals, or insects (Sultana et al. 2024; WHO 2024). For example, Sen et al. (2017) reported that recent flash floods in Nepal, Bangladesh, and India, considered the worst in a hundred years, resulted in more than 1,400 deaths, destroyed over 7,000 schools, submerged large areas of land, and left nearly half a million people in Nepal without enough food. Climate change can also affect reproductive health and increase financial stress due to rising healthcare costs. These problems are often made worse when multiple climate-related events happen at the same time or in a short period, such as heatwaves followed by floods. Such situations put pressure on health services, interrupt basic needs like clean water and electricity, and reduce the ability of health systems to respond quickly and fairly (WHO 2024). As a result, the health risks linked to climate change are growing more serious and affecting vulnerable groups the most.

Climate change also indirectly affects health and well-being through its influence on social, economic, and environmental conditions that determine how people live, grow, work, and age (Issa et al. 2023; Palmeiro-Silva et al. 2023). These include factors such as gender inequality, human rights, education, employment, poverty, food and water security, housing, and access to health services. As a result of climate change, these existing challenges are made worse, acting as a multiplier of risk (Sultana et al. 2024). Damage to essential infrastructure, rising unemployment, increased competition for limited resources, and the threat to basic needs such as food and clean water all contribute to greater vulnerability, particularly among disadvantaged groups (Javan et al. 2024). Air pollution, for example, can lead to respiratory illnesses, while disruptions to water systems may result in a rise in water-borne diseases. Changes in ecological conditions also affect the spread of vector-borne and zoonotic diseases (Kamel Buolos and Wilson 2023). In 2015, one of the deadliest heatwaves in history struck India and Pakistan, causing the deaths of approximately 3,500 people (Sen et al. 2017). Exposure to extreme heat can result in dehydration, heat exhaustion, and heat stroke, while also worsening cardiovascular and respiratory conditions. The already poor air quality in regions such as South Asia becomes even more harmful under extreme heat conditions. Furthermore, climate change worsens food insecurity, leading to malnutrition and other related health problems. Over time, environmental stress and displacement can contribute to mental health issues and increase the occurrence of noncommunicable diseases (NCDs) (Vaidya et al. 2024). The effect of climate change on mental health is a significant but often overlooked aspects of its overall burden on humanity (Vaidya et al. 2024).

Finally, the capacity of health systems and facilities to respond to these challenges is a critical determinant of outcomes. Climate change can damage healthcare infrastructure, disrupt services, and place additional strain on health personnel. Without adequate adaptation planning, these disruptions can compromise both the accessibility and quality of healthcare, particularly in resource-limited settings.

2.3 Geographical Variability of Climate-Induced Health Risk

Geographical factor is one of the key vulnerability factors, alongside demographic, biological factors and health status, sociopolitical, socioeconomic, health system capacity and, gender and equity. Geographical location is a key determinant in shaping the degree of exposure, sensitivity, and adaptive capacity of populations to climate-induced health risks. Variations in topography, climate zones, land use, and population distribution result in differing levels of hazard exposure and unequal access to resources for coping with heath threats. According to the IPCC (2014), geography directly influences the types of climate-related stressor encountered such as floods in low-lying deltas, droughts in inland plains, or heat stress in urban heat islands and how populations experience and respond to them (Fig. 2.2). The World Health Organization (WHO 2023) also emphasises that climate-related health outcomes are not evenly distributed but reflect underlying spatial inequalities in vulnerability and health system readiness.

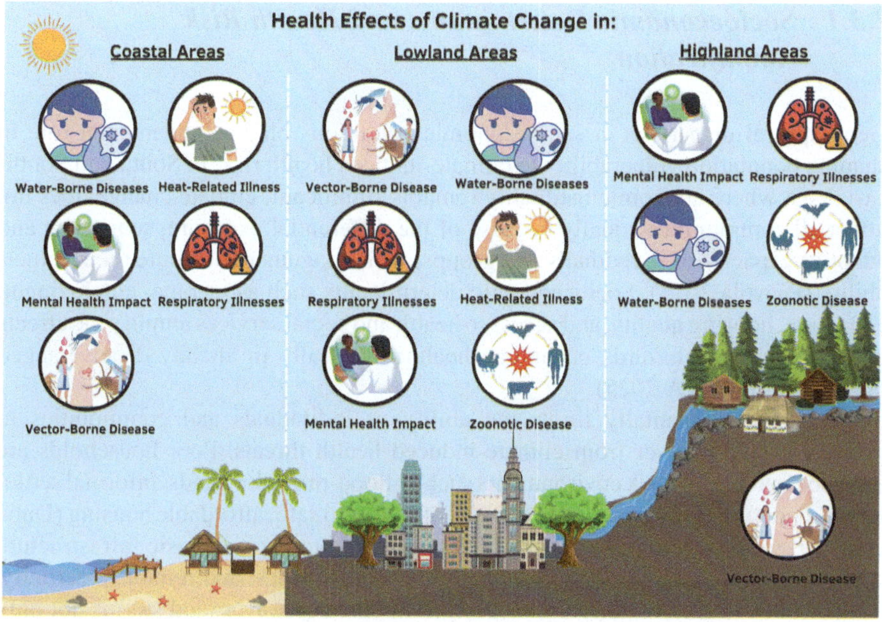

Fig. 2.2 Health effects of climate change to public health in different areas

This section explores the geographical variability of climate-induced health risks by examining how spatial factors intersect with broader determinants to influence different public health issues. It is organized into five interrelated subsections: (1) socioeconomic determinants and health risk amplification, which highlights how poverty, inequality, and limited access to services can heighten vulnerability in certain areas; (2) geographical exposure and environmental hazards, which focuses on how physical location and environmental conditions shape exposure to climate threats such as floods, heatwaves, and air pollution; (3) health impacts of climate-induced migration and displacement, which discusses the health challenges faced by populations uprooted by environmental stressors; (4) climate-sensitive disease burdens and regional inequalities, which examines the uneven distribution of vector-borne, waterborne, and other climate-sensitive diseases across regions; and (5) institutional response gaps and policy inequality, which addresses the disparities in adaptation efforts, infrastructure, and health system readiness across different geographical and administrative areas. Together, these dimensions illustrate how climate change affects public health in complex and place-specific ways.

2.3.1 Socioeconomic Determinants and Health Risk Amplification

Geographical disparities in socioeconomic conditions play a fundamental role in shaping population vulnerability to climate-induced health risk. In South and Southeast Asia, where economic inequality remains significant, climate change does not affect all communities equally because of the differential exposure, sensitivity and adaptive capacity of individuals and groups, which together constitute their vulnerability (Paavola 2017). Socioeconomic determinants such as income, employment, education, housing quality, and access to health and social services amplify the effects of climate-related hazards on human health, especially in already disadvantaged populations (USEPA 2025).

Poverty fundamentally limits the ability of individuals and communities to respond to and recover from climate-induced health threats. Poor households are often located in high-risk environments such as flood-prone lowlands, informal settlements, or deforested slopes due to the lack of access to safe, affordable housing (Dube et al. 2018; Hussainzad et al. 2024). These areas typically lack basic infrastructure such as piped water, drainage systems, sanitation facilities, and reliable electricity. As a result, extreme weather events such as floods, heatwaves, and storms not only pose direct health threats (e.g., injury, heat stroke, drowning) but also increase the risk of infectious diseases due to contamination of water sources and poor hygiene. This physical exposure, when combined with poor housing quality, limited mobility, and a lack of savings or insurance, severely restricts the capacity of residents to respond to and recover from climate shocks (Ebi and Hess 2020; Ebi et al. 2021; Sultana et al. 2024). Consequently, these populations experience a greater incidence of injury, waterborne diseases, and long-term mental health impacts.

Employment type is a key determinant of climate-related health risks. Individuals employed in the informal sector or in climate-sensitive industries such as agriculture, fisheries, and construction face heightened exposure to environmental hazards, including heatwaves, air pollution, and vector-borne diseases (Mahmood et al. 2023). Limited access to occupational protections, social safety nets, and health benefits further increases their vulnerability. Extended heat exposure in occupations such as bakeries, foundries, and laundries can result in dehydration, heatstroke, and worsening of cardiovascular or respiratory conditions, particularly among those with pre-existing illnesses (ILO 2024). Climate change also places additional strain on emergency services and healthcare systems, with frontline responders, including firefighters, encountering progressively hazardous working conditions (ILO 2024). These impacts are unevenly distributed, disproportionately affecting the working poor, informal and seasonal workers, and employees in micro and small enterprises (ILO 2024).

Education levels further influence health risk. Communities with lower levels of education may have limited awareness of climate-health linkages and inadequate knowledge of adaptive practices, such as vector control or the use of early warning systems (Reid 2019; Xu et al. 2023). In such areas, misinformation or delayed health-seeking behaviours during climate-sensitive disease outbreaks such as dengue fever or chlore can worsen health outcomes. On the other hand, communities with higher educational attainment are more likely to adopt risk-reducing behaviours and make informed decisions during emergencies.

Moreover, socioeconomic marginalisation often intersects with other forms of vulnerability, including gender, age, and disability. For instance, marginalised ethnic groups, Indigenous communities, women, children, the elderly, and people with disabilities often face structural disadvantages that hinder their ability to prepare for or respond to climate-related events. In Southeast Asia, for example, women in rural farming communities bear a disproportionate burden of food insecurity during droughts, as they are often responsible for household food production but have limited land tenure or access to credit. Similarly, in South Asia, caste-based and gender-based disparities intersect with climatic shocks to create differential health outcomes during periods of water scarcity or disease outbreaks (Thomas et al. 2018). According to Ngcamu (2023), in Bangladesh, the climate victims are mostly women who are poverty stricken. Food hierarchies have a likelihood to make women (possible when pregnant or breastfeeding) and children susceptible to the increased risks of malnutrition and water shortages during and after disasters (Chowdhury et al. 2024). In addition, older adults and persons with disabilities may struggle with evacuation or accessing care during emergencies due to mobility or support challenges (Ngcamu 2023). These overlapping vulnerabilities compound the health risks faced by certain groups, especially in under-resourced settings.

2.3.2 Geographical Exposure and Environmental Hazards

Geographical exposure is a critical determinant of climate-related health risks, particularly in a region as ecologically and demographically diverse as South and Southeast Asia. The varying topographies and land-use patterns, ranging from densely populated coastal cities to vulnerable rural heartlands shape the scale and type of climate impacts communities face. These impacts include sea-level rise, storm surges, floods, droughts, glacial melt, landslides, and extreme heat, all of which interact with socioeconomic and infrastructural conditions to influence health outcomes.

Low-lying coastal areas, floodplains, deltas, and small island states are among the most vulnerable to sea-level rise and storm surges, which can lead to displacement, saltwater intrusion, and outbreaks of waterborne diseases such as cholera and typhoid (Ehsan et al. 2022). Coastal cities such as Bangkok, Jakarta, and Manila face compounded risks from storm surges, extreme rainfall, and tidal flooding. These hazards contribute to land loss through erosion and inundation, disrupt transportation and healthcare infrastructure, and lead to large-scale displacement (Ruan et al. 2024). Floodwaters often overwhelm sanitation systems, increasing the risk of waterborne diseases such as cholera, dysentery, and typhoid. Salinization of freshwater sources due to sea-level rise further reduces the availability of clean drinking water.

In addition, the intrusion of saltwater into agricultural lands negatively affects food production, thereby intensifying public health risks such as malnutrition, especially among children and the elderly (Chowdhury et al. 2024; Howkins et al. 2024).

Similarly, extreme heat events have become more frequent and intense in urban areas, particularly in densely populated cities with limited green spaces and poor air circulation. The Urban Heat Island (UHI) effect, intensified by concrete infrastructure and reduced vegetation, contributes to heat-related illnesses such as heatstroke, cardiovascular stress, and exacerbation of respiratory conditions (Piracha and Chaudhary 2022; Yadav et al. 2023). In cities like Delhi, Bangkok, and Kuala Lumpur, high temperatures disproportionately affect elderly individuals, outdoor workers, and those without access to cooling systems—populations that are often geographically and economically marginalised. In contrast, arid and semi-arid regions face heightened vulnerability to droughts and water scarcity, which can lead to food insecurity, undernutrition, and increased prevalence of diarrheal diseases (Chowdhury et al. 2024). In Indonesia and India, prolonged dry spells have reduced agricultural yields, stressed freshwater resources, and driven rural–urban migration, further intensifying the burden on health and urban infrastructure (Sarkar et al. 2022; Ngadi et al. 2023). The collapse of local water systems also forces communities to rely on unsafe water sources, raising the incidence of gastrointestinal diseases, especially among children.

Another example of geographically contingent risk is the expansion of vector-borne diseases in previously unaffected areas due to shifting temperature and precipitation patterns (Eldos et al. 2025). In mountainous regions such as the Nepalese mid-hills, warming temperatures are enabling the northward and upward spread of malaria and dengue vectors, placing new populations at risk who may lack both immunity and public health preparedness (Acharya et al. 2025). These transitions highlight how biophysical geography and climate variability interact to reconfigure disease ecologies, often ahead of public health adaptation efforts.

Importantly, the interplay between environmental hazard exposure and socioeconomic context determines the scale and severity of health outcomes. Populations living in informal settlements, peri-urban areas, or remote hinterlands often bear a disproportionate burden due to inadequate housing, lack of access to emergency services, and insufficient public health infrastructure. Thus, geographical exposure to climate-sensitive hazards is not just a function of location but of vulnerability layered across place, shaped by historical development patterns, governance structures, and land-use decisions.

2.3.3 Health Impacts of Climate-Induced Migration and Displacement

According to the United Nations High Commissioner for Refugees (UNHCR), approximately 32 million individuals were displaced by weather-related hazards only in 2022, representing a 41% increase compared to 2008 (UNHCR 2022). Climate-induced migration and displacement represent some of the most profound yet under-addressed pathways through which climate change impacts human health. As environmental stressors such as sea-level rise, floods, droughts, and extreme weather events intensify, they increasingly undermine the habitability of certain regions, forcing individuals and communities to move, either temporarily or permanently. These movements, whether internal or cross-border, voluntary or forced, are not only demographic and socioeconomic challenges but also major public health concerns (Issa et al. 2023). The health effects of displacement are complex, intersecting with disruptions to shelter, food, water, sanitation, healthcare access, and psychosocial well-being (Shultz et al. 2018). Overcrowding, poor ventilation, inadequate shelter, lack of safe food and water, reduced access to health services, and the loss of social network, all of which contribute to an increased risk of infection in disaster-related informal settlements (Segala et al. 2025). Common infections in these settings include water- and food-borne diseases such as cholera, dysentery, typhoid fever, and hepatitis A and E; diseases associated with overcrowding such as measles, meningitis, acute respiratory infections, and tuberculosis; and vector-borne illnesses such as malaria and dengue (Sakapaji 2023). Furthermore, factors like social instability, increased sexual violence, and restricted access to barrier contraception contribute to a rise in sexually transmitted infections (Segala et al. 2025).

In South and Southeast Asia, climate-related displacement is already a tangible reality. Bangladesh, for instance, experiences recurrent monsoonal flooding and coastal erosion, displacing hundreds of thousands annually. Many of these internally displaced persons (IDPs) relocate to overcrowded urban slums in Dhaka, where they face heightened exposure to unsanitary living conditions, air and water pollution, limited access to healthcare, and nutritional deficiencies (Displacement Solutions 2012; Sakapaji 2023). The health system, often already stretched, struggles to meet the surge in demand, especially for maternal and child health, infectious disease control, and mental health services.

The health vulnerabilities of displaced populations are further exacerbated by the conditions under which migration occurs. Displacement often disrupts continuity of care for those with chronic illnesses or ongoing medical needs, such as tuberculosis, HIV, or diabetes. In the case of climate refugees from typhoon-affected regions of the Philippines, for example, temporary shelters often lack adequate medical facilities, exacerbating disease transmission risks and delaying treatment (Salazar et al. 2017). Moreover, the stress of displacement itself, marked by trauma, uncertainty, and loss, contributes to a growing burden of mental health disorders, including anxiety, depression, and post-traumatic stress disorder (PTSD), particularly among women, children, and the elderly (Kalukusu 2025).

Environmental degradation and livelihood loss also act as slow-onset drivers of migration, which carry long-term health implications. For instance, prolonged droughts in rural India and Indonesia have eroded agricultural productivity, compelling youth and working-age adults to migrate to cities in search of alternative income (Sarkar et al. 2022; Ngadi et al. 2023). This demographic shift often fragments families and leads to the feminisation of poverty in rural areas, with women and children left behind in increasingly resource-scarce and health-vulnerable environments (Kalukusu 2025). Simultaneously, migrants arriving in peri-urban areas often inhabit marginal lands such as floodplains or polluted industrial zones, where exposure to environmental hazards is intensified.

2.3.4 Climate-Sensitive Disease Burdens and Regional Inequalities

Climate change has reshaped the global burden of disease by altering the distribution, transmission dynamics, and intensity of a wide range of health outcomes. Diseases that are sensitive to climatic factors such as temperature, precipitation, and humidity are emerging or intensifying in new areas (Bahrami et al. 2024). At the same time, previously endemic regions are experiencing changes in disease seasonality or severity.

Vector-borne diseases such as dengue, malaria, chikungunya, and zika are particularly sensitive to climate variability (Ebi and Hess 2020). Rising temperatures and altered rainfall patterns have expanded the geographical range of disease vectors like Aedes aegypti and Anopheles mosquitoes into areas that were previously non-endemic, including highland and temperate regions (Ebi and Hess 2020; Meier et al. 2022). For example, parts of Nepal that were once too cold for dengue transmission have reported increasing incidence rates in recent years (WHO 2024). Wibawa et al. (2024) document a dramatic surge in dengue cases, from 505,430 in 2000 to over 5.2 million in 2019, with 70% of the burden occurring in Asia. This rise has been driven not only by climatic conditions but also by rapid urbanisation, global trade, population mobility, and breakdowns in vector control programs. Flood events and unplanned settlements exacerbate the proliferation of mosquito habitats, creating ideal conditions for rapid disease transmission (Ebi and Hess 2020). In Malaysia, dengue outbreaks have surged in peri-urban zones where unplanned development and inadequate drainage systems intersect with climatic conditions favourable to mosquito breeding (Abdul Hadi et al. 2025). These ecological changes translate directly into growing public health burdens, which place increasing pressure on healthcare infrastructure and expose geographical disparities in disease prevention and control.

In addition to vector-borne diseases, climate change influences the incidence of waterborne and foodborne illnesses. Extreme rainfall and flooding can overwhelm sanitation systems, leading to the contamination of drinking water with pathogens such as Vibrio cholerae, Salmonella, and E. coli (Balta et al. 2024). In South and Southeast countries, recurrent flooding events have been associated with outbreaks of diarrhoeal diseases, particularly among children under five (Wang et al. 2023; Li et al. 2025; Yirdaw et al. 2025). Conversely, drought conditions exacerbate water scarcity and lead to the use of unsafe water sources, further increasing the risk of gastrointestinal infections. These conditions disproportionately affect rural and low-income communities, where access to safe water and adequate healthcare is already limited.

Respiratory diseases are also influenced by climate-sensitive factors. Increased exposure to airborne particulate matter during prolonged droughts or wildfire events has been linked to rising rates of asthma and chronic obstructive pulmonary disease (COPD). Urban heat islands exacerbate respiratory and cardiovascular stress, particularly among older adults and those with pre-existing health conditions (Piracha and Chaudhary 2022; Yadav et al. 2023). In megacities, the health effects of heatwaves are magnified by poor air quality, high humidity, and limited green spaces. These conditions create compounded health burdens for urban populations, especially those living in informal settlements without adequate housing or ventilation (Segala et al. 2025).

Climate-sensitive disease burdens are deeply intertwined with regional inequalities in healthcare access, surveillance systems, and public health preparedness. Areas with limited disease monitoring capacity often fail to detect outbreaks in time to prevent widespread transmission. Furthermore, marginalised populations such as migrants, indigenous groups, and those living in ecologically fragile zones frequently fall outside the scope of national health protection frameworks (Salazar et al. 2017; Makharia et al. 2023).

2.3.5 Institutional Response Gaps and Policy Inequities

Institutional capacity and policy coherence are critical for effective public health adaptation to climate change. Across South and Southeast Asia, regional disparities in governance structures, resource allocation, infrastructure, and administrative capacity have led to spatially uneven public health responses to climate-induced challenges. A major institutional challenge is the limited integration of climate change into health planning, particularly at the sub-national and local levels. In many countries, national policies such as National Adaptation Plans (NAPs) and Nationally Determined Contributions (NDCs), acknowledge the health impacts of climate change, but implementation tends to be centralised, with insufficient support for district-level execution (Paz-Soldán et al. 2023). As a result, rural, coastal, and mountainous regions, where health risks from flooding, vector-borne diseases, or heatwaves are most acute, are often left out of coordinated adaptation responses. For example, while capital cities may benefit from advanced climate information systems, hospitals and clinics in deltaic communities and highland areas may lack even basic disease surveillance tools or reliable power supply.

Urban–rural and core-periphery divides also reflect deep-rooted policy inequities. Urban centres typically receive prioritised investments in early warning systems, emergency health infrastructure, and public communication networks. In contrast, peripheral areas such as the flood-prone areas and remote villages often face delayed or insufficient government responses during climate-related disasters (Palmeiro-Silva et al. 2023). This geographic disparity in institutional readiness results in higher mortality, longer recovery times, and more severe health burdens in under-resourced locations. Furthermore, institutional fragmentation across sectors and administrative boundaries limits the capacity for coherent action. Health and climate are frequency treated as separate domains, and coordination among ministries (e.g., health, environmental, disaster management) is often weak. These gaps are especially problematic in transboundary river basins and ecosystems, where floods, disease vectors, and pollution cross provincial or national borders (Xu et al. 2023) The absence of coordinated response mechanism in such areas leads to inconsistent health surveillance, delayed epidemic response, and uneven distribution of aid and medical supplies (Xu et al. 2023). For example, in many rural or low-capacity regions, health data collection is manual, underreported, and lacks real-time capacity. This undermines the early detection of climate-sensitive diseases like dengue, cholera, and leptospirosis. Without geographically disaggregated data, it becomes difficult to track disease outbreaks linked to rainfall, temperature, or flooding, particularly in regions that are already under stress due to limited healthcare access or recurring climate hazards.

Climate-induced infrastructure disruption further exposes geographic disparities in institutional preparedness (Tesfaye et al. 2025). Extreme events such as floods or landslides often damage roads, bridges, and health facilities, which cutting off access to healthcare in isolated communities. The lack of resilient infrastructure in these areas is not merely an environmental issue but a reflection of institutional neglect, where resource allocation is skewed towards politically or economically strategic locations. In addition, institutional inadequacies also impact mental health outcomes, particularly among healthcare workers and vulnerable populations in high-risk geographies. Health professionals stationed in disaster-prone areas are increasingly reporting stress, anxiety, and burnout due to repeated exposure to crises, poor working conditions, lack of support (Tesfaye et al. 2025). Feeling of abandonment and helplessness are magnified in regions where institutional response is delayed or minimal, leading to long-term psychological effects including depression, moral injury, and post-traumatic stress disorder (PTSD) (Tesfaye et al. 2025). Furthermore, the indirect social and economic disruptions caused by climate change, such as displacement, supply-chain issues, and strain on health systems, may compound the mental health challenges facing health workers.

Policy processes often overlook geographically marginalised and socioeconomically disadvantaged communities. Communities in informal settlements, Indigenous territories, or remote islands are frequently excluded from planning and decision-making, resulting in adaptation strategies that fail to consider local knowledge or context-specific risks. For example, traditional knowledge on flood risk management disaster in Indonesia and Sri Lanka is rarely incorporated into formal climate-health planning, despite its potential value (Karunarathne and Lee 2019; Troglić et al. 2019; Azizah et al. 2025). In addition, climate change and health literacy-related policy planning and research are predominantly shaped by institutions in the Global North, which may limit their relevance and applicability in different socio-political context such as Malaysia (Pandya-Wood et al. 2024). At the institutional level, geographical, political, and policy-related factors significantly influence a country's capacity to respond to and recover from climate-related shocks. According to Palmeiro-Silva et al. (2023), institutional resilience is closely tied to the availability of financial resources and political commitment to climate adaptation and mitigation. However, in South Asia, the lack of a unified or standardised framework to assess vulnerability and adaptive capacity has led to fragmented approaches, making it difficult to systematically address susceptibility to climate hazards across the region (Palmeiro-Silva et al. 2023).

Funding mechanisms further reflect geographic and institutional inequity. Access to international climate finance, such as from the Green Climate Fund (GFC), often depends on administrative capacity and technical expertise, resources that are concentrated in national ministries or major cities (Omukuti et al. 2022). Subnational governments and local health authorities in highly vulnerable but under-resourced regions are often unable to compete for or manage such funding, despite their higher exposure to climate-health risks.

2.4 Application of Geospatial Techniques for Analysing Climate-Health Vulnerabilities

Understanding how geography shapes the health impacts of climate change requires tools that can analyse the spatial variation of environmental hazards, population vulnerabilities, and health outcomes. In this context, the application of geospatial techniques such as geographic information systems (GIS), remote sensing, spatial modelling has been widely applied in South and Southeast Asia (Nepal et al. 2021; Kamel Buolos and Wilson 2023). By utilising geospatial techniques, researchers and policymakers can visualize, analyse, and predict climate-induced health vulnerabilities across different regions (Nepal et al. 2021).

Geospatial techniques help translate complex interactions between climate variables and health outcomes into spatially explicit formats. This allows for the identification of high-risk regions, visualisation of exposure pathways, and integration of multiple layers of data, including environmental indicators (e.g., temperature, rainfall), socio-economic metrics (e.g., population density, poverty), and health information (e.g., disease incidence, hospital access). By doing so, they reveal the geographic disparities in public health vulnerability, making it possible to design more equitable and targeted interventions (Kamel Buolos and Wilson 2023).

There are three key applications of geospatial techniques in managing climate-health vulnerability; (1) mapping climate-induced health risks, (2) identifying vulnerabilities hotspots, and (3) integrating geographic data into policy and planning.

2.4.1 Mapping Climate-Induced Health Risks

Mapping climate-induced health risks involves the spatial representation of environmental hazards and their potential impact on human health. Advances in geographical technology have significantly improved our capacity to detect, monitor, and predict these risks. Recent innovations such as high-resolution satellite imagery, synthetic aperture radar (SAR), artificial intelligence (AI)-powered predictive mapping, drone surveillance, the Internet of Things (IoT), and sensor networks have made data collection more accurate and timelier. These tools allow for the integration of environmental, climatic, and health data to generate dynamic maps that support real-time surveillance and forecasting of climate-related health threats (Hawchar et al. 2020; Berrang-Ford et al. 2021). This technological advancement directly supports targeted intervention by allowing for the identification of geographic hotspots of vulnerability. For example, a study in Kalimantan, Indonesia by Heo et al. (2023) demonstrated the use of a Random Forest machine learning to integrate climatic, topographic, vegetation, and anthropogenic data. This approach produced high-resolution maps identifying both current and projected hazard zones. Although the study focused on environmental risks such as droughts and forest fires, the same mapping approach can be applied to anticipate related health risks, such as respiratory illnesses due to haze exposure and heat-related morbidity under future climate scenarios. These spatial analyses enable public health agencies to allocate resources with precision, ensuring that interventions are deployed where they are most needed (Heo et al. 2023).

Beyond identifying high-risk zones for intervention, these technologies facilitate real-time monitoring of dynamic environmental and epidemiological variables. IoT-enabled sensor networks, for instance, continuously track microclimatic conditions such as humidity, surface temperature, and rainfall, which influence the breeding and transmission patterns of vectors like mosquitoes (Miller et al. 2023; Essamlali et al. 2024; Nemade et al. 2024). Drones equipped with thermal and multispectral sensors complement satellites-derived observations by offering low-attitude, localized surveillance, even in areas with cloud cover or poor accessibility (Manase et al. 2025). SAR technology, unaffected by weather conditions, allows for uninterrupted monitoring of flood-prone or coastal zones. This flow of real-time data supports early warning systems that can forecast the likelihood of disease outbreaks or heatwaves, enabling proactive public health responses such as issuing heat advisories or mobilizing mobile medical units (Miller et al. 2023; Nemade et al. 2024). Such timely interventions are vital in reducing morbidity and mortality in vulnerable communities.

Furthermore, mapping technologies play a crucial role in post-disaster recovery planning. In the immediate aftermath of floods, cyclone, or landslides, the use of drones and high-resolution satellite imagery allows for rapid assessment of infrastructure damage, displacement patterns, and accessibility of healthcare facilities (Iqbal et al. 2023). SAR imaging, particularly useful in inundated or cloud-covered areas, helps delineate the extent of physical damage and identifies zones cut off from critical services. This information enables emergency response teams to prioritize areas for restoration of essential services like potable water, sanitation, and healthcare delivery. Additionally, AI-assisted damage classification supports the efficient deployment of relief operations, ensuring that healthcare systems are not overwhelmed and that secondary outbreaks, such as cholera or respirator infections, are promptly addressed (Albahri et al. 2024; Essamlali et al. 2024; Morchid et al. 2024).

Lastly, these geospatial and digital technologies contribute to policy and preparedness. Spatial models that overlay future climate projections with demographic, epidemiological, and infrastructural data allow decision-makers to forecast health burdens under various climate scenarios. For instance, predictive mapping can reveal the potential expansion of malaria zones or the areas likely to experience repeated heatwaves, informing long-term planning for healthcare system strengthening, infrastructure development, and climate adaptation policies (WHO 2021; IPCC 2022). Such evidence-based planning is particularly important for urban planning in coastal cities and for ensuring healthcare access in rural, climate-sensitive areas. By supporting spatially explicit risk assessments, geospatial tools empower policymakers to prioritize investments in health and resilience based on the distribution and intensity of projected risks (Asaaga et al. 2024).

2.4.2 Identifying Vulnerabilities Hotspots

One of the key benefits of geospatial analysis is its ability to identify climate-health vulnerability hotspots, areas where populations face the highest risks due to environmental, social, and economic factors. Vulnerability hotspots are typically characterized by a combination of high exposure to climate hazards, limited adaptive capacity, and pre-existing health disparities (Bharadwaz et al. 2025). For example, densely populated urban slums often lack adequate infrastructure to cope with extreme heat or flooding, making residents more susceptible to heat stress, respiratory diseases, and waterborne infections (Damte et al. 2023). Recent geospatial research by Baqa et al. (2025) in Karachi, Pakistan demonstrates the effective use of Local Climate Zone (LCZ) classification combined with high-resolution satellite (SDGSAT-1) to assess heat-related health risks at the community level. By applying Crichton's Risk Triangle, incorporating hazard, exposure, and vulnerability, the study mapped vulnerability hotspots across 1,026 census wards. Areas such as LCZ2 (compact mid-rise), LCZ3 (compact low-rise), and LCZ7 (lightweight low-rise/slum areas) were identified as high-risk due to their dense population, poor infrastructure, and limited vegetation. The study also performed hotspot analysis using Optimized Hot Spot Analysis (OHSA) in ArcGIS to identify statistically significant clusters of high risk (Baqa et al. 2025). This spatial identification of risk zones offers actionable insights for targeted adaptation strategies, making it a valuable model for other South and Southeast Asian cities facing similar urban heat challenges.

Similarly, rural communities with poor healthcare access may struggle to respond to vector-borne disease outbreaks that are exacerbated by climate change. By mapping these vulnerabilities, authorities can prioritize interventions in high-risk areas, such as improving healthcare facilities, strengthening early warning systems, and implementing climate-resilient infrastructure projects (Rahaman et al. 2023). A recent five-year study in Selangor, Malaysia, applied techniques to identify persistent dengue hotspots across urban districts using ArcGIS-based spatial analysis, including the Getis-Ord Gi statistic and inverse distance weighted (IDW) interpolation (Mohd Hardy Abdullah et al. 2025). Over 3,000 hotspots were mapped, with outbreak linked to seasonal rainfall and driven by urban density, poor waste management, and inadequate drainage. The study demonstrates how geospatial techniques support targeted vector control and highlight the need for sustained surveillance in urbanising areas (Mohd Hardy Abdullah et al. 2025). In addition, identifying hotspots also enables targeted policy responses that address both environmental and social determinants of health, ensuring that vulnerable populations receive the necessary support to adapt to climate change impacts.

2.4.3 Integrating Geographic Data into Policy and Planning

Transforming geographic data into impactful policies is critical for improving climate adaptation and public health outcomes. The use of geographic information systems (GIS), satellite imagery, predictive models, and sensor networks, and artificial intelligence (AI) offers decision-makers powerful tools to drive targeted interventions. These tools allow planners and authorities to identify vulnerable areas, assess risk levels, and plan interventions based on reliable spatial data. For example, urban planning authorities can use geospatial information to guide the development of heat mitigation strategies such as expanding urban greenery, improving ventilation in residential areas, and designing transportation systems that reduce exposure to air pollution. In the health sector, geographic data helps in mapping the spread of diseases, locating gaps in healthcare access, and planning for future infrastructure needs (Ansah et al. 2024).

Geographic data integration delivers cross-sector benefits, enabling collaboration between health services, environmental agencies, urban development authorities, and disaster response teams. Shared geospatial information improves operational coordination during emergencies and ensures timely and efficient responses. During floods, landslides, or heatwaves, spatial analysis helps locate at-risk communities, optimize evacuation routes, and guide the deployment of emergency resources (Kamel Buolos et al. 2023). In the context of disaster preparedness, mapping flood plains, heat islands, or disease vector hotspots can improve the positioning of healthcare infrastructure and emergency response facilities. These examples illustrate how geographic intelligence strengthens resilience beyond a single sector and supports whole-of-government action (Neta et al. 2022).

Are we turning maps into action to protect vulnerable populations? This question underlines a critical gap between data generation and policy execution. While many countries have made strides in collecting and analysing geospatial data, challenges remain in translating insights into concrete policy decisions. To close this gap, governments must invest in the institutional capacity needed to interpret and apply geographic data across multiple levels of planning. This includes empowering local authorities, integrating spatial data into national development strategies, and ensuring that policies are informed by both scientific evidence and community realities (FGDC 2020). Geographic data must not remain confined to reports, it should be a central element in shaping equitable, inclusive, and forward-looking health and climate adaptation policies (Giest 2023).

In addition to guiding immediate responses, geographic data informs long-term preparedness and strategic planning. By linking environmental trends with health outcomes, spatial-climate analysis supports forecasting models that simulate future climate-health scenarios. These projections can inform policy development in areas such as urban expansion, water resource management, and disease surveillance systems (Kamel Buolos et al. 2023). In this context, geographic data serve as a critical driver of long-term of transformation by enabling evidence-based governance that anticipates climate-related risks, informs the development of resilience infrastructure, and protects the health and well-being of the most vulnerable populations (Puntub et al. 2024).

2.5 Conclusion

This chapter has provided a comprehensive synthesis of how geography influences the health impacts of climate change, with a specific focus on spatial variability and vulnerability in South and Southeast Asia. By integrating current knowledge from interdisciplinary literature, the chapter conceptualises the key geographical, socioeconomic, environmental, and institutional determinants that mediate climate-health risks. Rather than presenting original spatial analysis, the chapter's contribution lies in organising and interpreting current knowledge to highlight the importance of spatial context in understanding climate-health relationships.

Climate-induced health risks are not uniformly distributed; they are shaped by local conditions such as infrastructure, governance, socio-economic status, and environmental exposure. This review has shown that urban slums, coastal settlements, agricultural hinterlands, and displaced communities often bear a disproportionate burden of climate-health vulnerabilities. These findings underscore the need for responses that are sensitive to geographic disparities and informed by both empirical data and local realities. In addition, by placing geography at the centre of the climate-health discussion, this chapter advances the field in several keyways. It provides a spatial framework for assessing vulnerability, encourages integration of geospatial tools into health and climate planning, and emphasises the relevance of spatially disaggregated data for policy development. Geospatial techniques such as GIS, remote sensing, and spatial modelling are increasingly used to map health risks, identify vulnerability hotspots, and support evidence-based decision-making in the region.

Future research should aim to translate these conceptual insights into practice by linking geospatial, health, and climate datasets through empirical studies. There is also a pressing need for participatory approaches that incorporate local and Indigenous knowledge systems into mapping and policy development. Strengthening interdisciplinary capacity, particularly in low- and middle-income countries, will be vital to ensure that climate-health adaptation strategies are not only spatially informed but also equitable and inclusive.

Acknowledgements The authors extend their appreciation to all individuals and institutions whose work formed the basis of this review.

Funding: The study was not supported by any fundings.

Conflict of Interest: The authors declare that there are no conflicts of interest related to this work.

Informed consent: Not applicable.

Authors Contribution: All authors contributed equally to the conceptualisation, literature review, analysis, and writing of this manuscript.

References

Abdul Hadi Z, Che Dom N, Ishak AR, Salleh SA, Abdullah S, Precha N (2025) Geospatial dynamics of dengue fever density in Kuantan, Malaysia: GIS-based approach. Geogr Environ Sustain 54–64. https://doi.org/10.24057/2071-9388-2025-3298

Acharya BK, Khanal L, Dhimal M (2025) Increased thermal suitability elevates the risk of dengue transmission across the mid hills of Nepal. PLoS ONE 20(4):e0322031. https://doi.org/10.1371/journal.pone.0322031

Albahri AS, Khaleel YL, Habeeb MA, Ismael RD, Hameed QA, Deveci M, Homod RZ, Albahri OS, Alamoodi AH, Alzubaidi L (2024) A systematic review of trustworthy artificial intelligence applications in natural disasters. Comput Electr Eng 118(Part B):109409. https://doi.org/10.1016/j.compeleceng.2024.109409

Ansah EW, Amoadu M, Obeng P, Sarfo JO (2024) Health systems response to climate change adaptation: a scoping review of global evidence. BMC Public Health 24. https://doi.org/10.1186/s12889-024-19459-w

Asaaga FA, Tomude ES, Rickards NJ, Hassall R, Sarkar S, Purse BV (2024) Informing climate-health adaptation options through mapping the needs and potential for integrated climate-driven early warning forecasting systems in South Asia—a scoping review. PLoS ONE 19(10):e0309757. https://doi.org/10.1371/journal.pone.0309757

Azizah C, Lizar CA, Robo S, Zuraihan, Arsyani I, Iqbal M, Muharramsyah R, Ismahadi (2025) Indigenous wisdom in flash flood adaptation and mitigation: insights from the Gayo highlands, Indonesia. EGUsphere. https://doi.org/10.5194/egusphere-2024-3774

Bahrami G, Ghavidel N, Mahmoodi Z, McDermott R, Seddighi H (2024) The health effects of climate change: identifying strategies, policies, and knowledge gaps: an umbrella review. Int J Environ Health Res 1–18. https://doi.org/10.1080/09603123.2025.2498624

Balta I, Lemon J, Murnane C, Pet I, Vintila T, McCleery D, Callaway T, Douglas A, Stef L, Corcionivoschi N (2024) The one health aspect of climate events with impact on foodborne pathogens transmission. One Health 19:100926. https://doi.org/10.1016/j.onehlt.2024.100926

Baqa MF, Lu L, Guo H, Song X, Alavipanah SK, Nawaz-ul-Huda S, Li Q, Chen F (2025) Investigating heat-related health risks related to local climate zones using SDGSAT-1 high-resolution thermal infrared imagery in an arid megacity. Int J Appl Earth Obs Geoinf 136:104334. https://doi.org/10.1016/j.jag.2024.104334

Berrang-Ford L, Sietsma AJ, Callaghan M, Minx JC, Scheelbeek PFD, Haadaway NR, Haines SAH, Dangour AD (2021) Systematic mapping of global research on climate and health: a machine learning review. Lancet Planet Health 5(8):E514–E525. https://doi.org/10.1016/S2542-5196(21)00179-0

Bharadwaz GSVSA, Pal I, Mozumder C, Ahmad MM (2025) Assessing dynamic coastal vulnerability to climate hazards: a geospatial approach in Kakinada District, Andhra Pradesh, India. Environ Sustain Indic 26:100622. https://doi.org/10.1016/j.indic.2025.100622

Chaudhry D (2024) Climate change and health of the urban poor: the role of environmental justice. J Clim Chang Health 15:100277. https://doi.org/10.1016/j.joclim.2023.100277

Chowdhury NR, Moname EJ, Al Azad G, Hani U, Nazmin F, Ferdaus F (2024). Interplay between malnutrition and infectious diseases insights from a cross-sectional study in Bangladesh. Asia Pac J Med Innov 1(2):41–47. https://doi.org/10.70818/apjmi.2024.v01i02.07

World Health Organization (WHO) 12 October 2023. Climate change. https://www.who.int/news-room/fact-sheets/detail/climate-change-and-health

Damte E, Manteaw BO, Wrigley-Asante C (2023) Urbanization, climate change and health vulnerabilities in slum communities in Ghana. J Clim Chang Health 10:100189. https://doi.org/10.1016/j.joclim.2022.100189

Displacement Solutions (2012) Climate displacement in Bangladesh. The need for urgent housing, land and property (HLP) rights solutions. https://unfccc.int/files/adaptation/groups_committees/loss_and_damage_executive_committee/application/pdf/ds_bangladesh_report.pdf

Dube E, Mtapuri O, Matunhu J (2018) Flooding and poverty: two interrelated social problems impacting rural development in Tsholotsho district of Matabeleland North province in Zimbabwe. Jamba 10(1):455. https://doi.org/10.4102/jamba.v10i1.455

Ebi KL, Hess JJ (2020) Health risks due to climate change: inequity in causes and consequences. Health Aff 39(12):2056–2062. https://doi.org/10.1377/hlthaff.2020.01125

Ebi KL, Vanos J, Baldwin JW, Bell JE, Hondula DM, Errett NA, Hayes K, Reid CE, Saha S, Spector J, Berry P (2021) Extreme weather and climate change: population health and health system implications. Annu Rev Public Health 42:293–315. https://doi.org/10.1146/annurev-publhealth-012420-105026

Ehsan S, Begum RA, Maulud KNA (2022) Household external vulnerability due to climate change in Selangor coast of Malaysia. Clim Risk Manag 35:100408. https://doi.org/10.1016/j.crm.2022.100408

Eldos HI, Tahir F, Athira UN, Mohamed HO, Samuel B, Skariah S, Al-Ghamdi SG, Al-Ansari T, Sultan AA (2025) Mapping climate change interaction with human health through DPSIR framework: Qatar perspective. Heliyon 11(3):242455. https://doi.org/10.1016/j.heliyon.2025.e42455

Essamlali I, Nhaila H, El Khaili M (2024) Advances in machine learning and IoT for water quality monitoring: a comprehensive review. Heliyon 10(6):e27920. https://doi.org/10.1016/j.heliyon.2024.e27920

Federal Geographic Data Committee (FGDC) (2020) National spatial data infrastructure strategic plan 2021–2024: Reston, Virginia, USA, Federal Geographic Data Committee, p 17. https://www.fgdc.gov/nsdi-plan/nsdi-strategic-plan-2021-2024.pdf

Giest S (2023) Policy capacity mechanism for addressing complex, place-based sustainability challenges. In: Background paper for the OECD-EC high-level expert workshop series on place-based policies for the future, workshop 5, 15 September 2023. https://www.oecd.org/regional/place-based-policies-for-the-future.htm.

Greeham SV, Ferranti EJS, Jones S, Zhong J, Grayson N, Needle S, Acton WJF, Mackenzie AR, Bloss WJ (2024) An open access approach to mapping climate risk and vulnerability for decision-making: a case study of Birmingham, United Kingdom. Clim Serv 36:100521. https://doi.org/10.1016/j.cliser.2024.100521

Hawchar L, Naughton O, Nolan P, Stewart MG, Ryan PC (2020) A GIS-based framework for high-level climate change risk assessment of critical infrastructure. Clim Risk Manag 29:100235. https://doi.org/10.1016/j.crm.2020.100235

Heo S, Park S, Lee DK (2023) Multi-hazard exposure mapping under climate crisis using random forest algorithm for the Kalimantan Islands, Indonesia. Sci Rep 13:13472. https://doi.org/10.1038/s41598-023-40106-8

Howkins J, Schmidt DN, Thomas J, Hayward R, Eunice Lo YT, Neal J, Lewis J, Carruthers E, Coleborn S, Murray V, Oliver I (2024) Mapping vulnerability to climate-related hazards to inform local authority action in adaptation: a feasibility study. Public Health Pract 8:100549. https://doi.org/10.1016/j.puhip.2024.100549

Hussainzad EA, Gou Z (2024) Climate risk and vulnerability assessment in informal settlements of the Global South: a critical review. Land 13(9):1357. https://doi.org/10.3390/land13091357

Intergovernmental Panel on Climate Change (IPCC) (2014) Climate change 2014. Impacts, adaptation, and vulnerability. Summary for policymakers. https://www.ipcc.ch/site/assets/uploads/2018/03/ar5_wgII_spm_en-1.pdf

Intergovernmental Panel on Climate Change (IPCC) (2022) IPCC report 2022: How climate change impacts vulnerable countries. https://idronline.org/article/climate-emergency/ipcc-report-2022-how-climate-change-impacts-vulnerable-countries/?utm_source=Google&utm_medium=Grants&utm_campaign=Dalit_History_06&gad_source=1&gad_campaignid=22591559224&gbraid=0AAAAADCozDxEdBM8CvjYenHx_plVr-o1M&gclid=EAIaIQobChMI1bqYyeqQjgMV8qVmAh3tjiIHEAAYASAAEgID9PD_BwE

International Labour Organization (ILO) (2024) Ensuring safety and health at work in a changing climate. Global report. https://www.ilo.org/sites/default/files/2024-07/ILO_SafeDay24_Report_r11.pdf

Iqbal U, Riaz MZB, Zhao J, Barthelemy J, Perez P (2023) Drones for flood monitoring, mapping and detection: a bibliometric review. Drones 7(1):32. https://doi.org/10.3390/drones7010032

Issa R, Sarsour A, Cullip T, Toma S, Ruyssen I, Scheerens C (2023) Gaps and opportunities in the climate change, migration and health nexus: insights from a questionnaire based study of practitioners and researchers. J Migr Health 7:100171. https://doi.org/10.1016/j.jmh.2023.100171

Javan K, Altaee A, BaniHashemi S, Darestani M, Zhou J, Pignatta G (2024) A review of interconnected challenges in the water-energy-food nexus: urban pollution perspective towards sustainable development. Sci Total Environ 912:169319. https://doi.org/10.1016/j.scitotenv.2023.169319

Kalukusu AT (2025) Health impacts of climate-induced migration. Res Inven J Public Health Pharm 4(1):17–23. https://doi.org/10.59298/RIJPP/2025/411723

Kamel Buolos MN, Wilson JP (2023) Geospatial techniques for monitoring and mitigating climate change and its effects on human health. Int J Health Geogr 22:2. https://doi.org/10.1186/s12942-023-00324-9

Karunarathne AY, Lee G (2019) Traditional social capital and socioeconomic networks in response to flood disaster: a case study of rural areas in Sri Lanka. Int J Disaster Risk Reduct 14:101279. https://doi.org/10.1016/j.ijdrr.2019.101279

Li A, Toll M, Bentley R (2023) Mapping social vulnerability indicators to understand the health impacts of climate change: a scoping review. Lancet Planet Health 7:e925–e937

Li P, Tong M, Wu J, Meng X, Fan X, Sun X, Li J, Xue T (2025) Flood exposure contributes to under-five mortality in low- and middle-income countries. One Earth 8(5):101305. https://doi.org/10.1016/j.oneear.2025.101305

Lv Y, Sarker MNI, Radin Firdaus RB (2024) Disaster resilience in climate-vulnerable community context: conceptual analysis. Ecol Ind 158:111527. https://doi.org/10.1016/j.ecolind.2023.111527

Mahmood R, Zhang L, Li G, Roy NR, Rawnaq N, Yan M, Dong Y, Chen B (2023) Geospatial assessment of intrinsic resilience to the climate change for the central coast of Bangladesh. Clim Risk Manag 40:100521. https://doi.org/10.1016/j.crm.2023.100521

Makharia GK, Sadeghi A, Leddin D, Costello A (2023) Impact of climate change on vulnerable populations. Gut 72:2201–2204. https://doi.org/10.1136/gutjnl-2023-331195

Manase A, Manyevere A, Abd Elbasit MAM, Mashamaite C (2025) The use of UAV-based systems in monitoring forest health: potentials and challenges. Sci African 28:e02724. https://doi.org/10.1016/j.sciaf.2025.e02724

Meier D, Patkee P, Strange A (2022) The risk of a lifetime: mapping the impact of climate change on life and health risks. Swiss Re Institute. https://www.swissre.com/dam/jcr:b851c59c-d3e0-4f87-817c-62691b985bc5/2022-11-sri-expertise-publication-impact-of-climate-change-life-health-risks.pdf

Miller M, Kisiel A, Cembrowska-Lech D, Durlik I, Miller T (2023) IoT in water quality monitoring—are we really here? Sensors 23(2):960. https://doi.org/10.3390/s23020960

Mohd Hardy Abdullah NA, Che Dom N, Salleh SA, Salim H, Precha N, Dapari R (2025). Spatiotemporal dynamics of dengue hotspots in an urbanizing landscape: a five-year analysis in Selangor, Malaysia. Clin Epidemiol Glob Health 32:101966. https://doi.org/10.1016/j.cegh.2025.101966.

Morchid A, El Alami R, Raezah AA, Sabbar Y (2024) Applications of internet of things (IoT) and sensors technology to increase food security and agricultural sustainability: benefits and challenges. Ains Shams Eng J 15(3):102509. https://doi.org/10.1016/j.asej.2023.102509

Napoli CD, McGushin A, Romanello M, Ayeb-Karlsson S, Cai W, Chambers J, Dasgupta S, Escobar LE, Kelman I, Kjellstrom T, Kniveton D, Liu Y, Liu Z, Lowe R, Martinez-Urtaza J, MeMichael C, Moradi-Lakeh M, Murray KA, Rabbaniha M, Semenza JC, Shi L, Tabatabaei M, Trinanes JA, Vu BN, Brimicombe C, Robinson EJ (2022) Tracking the impacts of climate change on human health via indicators: lessons from the Lancet countdown. BMC Public Health 22:663. https://doi.org/10.1186/s12889-022-13055-6

Nemade B, Maharana KK, Kulkarni V, Mondal S, Pradeep Ghantasala GS, Al-Rasheed A, Getahun M, Soufiene BO (2024) IoT-based automated system for water-related disease prediction. Sci Rep 14:29483. https://doi.org/10.1038/s41598-024-79989-6

Nepal S, Tripathi S, Adhikari H (2021) Geospatial approach to the risk assessment of climate-induced disasters (drought and erosion) and impacts on out-migration in Nepal. Int J Disaster Risk Reduct 59:102241. https://doi.org/10.1016/j.ijdrr.2021.102241

Neta G, Pan W, Ebi K, Buss DF, Castranio T, Lowe RM, Ryan SJ, Steward-Ibarra AM, Hapairai LK, Sehgal M, Wimberly MC, Rollock L, Lichtveld M, Balbus J (2022) Advancing climate change health adaptation through implementation science. Lancat Planet Health 6(11):e909–e918. https://doi.org/10.1016/S2542-5196(22)00199-1

Ngadi N, Zaelany AA, Latifa A, Harfina D, Asiati D, Setiawan B, Ibnu F, Triyono T, Rajagukguk Z (2023) Challenges of agriculture development in Indonesia: rural youth mobility and aging workers in agriculture sector. Sustainability 15(2):922. https://doi.org/10.3390/su15020922

Ngcamu BS (2023) Climate change effects on vulnerable populations in the Global South: a systematic review. Nat Hazards 118:977–991. https://doi.org/10.1007/s11069-023-06070-2

Nilsson M, Sie A, Muindi K, Bunker A, Ingole V, Ebi KL (2021) Weather, climate, and climate change research to protect human health in sub-Saharan Africa and South Asia. Glob Health Action 14:1984014. https://doi.org/10.1080/16549716.2021.1984014

Omukuti J, Barrett S, White PCL, Marchant R, Averchenkova A (2022) The green climate fund and its shortcomings in local delivery of adaptation finance. Climate Policy 22(9–10):1225–1240. https://doi.org/10.1080/14693062.2022.2093152

Paavola J (2017) Health impacts of climate change and health and social inequalities in the UK. Environ Health 16(Suppl 1):113. https://doi.org/10.1186/s12940-017-0328-z

Palmeiro-Silva YK, Lescano AG, Flores EC, Yamileth Astorga E, Rojas L, Chavez MG, Mora-Rivera W, Hartinger SM (2023) Identifying gaps on health impacts, exposures, and vulnerabilities to climate change on human health and wellbeing in South America: a scoping review. Lancet RegNal Health—Americas 26:100580. https://doi.org/10.1016/j.lana.2023.100580

Pandya-Wood R, Azhari A, Johar H, Johns-Putra A, Muhamad N, Su TT (2024) Systematic review of climate change induced health impacts facing Malaysia: gaps in research. Environ Res Health 2:032002. https://doi.org/10.1088/2752-5309/ad6208

Parums DV (2024) A review of the increasing global impact of climate change on human health and approaches to medical preparedness. Med Sci Monit 30:e945763–e945768. https://doi.org/10.12659/MSM.945763

Paz-Soldán VA, Valcarcel A, Canal-Solis K, Miranda-Chacon Z, Palmeiro-Silva YK, Hartinger SM, Suárez-Linares AG, Falla-Valdez V, Intimayta-Escalante C, Lehoucq M, Pretell A, Castillo-Neyra R (2023) A critical analysis of national plans for climate adaptation for health in South America. Lancet RegNal Health 26:100604. https://doi.org/10.1016/j.lana.2023.100604

Piracha A, Chaudhary MT (2022) Urban air pollution, urban heat island and human health: a review of the literature. Sustainability 14(15):9234. https://doi.org/10.3390/su14159234

Puntub W, Greiving S, Birkmann J (2024) Framework for collaborative local climate adaptation scenario development – nexus between climate resilience, public health service and spatial planning. Int J Clim Chang Strat Manag 17(1):311–332. https://doi.org/10.1108/IJCCSM-09-2023-0110

Rahaman MA, Kalam A, Al-Mamun M (2023) Unplanned urbanization and health risks of Dhaka City in Bangladesh associations between urban environment and public health. Front Public Health 11:1269362. https://doi.org/10.3389/fpubh.2023.1269362

Reid A (2019) Climate change education and research: possibilities and potentials versus problems and perils. Environ Educ Res 25:767–790. https://doi.org/10.1080/13504622.2019.1664075

Ruan X, Sun H, Shou W, Wang J (2024) The impact of climate change and urbanization on compound flood risks in coastal areas: a comprehensive review of methods. Appl Sci 14(21):10019. https://doi.org/10.3390/app142110019

Sakapaji SC (2023) Climate-induced migration a new normal? A systematic research analysis of the climate-induced migration crisis in Bangladesh. Eur J Theor Appl Sci 1(4):463–489. https://doi.org/10.59324/ejtas.2023.1(4).42

Salazar MA, Law R, Pesigan A, Wikler V (2017) Health consequences of typhoon Haiyan in the Eastern Visayas region using syndromic surveillance database. PLoS Curr 9. https://doi.org/10.1371/currents.dis.4a3d3b4474847b2599aa5c5eefe3a621

Sarkar B, Dutta S, Singh PK (2022) Drought and temporary migration in rural India: a comparative study across different socio-economic groups with a cross-sectional nationally representative dataset. PLoS ONE 17(10):e0275449. https://doi.org/10.1371/journal.pone.0275449

Sarkar M, Paul S, Garai J (2024) Climate-induced coastal occupational vulnerability and livelihood insecurity: insights from coastal Bangladesh. Prog Disaster Sci 24:100382. https://doi.org/10.1016/j.pdisas.2024.100382

Segala FV, Guido G, Stroffolini G, Masini L, Cattaneo P, Moro L, Motta L, Gobbi F, Nicastri E, Vita S, Latta R, Otranto D, Locantore P, Occa E, Putoto G, Saracino A, Gennaro FD (2025) Insights into the ecological and climate crisis: emerging infections threatening human health. Acta Trop 262:107531. https://doi.org/10.1016/j.actatropica.2025.107531

Sen B, Dhimal M, Latheef AT, Ghosh U (2017) Climate change: health effects and response in South Asia. BMJ: Br Med J 359. https://www.jstor.org/stable/26950675.

Sentian J, Payus CM, Herman, Kong VWY (2022) Climate change scenarios over Southeast Asia. https://www.apn-gcr.org/bulletin/article/climate-change-scenarios-over-southeast-asia/. Accessed on 23 May 2025

Shultz JM, Rechkemmer A, Rai A, McManus KT (2018) Public health and mental health implications of environmentally induced forced migration. Disaster Med Public Health Prep 1–7. https://doi.org/10.1017/dmp.2018.27

Sultana F, Petzold J, John S, Muehlberger V, Scheffran J (2024) Systematic mapping of climate change impacts on human security in Bangladesh. Climate 12(9):141. https://doi.org/10.3390/cli12090141

Tesfaye AH, Prior J, McIntyre E (2025) Impact of climate change on health workers: a scoping review. J Public Health (Berl). https://doi.org/10.1007/s10389-025-02418-z.

Thomas K, Hardy RD, Lazrus H, Mendez M, Orlove B, Rivera-Collazo I, Roberts JT, Rockman M, Warner BP, Winthrop R (2018) Explaining differential vulnerability to climate change: a social science review. Wires Clim Change 10(2):e565. https://doi.org/10.1002/wcc.565

Tiwari I, Tilstra M, Campbell SM, Nielsen CC, Hodgins S, Vargas ARO, Whitfield K, Sapkota BP, Yamamoto SS (2022) Climate change impacts on the health of South Asian children and women subpopulations—a scoping review. Heliyon 8(10):e10811. https://doi.org/10.1016/j.heliyon.2022.e10811

Troglić RŠ, Wright GB, Duncan MJ, van der Homberg MJC, Adeloye AJ, Mwale FD, Mwafulirwa J (2019) Characterising local knowledge across the flood risk management cycle: a case study of Southern Malawi. Sustainability 11(6). https://doi.org/10.3390/su11061681

UNHCR (2022) Global trends report 2023. United Nations high commissioner for refugees. https://www.unhcr.org/global-trends-report-2022.

USEPA (2025) Climate change and the health of socially vulnerable people. https://www.epa.gov/climateimpacts/climate-change-and-health-socially-vulnerable-people. Accessed on 23 May 2025

Vaidya R, Bhatti VK, Bobdey S (2024) Climate change and health: an impending crisis and a public health challenge. Med J Armed Forces India 9, 80(3), 249–251. https://doi.org/10.1016/j.mjafi.2024.04.007

Wang P, Asare EO, Pitzer VE, Dubrow R, Chen K (2023) Floods and diarrhea risk in young children in low- and middle-income countries. JAMA Pediatr 177(11):1206–1214. https://doi.org/10.1001/jamapediatrics.2023.3964

Wibawa BSS, Wang Y-C, Andhikaputra G, Lin Y-K, Hsieh L-HC, Tsai K-H (2024) The impact of climate variability on dengue fever risk in central java, Indonesia. Clim Serv 33:100433. https://doi.org/10.1016/j.cliser.2023.100433

World Health Organization (WHO) (2021) SID Summit for Health: for a healthy and resilient future in small island developing states. https://cdn.who.int/media/docs/default-source/sids-summit/sids-climate-policy-brief.pdf?sfvrsn=c1536f14_5

World Health Organization (WHO) 12 October 2023. Climate change. https://www.who.int/news-room/fact-sheets/detail/climate-change-and-health

World Health Organization (WHO) 26 September 2024. Vector-borne diseases. https://www.who.int/news-room/fact-sheets/detail/vector-borne-diseases. Accessed on 1 Feb 2025

Xu R, Yu P, Liu Y, Chen G, Yang Z, Zhang Y, Wu Y, Beggs PJ, Zhang Y, Boocock J, Ji F, Hanigan I, Jay O, Bi P, Vargas N, Leder K, Green D, Quail K, Huxley R, Jalaludin B, Hu W, Dennekamp M, Vardoulakis S, Bone A, Abrahams J, Johnston FH, Broome R, Capon T, Li S, Guo Y (2023) Climate change, environmental extremes, and human health in Australia: challenges, adaptation strategies, and policy gaps. Lancet RegNal Health—West Pac 40:100936. https://doi.org/10.1016/j.lanwpc.2023.100936

Yadav N, Rajendra K, Awasthi A, Singh C, Bhushan B (2023) Systematic exploration of heat wave impact on mortality and urban heat island: a review from 2000 to 2022. Urban Climate 51:101622. https://doi.org/10.1016/j.uclim.2023.101622

Yirdaw G, Mekonen H, Assaye BT, Amare GA, Yenew C (2025) Prevalence of acute diarrhea and its risk factors among under five children in flood affected Dasenech District, Southern Ethiopia: a cross-sectional study. Sci Rep 15:16980. https://doi.org/10.1038/s41598-025-02120-w

Zain A, Sadarangani SP, Shek LP-C, Vasoo S (2024) Climate change and its impact on infectious diseases in Asia. Singapore Med J 65(4):211–219. https://doi.org/10.4103/singaporemedj.SMJ-2023-180

Chapter 3
Role and Impact of Holistic One Health Approaches for Protecting Public Health in the Era of Climate Change in South Asia

Muhammad Mubashar Beig, Waqas Ahmad, Hassnain Muhammad, Sarmad Zia, and Saqib Mehmood

Abstract Climate changes, decrease forests, rapid urban growth, and global warming facilitate the spread of animal borne diseases and antibiotic resistance in South Asia. Under variable weather conditions and ecosystem imbalance, the animals tend to lose their natural niches and rely on human populations and related societies, which in turn cause transmission of infectious diseases from wild life to humans. To address these health concerns and challenges associated with interconnectedness among animals, humans, and environment, One Health approach has been rated as most suitable strategy to tackle these multidisciplinary issues. Surveillance systems, artificial intelligence (AI) tools, and early disease detection methods facilitate the development of early disease warning signs to protect life on land from emerging and reemerging infectious diseases in South Asia or lower middle-income countries. In these countries, the interdepartmental research collaboration is weak and stakeholders related to climate changes and environment sector work in silos under budgetary constraints. South Asia, therefore, must empower and operationalize the holistic One Health approach not only at district or regional levels, but also at national levels in the form of legislative policy to sustain global health and

---

M. M. Beig
Institute of Microbiology, University of Veterinary and Animal Sciences, Lahore, Pakistan
e-mail: mubashar.beig@uvas.edu.pk

W. Ahmad (✉)
Department of Clinical Sciences, University of Veterinary and Animal Sciences, Lahore, Narowal Campus, Pakistan
e-mail: waqas.hussain@uvas.edu.pk

M. Hassnain
University of Veterinary and Animal Sciences, Lahore, Narowal Campus, Narowal, Pakistan

S. Zia
University of Veterinary and Animal Sciences, Lahore, Pakistan

S. Mehmood
Department of Pathology, University of Veterinary and Animal Sciences, Lahore, Pakistan

P. Ghosh et al. (eds.), *Climate Crises and Public Health Challenges*, Disaster Risk Reduction, https://doi.org/10.1007/978-981-95-9709-3_3

tackle emerging issues of climate changes and health emergencies. Developing technologies, raising community awareness, and community-centered approaches will protect climatic issues from future disease threats. The South Asian region can mitigate climate-sensitive diseases, stabilize ecosystems, and secure long-term public health goals using One Health based research and developmental projects. Success pertaining to good climatic conditions hinges on trans-disciplinary collaboration, data sharing, mutually agreed environmental policy, and equitable resource allocation. With coordinated actions, South Asia can transform its vulnerability into resilience, safeguarding both populations and ecosystems from escalating climate-health risks.

Keywords Climate change · Public health · One Health · South Asia · Zoonotic diseases

3.1 Introduction

Public health is vulnerable to biological and environmental threats due to fluctuating climate changes around the globe. One of the most affected areas is South Asia, which is grappling with a multitude of climatic and socio-economic conditions due to variations in temperatures, weather fluctuations, and the rising sea levels (Mirza 2011). Alongside all these, the air and water quality have worsened, which in turn affects the overall human health, food security, and disease outbreaks in the region. Due to global warming, South Asia is facing a continuous increase of heatwaves, which aggravates the frequency of heat strokes, asthma, dehydration, and cardiovascular diseases. Displacement due to coastal flooding further worsens sanitation, nutrition, and subsequently put people at higher risk of life threatening diseases. Moreover, reoccurring cyclones, prolonged drought, flooding, and the overall deterioration of the air and water quality makes this region more prone to several respiratory diseases (Elahi 2016).

South Asia possesses 2 billion people, while majority of them are living in urban areas in high population density and climate-sensitive rural communities (Dodman 2009). The immediate solutions of public health challenges associated with climatic changes are often beyond the capacity of the existing health care infrastructure of any country in the said region. Higher burden of respiratory diseases, including chronic obstructive pulmonary disease (COPD) and asthma have been linked to heat stress and air pollution (Hansel, McCormack, and Kim 2016). Flooding and water contamination have sparked outbreaks of cholera, dengue, and other vector-borne diseases. Deforestation and loss of biodiversity are responsible for the re-emergence and emergence of new zoonotic diseases like Nipah virus, avian influenza, and COVID-19 (Acosta-España et al. 2024). The climate-related extreme weather events displace millions of people, leading to psychological or mental distress and a greater health burden on healthcare systems. Climate changes are promoting health inequities by putting more pressure on vulnerable populations of those regions where communities lack resources and resilience to fight back (Ayub et al. 2024). Most of the climate adaptation policies are directed towards economic and infrastructural resilience, and do not consider the direct and indirect health consequences. Integrative and preventative approaches that target root causes of climate-related health crises are critical in protecting human and environmental health (Gasparri et al. 2022).

There are three main ways trough which climate change affects public health: (1) direct climate change-related exposures, like the rise in heat waves; (2) indirect climate change-related exposures, like water scarcity and the rise in infectious disease transmission due to the impact of climate change on vectors; and (3) societal and economic disruptions that may have an impact on health through population displacement, conflict, disruptions of safe food and water supplies, and governments' failure to ensure environmental quality (Daily and Ehrlich 1996).

The significance of addressing these climate-related health concerns cannot be overemphasized. While there is increasing awareness of the connection between climate change and public health, but there is still limited evidence on the efficacy of integrated One Health approaches in addressing these issues. The One Health concept, which advocates the collaboration of human, animal, and environmental health sectors for the well being of public health, has not been comprehensively studied, applied, and researched in South Asia (Qureshi et al. 2024). In Southeast Asia (ASEAN), regional initiatives such as the ASEAN One Health Network has been highlighted at various forums to respond cross-sectoral issues presented by climate-induced health emergencies (Hansen 2023). Increasing awareness of One Health approaches benefits global and regional levels (e.g., ASEAN), there is still a lack of knowledge on how well these strategies can be applied in various South Asian settings. Further studies are required on how One Health frameworks can be tailored specifically to combat climate-related public health risks in various socio-economic environments in South Asia (Coker et al. 2011).

This chapter aims to investigate the direct and indirect effects of climate changes on public health in South Asia. It will describe the existing gaps in climate resilience policy, recommended public health interventions, potential impediments, and policy issues in the application of One Health interventions at the interface of human–animal–environment. This chapter also describes policy guidelines to implement One Health regulations in national and regional climate resilience plans. The research outcomes will add value to the body of literature through the integration of climate science, public health, and biodiversity conservation, providing a multi-disciplinary approach that integrates environmental conservation with veterinary and public health security.

3.2 Public Health Consequences of Variable Climate in South Asia

Food security and health are the two aspects of human security in South Asian nations which are anticipated to be significantly impacted by climate changes (Akram and Hamid 2014). The climate change and climate sensitive health risks have a complex relationship that determine the burden of vulnerability, health system capacity, and resilience. Through various exposure pathways, climate change impacts human health by influencing demographic characteristics, geographic location, biological health status, and socioeconomic conditions which include poverty and accessibility to resources. These factors determine both an individuals or community's level of susceptibility and extent of exposures to climate related health risks (Shrestha et al. 2017). Extreme weather events (floods, storms), climate induced displacement and disasters, heat stress, deteriorated air quality, changes in water quality and quantity, food security issues and changes in vector distribution and ecology are the key exposure pathways (Butsch et al. 2023). These pathways end up resulting in a bunch of negative health outcomes ranging from injury to death due to malnutrition, food borne diseases, non communicable diseases (heart conditions due to stress of heat), and psychosocial health problems (Raza et al. 2023; White et al. 2023).

According to data in this year's study conducted by Lancet, the fast-changing climate is posing record-breaking risks to people's survival, well-being, and health everywhere in the world. The number of days of intense precipitation increased in 61% of the world's geographical area between 1961–1990 and 2014–2023, thereby increasing the risk of floods, infectious diseases, and contaminated water. In 2023, 48% of the world's geographical area experienced at least one month of severe drought, making it the second-largest afflicted area since 1951 (Romanello et al. 2024). The average annual temperature in South Asia is expected to rise which make it extremely vulnerable to climate change (Dimitrova et al. 2021). Temperatures in the area have risen by 1.0–3.0 °C over Western Afghanistan and Southwestern Pakistan and by 1.0 °C to 1.5 °C over portions of India, Sri Lanka, Pakistan, and Nepal between 1950 and 2010 (Mani et al. 2018).

We find that warming temperatures greatly impact vector-borne disease transmission and that moderate warming (30–32 °C) increases vectorial capacity and reduces the extrinsic incubation period, but results in reduced mosquito survival. Mosquito larvae quickly mature with rising temperatures, while adult mosquitoes digest blood faster and feed more frequently (Karmakar and Pradhan 2020). Dengue vector abundance has increased by 9.5% over the previous 110 years, and additional increases are anticipated, especially in heavily populated areas like South Asia (Zain et al. 2024). The distribution of malaria cases across Pakistan from 2004 to 2018 depict the temperature changes. Warmer regions like Sindh, Baluchistan, and some areas of Khyber Pakhtunkhwa have reported high malaria cases because of favourable temperature that encourage mosquito breeding (Fazal and Hotez 2020). Higher precipitation creates favourable breeding conditions for mosquitoes and other vectors, and increases the number of vector-borne diseases such as malaria, dengue and chikungunya. Malaria is reported in around 1 million people in Pakistan annually with major outbreaks in Baluchistan and Khyber Pakhtunkhwa and dengue and chikungunya have also been cases in considerable number of patients last year alone 30,000 cases of chikungunya were reported in Karachi in 2011 (Saeed et al. 2023). Each year India records over 120,000 dengue cases, more than 7 million malaria cases in 2018 and periodic chikungunya outbreaks in south and central regions (Cecilia 2014). In 2019, Bangladesh experiences more than 100,000 dengue cases, a steady reduction in malaria, while cases of chikungunya are on the rise especially in urban centres such as Dhaka (Zaman and Mitra 2024).

The severe rise in temperature range to 45 °C and 40 °C in various parts of Pakistan sprawled the heatwave and caused various deaths (Karim et al. 2024). In India, more than 1,200 people also died from extreme heat. Current trends could lead to over 20% increase of heat related mortality in the region by 2050 particularly affecting the elderly and people with pre-existing conditions (Cole et al. 2023).

Because of climate change, floods are becoming more frequent which greatly increase the spread of waterborne diseases. In 2010, typhoid cases increased by over 40%, during the incidents of floods in Pakistan, while 10,000 cholera cases were recorded during the 2021 monsoon floods in Bangladesh (Noureen et al. 2022; Jung et al. 2023). Similarly in Pakistan, the incidence of typhoid cases rose from 5,000 cases/year in 2000 to more than 20,000 during 2020 (Alied et al. 2024). Floods and droughts combine with high water stress in 74% of South Asia's population to make water quality worse and elevate the chances for infectious pathogens such as E. coli, Salmonella, and Shigella. These indicate the considerable bearing of environmental stress on the growing burden of waterborne diseases in the region (Zain et al. 2024).

Global warming results in displacement of millions of people in South Asia. Disappointed people have higher levels of anxiety and depression owing to losing their houses and sources of income. For instance, the prevalence rate of mental health disorders among the displaced people was 50% higher than those people who were not displaced (Cianconi et al. 2020). Recent studies showed that people facing natural disasters have higher chances of being diagnosed with mental disorders. Rates of post traumatic stress disorder (PTSD) may rise by 30% in displaced populations. Every year, millions of people in South Asia are affected and forced to flee by natural disasters (Kunnathepeedikayil 2020). Floods in Kashmir during 2014 affected over 20,0000 people and increased prevalence of mental health problems among the affected population (Wells et al. 2023).

3.3 The One Health Approach

Integrated monitoring and surveillance systems that screen health trends across the human–animal–environment interface are fundamental characteristics of the One Health approach. These systems are essential for the early identification of zoonotic diseases that are becoming increasingly affected by climate changes. For instance, high rise in temperature and precipitation changes may expand the range of disease vectors, making malaria and dengue fever more transmissible (Destoumieux-Garzón et al. 2018). Integrating data from many sectors enables a better understanding of the multifactorial nature of health risks, leading to timely interventions (Destoumieux-Garzón et al. 2018).

Another key component of the One Health approach is resource sharing among sectors. Better coordination of human, veterinary, and environmental health resources will likely result in better health outcomes. For example, joint funding opportunities can provide resources for research and interventions at the interface of these sectors to address health-related issues (Bakiika et al. 2023). In addition, cross-sectoral policy-making is critical for developing integrated solutions to the health risks posed by climate change. The national policies of any country need to ensure that they address multisectorial issues of agriculture, wildlife, and urban areas to secure healthy and planet-friendly production practices which will reduce health risks associated with environmental change (Acharya et al. 2020).

A few of the many global programs that demonstrate the successful use of the One Health approach include the response to tackle rift valley fever (RVF) through integrated surveillance systems across the African Union in collaboration with the World Organization for Animal Health (Samuel, Flores, and Frisancho 2020). This initiative highlights the need for international cooperation and data sharing to address zoonotic diseases more effectively. Kenya's One Health Initiative also aimed to strengthen the capacity for surveillance and response to zoonotic diseases. The initiative has subsequently enhanced the country's capacity by integrating veterinary and human health data to enable quicker and more effective responses in dealing outbreaks of infectious diseases (Munyua et al. 2019).

3.4 Role of One Health in Addressing Climate Change and Health in South Asia

3.4.1 Integrated Data Systems for Climate and Health Surveillance

The establishment of integrated data systems to monitor climate impacts on human and animal health is crucial for the control and prevention of infectious diseases in South Asia. Stakeholders can use the climate data layer to comprehend trends and anticipate outbreaks with greater precision by connecting climate data to human health surveillance. Real-time data on temperature and rainfall can help to inform public health officials about potential disease outbreaks and possible associations with vector populations (Mahrous et al. 2020). These integrated systems will not only improve preparedness, but also support targeted interventions that can significantly reduce health risks associated with climate variability (Yasobant et al. 2019). Undoubtedly, up-to-minute data are extremely useful in outbreak prediction or assessment of ecosystem changes. Equitable access to reliable epidemic information can forecast new health threats, especially in countries where climate change is paving the way for vectors and pathogens (Gifford et al. 2018). For instance, mobile health technologies have shown promise in rural South Asia to conduct survey pertaining to emerging diseases (Aptriana et al. 2022).

3.4.2 Role of Ecosystem Management in Preventing Zoonoses Through One Health

Ecosystem management is an important part of the One Health approach, particularly in preventing zoonoses. Habitat protection and wildlife management can help reduce the risks of disease spillover from animals to humans. For example, conserving natural habitats and biodiversity can reduce these types of human-wildlife interactions that promote the transmission of zoonotic pathogens (Bäck and Calltorp 2015). As such, land-use policies that directly affect ecosystem health also play a pivotal role in disease prevention. South Asian countries are beginning to adopt land-use strategies that reduce exposure to pathogen vectors. For example, integrated pest management practices in agriculture reduce vector-borne disease prevalence by reducing harmful pesticides and encourage sustainable agriculture (Makino-Oi et al. 2016). These policies can also increase overall ecosystem health.

3.4.3 Sustainable Agriculture and Animal Health in One Health for Climate Resilience

Climate change is a crucial context for sustainable agriculture, some of which are related to the One Health approach. South Asia can promote sustainable food security by adopting practices which are conducive to climate resilience, such as crop diversification and organic farming, thereby reducing the health risks associated with chemical inputs (Braam et al. 2023). Healthy soil and water resources are further strengthened by sustainable farming practices. Similarly, animal health is an important dimension of food security within the paradigm of One Health core competencies. Healthy livestock populations are essential to the general economic well-being of rural communities and countries because they provide germ free meat and milk. Animal vaccination and disease preventive measures can be implemented in animal populations for biosecurity, human health, and well-being to decrease potential risk factors for zoonotic disease and animal food (Shen et al. 2021).

3.4.4 Integrated Water Management to Prevent Waterborne Diseases

Water management is vital to limit contamination and fight waterborne diseases, especially in South Asia, where there is a continuous struggle in accessing clean drinking water. The One Health concept underlines the demand to adopt harmonize approaches in water management, where human and ecosystem health are viewed in tandem and share same water resources (O'Shanahan et al. 2022). Better sanitation and adequate water access can prevent waterborne diseases exacerbated by climate change in many countries. Several initiatives in India, Bangladesh, and Pakistan have demonstrated the importance of integrated water management for health outcomes. For instance, community-driven sanitation initiatives in India have increased access to clean water and decreased water-related diseases (Yasobant et al. 2021). Similarly, integrated management practices in Bangladesh have shown significant health benefits for at-risk populations through improved water quality (Yasobant et al. 2021).

3.4.5 Policy Integration and Funding Mechanisms

One Health principles should be integrated into national public health policies for the effective implementation of adaptive measures to climate change. It also involves the revision of health policies to include implications of climate changes, national One Health action plans, and coherence with the international frameworks such as the Sendai Framework for Disaster Risk Reduction and the Paris Agreement (Mettenleiter et al. 2023). More specifically, sufficient financial budget is also necessary to create a sustainable funding mechanism to build public health infrastructure. There are funding mechanisms available such as climate adaptation funds, public private partnerships, and international grants that governments in South Asia can explore to support One Health based strategies (Alenga 2023).

3.4.6 Innovative Research to Develop Climate-Health Linkages in One Health Framework

Research and innovations are key components to build an integrated One Health approach that minimizes health impacts of climate change. Transitional research that integrates epidemiology, climatology, ecology, and social sciences to generate rich insights in understanding how climate change alters disease frequencies, especially of vector-borne and zoonotic diseases across South Asia (Mumford et al. 2023). Joint multidimensional application of predictive modeling using geographical information system and genomic surveillance enable the healthcare professionals to design improved early warning systems and to predict outbreaks more effectively, while conducting real-time public health interventions. Nature-based solutions involving ecosystem restoration and sustainable agriculture can help to reduce climate-sensitive health risks to strengthen environmental resilience (Seddon et al. 2020). Innovative technological solutions for disease surveillance systems such as mobile health applications to capture event based surveillance data are often integrated with tele-sensing technology. These tele-sensing technologies are usually equipped with AI analytics in recent years that enable fingerprinting of disease vectors and allow transfer of data in real time between health systems of humans, animals, and environment (Kumar et al. 2023). Research that addresses robust climate-resilient solutions in tandem with community-oriented adaptive strategies and effective public health education approaches to ensure the strength of public health responses will lead to better health outcomes. Translating research into effective strategies at the local level depends on networks of collaborative research studies to inform policy change and active engagement with stakeholders (Wamsler 2017).

3.4.7 Strengthening Public Health Infrastructure in Climate Vulnerable Areas

Enhancing public health infrastructure in climate vulnerable regions is crucial to reduce the rising health threats from climate changes, extreme weather occurrences, vector borne ailments, and water borne diseases. This climate variability does affect some populations in South Asia far more than others, and also there can be no disaster preparedness and response without resilient health systems (Kurebwa and Kurebwa 2025) as shown in Fig. 3.1. It needs to build climate resilient healthcare systems that can withstand the storms or extreme weather events so that staff and patients alike will continue to enjoy consistent access to safe, clean water, sanitation, and medical supply chains in the aforementioned cases of emergency. Integrated disease surveillance systems also have the potential to complement early warning signs and enhance outbreak prediction and response (Bolton et al. 2020).

3.4.8 Capacity Building and Trained Workforce Development

Capacity building and workforce development also play an important role in building the capacity of healthcare workers to manage the climate sensitive health risks. Public health resilience can be enhanced through training programs on climate change adaptation, emergency response training as well as One Health approaches (Sharma 2024). To increase public awareness and encourage preventive measures of the public, community-based health interventions and effective communication strategies are very important. South Asian countries can create adaptive public health systems that

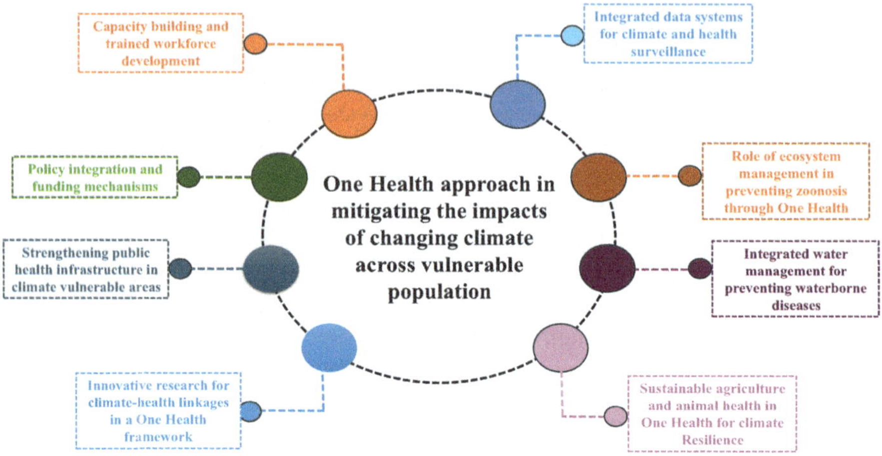

Fig. 3.1 One health approach: integrating climate change mitigation strategies for vulnerable populations

can respond to those climate change challenges through integration of cross sectoral collaboration and sustainable financing mechanisms (Sharma 2024).

3.5 Case Studies and Impacts: One Health Perspectives on Regional Climatic Changes in South Asia

3.5.1 India (Combating Vector-Borne Diseases with Integrated Approaches)

Vector-borne diseases (VBDs) like malaria, dengue fever, and chikungunya fever are major public health problems in India and developing world. Therefore, there is use of Intensified Vector Management, which uses a package of vector control methods in order to control these diseases. The National Vector Borne Disease Control Programme is to exterminate malaria by the year 2030 and the strategy used include community health, surveillance, and technology (Dhillon 2008).

In 2020, pilot project was conducted by using drones to survey breeding sites of mosquitoes in urban areas of India. This is in line with a larger framework to improve climate sensitivity through data acquisition, and a synchronization of vector control interventions. In addition to reduce overall mortality from acute diseases, the program also develops sustainable health systems to meet new challenges resulting from climate changes associated with vectors, seasonal temperature, and rainfall (Gaurav et al. 2024).

India has now started providing some grants for research on impact of climate change on VBDs and to create partnerships with health workers, ecological researchers, and policy makers. Inclusive to such main goals, there is a perspective to construct interlinking models that will be able to withstand alterations in climate conditions and enhance overall health (Ryan et al. 2023).

3.5.2 Bangladesh (Water and Sanitation Management for Public Health)

Bangladesh is among the most threatened countries by the current crises of flooding and cyclones as they worsen waterborne diseases. The country has embraced the community-based water and sanitation management practices in the efforts to improve the health sector endurance. After the devastating flooding in 2020 among large portions of the districts, some communities started opening up more projects to provide the communities in the rural areas with safe water and sanitation systems. There were measures such as putting up of water harvesting facilities and the establishment of flood-proof latrines. These measures have gone a long way in the transmission of post-flood incidences of cholera and other waterborne diseases (Sivakumar and Stefanski 2011). Furthermore, the Government of Bangladesh has included climate change adaptation measures in the country's health policies by investing in health infrastructure. This way the government is encouraging people and particularly the public health department to participate in health related events (European Foundation for South Asian Studies 2023).

3.5.3 Nepal and Bhutan (Wildlife Conservation for Man's Well Being)

Conservation of wildlife in Nepal and Bhutan have concern with prevention of zoonotic diseases which are the major factors causing diseases transmission between animals and humans. It has been established by both countries that measures aiming to preserve the biodiversity will help to address challenges related to human wildlife conflicts. Wildlife corridors are effective when implemented in Bhutan to reduce instances of human-wildlife conflict while conserving wildlife. These corridors enable animals to move safely across the landscape without causing any risks of zoonotic disease transmission to other wildlife and human beings (Lham et al. 2019). Likewise, Nepal's One Health operating model synchronizes wildlife protection by tracking the rate of zoonotic diseases in regions with abundant biological diversity (Acharya et al. 2019). They have also spent funds supporting community education, which aims at informing the people on risks of zoonotic diseases and ways of reducing contact risks with wildlife. In allowing coexistence between humans and wildlife conservation, Nepal and Bhutan are preparing for future possible disease outbreak handling models associated with climate change (Nepal 2012; Tse-Ring et al. 2010).

3.5.4 Pakistan (Leveraging Animal Health into the Public Animal Health System)

One Health approach has not been well operationalized in Pakistan yet due to limited advocacy and collaboration among multi disciplines. Recently, the Pakistan One Health Alliance (POHA) in collaboration with the Ministry of National Health Services Regulations and Coordination, Government of Pakistan has initiated a public–private partnership to develop a project called "One Health Work Force Development" which will flourish concepts, themes, and core competencies in various stakeholders related to health of animals, humans, environment, climate, and ecosystem. Another policy brief known as "Investing in Animal Health to Support One Health in Pakistan" was also prepared by Brooke Pakistan and Ministry of National Food Security and Research in May 2023 (Brooke 2023). This call therefore serves as a reminder as to the Board's lack of investment in Animal Health Services which remain underdeveloped. The brief established that 8,118,319 rural families rely on livestock products to sustain their households contributing between 35 and 40 percent of their income. However, there is always a low supply of veterinary doctors, and on average the population of animals per vet is 100,000 (Council et al. 2013).

The Food and Agriculture Organization (FAO) joined with the Ministry of National Food Security & Research and USAID to improve the animal and the public health systems in Pakistan. This particular workshop was centered on managing new zoonotic disease threats using the One Health concept (Pakistan 2024). To improve veterinary service delivery, particularly to rural clients, Pakistan has set up the Mobile Veterinary Dispensaries (MVDs). These services focus in delivering initial and basic health care to farmers specifically on diseases and its appropriate management in their own surroundings. The MVDs play a critical role in keeping surveillance system intact to monitor disease outbreaks in livestock, vaccinating for contagious diseases, and educating farmers regarding appropriate farm animal handling (Rasul et al. 2023).

The National Programme for Animal Disease Surveillance and Control was launched with an aim to increase the capacity of identifying and locating specific animal diseases and their correlation with extreme climate conditions in Pakistan. On disease surveillance, this program aimed at enhancing disease early detection and appropriate response mechanisms (Saleem et al. 2023). By implementing robust data collection methods regarding climate change and animal diseases, this initiative aims to decrease the frequency of the transboundary diseases, increase speed during disease outbreaks, and strengthen animal health care measures during extreme weather conditions (Usman et al. 2023).

3.6 Challenges and Barriers in Implementation of One Health in South Asia

3.6.1 Policy and Institutional Fragmentation

One of the biggest barriers in adopting One Health approaches in South Asia is the way policies and institutions have currently been set up. The problems of health policies, agriculture and environmental sectors do not align and therefore provide inefficient solutions for health crises. Various sectors of public health are frequently quite compartmentalized so the efforts do not converge, and hence do not adequately address the complex nature of health challenges (ESCAP 2020). There has been little cross talks among different ministries within South Asian region which are responsible for public health, veterinary services, and management of the environment (Mor 2023).

In India, the Ministry of Health addresses human health concerns, whereas Ministry of Agriculture is responsible for livestock health, even though the two areas interact in practice, but have not been well instanced to cooperate in written. This separation is undesirable as it might fail to confined early disease diagnosis especially when the disease are zoonotic in nature (Paul et al. 2020). In Bangladesh, measures required in preventing zoonotic diseases such as avian influenza have stressed the need for closer collaboration between sectors. While there have been efforts to put in place systems that link the human and animal health sectors. The frequent disease-specific One Health approaches have constrained other comprehensive approaches that would improve the health system's general advancement (Dahal et al. 2017b).

3.6.2 Funding Constraints and Economic Challenges

Lack of funds is the other important problem that hampers proper implementation of One Health endeavors in South Asia. Most of the countries in the region depend on international donors and aids to largely combat the health problems and environment initiatives (Dahal et al. Ewald 2017a). Projects developed to address zoonotic diseases are often interrupted by change in donor interests or withdrawal of support (Braam 2023). Funding allocations for One Health projects in Nepal also have a problem of budgeting for the country's public health stability. Despite attempts to promote the incorporation of animal health into the public health approaches, including the development of continuing professional development programs for veterinarians on zoonotic disease prevention and control, these strategies have been poorly funded to support expansion throughout the country (Malla et al. 2014).

3.6.3 Cultural and Social Barriers

Socioeconomic and cultural barriers are many relevant practices, which are directly or indirectly associated with broader health outcomes in societies. In Bhutan, despite a clear political engagement in One Health, the traditional view regarding livestock management is a limiting factor. Some farms may fail to vaccinate their animals due to some traditional beliefs on how to handle animals. Outreach in the form of community engagement campaigns that do not transgress cultural sensibilities while advocating for scientific knowledge are not available and usually inadequately supported or carried out (Namgyal et al. 2023). Rabies control programmes have improved Sri Lanka's intersectoral mechanisms, but still few owned animals skip vaccination. The use of community engagement activities involving society leaders are necessary, but necessitating large investment on outreach activities is difficult to complete (Bangert et al. 2017).

3.6.4 Technological and Infrastructure Limitations

There is dire need to have technological access and infrastructure to effectively work on One Health programs. It is common to see that most parts in South Asia lack proper health care facilities, which help in the delivery of health care services. This is particularly a huge challenge in rural India where there is very little use of technology to screen zoonotic diseases. Many regions are geographically located far from populated areas and experience low internet connectivity or signals from mobile phones to help support real time disease surveillance. This greatly limits outbreak control measures and also constrains the collection of essential data needed in the planning of public health interventions (Sekar et al. 2011).

3.6.5 Country-Wise Challenges to Implement One Health Policy

Country	Challenges	Details
Bangladesh (Ghosh et al. 2024)	Coordination and legislative gaps	Poor coordination among stakeholders and absence of appropriate legislation for intersectoral collaboration
	Resource limitations	A lack of long-term funding for research, surveillance, and logistical backup

(continued)

(continued)

Country	Challenges	Details
	Awareness and training deficits	Lack of awareness on One Health from stakeholders and shortage of professional training programs
India (Asaaga et al. 2021)	Fragmented intersectoral collaboration	Silos separation in human, animal, and environmental sectors
	Policy and implementation challenges	Zoonoses policies are in place, but inconsistent implementation at state and district levels
Nepal (Acharya et al. 2019)	Institutional and coordination deficiencies	Lack of a national One Health strategy and weak mechanisms for intersectoral coordination
	Resource and capacity constraints	Poor surveillance systems for zoonotic diseases, limited funding and a weak human resources base
Bhutan (Dahal, Upadhyay, and Ewald 2017b)	Policy development and implementation	Lack of legislative will to convert policies into actual programs owing to scarce resources
	Need for international partnerships	Reliance on international technical support and capacity-building programs
Pakistan (Abdullah and Shaikh 2025)	Ad hoc collaboration	Collaboration between sectors is variable and often is in response to national health crises
	Political and structural barriers	Fragmentation of public health efforts due to political instability and devolution of powers
Sri-Lanka (Kolla 2020)	Emerging collaborative efforts	Limited institutionalized One Health culture; Collaborations operate based on individual networks
	Resource and training needs	Lack of training programmes and resources required to deliver One Health initiatives

3.7 The Future of One Health in South Asia's Climate and Public Health Strategy

3.7.1 Opportunities for Innovation and Technology in One Health

It is crucial to use surveillance data to forecast prevalence of diseases under a variety of assumptions about future climatic trends, the characteristics of human and animal populations, changes in land use, and other variables that might be significant for certain diseases. AI and remote sensing technologies are emerging bright frontiers for surveillance and in the context of South Asia, these technologies can be integrated into One Health surveillance systems to increase its efficiency for disease monitoring and management (Zhang et al. 2024). Machine learning algorithms can improve human analysis by processing surveillance data and automating the identification of outbreaks of zoonotic diseases (Bezbaruah et al. 2023). Artificial intelligence is now being used to track and forecast the spread of COVID-19 (Allam et al. 2020). Remote sensing has been applied extensively in areas including resource analysis, environmental monitoring, weather forecasting, ocean monitoring, and land management (Gorji et al. 2017; West et al. 2019).

Natural language processing to collect insights from unstructured text-based reports and deep neural networks to combine and evaluate genetic, epidemiological, and meteorological data are being used in One Health surveillance (Bellinger et al. 2017; Ozturk et al. 2020). The use of IoT-based sensors in disease surveillance enables continuous and real-time monitoring of environmental parameters, animal health, and human activity. These systems are made up of a variety of components, such as biosensors, environmental sensors, agricultural sensors, wearable technology, and backend analytical platforms (Chude-Okonkwo 2022). Geographic information systems (GIS) is a multidisciplinary discipline that arose in the late 1960s as computer technology advanced (Cordes et al. 2020). Over the past three decades, significant advancements have been achieved in the monitoring of meteorological, climatic, environmental, and human variables that decrease or resurgence of vector-borne illnesses through the development of geographical information systems and earth observation satellites (Ceccato et al. 2018).

3.7.2 Global Partnerships and South–South Cooperation

The global scope of climate change and its associated concerns necessitate the broadest possible collaboration. To combat climate change and its impact on public health, international development cooperation or partnerships are crucial, especially to aid efforts of developing nations in South Asia. In the context of climate change, South–South collaboration enhances conventional North–South cooperation by expanding the scope and diversity of development partnerships that allow poor nations to pursue their own national sustainable development priorities and goals (Kim and Lee 2022). A shift towards more climate cooperation on the ground is reflected in South-South collaboration related to climate change (Stamm 2023). Climate change and its impact on One Health is one of the important problems impacting all SAARC (South Asian Association for Regional Cooperation) members, which brings overlapping interests together to create regional collaboration to address anticipated climate change challenges. The cross-border partnership will be successful in leveraging health and technological resources to strengthen climate resilience (Barrech et al. 2023).

International cooperations play a part in policy alignment, knowledge sharing, and finance. Recently, International Development Association has raised 24 billion dollars for developing countries including South Asia to address the effects of climate change and modernize healthcare services (Group 2024). The South-East Asia Regional Health Emergency Fund serves to bridge the gap between immediate emergency needs and formal disaster relief funding, enabling rapid financial support for health interventions in response to crises. Established by WHO in 2007, this institute allows member states to access funds within 24 h following a declared emergency, ensuring timely assistance for lifesaving healthcare and critical gaps during health emergencies across the region (Salvador et al. 2023).

3.7.3 Long-Term Vision for Climate-Resilient Health Systems

A climate resilient health system is an opportunity for sustainable human development. The objective of climate resilient health system is to strengthen medical facilities to safeguard and promote the health of their target populations in a changing and uncertain climate. A climate change resilient health system requires good governance, evidence-based policymaking, strategic planning, and cooperation with a range of stakeholders (Mosadeghrad and Shahsavani 2023). Additionally, national and local inter sectoral coordination should be improved. However, it has been noted that multi-sectoral partnerships and the execution of plans are difficult to implement in the long run (Organization 2024). Another strategy to increase institutional capacity to deliver better health services is to involve a variety of stakeholders (Sen et al. 2017). By implementing "health in all policies" and a multi-sectoral collaborative approach, the health sector is in charge of addressing environmental determinants of health (Organization 2015).

The technology and infrastructure used in healthcare should be sustainable and climate resilient. This can be achieved by replacing vulnerable healthcare facilities (Schnitter et al. 2019), and by updating the infrastructure. Making health systems climate resilient requires sufficient funds and resources. In developing nations, the annual cost of adapting to climate change is around USD 70 billion, and by 2030, it is expected to reach USD 140–300 billion (Neufeldt et al. 2021). As a result, funding should be allocated to assist strategies for health adaptation to climate change and to increase the resilience of the health system. To fund the health system, new inter-sectoral and partnership approaches should be employed. For example, in order to promote climate-resilient health initiatives, the Bangladesh Government created the Climate Change Trust Fund in 2010 (Rahaman et al. 2019). Designing a health system should take into consideration the global One Health projects and is pertinent to nutrition, pollution control, zoonotic disease prevention, and stringent surveillance systems. It is anticipated that this would lead to improved management of global health risks (Ganesh et al. 2020).

3.8 Conclusion

The operational, technical, and legislative One Health principles provide a revolutionary tool to deal with shared health crises of not only climate change, but also related public health, animal health, and environmental health issues in South Asia, which close key gaps in existing sectoral interventions. With regional case studies in India, Bangladesh, Nepal, Bhutan, and Pakistan, integrated surveillance, artificial intelligence-based instruments, and inter-sectoral cooperation can produce considerable resistance to climate-sensitive illness and serve as a blueprint for ecosystem management. Credible published literature supports the establishment of One Health in policy-making to sustain global health security, food security, but South Asian countries need to overcome health inequities, limited funds, legislative failure, limited cross-sectorial coordination, and lack of transparency.

Authors' Contributions Waqas Ahmad and Muhammad Mubashar Beig jointly conceptualized the chapter, conducted the core literature review, and led the critical revision of the manuscript. Muhammad Hassnain, Sarmad Zia, and Saqib Mehmood contributed equally to the writing and drafting of the manuscript. All authors reviewed and approved the final version of the chapter.

Funding The research was not supported by any funding.

Conflicts of Interest On behalf of all authors, the corresponding author declares that there is no conflict of interest.

Informed Consent Informed consent was obtained from all individual participants included in the study.

References

Abdullah MA, Shaikh BT (2025) Pathways to one health: enhancing inter-sectoral collaboration in Pakistan. EcoHealth 1–9

Acharya KP, Karki S, Shrestha K, Kaphle K (2019) One health approach in Nepal: scope, opportunities and challenges. One Health 8:100101

Acharya KP, Acharya N, Phuyal S, Upadhyaya M, Lasee S (2020) One-health approach: a best possible way to control rabies. One Health 10:100161

Acosta-España JD, Romero-Alvarez D, Luna C, Rodriguez-Morales AJ (2024) Infectious disease outbreaks in the wake of natural flood disasters: global patterns and local implications. Infez Med 32:451

Akram N, Hamid A (2014) Climate change and human security in South Asia. Sci Int 26:801–808

Alenga D (2023) A public–private partnership grand bargain for the green climate fund. Fisc Policy Instrum Green Dev

Alied M, Salam A, Sediqi SM, Kwaah PA, Tran L, Huy NT (2024) Disaster after disaster: the outbreak of infectious diseases in Pakistan in the wake of 2022 floods. Ann Med Surg 86:891–898

Allam Z, Dey G, Jones DS (2020) Artificial intelligence (AI) provided early detection of the coronavirus (COVID-19) in China and will influence future Urban health policy internationally 1:156–65

Aptriana CD, Sudarnika E, Basri C (2022) Nationally and locally-initiated One Health approach in controlling rabies in West Kalimantan, Indonesia. VetY World 15:2953

Asaaga FA, Young JC, Oommen MA, Chandarana R, August J, Joshi J, Chanda MM, Vanak AT, Srinivas PN, Hoti SL, Seshadri T, Purse BV (2021) Operationalising the "One Health" approach in India: facilitators of and barriers to effective cross-sector convergence for zoonoses prevention and control. BMC Public Health 21:1517

Ayub I, Sani SS, Zaman A, Shahid MR, Bashir W, Ijaz M, Zafar S, Rafique A (2024) Prevalence of Toxoplasma Gondii in human and animals and its economic impacts: a review. Pak Vet J

Bäck MA, Calltorp J (2015) The Norrtaelje model: a unique model for integrated health and social care in Sweden. Int J Integr Care 15

Bakiika H, Obuku EA, Bukirwa J, Nakiire L, Robert A, Nabatanzi M, Robert M, Moses M, Achan MI, Kibanga JB (2023) Contribution of the one health approach to strengthening health security in Uganda: a case study. BMC Public Health 23:1498

Bangert M, Molyneux DH, Lindsay SW, Fitzpatrick C, Engels D (2017) The cross-cutting contribution of the end of neglected tropical diseases to the sustainable development goals. Infect Dis Poverty 6:1–20

Barrech D, Malik M, Baloch MS, Sohail J (2023) Regional cooperation on climate change in South Asia. Int J Spec Educ 38

Bellinger C, Mohomed Jabbar MS, Zaïane O, Alvaro J (2017) BMC Public Health. A systematic review of data mining and machine learning for air pollution epidemiology 17:1–19

Bezbaruah R, Ghosh M, Kumari S, Nongrang L, Ali SR, Lahiri M, Waris H, Kakoti BB (2023) Role of AI and ML in epidemics and pandemics. Bioinformatics Tools for Pharmaceutical Drug Product Development Kakoti, pp 345–69

Bolton A, Khatri K, Ashworth M, S'Akolo Fa'u S, Fusi KF, Williamson J, Ofanoa R (2020) Strengthening adaptation planning and action to climate-related health impacts in Pacific islands countries: Tonga. Global J Health Sci 12:1–55

Braam DH, Tasker A, Parravani E, Saville K (2023) A fair share? Animal health actors and resources in one health initiatives: a multisite case study in Ethiopia and Pakistan. CABI One Health ohcs202300015

Braam D (2023) Zoonotic disease dynamics in displacement: A multisite case study in Sindh, Pakistan and Mafraq, Jordan

Brooke Pakistan (2023) Calls for animal health investment at policymaker event in Pakistan. https://actionforanimalhealth.org/wp-content/uploads/2023/02/A4AH-Report_FINAL-2023.pdf

Butsch C, Beckers L-M, Nilson E, Frassl M, Brennholt N, Kwiatkowski R, Söder M (2023) Health impacts of extreme weather events–Cascading risks in a changing climate. J Health Monit 8:33

Ceccato P, Ramirez B, Manyangadze T, Gwakisa P, Madeleine C (2018) J Infectious diseases of poverty Thomson. Data and tools to integrate climate and environmental information into public health 7:65–75

Cecilia D (2014) Current status of dengue and chikungunya in India. WHO South-East Asia J Public Health 3:22–27
Chude-Okonkwo UK (2022) One Health-inspired early implementation of airborne disease spread mitigation protocols aided by IoT-based biosensor network. Int J Sens Netw 39:215–26
Cianconi P, Betrò S, Janiri L (2020) The impact of climate change on mental health: a systematic descriptive review. Front Psych 11:490206
Coker RJ, Hunter BM, Rudge JW, Liverani M, Hanvoravongchai P (2011) Emerging infectious diseases in Southeast Asia: regional challenges to control. Lancet 377:599–609
Cole R, Hajat S, Murage P, Heaviside C, Macintyre H, Davies M, Wilkinson P (2023) The contribution of demographic changes to future heat-related health burdens under climate change scenarios. Environ Int 173:107836
Cordes RJ, Friedman DA, Maron M (2020) Reimagining maps
Council, National Research, Policy, Global Affairs, Division on Earth, Life Studies, Board on Higher Education, Committee to Assess the Current, and Future Workforce Needs in Veterinary Medicine (2013) Workforce needs in veterinary medicine (National Academies Press)
Dahal R, Upadhyay A, Ewald B (2017a) One Health in South Asia and its challenges in implementation from stakeholder perspective. Vet Rec 181:626
Dahal R, Upadhyay A, Ewald B (2017b) One Health in South Asia and its challenges in implementation from stakeholder perspective. VetY Rec 181:626–726
Daily GC, Ehrlich PR (1996) Impacts of development and global change on the epidemiological environment. Environ Dev Econ 1:311–346
Dananjayan S, Marshall G (2020) Artificial Intelligence during a pandemic: the COVID-19 example. Int J Health Plan Manag 35:1260
Destoumieux-Garzón D, Mavingui P, Boetsch G, Boissier J, Darriet F, Duboz P, Fritsch C, Giraudoux P, Le Roux F, Morand S (2018) The one health concept: 10 years old and a long road ahead. Front VetY Sci 5:14
Dhillon GPS (2008) National vector borne disease control programme—a glimpse, p 639
Dimitrova A, Ingole V, Basagaña X, Ranzani O, Milà C, Ballester J, Tonne C (2021) Association between ambient temperature and heat waves with mortality in South Asia: systematic review and meta-analysis. Environ Int 146:106170
Dodman D (2009) Urban density and climate change. Analytical review of the interaction between urban growth trends and environmental changes, p 1
Elahi KM (2016) Climate change and health impacts in Bangladesh. Climate change and human health scenario in South and Southeast Asia 207–219
ESCAP UN (2020) Pathways to manage cascading risks and protect people in South Asia: key takeaways for stakeholders
European Foundation for South Asian Studies (EFSAS), Amsterdam (2023) Climate Change In South Asia: The cases of Pakistan, Bangladesh and India. https://www.efsas.org/publications/articles-by-efsas/climate-change-in-south-asia-the-cases-of-pakistan,-bangladesh-and-india/
Fazal O, Hotez PJ (2020) NTDs in the age of urbanization, climate change, and conflict: Karachi, Pakistan as a case study. PLoS Negl Trop Dis 14:e0008791
Ganesh C, Schmeltz M, Smith J (2020) Introduction climate change and the legal, ethical, and health issues facing healthcare and public health systems. J Law Smith Med Ethics 48:636–642
Gasparri G, Tcholakov Y, Gepp S, Guerreschi A, Ayowole D, Okwudili É-D, Uwandu E, Iturregui RS, Amer S, Beaudoin S (2022) Integrating youth perspectives: adopting a human rights and public health approach to climate action. Int J Environ Res Public Health 19:4840
Gaurav K, Rajendra B, Kuldeep S, Kumar GS, Sam J, Amit S, Manju R (2024) Addressing challenges in vector control: a review of current strategies and the imperative for novel tools in India's combat against vector-borne diseases. BMJ Public Health 2:e000342
Ghosh S, Hasan M, Nath N, Haider N, Jones DH, Islam M, Rahaman MM, Mursalin H, Mahmud N, Kamruzzaman M (2024) Achieving zero by 30 in Bangladesh through a One Health approach
Gorji T, Sertel E, Tanik (2017) Monitoring soil salinity via remote sensing technology under data scarce conditions: a case study from Turkey. A Ecological indicators Tanik, vol 74, pp 384–91

Group, World Bank (2024) An Open Letter on IDA21
Hansel NN, McCormack MC, Kim V (2016) The effects of air pollution and temperature on COPD. COPD: J Chronic Obstr Pulm Dis 13:372–379
Hansen E (2023) Regional climate-related mobility governance in Southeast Asia: The Role of the Association for Southeast Asian Nations
Jung Y-J, Khant NA, Kim H, Namkoong S (2023) Impact of climate change on waterborne diseases: directions towards sustainability. Water 15:1298
Karim I, Fatima M, Ali M (2024) Alarming heat wave-2024: the summer season in Pakistan. Pak J Med Health Sci 18:1–1
Karmakar M, Pradhan MM (2020) Climate change and public health: a study of vector-borne diseases in Odisha, India. Nat Hazards 102:659–671
Kim T, Lee SW (2022) The embedded conundrum of South-South and triangular cooperation: A prologue to shifting frontiers from collaboration to contention. Asian J Peacebuilding 10:1–13
Kolla R (2020) Defining opportunities and challenges in one health agency management in Sri Lanka. University of Saskatchewan
Kumar D, Hasan Y, Afroz S (2023) Mobile health monitoring system: a comprehensive review. Int J Res Publ Rev 4:1922–1954
Kunnathepeedikayil S (2020) The long-term mental health consequences of floods and landslides in Thrissur District, Kerala. Harvard University, India
Kurebwa J, Kurebwa NY (2025) Climate change and health resilient systems: strengthening public health in a changing world. In: Managing the health risks of climate change (IGI Global Scientific Publishing)
Lham D, Wangchuk S, Stolton S, Dudley N (2019) Assessing the effectiveness of a protected area network: a case study of Bhutan. Oryx 53:63–70
Mahrous H, Redi N, Nguyen TMN, Awaidy SA, Mostafavi E, Samhouri D (2020) One Health operational framework for action for the Eastern Mediterranean Region, focusing on zoonotic diseases. East Mediterr Health J 26:720–725
Makino-Oi A, Ishii Y, Hoshino T, Okubo N, Sugito H, Hosaka Y, Fukaya C, Nakagawa T, Saito A (2016) Effect of periodontal surgery on oral health-related quality of life in patients who have completed initial periodontal therapy. J Periodontal Res 51:212–220
Malla S, Dumre SP, Shakya G, Kansakar P, Rai B, Hossain A, Nair GB, John Albert M, Sack D, Baker S (2014) The challenges and successes of implementing a sustainable antimicrobial resistance surveillance programme in Nepal. BMC Public Health 14:1–7
Mani M, Bandyopadhyay S, Chonabayashi S, Markandya A (2018) South Asia's hotspots: The impact of temperature and precipitation changes on living standards (World Bank Publications)
Mettenleiter TC, Markotter W, Charron DF, Adisasmito WB, Almuhairi S, Behravesh CB, Bilivogui P, Bukachi SA, Casas N, Becerra NC (2023) The one health high-level expert panel (OHHLEP). One Health Outlook 5:18
Qader MMM (2011) Climate change, flooding in South Asia and implications. RegNal Environ Chang 11:95–107
Mor N (2023) Organising for one health in a developing country. One Health 17:100611
Mosadeghrad AM, Afshari M, Sharifi T, Keliddar I, Shahsavani A, Yunesian M, Ostadtaghizadeh A, Dehnavi H (2023) Strengthening the resilience of Iran's health system against climate change 22:479–82
Mumford EL, Martinez DJ, Tyance-Hassell K, Cook A, Hansen GR, Labonté R, Mazet JAK, Mumford EC, Rizzo DM, Togami E (2023) Evolution and expansion of the One Health approach to promote sustainable and resilient health and well-being: a call to action. Front Public Health 10:1056459
Munyua PM, Kariuki Njenga M, Osoro EM, Onyango CO, Bitek AO, Mwatondo A, Muturi MK, Musee N, Bigogo G, Otiang E (2019) Successes and challenges of the One Health approach in Kenya over the last decade. BMC Public Health 19:1–9

Namgyal J, Chedup K, Couloigner I, Cork S (2023) Community engagement in research & policy development: a one health approach to the control of ticks and tick-borne diseases in Eastern Bhutan. One Health Cases ohcs20230002

Nepal, Practical Solution Consultancy (2012) Assessment of climate change impacts and identifying adaptation and mitigation measures in protected areas of Nepal. Environ Sci Policy 21:24–34

Neufeldt H, Christiansen L, Dale TW (2021) Adaptation gap report 2020

Noureen A, Aziz R, Ismail A, Trzcinski AP (2022) The impact of climate change on waterborne diseases in Pakistan. Sustain Clim Chang 15:138–152

O'Shanahan K, O'Flynn J, McCarthy D, MacGabhann L (2022) Creating alchemy within an Irish mental health service. Ment Health Rev J 27:412–425

Organization, World Health (2015) Meeting of experts on health in all policies in South-East Asia. In: WHO Regional Office for South-East Asia

Organization, World Health (2024) Refugee and migrant health system review: challenges and opportunities for long-term health system strengthening in Czechia (World Health Organization)

Ozturk T, Talo M, Yildirim EA, Baloglu UB, Yildirim O, Acharya UR (2020) Automated detection of COVID-19 cases using deep neural networks with X-ray images. Comput Biol Acharya Med 121:103792

Pakistan, FAO (2024) FAO Pakistan hosts technical workshop to strengthen animal and public health systems. https://www.fao.org/pakistan/news/detail-events/en/c/1683357/

Paul MM, Jadeja NB, Sidhu N, Vanak AT (2020) Whither One Health in India. National Institute of Advanced Studies, Bangalore, vol 55, p 40

Qureshi MHF, Azam F, Shafique M, Aslam B, Farooq M, Rafique MK, Meraj MT, Ahmed I (2024) A one health perspective of pet birds bacterial zoonosis and prevention. Pak VetY J 44

Rahaman MA, Rahman MM, Hafizur S (2019) Pathways of climate-resilient health systems in Bangladesh. Confronting Climate Change in Bangladesh: Policy Strategies for Adaptation Rahman, and Resilience, pp 119–143

Rasul I, Khan AQ, Gulzar SA, Qayyum S, Hollywood D, Sheikh S, Iqbal H, Siddiqui S, Lodhi AA (2023) Punjab economic opportunities program

Raza A, Ahmad S, Ahmad M, Zain-Ul-Abedin M, Channo A, Subhan A, Beig MM, Fazilani SA, Irshad Z, Mujahid U, Khan AK (2023) Zoonotic diseases: emerging threats to public health and livestock production. Zoonosis, Unique Scientific Publishers, Faisalabad, Pakistan, vol 1, pp 74–88

Romanello M, Walawender M, Hsu S-C, Moskeland A, Palmeiro-Silva Y, Scamman D, Ali Z, Ameli N, Angelova D, Ayeb-Karlsson S, Basart S, Beagley J, Beggs PJ, Blanco-Villafuerte L, Cai W, Callaghan M, Campbell-Lendrum D, Chambers JD, Chicmana-Zapata V, Chu L, Cross TJ, van Daalen KR, Dalin C, Dasandi N, Dasgupta S, Davies M, Dubrow R, Eckelman MJ, Ford JD, Freyberg C, Gasparyan O, Gordon-Strachan G, Grubb M, Gunther SH, Hamilton I, Hang Y, Hänninen R, Hartinger S, He K, Heidecke J, Hess JJ, Jamart L, Jankin S, Jatkar H, Jay O, Kelman I, Kennard H, Kiesewetter G, Kinney P, Kniveton D, Kouznetsov R, Lampard P, Lee JKW, Lemke B, Li B, Liu Y, Liu Z, Llabrés-Brustenga A, Lott M, Lowe R, Martinez-Urtaza J, Maslin M, McAllister L, McMichael C, Mi Z, Milner J, Minor K, Minx J, Mohajeri N, Momen NC, Moradi-Lakeh M, Morrisey K, Munzert S, Murray KA, Obradovich N, O'Hare MB, Oliveira C, Oreszczyn T, Otto M, Owfi F, Pearman OL, Pega F, Perishing AJ, Pinho-Gomes A-C, Ponmattam J, Rabbaniha M, Rickman J, Robinson E, Rocklöv J, Rojas-Rueda D, Salas RN, Semenza JC, Sherman JD, Shumake-Guillemot J, Singh P, Sjödin H, Slater J, Sofiev M, Sorensen C, Springmann M, Stalhandske Z, Stowell JD, Tabatabaei M, Taylor J, Tong D, Tonne C, Treskova M, Trinanes JA, Uppstu A, Wagner F, Warnecke L, Whitcombe H, Xian P, Zavaleta-Cortijo C, Zhang C, Zhang R, Zhang S, Zhang Y, Zhu Q, Gong P, Montgomery H, Costello A (2024) The 2024 report of the <em>Lancet</em> Countdown on health and climate change: facing record-breaking threats from delayed action. Lancet 404:1847–1896

Ryan SJ, Lippi CA, Caplan T, Diaz A, Dunbar W, Grover S, Johnson S, Knowles R, Lowe R, Mateen BA (2023) The current landscape of software tools for the climate-sensitive infectious disease modelling community. Lancet Planet Health 7:e527–e536

Saeed A, Ali S, Khan F, Muhammad S, Reboita MS, Khan AW, Goheer MA, Khan MA, Kumar R, Ikram A, Jabeen A, Pongpanich S (2023) Modelling the impact of climate change on dengue outbreaks and future spatiotemporal shift in Pakistan. Environ Geochem Health 45:3489–3505

Saleem F, Faraz A, Irtaza A, Ishaq HM, Ilyas MF, Khalid RH, Soomro KA, Sufiyan A, Ashraf M, Ghazali MH (2023) Policies to control zoonotic disease transmission in Pakistan. Zoonosis, Unique Scientific Publishers, Faisalabad, Pakistan, vol 1, pp 348–360

Samuel J, Flores W, Frisancho A (2020) Social exclusion and universal health coverage: health care rights and citizen-led accountability in Guatemala and Peru. Int J Equity Health 19:1–9

Schnitter R, Verret M, Berry P, Chung Tiam Fook T, Hales S, Lal A, Edwards S (2019) An assessment of climate change and health vulnerability and adaptation in Dominica. Int J Environ Res EdwS Public Health 16:70

Seddon N, Chausson A, Berry P, Girardin CAJ, Smith A, Turner B (2020) Understanding the value and limits of nature-based solutions to climate change and other global challenges. Philos Trans R Soc B 375:20190120

Sekar N, Shah NK, Abbas SS, Kakkar M (2011) Roadmap to Combat Zoonoses in India Initiative. Research options for controlling zoonotic disease in India, 2010–2015. PloS one 6:e17120

Sen B, Dhimal M, Latheef AT, Ghosh U (2017) Climate change: health effects and response in South Asia. bmj 359

Sharma S (2024) Training framework for frontline health workers on climate change and health. In: Healthcare administration and managerial training in the 21st century (IGI Global)

Shen N, Jankowicz D, Strudwick G (2021) Patient and family engagement approaches for digital health initiatives: protocol for a case study. JMIR Res Protoc 10:e24274

Shrestha SL, Shrestha IL, Shrestha N, Joshi RD (2017) Statistical modeling of health effects on climate-sensitive variables and assessment of environmental burden of diseases attributable to climate change in Nepal. Environ Model Assess 22:459–472

Salvador EC, Buddha N, Bhola A, Sinha SK, Kato M, Wijesinghe PR, Samuel R, Naidoo D, Singh SK, Perera WS, Singh PK (2023) Health emergency risk management in world health organization—South-East Asia region during 2014–2013: synthesis of experiences. Lancet RegNal Health-Southeast Asia 18

Sivakumar MVK, Stefanski R (2011) Climate change in South Asia. Climate change and food security in South Asia, pp 13–30

Stamm A (2023) The sustainability transition requires extended and differentiated north-south cooperation for innovation. In: UNCTAD Background Paper. Geneva: United Nations Conference on Trade and Development.[cit. 2023–06–06]. Dostupné na: https://unctad.org/system/files/non-official-document/tir2023_background5_en. pdf

Tse-Ring K, Sharma E, Chettri N, Shrestha AB (2010) Climate change vulnerability of mountain ecosystems in the Eastern Himalayas. In: International centre for integrated mountain development (ICIMOD)

Usman M, Ali A, Rosak-Szyrocka J, Pilař L, Baig SA, Akram R, Wudil AH (2023) Climate change and livestock herders wellbeing in Pakistan: Does nexus of risk perception, adaptation and their drivers matter? Heliyon 9

Wamsler C (2017) Stakeholder involvement in strategic adaptation planning: transdisciplinarity and co-production at stake? Environ Sci Policy 75:148–157

Wells C, Petty C, Saggioro E, Cornforth RJ (2023) Pakistan climate change impact storylines based on existing literature. Zenodo 1–117

West H, Quinn N, Horswell M (2019) Remote sensing for drought monitoring & impact assessment: progress, past challenges and future opportunities. Remote Sens Environ 232:111291

White BP, Breakey S, Brown MJ, Smith JR, Tarbet A, Nicholas PK, Ros AMV (2023) Mental health impacts of climate change among vulnerable populations globally: an integrative review. Ann Glob Health 89:66

Yasobant S, Bruchhausen W, Saxena D, Falkenberg T (2019) One health collaboration for a resilient health system in India: learnings from global initiatives. One Health 8:100096

Yasobant S, Bruchhausen W, Saxena D, Falkenberg T (2021) Systemic factors for enhancing intersectoral collaboration for the operationalization of One Health: a case study in India. Health Res Policy Syst 19:1–14

Zain A, Sadarangani SP, Shek LP-C, Vasoo S (2024) Climate change and its impact on infectious diseases in Asia. Singap Med J 65

Zaman MS, Mitra AK (2024) Dengue in Bangladesh and neighboring countries: an overview of epidemiology, transmission, control, and prevention. IMC J Med Sci 18:012

Zhang L, Guo W, Lv C (2024) Modern technologies and solutions to enhance surveillance and response systems for emerging zoonotic diseases. Sci One Health 3:100061

Chapter 4
Protecting Gig Workers from Heatwaves: Assessing Impact and Coping Mechanism in Delhi

Aditi Madan and Arjun Dubey

Abstract Climate change has intensified the frequency and severity of heatwaves in South Asia, posing severe public health and occupational risks for workers engaged in informal and outdoor labour. Among them, platform-based gig workers represent a rapidly expanding but highly vulnerable segment of the urban workforce. Despite their growing presence in cities like Delhi, these workers often fall outside the purview of conventional labour protections. This research examines the health impacts, earnings disruptions, and coping strategies employed by gig workers during heatwaves, utilising primary data from a mixed-methods study comprising a survey conducted with 390 workers during the 2024 heatwave period. The findings reveal prolonged work hours in extreme heat, a high incidence of heat-related symptoms, minimal employer-provided safeguards, and reliance on individual coping mechanisms such as altered work schedules and personal expenditure on protective gear. It concludes by situating the findings within broader global concerns on urban climate vulnerability, contributing to Sustainable Development Goals 3, 8, 10, and 13, and advocating for inclusive, equitable, and climate-adaptive planning.

Keywords Heatwaves · Gig economy · Heat stress · Gig work · Occupational health risk · Climate vulnerability · Coping mechanisms

A. Madan (✉) · A. Dubey
Institute for Human Development, Delhi, India
e-mail: aditimadanindia@gmail.com

A. Dubey
e-mail: arjun.dubey@ihdindia.org

P. Ghosh et al. (eds.), *Climate Crises and Public Health Challenges*, Disaster Risk Reduction, https://doi.org/10.1007/978-981-95-9709-3_4

4.1 Introduction

4.1.1 Background of the Study

India is currently experiencing a disturbing rise in the frequency and intensity of heatwaves, making extreme heat one of the most urgent public health and labour challenges of our time. Temperatures in several regions now routinely exceed 45 °C during the summer months, with Delhi recording multiple red-alert days in recent years. Heatwaves, defined as periods of excessively hot weather, have caused over 24,223 deaths in India between 1992 and 2015 (ORF 2023). Climate change has intensified these heatwaves, making them more frequent and severe. For instance, the early heatwaves of 2022 were observed to be 30 times more probable due to climate change (University of Cambridge 2022).

For millions of workers whose livelihoods depend on prolonged outdoor activity, heatwaves pose severe health risks and economic instability. Among these vulnerable groups, gig workers—particularly those in the food delivery and quick-commerce sectors—stand out due to the nature of their employment. They are more vulnerable to extreme health than other traditional workers. In India, the number of gig workers has increased exponentially over the last few years, from approximately 7.7 million in 2021 to projections exceeding 23 million by 2030 (NITI Aayog 2023). Unlike salaried workers, gig workers lack formal labour protections, work long hours outdoors, and rely on performance-based pay, often under pressure to meet algorithmic targets. Their marginal position in India's labour ecosystem raises important questions about occupational safety and climate resilience. Despite their large and growing presence, gig workers have limited recognition under current Indian labour laws, raising concerns about their working conditions (People's Association in Grassroots Action and Movement (PAIGAM 2023)). With the gig economy projected to reach 23.5 million workers by 2030 (NITI Aayog 2023), addressing their exposure to extreme weather conditions has become a policy imperative.

4.1.2 Evidence in the Previous Literature and Research Gaps

Recent studies have begun to highlight the intersection of climate risks and informal employment. The International Labour Organisation (ILO 2019) notes that heat stress is already reducing global working hours and productivity. In India, Das and Prakash (2022a, b) emphasise that informal and outdoor workers face disproportionate climate vulnerabilities due to limited access to cooling, healthcare, and employer support. Research from the UK and Southeast Asia has also revealed how gig workers, in particular, are compelled to work during heatwaves despite evident health hazards (Howarth et al., 2023; Eco-Business 2022). Climate change is increasingly influencing labour dynamics, and gig workers, who often work outdoors for extended hours, are facing serious public health challenges as a result (Sharma et al., 2022) including risks of dehydration, heat exhaustion, and heat strokes (ILO 2019). These workers face a precarious choice between protecting their health and securing their income, often compromising both (Robinson et al. 2023).

Research has highlighted the challenges faced by gig workers in managing their health during extreme weather conditions. For instance, a study by the Observer Research Foundation (2023) emphasised that gig workers often prioritise their job over personal well-being due to the nature of their contracts and performance ratings. This lack of flexibility and security further impacts their health, especially during heatwaves. A report by the Indian Federation of App-based Transport Workers indicates that many gig workers lack access to basic health services or heat-protective measures provided by their employers.

A study in Hyderabad, India, revealed that more than 50% of platform workers experienced heat exhaustion and nearly one-third suffered from heatstroke during severe heatwaves (HeatWatch and TGPWU 2024). Similar trends have been seen globally. A study from China reported that during heatwaves, gig workers faced increased workloads while suffering delayed payments and serious health issues, with delivery delays rising nearly 22% and income gains barely offsetting heightened health costs (Ge et al. 2024). In the United States, delivery and package drivers reported dizziness, kidney injuries, and even deaths owing to extreme heat, prompting renewed calls for national heat protection standards (Baker 2023).

Studies have shown that without significant changes in policy and practice, the frequency and severity of heatwaves are expected to increase, exacerbating existing vulnerabilities (Dasgupta et al. 2021). Despite these insights, few empirical studies have focused specifically on the experiences of gig workers in Indian cities during heatwaves. This is particularly relevant for India's burgeoning gig economy, where the rapid expansion of app-based, outdoor labour has outpaced the development of regulatory and protective frameworks for workers.

This gap is especially pronounced in the context of urban gig work, where flexible labour models collide with rigid environmental conditions. There is limited evidence on how gig workers perceive, experience, and respond to rising heat stress. Moreover, institutional responses—both from platform companies and urban governments—remain largely undocumented, leaving a critical blind spot in India's efforts to build climate-resilient cities and inclusive work environments.

In response to this research gap, this study investigates the impact of heatwaves on the health, earnings, and coping strategies of gig workers in Delhi. It asks three core questions: How do gig workers experience heat-related challenges? What employer provisions and individual coping mechanisms are in place to address these challenges? And what recommendations emerge from these patterns? This study also contributes towards achieving SDG 3 (well-being), SDG 8 (decent work and economic growth), SDG 10 (reducing inequalities), and SDG 13 (climate action).

4.1.3 Objectives

This study aims to assess the impact of heatwaves on gig workers in Delhi by examining their health vulnerabilities, income fluctuations, productivity, and coping mechanisms. Specifically, it examines the lived experiences of gig workers during heatwaves, the adequacy of employer provisions, and the individual coping mechanisms employed. The study offers recommendations to enhance the overall resilience of gig workers to future heatwaves.

4.2 Study Area

The present study was conducted in the National Capital Territory (NCT) of Delhi (Fig. 4.1), one of India's most populous and economically active urban agglomerations. Delhi lies in the semi-arid zone of North India and experiences a composite climate characterised by extremely hot summers, particularly from May to June, when temperatures frequently cross 45 °C. The city is located in the Indo-Gangetic Plain and is heavily affected by urban heat island effects due to dense construction, vehicular emissions, and limited green cover.

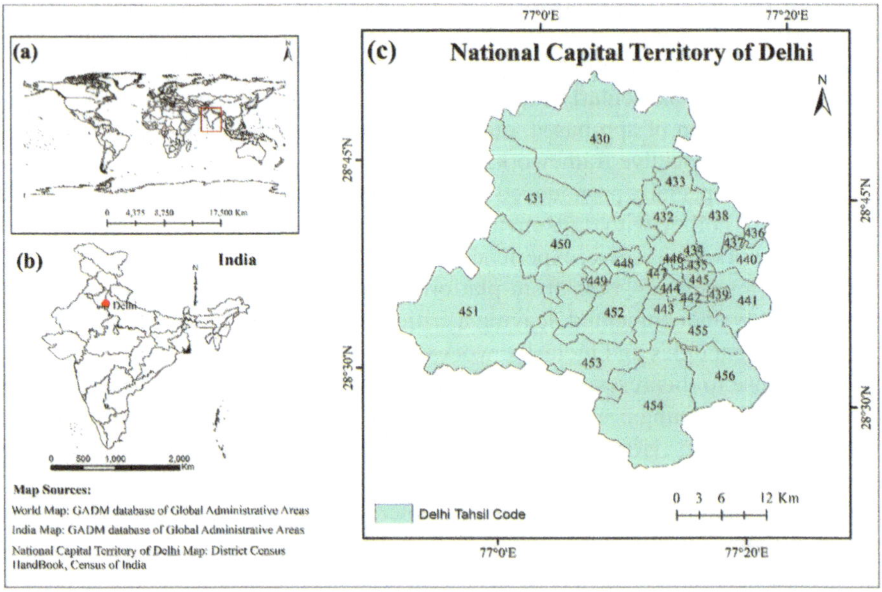

Fig. 4.1 Location of the study area, Delhi NCT

For this study, data collection was concentrated across five key urban hotspots in Delhi that serve as active operational zones for food delivery and quick commerce platforms—namely South Delhi, East Delhi, West Delhi, Dwarka, and the Central business district areas like Connaught Place and Karol Bagh. These locations were chosen due to their high concentration of gig workers, significant order volumes, and peak-time traffic congestion, which collectively intensify heat exposure for outdoor workers. The sites include malls, markets, residential societies, and EV charging stations—locations where delivery workers frequently gather or pause.

According to the 2011 Census, Delhi had a population of 16.7 million, which has now exceeded 20 million, making it one of the largest urban labour markets in South Asia. With a growing gig economy workforce—estimated to be over 3 lakhs in the capital alone by recent media and platform-specific reports[1]—Delhi offers a critical lens to study climate-health vulnerabilities in flexible labour models. The city's tropical steppe climate and prolonged heatwave alerts by the India Meteorological Department make it an ideal case for assessing occupational exposure and adaptive capacity.

4.3 Material and Methods

The field survey was conducted in Delhi between May 25 and June 15, 2024, a period marked by peak heatwave. A total of 390 gig workers were surveyed from quick commerce platforms (Blinkit and Zepto) and food delivery apps (Zomato and Swiggy). The selection of these platforms was strategic, given their significant market presence and the extensive outdoor activities involved in their operations.

A mixed-methods approach was employed to collect both quantitative and qualitative data through a semi-structured questionnaire, thereby providing a comprehensive understanding of the challenges faced by gig workers. It included 20 questions that addressed demographic details, working conditions, the impact of heatwaves on daily activities, and coping strategies (at both the employer and individual levels). A pilot survey was conducted to ensure the questionnaire was comprehensive yet could be completed within a short period and was easily understandable.

4.3.1 Sample Selection or Sampling Strategies

The study employed a mixed sampling strategy combining random selection with snowball sampling due to the absence of a formal or centralised database of gig workers in India. This strategy was designed to maximise reach and capture as many actively working gig workers as possible during the peak heatwave period. To ensure representativeness, respondents were randomly selected from identified delivery hotspots across the study area, including EV charging stations, outside malls

[1] Inside India's quick commerce boom: Earnings, growth, and gig workforce trends, Business Standard, 23 February 2025.

and commercial markets, and doorstep deliveries in societies. Early respondents were asked to refer other eligible gig workers, which enabled the study to reach individuals who might not have been present at typical delivery hotspots but were actively engaged in delivery services during the heatwave. This helped in addressing the hidden and mobile nature of the gig workforce, while minimising sampling bias.

4.3.2 Analytical Strategies

The analysis employed a combination of quantitative and qualitative techniques to identify patterns and associations. Descriptive statistics were used to build a respondent's profile and summarise key variables such as working hours, reported heat-related symptoms, access to employer support, and individual coping mechanisms. Further, cross-tabulations were carried out to explore associations between variables—such as working hours and heatwave impact—helping to understand how different groups of workers were affected in different ways. Qualitative analysis focused on understanding individual coping strategies adopted by gig workers in response to extreme heat. Open-ended survey responses, as well as photographs shared by respondents (with consent), were analysed thematically. These qualitative findings, combined with the results of quantitative analysis, helped identify specific adaptive practices, challenges, and unmet needs.

4.4 Results

Table 4.1 presents a consolidated view of the respondents' demographic and occupational characteristics, which are essential to understanding the socioeconomic context in which heatwave vulnerabilities emerge. While this profile is not the core objective of the study, it frames the structural conditions that influence coping ability, exposure, and access to institutional support during extreme weather events.

4.4.1 Overall Impact of Heatwave

A significant majority (75%) of respondents reported being affected by the heatwave. This indicates that heatwaves have a substantial impact on gig workers, likely due to the nature of their work, which often involves being outdoors for extended periods as shown in Fig. 4.2.

Out of all full-time workers, 78% reported being affected by the heatwave due to their longer working hours and greater exposure to outdoor conditions, resulting in significant drops in productivity and earnings. The high percentage indicates the significant exposure full-time gig workers have to extreme weather conditions, as

Table 4.1 Socio-demographic and work pattern of respondents

Variables	Category	Frequency	Percentage (%)
Gender	Male	100	100
Age group	<20	45	12
	20–25	85	22
	25–30	75	19
	30–35	88	23
	>35	97	25
Education level	Up to 10th	52	13
	10th or 12th grade	258	66
	Graduate	49	13
	Pursuing graduation	28	7
	Postgraduate	3	1
Accommodation type	Rented	237	61
	Owned	153	39
State of origin	Delhi-NCR	195	50
	Uttar Pradesh	100	26
	Bihar	58	15
	Others	37	9
Employer affiliation	Zomato	194	50
	Swiggy	94	23
	Blinkit	80	21
	Zepto	22	6
Type of employment	Full-time	268	69
	Part-time	122	31
Years of experience	<1 year	132	34
	1–3 years	102	26
	>3 years	156	40
Average working hours	>8 h/day	290	74
	4–8 h/day	81	21
	<4 h/day	19	5
Average earnings/day	₹900–1300	157	41
	₹700 or less	103	26
	₹700–900	93	24
	₹1300 or more	37	9
No. of orders/day	>30	25	6
	20–30	177	45
	10–20	160	41
	<10	28	7

(continued)

Table 4.1 (continued)

Variables	Category	Frequency	Percentage (%)
Insurance coverage	Yes	161	41
	No	77	20
	Unaware	152	39

Source Author's field survey (May–June, 2024);
Note Earnings refer to total income without any deductions

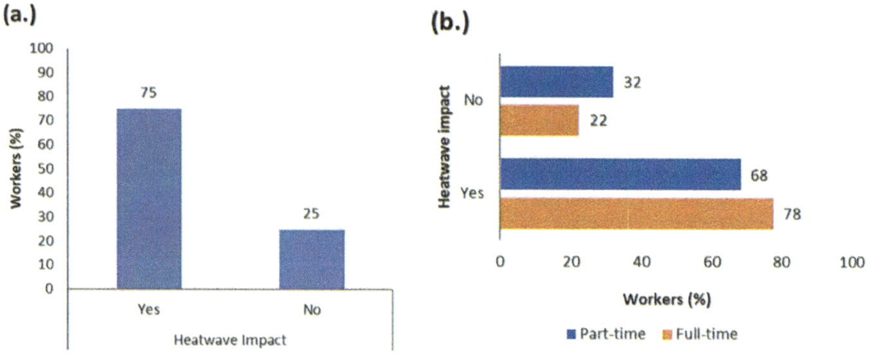

Fig. 4.2 Impact of heatwave among gig workers in Delhi; **a** Overall impact of heatwave and **b** Impact of heatwave by type of workers

they often work longer hours and are more consistently on the move throughout the day.

Full-time workers may also rely more heavily on personal coping mechanisms due to inadequate employer provisions. Out of all the part-time workers, 68% reported being affected by the heatwave. Potentially shorter shifts, might mitigate some of the heatwave effects but still face substantial health risks. They are not as severely impacted as full-time workers in terms of productivity and earnings loss, but the lack of employer support still poses significant risks.

A distinct pattern emerges when examining the impact of heatwaves and daily working hours as shown in Fig. 4.3. Among workers who reported working less than 4 h a day, 57.9% were affected by the heatwave, compared to 65.4% among those working 4–8 h, and 80.4% among those working 8–10 h. The proportion further increased to 90.7% for those working 10–12 h. Interestingly, this percentage falls for those working more than 12 h, possibly indicating that these workers might avoid working in the peak heat hours or have grown more resilient to the heat.

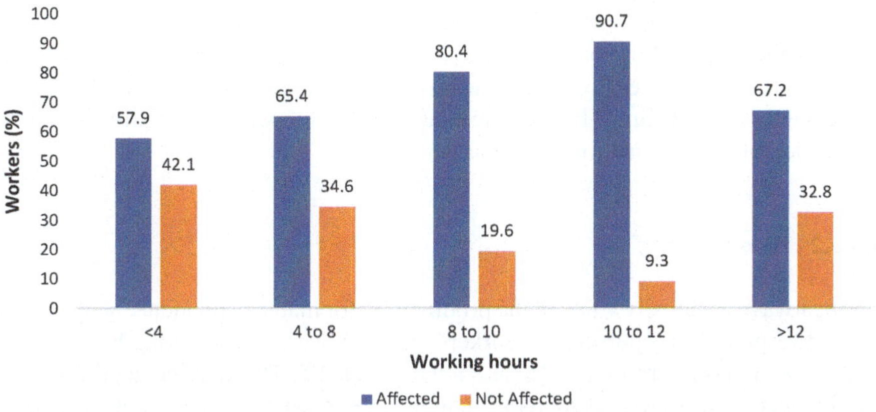

Fig. 4.3 Self-reported heatwave impact by daily working hours among gig workers in Delhi

4.4.1.1 Health

The heatwave has significant health impacts on gig workers, with common symptoms including exhaustion (42%), headaches (38%), and dizziness (25%) as shown in Fig. 4.4. These symptoms can significantly impair a worker's ability to perform their duties efficiently. Other health issues reported include body aches (18%), diarrhoea (21%), dizziness (25%), loo (15%) all of which can further exacerbate the challenges faced by gig workers, making it difficult to maintain their work schedules and performance. Physical symptoms such as exhaustion, headaches, dizziness (indicating dehydration and stress) significantly impair the ability of workers to perform their duties, leading to decreased productivity and potential loss of income.

Around 67% of full-time workers reported experiencing health impacts due to heatwaves. This high percentage indicates that full-time gig workers are heavily exposed to extreme heat conditions, which can lead to serious health issues such as

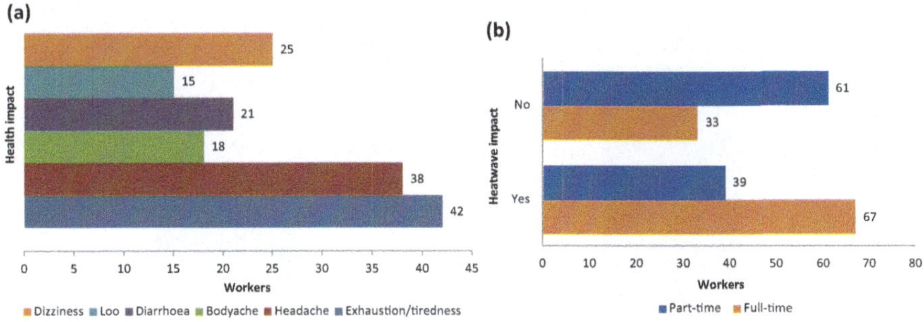

Fig. 4.4 Type of health risk faced by **a** Overall and **b** Part-time or full-time gig workers due to heatwave in Delhi

dehydration, heat exhaustion, heat stroke, and exacerbation of pre-existing conditions, making them highly vulnerable to heat-related health problems. Around 39% of the part-time workers reported health impacts due to heatwaves. While this percentage is significantly lower than that of full-time workers, it still highlights a considerable health risk for part-time workers.

4.4.1.2 Productivity

The heatwave negatively affected the productivity for many respondents, with around 17% of respondents reporting it. Workers reported decreased working hours (13%) and a drop in daily earnings (11%) as shown in Fig. 4.5. This decline in productivity can be attributed to several factors, including the need to take frequent breaks to avoid heat exhaustion, shorten their shifts or avoid working during peak heat hours to manage physical strain and health risks associated with extreme temperatures. This results in reduced efficiency and fewer working hours during peak heat periods. The decrease in earnings further exacerbates the financial instability of gig workers, who often rely on daily wages to meet their basic needs.

Workers often need to take frequent breaks, shorten their shifts, or avoid working during peak heat hours to manage the physical strain and health risks associated with extreme temperatures. This reduction in working hours directly translates to a loss of income, exacerbating the financial instability faced by many gig workers. For instance, some of them chose to work at night as a strategic choice to avoid the intense daytime heat. However, it highlights the trade-offs gig workers make to mitigate heat exposure while maintaining their income.

Workers earning higher daily incomes were more likely to report being affected by the heatwave. Among those earning 1300–1500 rupees per day, 96% reported adverse effects, compared to only 50% in the lowest earnings group (less than 500) as shown in Fig. 4.6. This trend may reflect the fact that higher earnings often correspond to longer work hours or greater physical exposure, thereby increasing vulnerability during extreme heat.

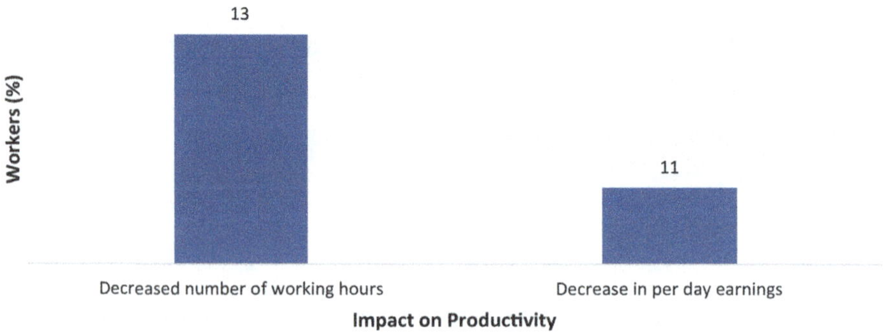

Fig. 4.5 Self-reported impact of heatwave on work productivity among gig workers

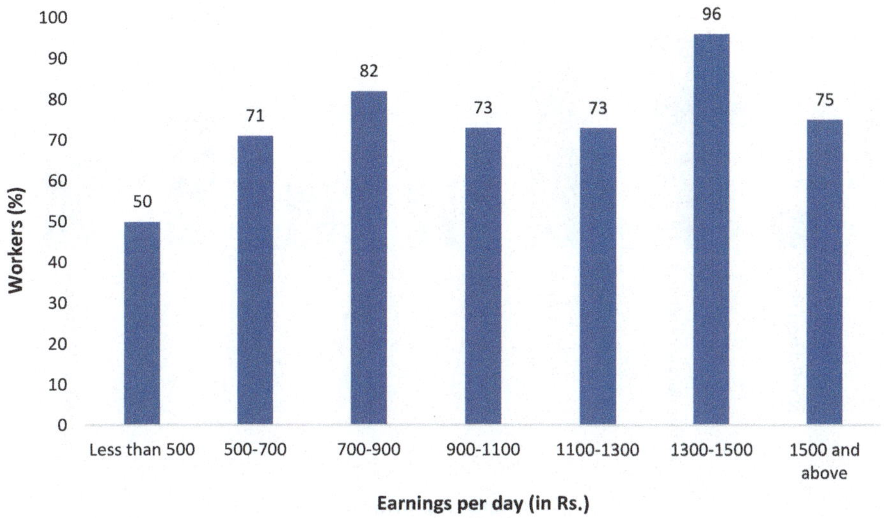

Fig. 4.6 Self-reported heatwave impact by daily earnings among gig workers

4.4.2 Coping Strategies and Adaptation

4.4.2.1 Employer Provisions

Employer-provided provisions to mitigate the effects of heatwaves are limited and largely inadequate. Study findings reveal that only a small percentage of respondents remembered receiving notifications/alerts (3%) or had access to air coolers (5%) and toilets (5%) as shown in Fig. 4.7. The limited availability of water (10%) and hydration solutions like Glucose/ORS/buttermilk (7%) further highlights the gaps in employer support. While most companies lacked cooling facilities, a few dark stores had some basic provisions. One of the Blinkit workers noted, '*Some dark stores have few facilities, like a fan or fridge. We keep our bottle in the fridge to cool water*'. These deficiencies point to a critical need for comprehensive health and safety policies that address the challenges posed by extreme weather conditions.

4.4.2.2 Individual Mechanisms

Gig workers employ various strategies to cope with the heat as shown in Fig. 4.8. The most common methods include carrying water (90%), wearing a helmet/cap (77%), and requesting water from restaurants or customers (71%). These measures help mitigate the immediate effects of heat exposure and are essential for maintaining hydration and protecting against direct sun exposure. For instance, a Swiggy

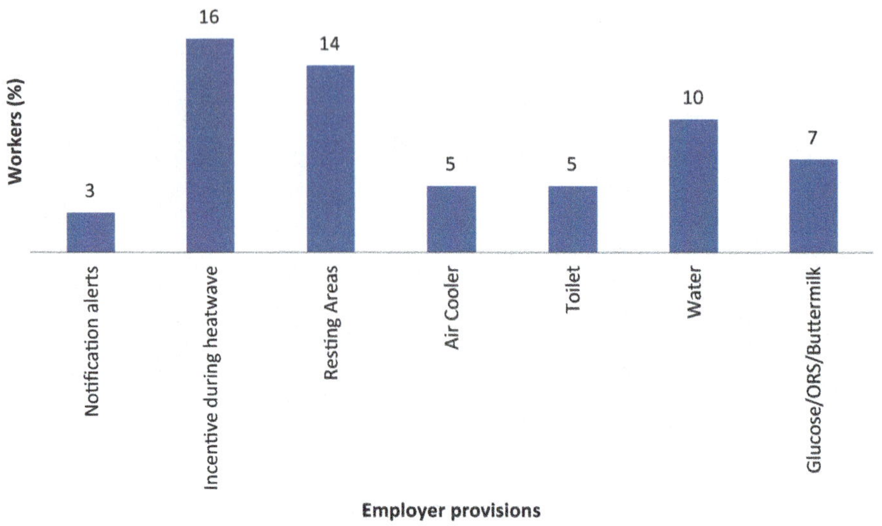

Fig. 4.7 Distribution of employer-provided heatwave provisions among gig workers in Delhi (N-390)

worker explained the added burden of waiting for customer approvals in harsh conditions: '*Have to wait too long in the hot weather for customer approvals in high-rise apartments. Sometimes, guards don't allow us to use the lift, and we wait in the heat*'.

Workers also adopt protective clothing, such as cloth around the face (59%) and arms (30%), to shield themselves from the sun. A smaller percentage avoids working

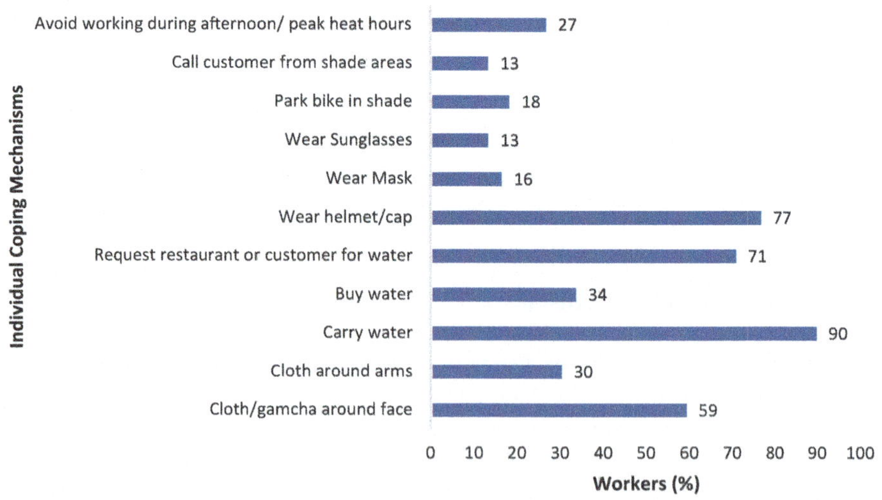

Fig. 4.8 Individual coping mechanisms adopted by gig workers during heatwave (N-390)

during peak heat hours (27%), indicating that while some workers can adjust their schedules, many cannot due to financial pressures or job requirements. However, the reliance on personal coping strategies highlights the inadequacy of employer-provided support.

Full-time workers might rely more heavily on personal coping mechanisms due to prolonged exposure to heat and inadequate employer provisions. This reliance can lead to increased fatigue and health issues. Part-time workers might not be as severely impacted due to shorter working hours, but the lack of employer support still poses significant risks.

Figure 4.9 shows how people are using gamcha, a cotton cloth, to shield their face and neck from the sun and how it helps absorb sweat, thus cooling the skin and preventing dehydration. The combination of a helmet (head protection) and a gamcha illustrates the layered approach to sun protection adopted by many gig workers. It shows an insulated delivery bag designed to keep deliveries cool, but the worker carrying it can't keep cool in his uniform given the synthetic fabric used in the uniform material that ends up trapping heat. Some also preferred to wear cotton clothing under the uniform for protection against heat. Several respondents shared the intensity of their experiences. '*My body burns and my hands have blisters riding the bike during such hot weather*', reported one of the Swiggy workers. A Zomato delivery executive shared, '*I stopped working from 12 to 4 pm as I fell sick with diarrhoea and low BP, and since then, I feel very weak to work for long hours*'. Another executive, also from Zomato, added, '*I stay offline in afternoon hours unless there is an incentive*'.

Image 2 illustrates a high surge announcement by Swiggy (Fig. 4.10), a food delivery platform, offering significant financial incentives for gig workers to work during peak heat hours on June 5, 2024. High incentives to work on peak heat days create economic pressure, pushing workers to endure harsh conditions to maximise their earnings during the hottest parts of the day. This can lead to reduced breaks and inadequate hydration, further exacerbating health risks. Despite the financial gains, the health implications can be severe, especially for workers who may not have adequate protective measures or access to hydration to counter the intense heat during peak hours, especially when the employer's incentive structure encourages continuous work with minimal breaks. As one of the workers from Swiggy puts it, '*It is very hot, but I am forced to work to earn for my family. People don't move out during the afternoon, as per the news, but we are made to work to earn an incentive*'. This tension between livelihood and health underscores the structural vulnerabilities faced by gig workers during extreme weather events.

4.5 Discussion

This study investigated the impact of heatwaves on gig workers in Delhi, focusing on their health experiences, coping mechanisms, and access to employer and policy support. Using a mixed-methods approach, including a structured survey of 390

Fig. 4.9 Coping mechanisms adopted by gig workers

workers and qualitative quotes, the study documented how rising temperatures are reshaping labour patterns, productivity, and occupational health in the urban informal sector. The findings contribute to ongoing conversations on around climate resilience, decent work, and occupational health in urban economies of the Global South, aligning with SDG 3, SDG 8, SDG 10, and SDG 13.

4.5.1 Key Findings

The study confirms earlier findings that gig workers face disproportionate climate risks due to the unregulated, flexible nature of their work (Ahmed and Singh 2022; ILO 2019). Around 75% of the respondents reported being affected by the heatwave. The high percentage indicates the significant exposure gig workers have to extreme weather conditions, as they often work longer hours and are more consistently on the move throughout the day. Occupational heat exposure has become a growing concern, particularly for gig workers in South Asia, who spend a significant portion of their workday outdoors (Patel and Choudhary 2022).

One key finding highlights the significant health risks faced by gig workers, with symptoms such as dehydration, dizziness, and fatigue being widely reported. Such symptoms significantly impair a worker's ability to perform their duties, leading to decreased productivity and a potential loss of income. These experiences align

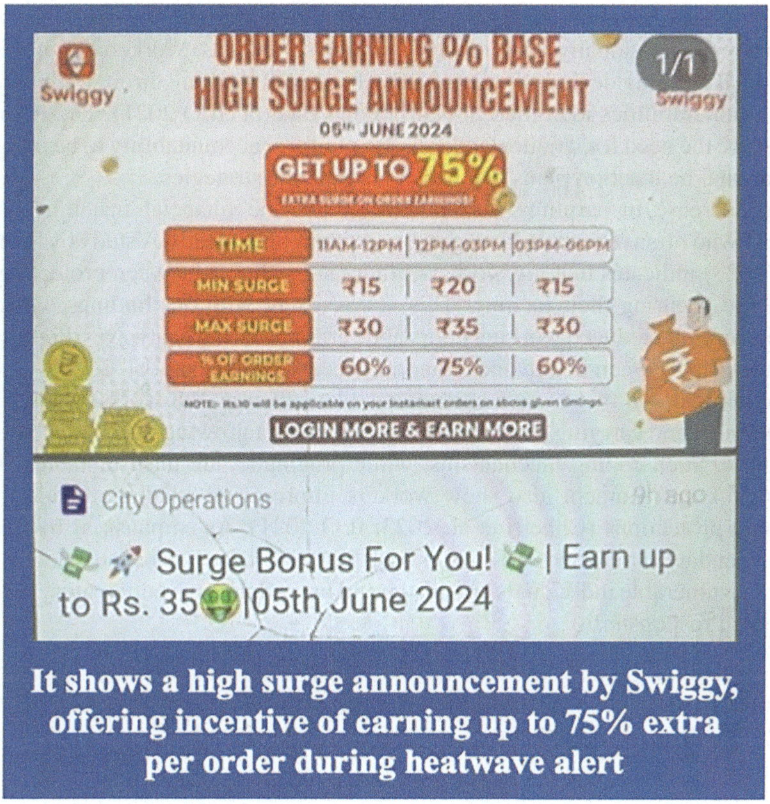

Fig. 4.10 Surge announcement by employer

with findings from similar studies in Ahmedabad (Azhar et al. 2017) and Dhaka (Khan et al. 2021), where outdoor workers reported an increased susceptibility to heat illnesses. The fact that 74% of respondents worked over 8 hours a day mirrors concerns raised by Kjellstrom et al. (2016), who demonstrated a strong correlation between prolonged exposure and decreased labour productivity. Workers in cities like Manila and Nairobi have also shown similar health challenges, particularly among riders in the food delivery and courier sectors (ILO 2021). This finding confirms the need for integrating occupational heat exposure into urban health frameworks (Das and Prakash 2022a, b; ILO 2019).

Another critical insight from the study is the inadequacy of institutional support from platform companies. Fernández et al. (2023) emphasised that the intersection of climate change with occupational health risks is increasingly evident, especially for informal workers who lack the protections of traditional employment models. Despite the hazardous conditions, only 41% of respondents reported access to employer-provided insurance or hydration provisions. This aligns with findings from Narain et al. (2022) on informal workers in Kolkata and Delhi, and is further supported by global studies, including in Lagos (Aderemi and George 2020), which show that platform workers globally operate with minimal safety nets. Workers' quotes in this study, such as being denied lift access in high-rises or working through illness, echo similar vulnerabilities identified in Nairobi and Jakarta (ILO 2021). These findings underscore the need for labour standards and platform accountability to be integrated into national heat action plans and urban governance strategies.

The decrease in earnings further exacerbates the financial instability of gig workers, who often rely on daily wages to meet their basic needs. A study by Robinson et al. (2023) indicates that gig workers often face a choice between protecting their health and securing their income. This is consistent with our findings, where gig workers reported reduced working hours and earnings during heatwaves, highlighting the trade-offs between health and economic needs (Robinson et al. 2023).

The comparison of adaptive strategies—like self-imposed work breaks, altered shift timings, and carrying personal water—suggests a growing individual burden of adaptation. Such coping mechanisms, while pragmatic, are unsustainable. Studies in Bangkok and Karachi also show workers improvising adaptations due to lack of formal protections (Chhetri et al. 2023; ILO 2021). As emphasised by SDG 13 and the Sendai Framework for Disaster Risk Reduction, such risks must not be borne solely by vulnerable individuals but must be addressed through anticipatory planning and shared responsibility.

4.5.2 Recommendations

This study recommends implementing climate-responsive labour policies and occupational safety protocols for the gig economy. Specific measures include the inclusion of gig workers under the purview of India's Code on Occupational Safety, Health, and Working Conditions (OSH Code), city-level mandates for rest shelters and hydration zones, and incentives for platform companies that adopt heat-safe delivery practices. Globally, cities like Paris and Los Angeles have begun experimenting with heat-protection mandates for outdoor workers (World Bank 2022). Delhi, too, can pilot policy interventions guided by its Heat Action Plan and align them with NDMA (2019) recommendations. Such steps will also help meet SDG 10 and reduce urban income inequalities exacerbated by climate risks.

Based on the findings, the following suggestions (Fig. 4.11) are recommended:

4.5.2.1 Food Delivery/Quick Commerce Companies

To enhance worker safety and reduce heat-related risks, delivery platforms must make operational changes that are both climate-responsive and worker-centric.

(a) Remove Cash on Delivery (COD) option for eliminating COD can streamline payment processes, reducing the time workers spend waiting for payments.

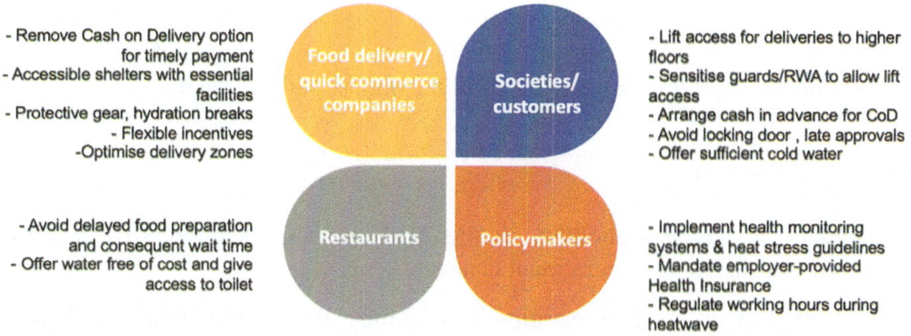

Fig. 4.11 Stakeholder-specific recommendations to improve heatwave preparedness for gig workers in India (*Source* Diagram created based on the results of the field survey, 2024)

This can help in avoiding unnecessary delays and reducing the time spent in potentially hazardous conditions.

(b) Accessible shelters with essential facilities such as fans or water coolers can offer much-needed relief to gig workers. These shelters can serve as rest points where workers can cool down, hydrate, and recover from the heat. Ensure that rest points are strategically located within delivery zones to provide workers with much-needed breaks from the heat.

(c) Provide additional protective gear, regular hydration breaks, and access to shaded rest areas.

(d) Offer flexible incentives that do not necessarily require working during the hottest part of the day can help balance financial benefits with health considerations. While high financial incentives can drive increased productivity and meet demand, they must be balanced with the health and well-being of the workers. Employers should ensure that the financial benefits do not come at the cost of workers' health and safety.

(e) Optimise delivery zones to help reduce the strain on workers as long-distance deliveries and slower travel speeds on e-bikes increases exposure to heat and fatigue.

4.5.2.2 Policymakers

Government interventions must evolve to formally recognise gig workers and embed heat-resilience into urban labour protections.

(a) Implement health monitoring systems to ensure workers are not overexerting themselves during peak heat times.

(b) Enforce comprehensive heat stress guidelines specific to gig workers.

(c) Given the high health risks associated with heatwaves gig economy platforms should provide health insurance to all their workers through mandatory policies.

(d) Regulations should be introduced to limit the working hours of gig workers during extreme heat conditions. Flexible working schedules should be mandated, allowing workers to avoid the hottest parts of the day.

4.5.2.3 Restaurants

Restaurants, as key stakeholders in the delivery chain, can improve working conditions through small operational changes.

(a) Avoid delayed food preparation to minimise the waiting time for delivery workers, reducing their exposure to heat. Efficient order management can enhance the overall delivery process and improve worker satisfaction.
(b) Offer free access to cold water and allowing use of restroom facilities can significantly improve the working conditions for gig workers. This not only helps in maintaining hydration but also ensures basic human needs are met.

4.5.2.4 Societies/Customers

Societies and individual customers play a vital role in shaping the delivery experience and reducing health risks for workers.

(a) Lifting access for deliveries to higher floors can significantly reduce the physical strain on gig workers, especially during extreme heat. It also ensures quicker delivery times and better efficiency.
(b) Sensitise security guards/residential welfare associations (RWA) to allow lift access to facilitate smoother operations and reduce unnecessary hardships for delivery personnel.
(c) Customers should arrange cash in advance for COD to expedite the payment process. This reduces the waiting time for delivery workers, especially during peak heat hours.
(d) Ensure doors are not locked and delivery personnel can access the premises quickly can prevent prolonged exposure to heat.
(e) Prompt approvals from customers can reduce the time delivery workers spend waiting in hot conditions.
(f) Provide adequate amounts of cold water can help gig workers stay hydrated and reduce the risk of heat-related illnesses.

By addressing issues related to payment processes, providing essential facilities, ensuring better cooperation from societies and customers, and improving restaurant operations, the overall efficiency and safety of gig workers can be significantly enhanced. By implementing these measures, companies can help ensure that the financial incentives do not come at the cost of the workers' health and well-being.

While the study offers empirical evidence from a critical urban sector, it has certain limitations. The data are limited to Delhi and are not generalisable to all Indian cities or other forms of gig work. However, the methodological approach—grounded surveys during peak summer and integration of worker narratives—offers a rich and replicable model. The study contributes to global knowledge by foregrounding the vulnerability of new-economy workers in climate-affected cities, offering a replicable framework for comparative studies in other Global South contexts.

4.5.3 Study Relevance and Limitations

The study was limited to Delhi and quick commerce food delivery platforms, which may not capture the diversity of gig work across cities and sectors. Additionally, the study relied on self-reported health symptoms, which can introduce subjective bias. However, the real-time nature of the data collection during Delhi's peak heat period (May–June 2024) offers rich, context-specific insights.

Globally, this study contributes to a growing body of literature linking climate adaptation to labour rights. It echoes findings from the UK (Howarth et al., 2023) and Southeast Asia (Eco-Business 2022), emphasising that gig work models, regardless of geography, are ill-equipped to deal with climate shocks. By embedding a worker-vulnerability lens into climate discourse, this study advocates for cross-sectoral policy shifts to ensure dignified, safe, and sustainable employment in a warming world.

4.6 Conclusion

This study highlights how the intensifying heatwave crisis intersects with the precarity of gig work in urban India. Through primary fieldwork in Delhi, it shows that gig workers—particularly those in food and quick-commerce delivery—face significant health, income, and productivity risks due to prolonged heat exposure and inadequate institutional safeguards. Their coping strategies remain largely individual and informal, reflecting a gap in both platform responsibility and policy support. By drawing attention to this often-overlooked population, the study contributes to the growing discourse on climate justice and labour rights. It offers a grounded assessment of how emerging labour models are vulnerable to environmental extremes and underlines the urgent need for integrating worker-centric protections into climate adaptation and urban policy frameworks.

The findings have broader relevance not just for India but for other developing countries experiencing parallel shifts in the future of work. As climate shocks become more frequent, this research calls for a reimagining of work environments that centre equity, resilience, and health. Future studies may expand to other cities and gig sectors to create a more comprehensive framework for climate-resilient urban employment.

Funding The research was not supported by any funding.

Potential Conflicts of Interest: On behalf of all authors, the corresponding author states that there is no conflict of interest. All images are original field photos with no identifiable individuals.

Informed Consent: Informed consent was obtained from all individual participants included in the study.

Authors Contribution: All authors have contributed equally to this work.

Data Availability Statement: The datasets during and/or analysed during the current study available from the corresponding author on reasonable request.

Appendix: Questionnaire

1. Gender
   (a) Male
   (b) Female
   (c) Prefer not to say
   (d) Other:
2. Age Group
   (a) Less than 20
   (b) 20–25
   (c) 25–30
   (d) 30–35
   (e) 35–40
   (f) Above 40
3. Education level
   (a) Below 10th class
   (b) 10th–12th
   (c) Pursuing Graduation
   (d) Graduate
   (e) Post Graduate and above
4. Type of accommodation in Delhi-NCR
   (a) Own
   (b) Rented
5. Which state/city you belong to?
   (a) Born in Delhi
   (b) Other:
6. Since how many years have you been staying in Delhi?
   (a) NA
   (b) Other:
7. No. of family members:

8. How many people earn/work in your family? (including you):
9. Company
   (a) Zomato
   (b) Blinkit
   (c) Swiggy
   (d) Zepto
10. Number of years/months of working
    (a) Less than 6 months
    (b) 6 months to 1 year
    (c) 1–3 years
    (d) 3–5 years
    (e) More than 5 years
11. What type of job is this for you?
    (a) Full-time
    (b) Part-time
12. What type of part-time jobs you undertake, if any? (Check all that apply)
    (a) NA
    (b) Studying
    (c) Small business
    (d) Private job
    (e) Government job
    (f) Startup
    (g) Other:
13. Why did you join this company? (Check all that apply)
    (a) Unemployed/ had no other option
    (b) To earn more than previous job
    (c) Additional income sources
    (d) Flexible work timings
    (e) Near to house
    (f) Other:
14. Number of working hours across the entire day
    (a) Less than 4 h
    (b) 4–8 h
    (c) 8–10 h
    (d) 10–12 h
    (e) More than 12 h
15. Average no. of orders per day:

16. Average Earnings per day (in Rs.) (Total Income)
    (a) Less than 500
    (b) 500–700
    (c) 700–900
    (d) 900–1100
    (e) 1100–1300
    (f) 1300–1500
    (g) 1500 and above
17. Number of days you work in a month
    (a) 1 week or less
    (b) 10–15 days
    (c) 15–20 days
    (d) 20–25 days (1 day a week off)
    (e) 25 and above
18. Type of insurance provided by your company? (Check all that apply)
    (a) No insurance provided
    (b) Health Insurance
    (c) Accident Insurance
    (d) Life insurance
    (e) Don't know/Unsure
    (f) Other:
19. How has hot weather/heatwave affected you? (Check all that apply)
    (a) Not affected
    (b) Exhaustion/tiredness
    (c) Headache
    (d) Bodyache
    (e) Diarrhoea
    (f) Body is burning
    (g) Decreased number of working hours
    (h) Decrease in per day earnings
    (i) Don't work during afternoon gig slot
    (j) Dizziness
    (k) Other:
20. What are the provisions that your company provides during heatwave or hot weather? (Check all that apply)
    (a) No provisions
    (b) Notification/guidelines (heatwave alerts and safety measures)
    (c) Normal Water
    (d) Cold water

(e) Glucose/Glucon-D/ORS etc.
(f) Buttermilk/chhach
(g) Incentive (additional compensation per order during heat hours/afternoon hours)
(h) Resting areas/stores
(i) Air Cooler
(j) First-aid box
(k) Toilet/bathroom facility
(l) Other:

21. What all precautions you take to protect yourself from heatwave? (Check all that apply)

    (a) Cloth/gamcha around face
    (b) Cloth around arms
    (c) Carry water
    (d) Buy water
    (e) Request restaurant or customer for water
    (f) Wear helmet/cap
    (g) Wear Mask
    (h) Wear Sunglasses
    (i) Park bike in shade
    (j) Call customer from shade areas
    (k) Avoid working during daytime/heat hours
    (l) Other:

22. REMARKS Any experiences to share or suggestions to give (eg. taking stairs as not allowed to use lifts in apartments)

References

Aderemi TA, George T (2020) Precarity and platform labour in Lagos: experiences of informal digital workers. Afr Rev Econ Finan 12(1):112–130

Ahmed R, Singh A (2022) Gig economy and climate change: examining heat vulnerability of gig workers in India. Int J Clim Change Strat Manag 14(1):45–60. https://doi.org/10.1108/IJCCSM-12-2021-0157

Azhar GS, Mavalankar D, Nori-Sarma A, Rajiva A, Dutta P, Jaiswal A, Hess JJ (2017) Heat-related mortality in India: excess all-cause mortality associated with the 2010 Ahmedabad heat wave. PLoS ONE 9(3):e91831. https://doi.org/10.1371/journal.pone.0091831

Baker A (2023) Extreme heat is endangering America's workers—and its economy. Time. https://time.com/6299091/extreme-heat-us-workers-economy/

Chhetri N, Rai RK, Aggarwal PK, Joshi PK (2023) Climate risks, adaptation strategies and food security in South Asia: a micro-level evidence. Clim Risk Manag 39:100518. https://doi.org/10.1016/j.crm.2023.100518

Das S, Prakash A (2022) Social inequality and climate change: the exacerbation of heat stress among informal workers. Environ Res Lett 17(7):074019. https://doi.org/10.1088/1748-9326/ac7b9a

Das R, Prakash P (2022) Health and safety of informal sector workers during extreme weather events in India. J Occup Health 64(1):e12345. https://doi.org/10.1002/1348-9585.12345

Dasgupta S, van Maanen N, Gosling SN, Piontek F, Otto C, Schleussner C-F (2021) Effects of climate change on combined labour productivity and supply: an empirical, multi-model study. Lancet Planet Health 5(7):e455–e465. https://doi.org/10.1016/S2542-5196(21)00170-4

Eco-Business (2022) Heatwave fee for Asia's gig workers does little to cool climate stress. Retrieved from https://www.eco-business.com/news/heatwave-fee-for-asias-gig-workers-does-little-to-cool-climate-stress/

Fernández C, López M, Pérez G (2023) Climate change and occupational health: a review of the risks for informal workers. Glob Environ Health J 18(2):115–126. https://doi.org/10.1016/j.gehj.2023.03.005

Ge R, Lu SF, Mao W (2024) The impact of heat hazard on gig economy workers: evidence from an on-demand food delivery platform. SSRN. https://doi.org/10.2139/ssrn.4993644

HeatWatch & Telangana Gig and Platform Workers Union (2024) Impact of extreme heat on gig workers: a survey report. https://tgpwu.org/

Howarth C et al (2023) How can the UK better adapt to heat in the workplace? London School of Economics and Political Science

ILO (2019). Working on a warmer planet: the impact of heat stress on labour productivity and decent work. Retrieved from https://www.ilo.org/global/publications/books/WCMS_711919/lang--en/index.htm

International Labour Organization (ILO) (2021) Ensuring decent work for platform workers in urban Asia: Emerging practices and policy options. ILO Regional Office for Asia and the Pacific, Bangkok

Khan MH, Ahsan MN, Uddin SMN (2021) Vulnerability of informal workers to climate extremes: evidence from street vendors in Dhaka. Environ Res Lett 16(11):114034. https://doi.org/10.1088/1748-9326/ac2d6f

Kjellstrom T, Freyberg C, Lemke B, Otto M, Briggs D (2016) Estimating population heat exposure and impacts on working people in conjunction with climate change. Int J Biometeorol 60(5):719–730. https://doi.org/10.1007/s00484-015-1065-1

Narain V, Anand M, Bhaskar A (2022) Informal work and climate change in India: narratives from Delhi and Kolkata. Environ Urban ASIA 13(2):273–290. https://doi.org/10.1177/09754253221109558

National Disaster Management Authority (NDMA) (2019) Guidelines for preparation of action plan–prevention and management of heat wave. Government of India. https://ndma.gov.in/sites/default/files/PDF/Guidelines/heatwaveguidelines2021.pdf

NITI Aayog (2023) India's booming gig and platform economy. Retrieved from https://www.niti.gov.in

ORF (2023) Heatwave vulnerability: the plight of gig workers in India. Retrieved from https://www.orfonline.org

Patel S, Choudhary K (2022) Occupational heat exposure in the gig economy: the emerging challenge in South Asia. J Occup Environ Med 64(9):705–710. https://doi.org/10.1097/JOM.0000000000002385

People's Association In Grassroots Action and Movement [PAIGAM] (2023) Report on the working conditions of gig workers in India. Retrieved from https://www.paigam.org

Robinson E, Khosla R, Harish S (2023) Heatwave vulnerability in the gig economy: an emerging risk. J Clim Risk Manag. https://doi.org/10.1016/j.crm.2023.100408

Sharma A, Singh P, Kumar R (2022) Climate change, occupational heat exposure, and health risks among outdoor workers in India. J Environ Public Health 2022:1–10

University of Cambridge (2022) Study on the increased probability of early heatwaves due to climate change. Retrieved from https://www.cam.ac.uk

World Bank (2022) Heat and labour: a global review of heat exposure, productivity losses, and worker protections. World Bank Group, Washington, D.C. https://documents.worldbank.org/en/publication/documents-reports/documentdetail/099042303152228692/p17577508a75df0b50a303034e03e49ecfa

Chapter 5
Ambient Air Pollution and Acute Respiratory Infection Among Children: A Review of Evidence from South Asia and Southeast Asia

Pin Cheng Wong, Azliyana Azhari, Mohd Shahrul Mohd Nadzir, Tin Tin Su, and Hamimatunnisa Johar

Abstract The climate crisis intensifies air pollution, which is one of the most significant environmental health risks in South Asia (SA) and Southeast Asia (SEA). Air pollution directly contributes to the prevalence and severity of acute respiratory infections (ARIs), further contributing to morbidity and mortality among young children across the region. The review aims to summarise the current evidence on the link between air pollution and ARI among children under five within SA and SEA regions and assess environmental and socioeconomic factors that modifies the ARI risks. This chapter reports a narrative evidence synthesis from meta-analyses and

P. C. Wong
Monash Medical Centre, Clayton, VIC, Australia
e-mail: PinCheng.Wong@monashhealth.org

A. Azhari
Monash Climate Communication Hub (MCCH), School of Arts and Social Sciences, Monash University Malaysia, Subang Jaya, Malaysia
e-mail: azliyana.azhari@monash.edu

M. S. Mohd Nadzir
Department of Earth Sciences and Environment, Universiti Kebangsaan Malaysia, Bangi, Selangor, Malaysia
e-mail: shahrulnadzir@ukm.edu.my

T. T. Su
Regional Hub for Asia Climate Change and Health (REACH), Jeffrey Cheah School of Medicine and Health Sciences, Monash University Malaysia, Subang Jaya, Malaysia
e-mail: TinTin.Su@monash.edu

H. Johar (✉)
Jeffrey Cheah School of Medicine and Health Sciences, Monash University Malaysia, Subang Jaya, Malaysia
e-mail: hamimatunnisa.johar@monash.edu

T. T. Su · H. Johar
Heidelberg Institute of Global Health (HIGH), Faculty of Medicine and University Hospital, Heidelberg University, Heidelberg, Germany

P. Ghosh et al. (eds.), *Climate Crises and Public Health Challenges*, Disaster Risk Reduction, https://doi.org/10.1007/978-981-95-9709-3_5

systematic reviews on the link between air pollution and ARIs in children under five years. Limited studies from SA and SEA were identified. The findings revealed that exposure to particulate matter ($PM_{2.5}$ and PM_{10}) and gaseous pollutants (sulphur dioxide (SO_2), ozone (O_3), nitrogen dioxide (NO_2), and carbon monoxide (CO)), significantly increases ARI risks and hospitalisations. Socioeconomic disparities play a significant role in determining the health outcomes associated with exposure to air pollutants. Key recommendations for future work are outlined to guide policy initiatives in mitigating the respiratory health impacts of air pollution in the young vulnerable populations.

Keywords Air pollution · Acute respiratory infection · Epidemiological studies · Children · Vulnerable populations

Abbreviations

Acute lower respiratory infection	ALRI
Air pollution index	API
Acute respiratory infection	ARI
Acute respiratory infections	ARIs
Association of Southeast Asian Nations	ASEAN
Carbon monoxide	CO
Demographic and Health Surveys	DHS
Excess risk	ER
Extreme weather events	EWE
International Classification of Diseases	ICD
Interquartile range	IQR
Low-middle income countries	LMICs
Lower respiratory tract infections	LRTI
Nitrogen dioxide	NO_2
Nitrogen oxides	NO_x
Odds ratio	OR
Ozone	O_3
Particulate matter	PM
Particulate matter with aerodynamic diameter less than 10 μm	PM_{10}
Particulate matter with aerodynamic diameter less than 2.5 μm	$PM_{2.5}$
Relative risk	RR
Reactive oxygen species	ROS
Respiratory syncytial virus	RSV
South Asia	SA
Southeast Asia	SEA
Socioeconomic status	SES
Sulphur dioxide	SO_2
Total suspended particles	TSP

United Nations International Children's Emergency Fund	UNICEF
Upper respiratory tract infection	URTI
World Health Organisation	WHO

5.1 Introduction

Anthropogenic emissions, leading to air pollution, are a major cause of global warming and pose a critical public health challenge in South Asia (SA) and Southeast Asia (SEA). The SA and SEA regions experience some of the highest levels of air pollution globally, thereby significantly contributing to the burden of disease (IQAir 2023). Children under five are particularly vulnerable because of their highly developing respiratory systems, higher ventilation rates, and greater exposure to pollutants in environments such as schools and crowded living conditions (UNICEF 2016). According to the United Nations International Children's Emergency Fund (UNICEF), almost 300 million children worldwide are exposed to toxic levels of outdoor air pollution, and nearly one million children die prematurely from pneumonia annually, with more than half of these deaths attributed directly to air pollution (UNICEF 2016, 2024; Friedrich 2017). It has been reported that air pollution accounted for approximately 709,000 deaths in young children aged five years and below, amounting to 15% of deaths in this age group, globally (State of Global Air 2024).

The climate crisis intensifies the impact of air pollution, creating a compounding threat to children's health through several mechanisms. Rising temperature causes the formation of ground level ozone and increases wildfire frequency which impacts the air quality to deteriorate significantly. Climate change-driven extreme weather events (EWE) like prolonged heat waves and El Niño trap pollutants near the ground and boost emissions from industry, transportation, and agricultural burning (Bolan et al. 2024). Studies indicate that an increase in temperature (during drier seasons) induces the seasonal surges of pollutants such as fine particulate matter ($PM_{2.5}$) and ozone (Amnuaylojaroen et al. 2023; Manager 2024). Moreover, SA and SEA countries not only rank among the highest levels of air pollution but also have some of the highest shares of children in their populations, making this demographic particularly vulnerable to the adverse effects of air pollution (IQAir 2023). Existing literature highlighted that ARIs are a leading cause of morbidity and mortality among children under five in SA and SEA. Countries in the SA region recorded high mortality rates for children under five, from air pollution with 164 deaths per 100,000 children. For instance, in 2021, the highest record was from India with 169,400 deaths, while Pakistan and Bangladesh recorded 68,100 deaths and 19,100 deaths, respectively (State of Global Air 2024). In SEA, the situation is similar, with high mortality rates seen throughout the region, with the three leading countries being Indonesia, Vietnam, and the Philippines, with 221,600 deaths, 99,700 deaths, and 98,209 deaths, respectively (State of Global Air 2024).

5.1.1 Economic Burden Associated with Air Pollution

Furthermore, the economic burden associated with ARI as a result of air pollution is substantial. In SA, economic losses of over 4% of gross domestic product (GDP) are seen in countries like Bangladesh as a result of air pollution, which resulted in productivity losses from health impacts of respiratory diseases (Khwaja et al. 2012). In SEA, the economic impacts are equally severe, where data from Association of Southeast Asian Nations (ASEAN) countries indicated the importance of implementing effective air quality policies, as failure could result in heavy economic losses by 2030, with Cambodia, Indonesia, and Thailand projected to incur losses of approximately 2.6–3.1% of each country's GDP in 2030 solely from air pollution-related health issues which include ARI (IIASA 2023).

5.1.2 Drivers of Air Pollution and Their Impact on Public Health in SA and SEA

SA and SEA experience disproportionately high levels of air pollution, resulting in detrimental public health consequences that are exacerbated by these regions' rapid urbanisation and industrialisation. Key drivers include emissions from industrial activities, traffic-related pollution, solid fuel combustion for cooking, deforestation, and agricultural burning (Jabbar et al. 2022; Gu et al. 2024). It is undeniable that these activities are important for the region's economic growth. However, the fast pace of development and a rapid growth in population in the region creates significant environmental challenges (Anderson 1979; Yeung 2022), contributing to environmental disparities, particularly affecting marginalised communities with lower socioeconomic status (SES) (Hajat et al. 2015). Socioeconomic disparities such as poverty, limited infrastructures, and low accessibility to healthcare services worsen the effects of air pollution on children's respiratory health. Low SES was associated with greater household pollutant exposure and restricted healthcare access, both of which exacerbate respiratory conditions like ARIs (George et al. 2024). Furthermore, transboundary haze has become a major environmental concern in SEA, further exacerbating the region's air quality crisis, with El Niño delaying the rainy season's ability to reduce $PM_{2.5}$ levels (IQAir 2023). Power generation, vehicle emissions, industrial activities, and open burning remain the primary sources of $PM_{2.5}$ within the regions (IQAir 2023). Cooperation among countries is required to address the impacts of transboundary haze, which necessitates further efforts in understanding the causes of air pollution, to develop effective solutions and policies to improve air quality and protect public health in the region.

5.1.3 *Acute Respiratory Infections (ARIs) Increase the Risk of Morbidity and Mortality Among Children Under 5 Years Old in SA and SEA*

ARIs increase the risk of morbidity and mortality, particularly among children under five in SA and SEA regions, contributing significantly to the overall disease burden among young children in these regions, (Anwar et al. 2021; UNICEF 2017) contributing to 25% of the cause of deaths (Shapiro 1998) with 4.5 million deaths annually due to pneumonia (Berman 1991). To date, the understanding of how air pollutants may affect the physiological development of children's lungs remains limited. Animal studies suggest that these conditions can significantly impact lung growth and function, particularly during critical developmental stages (Assessment 2009). Children's unique physiology and behaviours further heighten their vulnerability to air pollution. While it is typically anticipated that school-aged children experience greater exposure to air pollution due to increased outdoor activity and commuting (Wolfe et al. 2020), emerging evidence indicates that young children under five are also significantly exposed to air pollution due to spending more time outdoors during peak pollution periods (Cohen Hubal et al. 2000). Several mechanisms have been proposed to explain how air pollution increases the risk of respiratory infections (Bateson and Schwartz 2008). For instance, fine particulate matter such as $PM_{2.5}$, a major component of air pollution, poses a significant threat to respiratory health. Upon inhalation, $PM_{2.5}$ can penetrate deep into the lungs and enter the bloodstream, triggering systemic inflammation and oxidative stress (Checa and Aran 2020; Zhao et al. 2020). This process damages respiratory epithelial cells, compromising their ability to clear pathogens and pollutants from the airways (Miller 2020).

Several meta-analyses and systematic reviews have provided evidence that ambient air pollution is a significant risk factor for ARI in the general population (Nhung et al. 2017; Ibrahim et al. 2021; Kaspersen et al. 2024). However, studies specifically examining the associations between individual air pollutants and ARI in children under the age of 5, particularly within the SA and SEA regions remain limited (Luong et al. 2019; Tiwari et al. 2022). The World Health Organisation (WHO) emphasises that every child has the right to clean air, yet many live in areas where air quality exceeds international safety limits by significant margins (UNICEF 2024; WHO 2024).

5.1.4 *Research Gaps*

The number of literatures investigating the relationship between air pollution and ARIs among children in SA and SEA remains limited, which is mainly due to constraints in resources and infrastructure to support the research on public health in low- and middle-income countries (LMICs). The limited evidence also contributes to the insufficient understanding of how various air pollutants may impact children who are undergoing physiological development of their lungs. Further research into this concerning health issues is important to formulate targeted and effective interventions for mitigating the health risks of air pollution for young vulnerable populations. Organisations such as UNICEF and WHO are strong advocates of mitigating air pollution risks via policy changes and engagement with the community emphasising that protecting children's health should be the priority in climate action studies (UNICEF 2016, 2024; WHO 2024). Implementing an effective plan against the climate change crisis remains challenging, therefore, this study aims to synthesise available evidence on the association between air pollution and ARIs especially among children across the SA and SEA regions and seeks to highlight critical research gaps to guide future studies.

Therefore, there is a need for comprehensive approaches to tackle air pollution, despite significant challenges remaining in implementing effective solutions. To explore the research gaps, a research question for this literature review was formulated according to the PECO framework, with Participants (P) defined as children aged five years and under; Exposure (E) refers to ambient air pollution; Comparison (C) is not applicable due to the nature of observational study designs; and the Outcome (O) is Acute Respiratory Infection (ARI), as depicted in Table 5.1. Based on this framework, the research question is: '*In children aged 5 years and below, does outdoor air pollution increase the risk of acute respiratory infection (ARI)?*'. The PECO framework provides clear guidance in formulating research questions that specifically assess the association between exposures and outcomes, particularly in the context of environmental and health risks (Morgan et al. 2018) (Fig. 5.1).

Hence, the present review aims to synthesise the current evidence on the association between various air pollutants and ARI among children under five, specifically within SA and SEA regions, and additionally to investigate environmental and socioeconomic factors, and how they modify the risks of ARI in children. This review will further highlight crucial research gaps which would then contribute to the understanding of climate change induced environmental risks for children in these regions. By addressing these research aims, the study will provide valuable insights that can inform targeted interventions, policy development and public health strategies to mitigate the adverse effects of air pollution on children's respiratory health.

PECO components	Description
Participants (P)	Children aged five years and below
Exposure (E)	Ambient air pollution
Comparison (C)	Not applicable (due to observational study designs)
Outcome (O)	Acute respiratory infection (ARI)

Fig. 5.1 Components of a PECO framework

Table 5.1 Meta-analyses and systematic reviews on the association between air pollution and acute respiratory infections (ARI)

References	Study design	Sample population	Exposure	Health outcome	Main findings	Strengths and limitations	Conclusion	Confounding
Ziou et al. (2022)	Systematic review and meta-analysis of short-term studies of time-series, case-crossover, cohort, or cross-sectional studies	Population aged <19 years old Population from Switzerland, USA, Republic of Korea, Sweden, Germany, UK, Nigeria, Bangladesh Systematic review of 34 studies, 16 included in meta-analysis Sample size: N = 40–3,472,347	Short-term exposure to air pollutants: (i) $PM_{2.5}$ (ii) PM_{10}	Diagnosis or symptoms of acute upper respiratory tract infection	Group result: (i) $PM_{2.5}$ is associated with an increased risk of acute URTI, RR 1.010 (95% CI: 1.007 to 1.014) (ii) PM_{10} is associated with an increased risk of acute URTI, RR 1.016 (95% CI: 1.011–1.021)	Strengths: (i) The study includes of broad variety of epidemiological study designs (ii) PM of different size fractions, and different exposure time periods are considered (iii) Time-series and case-crossover studies had large sample sizes, small standard errors, with study duration over years, increasing the confidence of results (iv) Extensive critical evidence performed for each pollutant Limitations: (i) Possible reporting bias as studies included are published only in English, French or Spanish, full text studies (ii) Limited studies included for each PM of different sizes (iii) Most studies obtain data through healthcare services, which may have captured only the more serious URTIs (iv) Exposure assessment performed through monitoring or modelled estimates over cities or counties and does not measure more precise locations such as residences, school, and childcare locations (v) Limited evidence on long-term effects	TSP, $PM_{2.5}$, and PM_{10} and were significantly associated with hospitalisations for URTI in children <19 years	Time-series, case-crossovers: holidays, seasonal influenza epidemics, seasonality, temperature, humidity, day of the week Cohort and cross-sectional studies: age and sex of participants, tobacco smoke exposure, SES

(continued)

Table 5.1 (continued)

References	Study design	Sample population	Exposure	Health outcome	Main findings	Strengths and limitations	Conclusion	Confounding
Mehta et al. (2013)	Systematic review and meta-analysis of cohort, time-series, case-controls, case-crossover, cross-sectional, panel, ecologic and interventional studies	Children up to 12 years, findings stratified for under 5 years Population from Australia, New Zealand, Brazil, Vietnam, Mexico, USA, Chile, France, Netherlands, Czech Republic, Germany, Canada	Short-term and long-term exposure to a 10 μg/m^3 increase in concentration of air pollutants: (i) $PM_{2.5}$ (ii) PM_{10}	ALRI morbidity and mortality	For short-term studies on PM_{10} and ALRI morbidity in children under 5 years of age, no pooled result is available For long-term studies, the pooled RR of ALRI occurrence was 1.12 (95% CI 1.03–1.30) per 10 μg/m^3 increase in annual average $PM_{2.5}$ concentrations in children under 5 years of age	Strengths: (i) No substantial effects resulting from differences in outcome assessment from sensitivity analyses, strengthening the credibility of the summary estimate reported Limitations: (i) The definition of ALRI is not standardised across all studies (ii) Outcome assessments are based on maternal recall on ALRI among their children, leading to possibly poorer validity (iii) Possibility that ambient PM may have independent effects on ALRI which is difficult to estimate	Increased risk in acute lower respiratory infection occurrence per 10 μg/m^3 increase in annual average $PM_{2.5}$ concentration	Different sources of combustion (second hand smoke, solid fuel use within the household), regional variations of case fatality rates, timely access to adequate treatment, malnourishment, immunosuppression

Abbreviations Acute Lower Respiratory Infection (ALRI), Acute Respiratory Infection (ARI), Socioeconomic Status (SES), Total Suspended Particles (TSP), Upper Respiratory Tract Infection (URTI)

5.2 Methods

5.2.1 Search Strategy

The literature search was conducted using the umbrella review approach to compile available systematic reviews or meta-analyses from the main health and medical science databases, PubMed and Ovid MEDLINE. Firstly, a search was performed to identify available meta-analyses and systematic reviews on the association between air pollution and ARI among children. A flow diagram in accordance with the PRISMA guidelines is utilised for the first literature search. Then, suitable individual studies were extracted from the meta-analyses and systematic reviews. Additionally, a second search was performed, which focused on studies conducted in SA and SEA and included 13 individual studies (HEI Collaborative Working Group et al. 2012; Nathan et al. 2014; Abdul Rahman et al. 2017; Luong et al. 2017, 2020; Nhung et al. 2018, 2019; Suryadhi et al. 2020; Sherris et al. 2021; Thongphunchung et al. 2021; Balasubramani et al. 2022; Ibrahim et al. 2022; Abudureyimu et al. 2023; Seah et al. 2023; Adhikary et al. 2024), that were not part of the meta-analyses or systematic reviews. The complete search strategy, including specific keywords and search terms, is provided in Appendix 1.

5.2.2 Inclusion and Exclusion Criteria

Inclusion criteria of the studies were: (1) children aged 5 and below; (2) air pollution or particulate matter from an outdoor source; (3) acute respiratory infection as an outcome; (4) effect estimates such as relative risk (RR), odds ratio (OR), or excess risk (ER) were calculated between the exposure and outcome; (5) publications in the English language; (6) publications from/with data from SA and SEA. Exclusion criteria include: (1) air pollution from an indoor source; and (2) only an abstract available for the study.

5.2.3 Data Extraction

The studies were reviewed, and data was extracted from the studies into a table, summarising the author, year, study design, sample population, exposure, health outcomes, main findings, confounders and additional comments such as strengths and limitations.

5.2.4 Strategy for Data Synthesis

The results will be reported using a narrative synthesis grouped by studies from (1) meta-analyses and/or systematic reviews, (2) SA and SEA. The meta-analysis and systematic review findings will be summarised based on the types of air pollutants, adapting the format of an umbrella review.

5.2.5 Literature Search Process

Two studies published from 2013 to 2022, studying the association between ambient air pollution and ARI among children aged 5 and below, were extracted and included in Table 5.1. As shown in the flow diagram (Fig. 5.2), we identified two systematic reviews and meta-analyses as eligible: Ziou et al. (2022) and Mehta et al. (2013).

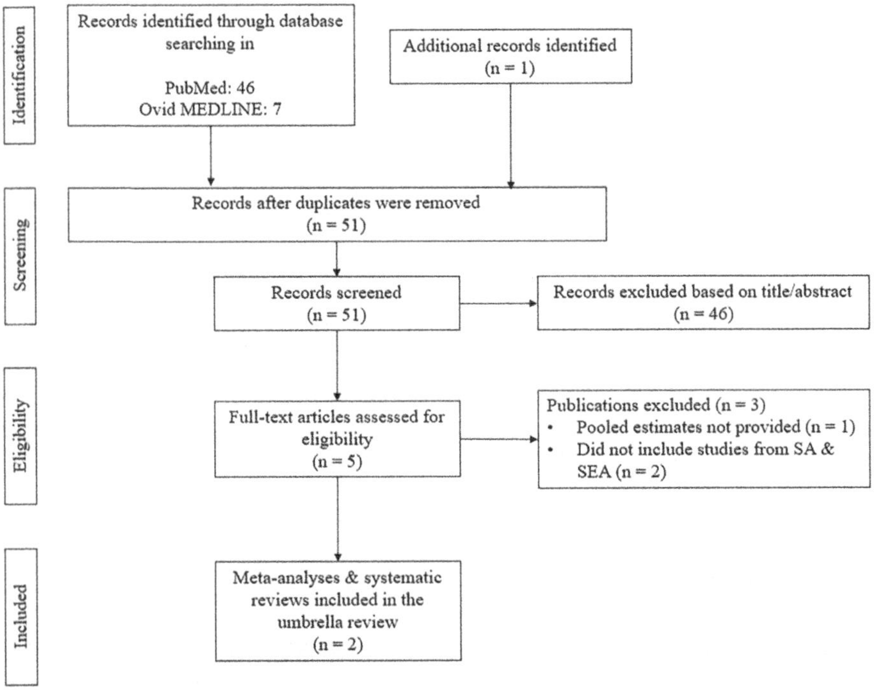

Fig. 5.2 Flow diagram of literature search for meta-analyses and systematic reviews. *Abbreviations* South Asia (SA), Southeast Asia (SEA)

5.3 Results

Table 5.1 summarises the two meta-analyses and systematic reviews on the association between air pollution and acute respiratory infections (ARI). Ziou et al. (2022) is a systematic review and meta-analysis of 16 studies, consisting of time-series, case-crossover and cohort studies, aiming to determine the relationship between exposure to $PM_{2.5}$ and PM_{10}, and the diagnosis or symptoms of acute upper respiratory tract infection (URTI). Mehta et al. (2013) is a systematic review and meta-analysis of 73 studies investigating the association between short- and long-term exposure to air pollutants PM_{10}, and $PM_{2.5}$, and ALRI morbidity and mortality.

Among the studies, only PM_{10} and $PM_{2.5}$ see significant positive associations with ARI among the population. Associations are seen with the exposure of PM_{10} and ARI or respiratory morbidity among children in both studies. For diagnosis or symptoms of acute URTI, relative risks of RR 1.016 (95% CI: 1.011–1.021) were reported (Ziou et al. 2022). Mehta et al. (2013) studied the short-term effects of PM_{10} on ALRI morbidity in children under 5 years, however, no pooled result was available for the study. For $PM_{2.5}$, Ziou et al. (2022) reported significant associations between exposure to $PM_{2.5}$ and acute URTI, RR 1.010 (95%CI: 1.007–1.014), while Mehta et al. (2013) reported a significant association between ALRI with exposure to $PM_{2.5}$ in long-term studies, pooled RR 1.12 (95% CI: 1.03–1.30).

5.3.1 South Asia (SA) and Southeast Asia (SEA) Studies

Table 5.2 presents twelve studies that fulfilled the inclusion criteria for this review, investigating the association between outdoor air pollution and ARI among children aged 5 and below, with a focus on the population of SA and SEA. The studies that are included consist of case-crossover studies (n = 7), time-series studies (n = 3), cross sectional studies (n = 2), a retrospective cohort study (n = 1), an ecological study (n = 1), and one study which utilised both case-crossover and time-series analysis. Eleven studies are from SEA, and four studies are from SA. Table 5.3 summarises the associations between each air pollutant with ARI in each SA and SEA study. From the literature search, fifteen studies from both SA and SEA were extracted, and there is a disparity between the number of studies conducted in SA (n = 4), and SEA (n = 11). Among the studies from SEA, five studies are from Vietnam, three consist of studies from Malaysia, and one study each from Singapore, Indonesia, and Thailand, whereas among the studies from SA, three are from India and one is from Bangladesh. Among the fifteen studies from SA and SEA, five studies included children aged up to 18 years old, two of them did not conduct subgroup analyses for children aged 0–5, while three of the studies did. However, the majority of the participants in these studies still consisted of children aged 5 and below, providing sufficient data to represent the population, which aligns with the inclusion criteria.

Table 5.2 Summary of studies investigating the association between air pollution and ARI in SA and SEA

References	Study design	Sample population	Exposure	Health outcome	Main findings	Strengths and limitations	Conclusion	Confounding
Abdul Rahman et al. (2017)	Case-crossover analysis	Children from urban regions of Klang Valley, Malaysia, admitted for ARI, from 2006 to 2010 Sample size Children aged under 14 years old, N = 54,542 <5 years old, N = 45,393 (83.22%) 58.09% Male 41.91% Female	Short-term exposure to air pollution, from January 2006 to December 2010: (i) PM_{10} (ii) CO (iii) O_3 (iv) NO_2	Hospitalisations for 16 ARI subtypes	(i) PM_{10} is significantly associated with 13 ARI subtypes (OR ranging from 1.037 to 1.115) (ii) CO is significantly associated with 4 ARI subtypes (OR ranging from 1.287 to 1.785) (iii) O_3 is significantly associated with 2 ARI subtypes (OR ranging from 1.007 to 1.010) (iv) NO_2 is significantly associated with 8 ARI subtypes (OR ranging from 1.012 to 1.026)	Limitations (i) Study takes into account only the secondary data analysis without self-monitoring from the site (ii) Data collected from major urban cities which are focal points for industries, residential and commercial areas which produce high traffic-related pollution. Not representative of the entire population as there is no data from rural areas (iii) Only short-term outcomes are verified	PM_{10} significantly increased the risk of 13 ARI subtypes, CO significantly increased the risk of 4 ARI subtypes, O_3 significantly increased the risk for 2 ARI subtypes, and NO_2 significantly increased the risk for 8 ARI subtypes	Age, nationality, gender, race, home environment, behavioural factors

(continued)

Table 5.2 (continued)

References	Study design	Sample population	Exposure	Health outcome	Main findings	Strengths and limitations	Conclusion	Confounding
Abudureyimu et al. (2023)	Cross-sectional study	Children under 5 years old, from the DHS in India, from 20th January 2015 to 4th December 2016 Sample size N = 55,118 52.3% Male 47.7% Female	Short-term exposure to $PM_{2.5}$ from 2015 to 2016, obtained from the atmospheric composition analysis group	AURI and ALRI based on symptoms experienced by the child two weeks preceding the survey	Every 10 μg/m^3 increase in levels of $PM_{2.5}$ is associated with: (i) AURI: OR 1.02 (95% CI 1.02–1.03) (ii) ALRI: OR 1.05 (95% CI 1.03–1.06) Short-term exposure to $PM_{2.5}$ at the highest quartile (94.15–133.70 μg/m^3) is associated with the greatest increase in AURI, OR 1.13 (95% CI 1.04–1.23), and ALRI, OR 1.44 (95% CI 1.28–1.62)	Strength (i) Survey received a high response rate of 98% (ii) Solid fuel use was adjusted as a confounder and was found not to modify the association between $PM_{2.5}$ exposure and AURI and ALRI Limitations (i) Cluster level of $PM_{2.5}$ utilised which may not reflect actual level of exposure (ii) Risk of disease misclassification from self-reported outcomes, due to the nature of the data collection from surveys (iii) Cross-sectional study limits the establishment of relationship between cause and effect (iv) The relationship between exposure and outcome may be underestimated due to high levels of $PM_{2.5}$ across India (v) Possible underestimation of lower SES groups does not fully reflect actual population	$PM_{2.5}$ exposure significantly increases the odds of AURI and ALRI among children under five in India, with stronger associations with ALRI	Adjusted confounders: child's age, child's sex, maternal education level, maternal smoking status, solid fuel use, population density

(continued)

Table 5.2 (continued)

References	Study design	Sample population	Exposure	Health outcome	Main findings	Strengths and limitations	Conclusion	Confounding
Adhikary et al. (2024)	Cross-sectional study	Children under 5 years old, from the DHS in India, from 2019 to 2021 Sample size N = 223,375 51.7% Male 48.4% Female	$PM_{2.5}$ levels from each cluster in India using geographic data from the Standard DHS dataset, from 2019 to 2021	ARI based on symptoms experienced by children two weeks prior to the survey	(i) Each 10 $\mu g/m^3$ increase in $PM_{2.5}$ levels is associated with increased odds of ARI, OR 1.23 (95% CI 1.19–1.27) (ii) A two-fold increase in the odds of developing ARI is seen with a change from the first quartile (2.5–34.4 $\mu g/m^3$) to the second quartile (34.5–51.5 $\mu g/m^3$) of $PM_{2.5}$, OR 2.06 (95% CI 1.60–2.66) (iii) A four-fold increase in the odds of developing ARI is seen with a change from the first quartile (2.5–34.4 $\mu g/m^3$) to the fourth quartile (78.3–128.9 $\mu g/m^3$) of $PM_{2.5}$, OR 4.45 (95% CI 3.37–5.87)	Similar strengths: high response rates and adjusted confounders, and limitations: cluster levels of PM2.5, self-reported outcomes, cross sectional study design, underestimation of exposure and outcome and SES levels as Abudureyimu et al. (2023), given both studies are similar in design and obtained from the same data source	Exposure to high levels of $PM_{2.5}$ significantly increases the odds of developing ARI among children under 5 in India. Exposure to greater levels of $PM_{2.5}$ is significantly associated with greater odds of developing ARI compared to lower levels	Child's age and sex, maternal education level, household wealth, type of residence, maternal smoking status, household cooking fuel type

(continued)

Table 5.2 (continued)

References	Study design	Sample population	Exposure	Health outcome	Main findings	Strengths and limitations	Conclusion	Confounding
Balasubramani et al. (2022)	Case-crossover study	Children under 5 years across India, utilising data from NFHS-4 and the Census of India, from January 2015 to December 2016	$PM_{2.5}$ collected from the Global Estimates and National Ambient Air Quality Monitoring Programme, from the period of 2013 to 2016	ARI in each of the 640 districts of India based on symptoms, obtained from NFHS-4	$PM_{2.5}$ exposure is significantly associated with ARI, RR 1.007 (95% CI 1.005–1.009)	Limitations (i) The results of the study are not causative and only reflect associations between different independent variables and ARI (ii) Possible information bias due to the nature of study, where data is obtained from surveys, conducted at a district-level (iii) Data obtained from multiple sources (i.e. NFHS-4, Census of India, satellite and ground-based stations) makes statistical analysis challenging	$PM_{2.5}$ exposure is significantly associated with ARI among children under 5 in India	Household variables: use of unclean cooking fuels, improved sanitation facilities, tobacco consumption Maternal characteristics: breastfeeding duration, number of children, maternal BMI Children-level variables: diarrhoea, low birth weight, immunisations Demographic variables: household size, rural or urban residence, illiteracy levels, income levels

(continued)

Table 5.2 (continued)

References	Study design	Sample population	Exposure	Health outcome	Main findings	Strengths and limitations	Conclusion	Confounding
Ibrahim et al. (2022)	Case-crossover analysis	Children from Bintulu, Malaysia admitted for respiratory diseases, from 2010 to 2019 Sample size <18 years, N = 16,764 Children aged 0–4 (N = 14,311) and 5–17 (N = 2453) 59.0% Male 41.0% Female	Daily ambient air pollution, from 1st January 2010 to 31st December 2019: (i) PM_{10} (ii) $PM_{2.5}$ (iii) SO_2 (iv) NO_2 (v) O_3 (vi) CO	Daily hospital admissions for childhood respiratory diseases (ICD-10 code: J0-J49)	In age group 0–4: For a 10 μg/m^3 increase of air pollutants: (i) $PM_{2.5}$ highest RRs were 1.095 (95% CI 1.001–1.197) at Lag 0–2 days (ii) SO_2 highest RRs were 1.192 (95% CI 1.029–1.381) at Lag 0–6 days (iii) No significant associations seen with PM_{10}, NO_2, CO, O_3	(i) Study performed in urban cities may not represent the entire population (ii) Exposure measured using average concentration from one fixed-site monitor, the possibility of systematic measurement errors (iii) Possibility of misclassification of patients as patients who are admitted are assumed to be living close by	Exposure to air pollutants $PM_{2.5}$ and SO_2 were significantly associated with increased risk of hospital admissions for respiratory diseases among children, with stronger association in younger children below 5 years SO_2 showed a stronger association compared with $PM_{2.5}$ Association not seen with exposure to PM_{10}, NO_2, CO, and O_3	SES, parents' education, nutritional status

(continued)

Table 5.2 (continued)

References	Study design	Sample population	Exposure	Health outcome	Main findings	Strengths and limitations	Conclusion	Confounding
HEI Collaborative Working Group et al. (2012)	Case-crossover and time-series analysis	Children under age of 5 in HCMC, Vietnam from 2003 through 2005 Sample size N = 15,708	Daily citywide 24-h average concentrations of air pollutants, from 2003 to 2005: i) PM_{10} ii) NO_2 iii) SO_2 iv) 8-h maximum average concentrations of ozone (O_3)	Hospitalisations for ALRI, defined as pneumonia or bronchiolitis	Excess risk of ALRI admissions (95% CI) per 10 $\mu g/m^3$ increase in air pollutants: *(A) Case-crossover analyses* (i) PM_{10}: – Dry season: ER 1.25 (−0.55 to 3.09) at Lag 0, ER 1.63 (−0.3 to 3.59) at Lag 1–6 – Rainy season: ER −3.11 (−1.42 to −4.76) at Lag 0, ER −2.43 (−0.69 to −4.13) at Lag 1–6 (ii) O_3: – Dry season: ER −0.79 (−2.67 to 1.13) at Lag 0, ER −0.65 (−2.61 to 1.35) at Lag 1–6 – Rainy season: ER −2.98 (−1.14 to −4.78) at Lag 0, −2.49 (−0.5 to −4.43) at Lag 1–6 (iii) NO_2: – Dry season: ER 8.50 (0.8–16.79) at Lag 0, ER 12.69 (4.07–22.02) at Lag 1–6 – Rainy season: ER −5.15 (−0.1 to −9.94) at Lag 0, ER −4.34 (0.81 to −9.23) at Lag 1–6 (iv) SO_2: – Dry season: ER 5.85 (0.44–11.55) at Lag 0, ER 5.90 (0.42–11.68) at Lag 1–6 – Rainy season: ER −2.13 (4.41 to −8.25) at Lag 0, ER −2.95 (3.6 to −9.09) at Lag 1-–6 *(B) Time-series analyses* (i) PM_{10}: – Dry season: ER 0.53 (−1.46 to 2.56), ER 1.2 (−0.6 to 3.04) with rainfall – Rainy season: ER −2.58 (−4.44 to −0.68), ER −3.06 (−4.71 to −1.38) with rainfall (ii) O_3: – Dry season: ER 0.47 (−1.56 to 2.54), ER -0.87 (−2.76 to 1.06) with rainfall - Rainy season: ER -2.33 (-4.26 to -0.36), ER -2.91 (-4.72 to -1.07) with rainfall (iii) NO_2: - Dry season: ER 12.62 (3.91–22.05), ER 8.24 (0.5–16.53) with rainfall - Rainy season: ER -3.29 (-8.51 to 2.23), ER −4.99 (−9.79 to 0.07) with rainfall (iv) SO_2: – Dry season: ER 4.21 (−1.37 to 10.10), ER 5.87 (0.45–11.57) with rainfall – Rainy season: ER 2.70 (−4.88 to 10.88), ER −1.72 (−7.87 to 4.84) with rainfall	Strengths (i) Large population size (ii) Comparing and contrasting results from case-crossover and time-series methods allows for informative results, and strengthened confidence in the results, especially for NO_2 (iii) Time-series design allows for evaluation of short-term effects, and accounts for variations in daily pollutant levels Limitations (i) Presence of unmeasured confounding variables	Exposure to air pollutants was associated with hospital admissions for ALRI during the dry seasons (From November to April), while an inverse association was seen during the rainy season Increased concentrations of NO_2, SO_2, and PM_{10} were associated with an increased risk of ALRI resulting in hospitalisations among young children in the dry season. The associations are seen strongest among SO_2 and NO_2	SES, weather conditions, seasonal variations, respiratory syncytial virus infection

(continued)

Table 5.2 (continued)

References	Study design	Sample population	Exposure	Health outcome	Main findings	Strengths and limitations	Conclusion	Confounding
Luong et al. (2020)	Cross-sectional study	Patients aged between 28 days, to less than 5 years old, admitted to Children's Hospital No. 2 in Vietnam for ALRI, from February 2016 to December 2017 Sample size N = 11,416 60.9% Male 39.1% Female	Concentrations of $PM_{2.5}$ in HCMC from February 2016 to December 2017	Daily hospital admissions for ALRI	$PM_{2.5}$ exposure is significantly associated with ALRI resulting in admissions after 3 days of exposure, ER 3.51% (95% CI: 0.96–6.12%) at Lag 3 Each 10 μg/m3 increase in $PM_{2.5}$ levels was significantly associated with an increased risk of ALRI admission, ER 5.25% (95% CI 2.00–8.60%) among male, and no significant associations with female children, at Lag 3	Strengths (i) Confounders are adjusted which includes temperature, day of the week, and relative humidity Limitations (i) Data of hospitalisations for ALRI were collected from only one hospital, which does not account for the total number of child admissions in the city (ii) Potential confounding between age and ICD codes not considered (iii) Ambient air pollution ($PM_{2.5}$) data was only collected from only one monitoring station (iv) Study lack data for the effect of other air pollutants (NO_2, SO_2, O_3) (v) Influenza outbreak not adjusted for	$PM_{2.5}$ exposure in HCMC is significantly associated with hospital admissions for ALRI among children under 5 years old	Age, breastfeeding, influenza outbreak

(continued)

Table 5.2 (continued)

References	Study design	Sample population	Exposure	Health outcome	Main findings	Strengths and limitations	Conclusion	Confounding
Luong et al. (2017)	Time-stratified case-crossover study	Children aged 28 days–5 years, admitted to Vietnam National Hospital of Paediatrics from September 2010–September 2011 Sample population N = 8934	Hourly average value of air pollutant concentrations (μg/m^3), from September 2010 to September 2011: (i) PM_{10} (ii) $PM_{2.5}$ (iii) PM_1	Hospital admissions for respiratory disease (ICD-10 code: J00-J99)	The relationship between air pollution and admissions for respiratory disease, at different time lags (i) PM_{10}: (a) Lag 0: OR range from 1.012 to 1.015 ($P < 0.05$) (b) Lag 1: OR range from 1.010 to 1.012 ($P < 0.05$) (c) Lag 2: OR range from 1.007 to 1.009 ($P < 0.05$) (d) Lag 3: OR range from 1.006 to 1.007 ($P < 0.05$) (ii) $PM_{2.5}$: (a) Lag 0: OR range from 1.022 to 1.017 ($P < 0.05$) (b) Lag 1: OR range from 1.013 to 1.016 ($P < 0.05$) (c) Lag 2: OR range from 1.009 to 1.012 ($P < 0.05$) (d) Lag 3: OR 1.009 ($P < 0.05$) (iii) PM_1: (a) Lag 0: OR range from 1.019 to 1.025 ($P < 0.05$) (b) Lag 1: OR range from 1.014 to 1.019 ($P < 0.05$) (c) Lag 2: OR range from 1.012 to 1.014 ($P < 0.05$) (d) Lag 3: OR 1.012 ($P < 0.05$)	Limitations (i) Hospital admission data for respiratory diseases only obtained from one major hospital and does not account for minor cases admitted to lower-level hospitals (ii) Factors that can be changed in a short period of time are not able to be controlled (e.g. air conditioning use, indoor and outdoor work) (iii) Ambient air pollution levels only collected from a single monitoring station and does not represent the entire population, possibility of bias (iv) The study was conducted around a short time period, which may miss out on other potential confounding factors (v) Unadjusted confounders such as humidity, influenza outbreaks, air conditioning use, indoor and outdoor work	A 10 μg/m^3 increase in the levels of PM_{10}, $PM_{2.5}$ or PM_1 increases the risk of hospital admission for respiratory disease, with an increased risk of 1.4%, 2.2% or 2.5%, respectively, for same day exposure	Temperature, sex, day of the week, other air pollutants (NO_2, SO_2, CO, O_3), humidity, influenza outbreaks, air conditioning use, indoor and outdoor work

(continued)

Table 5.2 (continued)

References	Study design	Sample population	Exposure	Health outcome	Main findings	Strengths and limitations	Conclusion	Confounding
Nathan et al. (2014)	Retrospective cohort study	Children from an urban city, Kuala Lumpur, Malaysia, admitted between 1st November 2010 and 30th November 2011 Sample size N = 391 Children aged between 28 days to 18 years old ≤1 years old N = 344 (88%) >5 years old N = 6 59.3% Male 40.7% Female	Particulate matter ≤10 μm (PM_{10}) levels, and API, from November 2010 to November 2011	Hospitalisations for LRTI, either pneumonia, empyema or bronchiolitis diagnosed clinically	(i) Unhealthy API and poor API not associated with admissions for LRTI (ii) No significant correlation between monthly PM_{10} levels and API with total admissions for LRTI and life-threatening LRTIs	(i) Study population is from the urban population, and may not represent the entire population (ii) Study population is small (iii) Lack of uniform methods to determine pneumonia aetiology (iv) Certain sociodemographic factors are not determined and included in the study as variables, which includes overcrowding, access to sanitary water, indoor air pollution, and accessibility to care	No significant association between PM_{10} or API and admissions for LRTI	Race, comorbidities (Prematurity, cardiac defects, metabolic or genetic conditions), hypoxia, dyspnoea, apnoea and lethargy, fever symptoms, previous lung infection, previous admissions, nursery care, positive nasopharyngeal culture, biochemical levels (C-reactive protein, bicarbonate neutrophils, and lymphocyte levels)

(continued)

Table 5.2 (continued)

References	Study design	Sample population	Exposure	Health outcome	Main findings	Strengths and limitations	Conclusion	Confounding
Nhung et al. (2019)	Time-series study	Children aged 0–5 years in Hanoi, hospitalised with ALRI between 2007 and 2016 Sample size N = 75,432 64.1% Male 35.9% Female	Hourly concentrations of air pollutant levels from 2007 to 2016: (i) PM_{10} (ii) $PM_{2.5}$ (iii) PM_1 (iv) SO_2 (v) NO (vi) NO_2 (vii) NOx (viii) CO (ix) O_3	Length of hospital stay	(i) An 86 $\mu g/m^3$ increment (IQR) in O_3 concentration is associated with lower odds of discharge, OR 0.95 (95% CI 0.92–0.98), with exposures one to four days prior to admission (ii) A 61.5 $\mu g/m^3$ (IQR) increase in PM_{10} significantly increases LOS in children aged 2–5 yrs, with a 6% decrease in odds of discharge (95% CI 1%–12%) (iii) Four-day average ambient concentrations of SO_2 is negatively associated with LOS in children aged 0–5, OR 1.069 (95% CI 1.011–1.103) (iv) A 4 day average NO concentration is negatively associated with LOS in children aged 0–5, OR 1.026 (95% CI 1.002–1.051) (v) A 4-day average CO concentration is negatively associated with LOS in children aged 0–5, OR 1.031 (95% CI 1.005–1.057) (vi) LOS is not significantly associated with other air pollutants: $PM_{2.5}$, PM_1, NO_2, and NO_x	Limitations (i) Information regarding air quality is obtained from only via a few fixed monitoring stations and does not represent the actual exposure levels in residential location of children, commuting modes and school locations (ii) Variables such as wind factor and speed were not adjusted (iii) Data collected from a national hospital and focusing on specific children with severe illness or pre-existing conditions, which limits the generalisability of results from other hospitals and settings	An IQR increase in O_3 concentrations decreases the odds of discharge from hospital for children with ALRI. PM_{10} is associated with decreased odds of discharge for children aged 2–5 years Higher SO_2 and CO concentrations decrease the length of hospital stay Increase in air pollutants such as NO_2, NO, NOx, SO2, and CO decrease the length of hospital stay	Adjusted confounders: temperature, humidity, wind speed, child's age and sex, place of residence, days of the week, time of discharge, hospital occupancy

(continued)

Table 5.2 (continued)

References	Study design	Sample population	Exposure	Health outcome	Main findings	Strengths and limitations	Conclusion	Confounding
Nhung et al. (2018)	Time-series study	Children aged 0–17 in Hanoi, Vietnam, admitted due to pneumonia, bronchitis and asthma, obtained from hospital admission records, from January 1, 2007 to December 30, 2014 Subgroup analysis for age group < 1 year and 1–5 Sample size: All age (0–17) N = 57,851 < 1 year N = 34,211 1–5 years N = 21,375	Short-term ambient air pollution, daily average obtained from January 2007 to December 2014: (i) PM_{10} (ii) $PM_{2.5}$ (iii) PM_1 (iv) SO_2 (v) NO_2 (vi) NOx (vii) CO (viii) 8-h average O_3 (ix) 24-h maximum O_3	Daily counts of pneumonia (ICD-10 code J12-J18), bronchitis and asthma (ICD-10 code J20, J21, J45)	Increase in hospitalisations due to a result of 7-day moving average of ambient air pollution: (i) Pneumonia (1–5 years) (a) PM_{10} (RR 1.056, 95% 1.012 to 1.102) (b) $PM_{2.5}$ (RR 1.063, 95% 1.011 to 1.117) (c) PM_1 (RR 1.077, 95% 1.018 to 1.139) (d) NO_2 (RR 1.100, 95% 1.041 to 1.162) (e) NO_X (RR 1.101, 95% 1.041 to 1.166) (f) CO (RR 1.066, 95% 1.005 to 1.130) (ii) Pneumonia (< 1 years) (a) NO_2 (RR 1.050, 95% 1.005 to 1.097) (iii) Bronchitis and asthma (1–5 years) (a) NO_2 (RR 1.079, 95% 1.004 to 1.160) (b) NO_X (RR 1.116, 95% 1.035 to 1.203) (iv) Bronchitis and asthma (< 1 years) (a) CO (RR 0.908, 95% 0.848 to 0.973)	Limitations (i) Study conducted only in a major tertiary hospital, which may miss out children admitted to other hospitals (ii) Outpatients are not included in the study (iii) Possibility of misclassification of diagnosis of the type of respiratory disease using ICD10 coding definition (iv) Air pollution data collected from only one fixed-site monitoring station (v) There are limited admissions due to bronchitis and asthma	All ambient air pollutants were significantly associated with pneumonia hospitalisations, except for SO_2 and O_3 in children aged 1–5 years, while only NO_2 is associated with pneumonia hospitalisations in children aged <1 years NO_2 and NO_X is significantly associated with hospital admissions for bronchitis and asthma in children aged 1–5 years, while CO is significantly associated with bronchitis and asthma hospitalisations in children aged <1 years Associations were stronger in children aged 1–5 years	Temperature, humidity, wind speed, influenza epidemics, day of the week, and holidays

(continued)

Table 5.2 (continued)

References	Study design	Sample population	Exposure	Health outcome	Main findings	Strengths and limitations	Conclusion	Confounding
Seah et al. (2023)	Ecological study	Daily influenza A and B infections in children <5 years old obtained from the women and children's hospital in Singapore, from 2009 to 2019 Sample size N = 7,329	Concentrations of air pollutants from 2009 to 2019: (i) $PM_{2.5}$ (ii) PM_{10} (iii) SO_2 (iv) NO_2 (v) CO	Positive cases of paediatric influenza infections, confirmed by laboratory, among children under 5-years of age, from 2009 to 2019	Influenza A (i) Increased SO_2 concentration was positively associated with influenza A infections at Lag 4, RR 1.009 (95% CI 1.001–1.078) (ii) Increased CO concentration was negatively associated with influenza A infections at Lag 2, RR 0.712 (95% CI 0.555–0.914) Influenza B (i) SO_2 significantly increases the risk of influenza B on the same day, RR 1.020 (95% CI 1.006–1.033), while non-linear positive associations are seen at lag 4 (ii) A minimal increase in CO concentrations up to 0.5 mg/m^3 increases the risk of influenza B, while a negative association is seen with levels beyond 0.5 mg/m^3 (iii) Increased NO_2 levels significantly increases influenza B infection at lag 5, RR 1.015, (95% CI: 1.005–1.025)	Limitations (i) Possibility of selection bias as symptomatic cases are more likely to be picked up where a nasopharyngeal swab was performed, compared to asymptomatic or mild cases (ii) Possible variables may alter outcomes, such as air conditioner or air purifier use, and duration stayed indoors	SO_2 exposure is associated with increased risk of influenza A and influenza B infection. The association is greater in influenza B compared to influenza A NO_2 exposure increases the risk of influenza B infection CO exposure increases the risk of influenza B at low concentrations, however, exhibits protective effects against influenza B at higher concentrations $PM_{2.5}$ and PM_{10} exposure is negatively associated with the risk of influenza infection	Adjusted confounders: Rainfall, temperature, long-term trend, seasons, absolute humidity, and days of the week

(continued)

Table 5.2 (continued)

References	Study design	Sample population	Exposure	Health outcome	Main findings	Strengths and limitations	Conclusion	Confounding
Sherris et al. (2021)	Case-crossover study	Children aged < 5 years in Kamalapur, Bangladesh between 2005 and 2014 from clinic visits Sample size N = 28,089 45.9% Female, 54.1% Male	Exposure to ambient $PM_{2.5}$, and $PM_{2.5}$ of different sources (biomass burning, brick kilns, soil dust, road dust, fugitive lead, sea salt, zinc manufacturing, vehicle emissions), collected between 1st January 2005, and 31st December 2014	Physician diagnosed pneumonia or URTI	A 10 μg/m^3 increase in two-day lagged $PM_{2.5}$: RR 1.032 (1.008 to 1.056) IQR increase in $PM_{2.5}$ attributed to: (i) Brick kilns: RR 1.062 (1.000–1.194) for pneumonia, RR 1.048 (0.967–1.267) for URTI (ii) Fugitive lead: RR 1.060 (0.982–1.101) for pneumonia (iii) Road dust: RR 0.936 (0.798–0.960) for URTI (iv) Zinc manufacturing: RR 0.951 (0.849–0.984) for URTI	Limitations (i) Not adjusted to different sources of air pollutants as the categories of air pollutants are composed of a source mixture. Hence, it is challenging to determine the impact of a specific source of pollutant (ii) $PM_{2.5}$ samples were collected approximately twice weekly and are unable to estimate all lag analyses and lag effects (iii) Unable to account for nonindependence, as a result of possible repeat visits by the same children, that could result in underestimated standard errors (iv) Distance between the study community and the site of $PM_{2.5}$ measurement is 4 km, potential exposure misclassification (v) Source specific $PM_{2.5}$ masses were not measured directly	An increased level of ambient $PM_{2.5}$ significantly increases pneumonia among children under 5 years, in urban Dhak. Greater associations were seen with $PM_{2.5}$ that has a higher mass percentage with different sources of $PM_{2.5}$ composition (brick kiln and fugitive lead emissions)	Adjusted confounders: temperature, relative humidity, precipitation Potential confounders not adjusted: other air pollutants (O_3, NO_2, CO, SO_2), different $PM_{2.5}$ sources (biomass burning generated from waste and cookstoves)

(continued)

Table 5.2 (continued)

References	Study design	Sample population	Exposure	Health outcome	Main findings	Strengths and limitations	Conclusion	Confounding
Suryadhi et al. (2020)	Time-series study	Indonesian children born between 2009 and 2011, aged 0–3 Sample population N = 4931 51.2% Male 48.9% Female	Concentrations of air pollutants from 2009 to 2011: (i) NO_2 (ii) SO_2	Self-reported ARI	Increased ARI risk for an IQR increase in: (i) Mean NO_2 exposure OR 1.18 (95% CI 1.08–1.28) (ii) Mean SO_2 exposure OR 0.98 (95% CI 0.90–1.06)	Limitations (i) NO_2 and SO_2 concentrations were only measured four times yearly, where the mean concentration is obtained from averaging the maximum and minimum concentrations of these 4 measurements (ii) No information regarding air pollution was available at individual-level, which does not reflect the actual level of exposure (iii) Child health outcomes were self-reported, risk of disease misclassification (iv) Information on the child's history and family history, such as asthma and allergies were not obtained, which may lead to potential residual confounding (v) The study is only generalised to urban populations	NO_2 exposure increased the risk for ARI in children aged 0–3 years in Indonesia, whereas SO_2 was not associated with ARI	SES, smoking

(continued)

Table 5.2 (continued)

References	Study design	Sample population	Exposure	Health outcome	Main findings	Strengths and limitations	Conclusion	Confounding
Thongphunchung et al. (2021)	Case-crossover study	Children aged 0–14 years visiting the outpatient department in Bangkok Metropolitan Region from January 2015 to December 2018 Age group stratified into preschool children (Aged 0–5 years, N = 280,362) and school children (Aged 6–14, N = 127,509)	Hourly concentrations of air pollutants from January 2015 to December 2018: (i) $PM_{2.5}$ (ii) PM_{10}	Daily number of outpatient department visits for respiratory diseases	The OR for total respiratory diseases per 10 μg/m^3 of: (i) PM_{10}: OR range from 1.0168 (1.0075–1.0262) to OR 1.0354 (1.0267–1.0442) (ii) $PM_{2.5}$: OR range from 1.0210 (1.0133–1.0288) to OR 1.0561 (1.0410–1.0715)	Strengths (i) Time-stratified case-crossover design minimises potential confounders, such as demographic factors, seasons, day of the week, and long-term trends Limitations (ii) 4 provinces of Bangkok Metropolitan Region are only included, as the remaining 2 provinces do not have sufficient air pollution data (iii) Air pollutant data is obtained from fixed monitoring stations as, which leads to misclassification of the level of exposure of participants	PM_{10} and $PM_{2.5}$ were significantly associated with an increased risk of respiratory diseases requiring outpatient visits among children aged 0–5 in Bangkok Metropolitan Region	Adjusted confounders: long-term trend, seasonality Non-adjusted confounders: day of the week, patient characteristics (age, race, sex, body mass index, diet, behavioural, and SESs)

Abbreviations Acute Lower Respiratory Infection (ALRI), Air Pollution Index (API), Acute Respiratory Infection (ARI), Acute Upper Respiratory Infection (AURI), Demographic and Health Surveys (DHS), Ho Chi Minh City (HCMC), International Classification of Diseases (ICD), Interquartile Range (IQR), Lower Respiratory Tract Infection (LRTI), Length of Stay (LOS), National Family Health Survey-4 (NFHS-4), Socioeconomic Status (SES), Upper Respiratory Tract Infection (URTI)

Table 5.3 Summary of the association between air pollution and ARI in SA and SEA for under 5 years

Author	Outcome	PM_1	$PM_{2.5}$	PM_{10}	CO	NO_2	NO	NO_x	O_3	SO_2	API
Abdul Rahman et al. (2017)	Hospitalisations for ARI subtypes	N/A	N/A	/	/	/	N/A	N/A	/	N/A	N/A
Abudureyimu et al. (2023)	AURI and ALRI	N/A	/	N/A	N/A	N/A	N/A	N/A	N/A	N/A	N/A
Adhikary et al. (2024)	ARI	N/A	/	N/A	N/A	N/A	N/A	N/A	N/A	N/A	N/A
(Balasubramani et al. (2022)	ARI	N/A	/	N/A	N/A	N/A	N/A	N/A	N/A	N/A	N/A
Ibrahim et al. (2022)	Hospitalisations for respiratory diseases (ICD-10 code: J0-J49)	N/A	/	X	X	X	N/A	N/A	X	/	N/A
HEI Collaborative Working Group et al. (2012)	Hospitalisations for ALRI	N/A	N/A	/	N/A	/	N/A	N/A	/	/	N/A
Luong et al. (2020)	Hospitalisations for ALRI	N/A	/	N/A	N/A	N/A	N/A	N/A	N/A	N/A	N/A
Luong et al. (2017)	Respiratory disease (ICD-10 code: J00-J99)	/	/	/	N/A	N/A	N/A	N/A	N/A	N/A	N/A
Nathan et al. (2014)	Hospitalisations for LRTI	N/A	N/A	X	N/A	N/A	N/A	N/A	N/A	N/A	X
Nhung et al. (2019)	Length of hospital stay	X	X	/	/	X	/	X	/	/	N/A
Nhung et al. (2018)	Daily counts of pneumonia (ICD-10 code: J12-J18), bronchitis and asthma (ICD-10 code: J20, J21, J45)	/	/	/	/	/	N/A	/	X	X	N/A
Seah et al. (2023)	Laboratory-confirmed Influenza A and B infections	N/A	/	/	N/A	/	N/A	N/A	N/A	/	N/A
Sherris et al. (2021)	Pneumonia or URTI diagnosis	N/A	/*	N/A	N/A	N/A	N/A	N/A	N/A	N/A	N/A

(continued)

Table 5.3 (continued)

Author	Outcome	PM_1	$PM_{2.5}$	PM_{10}	CO	NO_2	NO	NO_x	O_3	SO_2	API
Suryadhi et al. (2020)	Self-reported ARI	N/A	N/A	/	N/A	/	N/A	N/A	N/A	/	N/A
Thongphunchung et al. (2021)	Outpatient visits for respiratory diseases	N/A	/	/	N/A	N/A	N/A	N/A	N/A	N/A	N/A

Significant associations are shown as "/" and no significant associations as "X". Abbreviations: Not available (N/A), Acute Lower Respiratory Infection (ALRI), Acute Respiratory Infection (ARI), Lower Respiratory Tract Infection (LRTI), Air Pollution Index (API), Acute Upper Respiratory Infection (AURI), Upper Respiratory Tract Infection (URTI), International Classification of Diseases (ICD). *PM2.5 from various sources (biomass burning, brick kilns, soil dust, road dust, fugitive lead, sea salt, zinc manufacturing, vehicle emissions

5.3.1.1 *$PM_{2.5}$*

$PM_{2.5}$ was investigated in eleven of the studies. Ibrahim et al. (2022) revealed that a 10 µg/m^3 increase in $PM_{2.5}$ levels was associated with admissions for childhood respiratory diseases, defined by the International Classification of Diseases (ICD) code J0-J49, with a relative risk of RR 1.095 (95% CI 1.001–1.197). Abudureyimu et al. (2023) conducted a study in India based on the DHS from 2015 to 2016 which found that a 10 µg/m^3 increase in $PM_{2.5}$ levels increases the odds of acute upper respiratory infection (AURI), OR 1.02 (95% CI 1.02–1.03), and ALRI, OR 1.05 (95% CI 1.03–1.06), and the greatest odds of AURI, OR 1.13 (95% CI 1.04–1.23) and ALRI, OR 1.44 (95% CI 1.28–1.67) is seen with exposure to the highest quartile of $PM_{2.5}$ levels. Adhikary et al. (2024) continued the study by utilising data from the DHS from 2019 to 2021, and found that a 10 µg/m^3 increase in $PM_{2.5}$ significantly increases the odds of ARI, OR 1.23 (95% CI 1.19–1.27), and additionally noted linear association of ARI with higher levels of $PM_{2.5}$, where the odds is doubled when comparing the second quartile of $PM_{2.5}$ levels to the first, OR 2.06 (95% CI 1.60–2.66), and a quadrupling of odds when comparing the fourth quartile of $PM_{2.5}$ levels to the first, OR 4.45 (95% CI 3.37–5.87). Balasubramani et al. (2022) revealed that $PM_{2.5}$ exposure significantly increases the risk of ARI, RR 1.007 (95% CI 1.005–1.009). Additionally, Luong et al. (2020) discovered significant associations between daily hospital admissions for ALRI and exposure to $PM_{2.5}$, with excess risk of ER 3.51% (95% CI 0.96–6.12%) at Lag 3. The association was also different between sex, where male children more significantly affected, ER 5.25% (95% CI 2.00–8.60%) at Lag 3, while no association was seen among females (Luong et al. 2020). The risk of admissions for respiratory diseases (ICD code: J00-J99) was associated with exposure to $PM_{2.5}$ up to a Lag of 3 days, and the effect is greatest for same day exposure, and decreases with more days of exposure, with statistically significant OR ranging from 1.022 to 1.017 at Lag 0, OR from 1.013 to 1.016 at Lag 1, OR from 1.009 to 1.012 at Lag 2, and OR of 1.009 at Lag 3 (Luong et al. 2017). The time-series study by Nhung et al. (2019) noted no significant association between $PM_{2.5}$ exposure and length of hospital stay among children aged 0–5. Nhung et al. (2018) found significant associations between short-term exposure to $PM_{2.5}$ and an increased risk for pneumonia, RR 1.063 (95% CI 1.011–1.117). Seah et al. (2023) reported that $PM_{2.5}$ exposure is inversely associated with risks of influenza infection, however, no effect estimates were reported in the study. Sherris et al. (2021) investigated on different sources of $PM_{2.5}$ exposure, and their effects on an increased risk of pneumonia or URTI, which yielded significant positive associations with $PM_{2.5}$ from brick kilns RR 1.062 (1.000–1.194) for pneumonia, while $PM_{2.5}$ from road dust and zinc manufacturing is associated with lower risks of URTI, with relative risks of RR 0.936 (95% CI 0.798–0.960) and RR 0.951 (95% CI 0.849–0.984), respectively. Thongphunchung et al. (2021) conducted a case-crossover study and found significant associations between $PM_{2.5}$ and outpatient visits for respiratory diseases, with OR ranging from 1.0210 (95% CI 1.0133–1.0288) to 1.0561 (95% CI 1.0410–1.0715).

5.3.1.2 *PM_{10}*

Abdul Rahman et al. (2017) found a significant association between short-term exposure to PM_{10} and ARI, where PM_{10} increases the risk of 13 subtypes of ARI in Malaysia, OR ranging from 1.037 to 1.115. In a study looking at the association between air pollutants and ALRI admissions during dry and rainy seasons, PM_{10} was found to be negatively associated with ALRI during the rainy season, with excess risks of ER −3.11 (−1.42 to −4.76) at Lag 0, ER −2.43 (−0.69 to −4.13) at Lag 1–6, using a case-crossover analysis, and ER −2.58 (−4.44 to −0.68), ER −3.06 (−4.71 to −1.38) with rainfall, using a time-series analysis (HEI Collaborative Working Group on Air Pollution, Poverty, and Health in Ho Chi Minh City et al. 2012). Luong et al. (2017) discovered that $PM_{2.5}$ is significantly associated with admissions for respiratory diseases (ICD-10 code J00-J99) at different time lags, with greatest effects seen at same day exposure of PM_{10}, OR ranging from 1.012 to 1.015 at Lag 0, OR ranging from 1.010 to 1.012 at Lag 1, OR range from 1.007 to 1.009 at Lag 2, and OR ranging from 1.006 to 1.007 at Lag 3. Additionally, PM_{10} increases the length of hospital stay for children aged 2–5 that was exposed to the air pollutant, with a 6% decrease (95% CI 1%–12%) in the odds of patient discharge per interquartile range (IQR) increase in PM_{10} concentrations (Nhung et al. 2019). Furthermore, Nhung et al. (2018) found that PM_{10} increases the risk of pneumonia in children aged 1–5, with relative risk of RR 1.063 (95% CI 1.012–1.102). Seah et al. (2023) found that PM_{10} has a significant negative association with risks of influenza, however there were no effect estimates reported. A positive association of PM_{10} with outpatient visits for respiratory diseases was discovered by Thongphunchung et al. (2021) with odds ratio ranging from OR 1.0168 (95% CI 1.0075–1.0262) to OR 1.0354 (95% CI 1.0267–1.0442). Among the studies investigating PM_{10}, two studies found no significant association between PM_{10} exposure and an increase in admission for respiratory disease and lower respiratory tract infection (LRTI), respectively (Nathan et al. 2014; Ibrahim et al. 2022).

5.3.1.3 *PM_1*

PM_1 was investigated in three studies, where Luong et al. (2017) found a significant association between PM_1 exposure and hospital admissions for respiratory diseases (ICD-10 code: J00-J99), with the greatest effect seen at same-day exposure, and decreasing effect with more days after exposure, with OR ranging from 1.019 to 1.025 at Lag 0, OR ranging from 1.014 to 1.019 at Lag 1, OR ranging from 1.012 to 1.014 at Lag 2, and OR of 1.012 at Lag 3. Nhung et al. (2019) found that there are no significant associations between exposure of PM_1 and the length of hospital stay for children aged 5 and below, whereas PM_1 was found to contribute significantly to daily counts of pneumonia RR 1.077 (95% CI 1.018–1.139) in a separate study (Nhung et al. 2018).

5.3.1.4 CO

The effects of CO on ARI among children are investigated in five studies in the SA and SEA population. Abdul Rahman et al. (2017) found that CO significantly increases the risk of 4 different ARI subtypes, with OR ranging from 1.287 to 1.785. Ibrahim et al. (2022) did not find significant associations between CO exposure and respiratory diseases among children. CO was found to be negatively associated with length of stay in children aged 0–5, OR 1.031 (95% CI 1.005–1.057), showing that it increases the odds of discharge in the study by Nhung et al. (2019). In the time-series by Nhung et al. (2018), CO was found to be significantly associated with pneumonia, RR 1.066 (95% CI 1.005–1.130) among children aged 1–5 years, however was negatively associated with bronchitis and asthma among children aged less than one years old, RR 0.908 (95% CI 0.848–0.973). Additionally, Seah et al. (2023) found that CO is negatively associated with influenza A infections 2 days after exposure, RR 0.712 (95% CI 0.555–0.914).

5.3.1.5 *NO_2*

NO_2 was investigated in seven studies, and was found to significantly increase the risk of 8 subtypes of ARI, OR ranging from 1.012 to 1.026 in a case-crossover analysis by Abdul Rahman et al. (2017). A study by HEI Collaborative Working Group et al. (2012) found that NO_2 is positively associated with hospitalisations for ALRI during the dry season, and negatively associated with ALRI hospitalisations during rainy seasons. Using the case-crossover analysis, excess risks of 8.50 (0.8–16.79) at Lag 0, and 12.69 (4.07–22.02) at Lag 1–6, were seen during the dry season, while excess risk of ER −5.15 (−0.1 to −9.94) at Lag 0 was seen in the rainy season (HEI Collaborative Working Group et al. 2012). Meanwhile, utilising time-series analysis, the HEI Collaborative Working Group et al. (2012) noted that NO_2 increases ALRI admissions during the dry season, with the excess risk of 12.62 (95% CI 3.91–22.05) without rainfall, and excess risk of 8.24 (95% CI 0.53–16.53) with rainfall, and no associations were seen during rainy seasons. NO_2 is significantly associated with pneumonia among children aged 1–5, RR 1.110 (95% CI 1.041–1.162) and among children less than 1 year old, RR 1.050 (95% CI 1.005–1.097) (Nhung et al. 2018). The risk of bronchitis and asthma is also increased with exposure to NO_2, RR 1.079 (95% CI 1.004–1.160) (Nhung et al. 2018). Seah et al. (2023) found a significant association between influenza B infection and an increase in NO_2 concentration 5 days later, RR 1.015 (95% CI 1.005–1.025). Additionally, exposure to an IQR increase of NO_2 is significantly associated with an increased risk of ARI, OR 1.18 (95% CI 1.08–1.28) (Suryadhi et al. 2020). In two studies, NO_2 was not significantly associated with respiratory diseases and length of hospital stay, respectively (Nhung et al. 2019; Ibrahim et al. 2022).

5.3.1.6 NO

NO was investigated in only one study which looked at the association between air pollutants and length of hospital stay, and NO was found to be negatively associated with length of stay, with higher odds of discharge from hospital, OR 1.026 (95% CI 1.002–1.051) among children aged 0–5 years old (Nhung et al. 2019).

5.3.1.7 NO_x

The effects of nitrogen oxides (NO_x) on ARI were investigated in two studies. Exposure to NO_x is not associated with the difference in length of hospital stay among children aged 0–5 (Nhung et al. 2019). Whereas, NO_z significantly increases the risk of pneumonia, RR 1.101 (95% CI 1.041–1.166), as well as bronchitis and asthma, RR 1.116 (95% 1.035–1.203) in children aged 1–5 years old (Nhung et al. 2018). Nhung et al. (2018) found significant associations between NO_x and respiratory diseases, where NO_x increases the risk of pneumonia among children aged 1–5, RR 1.101 (95% CI 1.041–1.166), as well as increases the risk of bronchitis and asthma, RR 1.116 (95% CI 1.035–1.203).

5.3.1.8 O_3

Five studies investigated O_3 and its effects on ARI. Abdul Rahman et al. (2017) found that O_3 is significantly associated with two subtypes of ARI, with OR ranging from 1.007 to 1.010. During rainy seasons, O_3 was negatively associated with ALRI admissions, ER −2.98 (−1.14 to −4.78) at Lag 0, and ER −2.49 (−0.5 to −4.43) at Lag 1–6 with case-crossover analysis, while utilising time-series analysis, the excess risks were ER −2.33 (−4.26 to −0.36) without rainfall, and ER −2.91 (−4.72 to −1.07) with rainfall (HEI Collaborative Working Group et al. 2012). The time-series study by Nhung et al. (2019) showed that O_3 has a significant association with length of stay among children aged 0–5 years, where an IQR increase in O_3 concentration is associated with 5% lower discharge odds, OR 0.95 (95% CI 0.92–0.98). Meanwhile, two studies found no association between O_3 and respiratory diseases, pneumonia, bronchitis, and asthma (Nhung et al. 2018; Ibrahim et al. 2022).

5.3.1.9 SO_2

The effect of SO_2 on ARI among children was investigated in six studies, where Ibrahim et al. (2022) noted significant associations between SO_2 exposure and hospitalisations for respiratory diseases, with a relative risk of 1.192 (95% CI 1.029–1.381) seen at Lag 0–6 days. In the case-crossover analysis by HEI Collaborative Working Group et al. (2012), SO_2 was found to significantly increase ALRI admissions during the dry season, ER 5.85 (0.44–11.55) at Lag 0, and ER 5.90 (0.42–11.68) at Lag 1–6, while in the time-series analysis SO_2 significantly increases ALRI admissions with an ER of 5.87 (95% CI 0.45–11.57) during the dry seasons with rainfall. SO_2 is negatively associated with an increased length of hospital stay, with an OR 1.069 (95% CI 1.011–1.103) for odds of discharge (Nhung et al. 2019). Seah et al. (2023) found that SO_2 is positively associated with both influenza A and B, with relative risks of 1.009 (95% CI 1.001–1.078) for influenza A reports at Lag 4, and relative risks of 1.020 (95% CI 1.006–1.033) for influenza B reports on the same day, with greater association seen with influenza B. In contrast, two studies did not find any significant associations with SO_2. SO_2 was not significantly associated with pneumonia, bronchitis, and asthma in the study by Nhung et al. (2018), and does not increase the risk of ARI in the study by Suryadhi et al. (2020).

5.3.1.10 Air Pollution Index (API)

API and its effects on hospitalisations for LRTI among children was investigated by Nathan et al. (2014) where no significant associations were found.

5.4 Discussion

The findings of the narrative synthesis summarise the evidence regarding the link between ambient air pollution and ARI among children aged 5 years and younger, with a particular focus on SA and SEA. The majority of studies identified consistently report that short-term exposure to PM_{10}, $PM_{2.5}$, PM_1, NO_2, O_3, SO_2, and CO is associated with an increased incidence of ARI in this vulnerable population. Specifically, $PM_{2.5}$ is the most extensively studied air pollutant, featured in 13 out of the 17 total studies included in this review, with significant associations reported in 12 of these studies for $PM_{2.5}$. Other pollutants also demonstrate significant associations with ARI: PM_{10} in 9 out of 11 studies, NO_2 in 5 out of 7 studies, SO_2 in 5 out of 6 studies, O_3 in 3 out of 5 studies, CO in 3 out of 4 studies, PM_1 in 2 out of 3 studies, and NO_x in 1 out of 2 studies.

The extensive research on PM as a key component of air pollution highlights its significance, with the World Health Organisation (WHO) often using it as a proxy indicator to assess air quality and its health impacts (WHO 2021). Moreover, the WHO Global Burden of Disease Project has also recognised $PM_{2.5}$ exposure as a key indicator for evaluating the risk of LRTI in children, emphasising its significant contribution to paediatric respiratory diseases (Forouzanfar et al. 2015). Both short-term and long-term exposure to PM have been documented for their significant health risks, due to the ability of PM to penetrate deep into the circulatory system through lung infiltration, increasing morbidity and mortality associated with respiratory, cardiovascular, and cerebrovascular diseases (WHO 2021). Sources of PM differ depending on their diameters, with coarse particles of diameters between

2.5–10 μm arising in agricultural spaces, erosion dusts, roads, and mining activities, whereas fine particles such as $PM_{2.5}$ are derived from the combustion of fuels mainly from vehicles, power generation facilities, industries, waste burnings, and chemical reactions between gases (WHO 2021). The widespread presence of these sources contributes to significant exposure among populations. This coincides with data showing that between 2000 and 2019, both annual population-weighted exposed days and the population-weighted concentration of $PM_{2.5}$ had decreased in Europe and northern America, however, an increase in exposure is seen in the southern part of Australia, New Zealand, Asia, the Caribbean, and Latin America, highlighting the higher emissions in the Global South and Oceania (Yu et al. 2023).

5.4.1 Pathophysiological Mechanism Linking Air Pollution and ARI

Ambient air pollutants are linked to ARI through various pathophysiological mechanisms. Experimental studies suggest that PM contributes to ARI via oxidative stress, that results from the upregulation of reactive oxygen species (ROS) (Kadiiska et al. 1997; Kelly and Fussell 2015). The increase in ROS translates to an increase in inflammation, which can lead to tissue damage in the excess of oxidative stress, heightening the vulnerability of the lung tissues to respiratory infections (Mittal et al. 2014). The pathophysiology of lung injury caused by NO_2 is that NO_2 oxidises protective antioxidants within the lining fluid of the epithelial cells of the lungs, leading to extracellular damage (Yang and Omaye 2009; Petit et al. 2017). Meanwhile, inhalation of O_3 leads to peroxidation and ozonation of proteins and lipids in the epithelial fluid layer of the lung, resulting in lung damage from the production of free radicals, oxidised proteins and aldehydes (Yang and Omaye 2009; Groves et al. 2012). SO_2 can be converted to SO_2 radicals in low pH situations, leading to oxidative stress, and additionally SO_2 inhalation leads to apoptosis of polymorphic neutrophils, decreasing their migration into the lungs which increases lipopolysaccharide-induced acute lung injury (Yang and Omaye 2009; Ma et al. 2012). Carbon monoxide exerts its effects on lung injury via the competitive binding of oxygen on the haemoglobin, resulting in tissue hypoxia (Yang and Omaye 2009).

Children are more susceptible to air pollution due to a multitude of factors, including a developing respiratory and immunological system, and behavioural patterns. Nasal breathing is the primary mode of respiration in adults, offering several advantages, including the filtration of airborne particles and the regulation of humidity and temperature of the air entering the lungs (Bennett et al. 2003; Nikasinovic et al. 2003). In contrast, children predominantly breathe through their mouths, which bypasses the benefits associated with nasal breathing during early developmental years. This pattern increases their exposure to air pollutants, as the unfiltered air enters directly into their lungs. Additionally, children have a larger lung surface area relative to their body weight, necessitating greater volumes of air intake. This is achieved through deeper breaths and higher respiratory rates (Arcus-Arth and Blaisdell 2007). Such deep breathing allows pollutants to bypass initial deposition sites in the upper respiratory tract, leading to a higher concentration of harmful particles in the pulmonary regions (Ginsberg et al. 2005). Moreover, younger children tend to be more physically active, which further elevates their ventilation rates compared to adults (Sigmund et al. 2007). They also spend more time outdoors, especially during peak pollution times such as midday in summer when levels of O_3 and SO_2 are at their highest. Compounding this issue is the fact that children possess immature immune systems; their antibody levels decline during the first few months of life, rendering them more susceptible to infections. Consequently, infants rely heavily on vaccinations, breastfeeding, and natural infections to build immunity (Peltola 1993).

5.4.2 Limited Evidence of the Link Between Air Pollution and ARI Among Children in SA and SEA

Two meta-analyses were synthesised from the literature review to understand the association between air pollution and ARI in children. The earliest of these initiatives is a systematic review and meta-analysis conducted by Mehta et al. (2013), which analysed a total of 73 observational and interventional studies examining the effects of $PM_{2.5}$ and PM_{10} on ALRI morbidity and mortality. The majority of studies included in this review originated from Europe and North America, with only four studies from Asia included. The research focused on children up to the age of 12, with the findings specifically stratified for those under five years, aligning with the inclusion criteria of the study. Mehta et al. (2013) demonstrated high confidence in their risk estimates, as different sensitivity analyses revealed no substantial effects across different outcome assessments. The inclusion of diverse study designs broadens the insights gained from this systematic review. However, several limitations were present, notably, the definition of ALRI was not standardised across all studies, with some relying on maternal recall, which could lead to misclassification bias. Additionally, unadjusted confounders such as second hand smoke, household solid fuel use, regional differences in case fatality rates, access to healthcare, malnutrition, and immunocompetence may have influenced the results. Future research should investigate the effects of various combustion sources and include more studies from SA and SEA to enhance understanding of their impact on ALRI in children.

To address the limitations identified in previous systematic reviews and meta-analyses, Ziou et al. (2022) revisited the topic and published a comprehensive systematic review and meta-analysis of 16 studies consisting of various study designs such as time-series, case-crossover, cohort, and cross-sectional studies, from a diverse range of countries across America, Europe, Asia, and Africa. This approach allowed for the exploration of populations of different income levels. Significant associations were observed between exposure to $PM_{2.5}$ and PM_{10} and hospitalisations for URTI, with high heterogeneity among the pooled results for children up to 18 years old.

Although pooled effect estimates for children aged up to 5 years were unavailable from the meta-analysis, individual studies indicated an increased incidence of URTI linked to total suspended particles (TSP) among children in this age group (RR 1.12; 95% CI 1.00–1.24) in Europe (Braun-Fahrländer et al. 2012). Additionally, increased hospitalisations for URTI due to exposure to $PM_{2.5}$ (RR 1.015; 95% CI 1.004–1.027) and PM_{10} (RR 1.023; 95% CI 1.012–1.036) were reported in the USA (Darrow et al. 2014). Notably, a 30 μg/m^3 increase in $PM_{2.5}$ levels was associated with increased hospitalisations (RR 1.048; 95% CI 1.028–1.069), as was a 40 μg/m^3 increase (RR 1.083; 95% CI 1.046–1.122), both at Lag 7 (Kim et al. 2020). Confounders by study design were grouped and adjusted for, where holidays, influenza epidemics, seasonality, temperature, humidity, and day of the week were adjusted for time-series and case-crossover studies, and age, sex, smoking exposure and SES were adjusted for cohort and cross-sectional studies. The strengths of Ziou et al.'s (2022) study includes the inclusion of a wide array of epidemiological study designs with large sample sizes, small standard errors, and extended study durations. However, limitations persist due to the relatively small number of studies available for analysis and potential bias arising from data collection methods that primarily capture severe URTI through hospitals or physician registries. Moreover, long-term effects were not adequately addressed, and exposure assessments were not performed at precise locations within the studied populations. This suggests either a lack of sensors in close proximity to the populations or limitations in the availability of air pollution monitoring devices. Future studies should focus on long-term exposure effects to further elucidate the relationship between air pollution and respiratory morbidity in children.

5.4.3 Socioeconomic Disparities and Unequal Burden of $PM_{2.5}$Exposure and ARI Risk Among Children in SA and SEA

Although systematic reviews and meta-analyses have been conducted, big data analyses provide new evidence on the impact of $PM_{2.5}$ exposure on ARI among children under five in LMICs, including those in SA and SEA (Odo et al. 2022). Odo et al. (2022) utilised cross-sectional data from the Demographic and Health Surveys (DHSs) across 35 LMICs, and found that a 10 μg/m^3 increment in $PM_{2.5}$ increases the risk of ARI, OR 1.06 (95% CI: 1.05–1.08), with rural areas (OR_{Rural} 1.08; 95% CI 1.06–1.10) more greatly affected compared to urban areas (OR_{Urban} 1.03; 95% CI 1.01–1.06). One of the strengths of this study is that it featured a large sample size (N = 22,506), which enhances the reliability and generalisability of the findings. Additionally, subgroup analyses were performed on the association between $PM_{2.5}$ and ARI, highlighting that the risk of ARI is notably higher in rural and economically disadvantaged areas compared to urban wealthier regions, providing valuable insights into how SES modifies the relationship between air pollution and ARI. The increased odds of ARI seen with exposure to $PM_{2.5}$ in rural and poorer areas compared to urban and richer areas indicate that individuals in rural and lower-income regions disproportionately experience a greater burden of outdoor air pollution and its adverse health impacts. The evidence also revealed a critical gap in current studies whereby air pollution-related health impacts among diverse SES groups were not comprehensively assessed. Some studies indicated that higher SES groups have lower exposure to air pollutants compared to lower SES groups (Hajat et al. 2015). This discrepancy is present because lower SES groups often do not have access to resources that reduce their exposure to harmful pollutants, such as substandard housing, limited access to air filtration systems, increased time spent in outdoor work environments, and a greater reliance on public transportation contribute to this heightened vulnerability (Hajat et al. 2015). Recognising the role of SES in the association between air pollution and ARI is crucial, as it can influence both exposure levels and health outcomes, potentially mitigating the adverse effects of ARI. Odo et al. accounted for confounding factors including household air pollution, second hand smoke, lifestyle, sociodemographic, other climatic factors and pollutants (NO_2, O_3), providing strong evidence that air pollution independently contributes to ARI in children under five. Overall, this research offers valuable insights into the association between $PM_{2.5}$ exposure and ARI in LMICs while highlighting the modifying influence of SES on this relationship. These findings can inform future recommendations and policies aimed at addressing air pollution's health impacts on vulnerable populations. Nevertheless, as this study focused exclusively on $PM_{2.5}$ as an exposure variable, future research should consider including other air pollutants to provide a more comprehensive understanding of their health effects on children in LMICs.

Additionally, recent studies have shed light on mechanisms underlying the potential role of pollution sources, distinguishing between fire-related and non-fire-related $PM_{2.5}$ exposure (Li et al. 2023a, b) analysed data from 48 DHS studies across LMICs, and reported that a 1 $\mu g/m^3$ increment in fire-sourced $PM_{2.5}$ increases the risk of ARI by 3.2%, OR 1.032 (95% CI: 1.002–1.062) while a 5 $\mu g/m^3$ increment in non-fire-sourced $PM_{2.5}$ increases the risk of ARI by 3.1% OR 1.031 (95% CI: 1.024–1.037). This finding aligns with an experimental study demonstrating that fire-sourced PM is more toxic than non-fire-sourced PM, due to its association with increased inflammation of the lungs (Williams et al. 2013). The phagocytosis of fire-soured PM by macrophages activates nuclear factor kappa B, leading to neutrophilic inflammation of lungs through mechanisms such as transcription of genes leading to rapid death of macrophages, or secretion of neutrophil-attracting chemokines (Williams et al. 2013). This information is valuable as it can inform targeted interventions which prioritise specific fire-sourced air pollutants. Li et al.'s (2023a, b) study adjusted for various confounders including child sex, haemoglobin level, residence type, wealth status, maternal education and employment, breastfeeding practices, temperature variations, wind speed, and humidity levels. However, there was uncertainty in exposure assessment, due to factors like crop residue burning and geographical discrepancies among many countries included in the DHSs. Nevertheless, (Li et al. 2023a, b) study provides valuable insights into the detrimental effect of fire sources of $PM_{2.5}$, underscoring the urgent need for increased public awareness and targeted interventions to mitigate exposure to landscape fire smoke, particularly among vulnerable populations and LMICs that are disproportionately affected by air pollution. Effective air quality management requires stringent fire management practices, including the reduction of excessive burning of vegetation, agricultural wastes, and land clearing. Additionally, minimising exposure for populations at risk can be achieved through the use of face masks, the implementation of effective evacuation protocols, and the deployment of air cleaners. Alternatively, new methods for land clearing should be explored that move away from the reliance on fire.

5.4.4 Uneven Research Efforts and Seasonal Complexities in the Impact of Air Pollution on ARI Risk

More research initiatives were conducted in SEA compared to SA, despite SA almost tripling the population size of SEA, with population numbers of roughly 2.074 billion and 697 million, respectively, as of December 2024 (Worldometer 2024). Most of the studies indicate that exposure to air pollutants is linked to higher hospital admissions for ALRI during the dry season. However, a reduction in air pollution levels during rainy seasons was observed, which can be attributed to the natural cleansing effects of precipitation, which facilitates the removal of gaseous pollutants through processes such as absorption by raindrops (Shukla et al. 2008). This is further supported by a study conducted in China which found that precipitation events significantly improved air quality indices in 15 out of 18 cases due to the scavenging of particulates and dissolved gases from the atmosphere (Tian et al. 2021). However, it is also noted that certain precipitation events can lead to worsened air quality due to reduced atmospheric vertical convection caused by lower wind speeds during colder weather (Tian et al. 2021). Additionally, respiratory syncytial virus (RSV) infection confounded the association between air pollutant exposure and ALRI hospital admissions because the prevalence of RSV is typically higher during the monsoon seasons, which coincides with a decrease in air pollution levels. This seasonal variation means that while air pollutants may generally increase the risk of respiratory infections, the protective effect observed during the rainy season could be misleading due to the simultaneous rise in RSV infections, which are a significant cause of ALRI in children. Additionally, other factors such as increased humidity and lower temperatures during this season can influence RSV transmission dynamics, as increased humidity possibly increases the build-up of the virus in the human body, paired with increased virus transmissibility due to indoor crowding for sheltering (Cherian et al. 1990; Lee et al. 2023). This further complicates the relationship between air quality and respiratory health outcomes.

5.4.5 Urban-Centric Evidence of the Relationship Between $PM_{2.5}$and ARI Risk

Several studies particularly in Vietnam found that exposure to air pollutants like PM_{10}, $PM_{2.5}$, and NO_2 increased the risk of respiratory infections and hospital admissions in children, with some pollutants linked to shorter or longer hospital stays. However, these studies had limitations, including reliance on data from major cities and fixed monitoring stations, which may not reflect the true exposure levels and disparities across the country. Similarly, Malaysian studies on short-term exposure to outdoor air pollutants and respiratory diseases were all conducted in urban areas using hospital admission data from major government hospitals, which may not represent rural communities and low SES populations. Of note, one study found no significant association between PM_{10} and LRTI admissions (Nathan et al. 2014), possibly due to a small sample size and inconsistent outcome diagnosis, highlighting the need to accurately assess the impact of air pollution on respiratory health. Future studies in Malaysia could look into SES as a confounding variable, and subgroup analysis can be conducted with a larger sample size to better look into the potential modifying effects of SES. Comparable to the Malaysian studies due to the similar geographical and climatic conditions, Seah et al. (2023) conducted an ecological study in Singapore, which noted significant positive associations between SO_2 and influenza A, as well as SO_2, CO, and NO_2, and influenza B. The study found that CO levels at higher concentrations were negatively associated with the risk of influenza A. This study adds to the literature on potential mechanisms of how CO may affect influenza infections by inhibiting the production of pro-inflammatory cytokines, preventing excessive tissue damage and promoting a controlled immune response (Foresti et al. 2008). Additionally, research showed that each incremental increase in CO concentration significantly decreases the risk of admissions from respiratory tract infection, suggesting that a low level of CO enhances respiratory health by improving immune function and modulating inflammatory responses (Tian et al. 2013). Meanwhile, in Indonesia, Suryadhi et al. (2020) time-series study found significant associations between NO_2 and ARI among children aged 0–3 years in urban areas, though it lacked adjustments for socioeconomic factors and had potential misclassification bias. Similarly, a case-crossover study in Thailand by Thongphunchung et al. (2021) observed positive associations between $PM_{2.5}$, PM_{10}, and outpatient visits for respiratory diseases in children. Despite the heavy reliance on the API as the most commonly used indicator for reporting air pollution severity to the public, surprisingly, very few studies in the region have evaluated its impact on the risk of ARI (Nathan et al. 2014). This underscores the need for future research on the health impacts of API exposure, particularly among young children.

5.4.6 *Source-Specific PM Exposure and Prevalence of ARI Among Children in SA and SEA*

The studies conducted in SEA provided crucial information on the association between air pollution and respiratory health among children, especially PM. However, the studies in SEA shared one characteristic, which is that the actual composition and source of PM were not considered. This limitation was addressed by the study of Sherris et al. (2021) who conducted a case-crossover study in Bangladesh investigating the effect of different sources of $PM_{2.5}$ and URTI. Interestingly, the study is the only study among the fifteen which investigated different sources of $PM_{2.5}$, which provided valuable insights into the health effects of different sources of PM, especially brick kilns, fugitive lead, road dust, and zinc manufacturing, on ARI among children. Studies have shown that certain metals significantly correlate with oxidative potential, subsequently resulting in oxidative stress and tissue damage, particularly lead, zinc, copper, iron, and manganese (Bates et al. 2019). This potentially explains the greater health effects seen with $PM_{2.5}$ from fugitive lead and zinc manufacturing sources, which is crucial to develop strategies to mitigate air pollution, focusing on different sources of air pollutants. The study adjusted for temperature, humidity, and precipitation. However, limitations such as unadjusted variables and possible misclassification and exposure bias are still present in the study. Three studies conducted in India investigated $PM_{2.5}$, which noted significant associations with ARI. The three studies are similar in terms of data collection, as the exposure and outcome data are obtained from large nationwide surveys, which is advantageous as it provides strong evidence of the association between $PM_{2.5}$ exposure and ARI. However, similar limitations persist, most prominently the potential of disease misclassification as the outcome assessment is dependent on self-reported questionnaires.

The studies conducted in SA and SEA reveal several notable similarities regarding the pervasive issue of air pollution. Firstly, it is clear that air pollution poses a significant challenge in many countries within these regions, with measured concentrations frequently exceeding the recommended safe levels established by national and global standards. Research conducted in Malaysia, Vietnam, India, Bangladesh, and Thailand has consistently shown alarming levels of air pollutants, which correlate with the rising incidence of respiratory diseases among children in these nations (HEI Collaborative Working Group et al. 2012; Luong et al. 2017; Nhung et al. 2018; Ibrahim et al. 2022; Adhikary et al. 2024). The primary sources of these pollutants are multifaceted, including vehicle emissions, construction activities, industrial production, agricultural practices, mining operations, and the transboundary haze that SEA countries often experience (Aouizerats et al. 2015; S. Li et al. 2023a, b). As both SA and SEA continue to develop rapidly, they face increasing population pressures that contribute to higher vehicle volumes and greater demand for goods and services. This surge in economic activity inevitably leads to intensified production and transportation efforts, exacerbating air quality issues (Gu et al. 2024; World Bank 2023). The main difference between the studies is that one study from SA considered the effects of different sources of PM and acknowledged the confounding effects of other air pollutants, which are not considered in the other studies.

5.4.7 Recommendations for Future Studies

One significant limitation identified in many studies on air pollution and its health effects is the reliance on fixed monitoring sites, which may not accurately reflect the air quality experienced by populations living at varying distances from these locations. The actual exposure of air pollutants experienced by the population of interest from different locations is assumed to be the same, resulting in possible errors in estimating the relationship between air pollution and ARI. To minimise this limitation, future studies could increase the number of monitoring sites to capture greater variations in the quality of air in different locations and of different sociodemographic groups. This could be further achieved by implementing mobile monitoring units to broaden the areas of air quality collection. Additionally, time-series analysis could be utilised for future studies as it provides valuable information by recording levels of air pollution at consistent intervals over a certain period, giving a more accurate concentration of air pollutants exposed to the population, thereby strengthening the association. Furthermore, more research on the pathophysiology of how air pollutants impact respiratory health should be conducted, especially within the younger populations, to better understand the biological mechanisms which could be utilised to formulate better interventions aimed at reducing the specific risks associated with air pollution. The majority of the current literature focuses only on short-term outcomes of air pollution, while research into the possible long-term impacts of air pollution on children is scarce. Longitudinal studies can be conducted to track the health outcomes of children, providing insights into the chronic health impacts of air pollution. By addressing these research gaps and recommendations, we can deepen our understanding of air pollution's impacts on vulnerable populations such as children, where effective policies and strategies could be implemented to safeguard the well-being of the younger generations.

5.4.8 Existing Policy Landscape

The governments in the SA and SEA regions are currently facing public health challenges related to poor air quality. Therefore, in Southeast Asia, ASEAN adopted the Agreement on Transboundary Haze Pollution (AATHP) in 2002 to prevent, monitor and mitigate transboundary haze pollution through coordinated national and regional corporations (ASEAN 2015). Recently in 2023, the ASEAN Coordinating Centre for Transboundary Haze Pollution Control (ACC THPC) was established to strengthen regional initiatives through implementation framework, monitoring haze situations, and coordinating emergency responses (ASEAN 2024). The 2016 roadmap on ASEAN Cooperation towards Transboundary Haze Pollution Control (2016) (ASEAN 2018) was then revised in 2021, leading to a Second Roadmap for 2023–2030 to continue transboundary haze reduction efforts (ASEAN 2024). Air pollution in SEA poses significant economic challenges in multiple sectors, including healthcare expenditures, productivity loss, and diminished tourism (IIASA 2023). A study estimated that failing to implement robust air quality policies could lead to substantial economic losses by 2030, approximately US$ 1.2 billion for Cambodia, US$ 43.2 billion for Indonesia, and US$ 18.3 billion for Thailand (IIASA 2023).

In contrast to the ASEAN-led framework which have started since early 2000s, a cooperation among South Asian countries was only recently established through the Second Regional Science Policy Dialogue (SPD) in the Indo-Gangetic Plain and Himalayan Foothills (IGP-HF) that recommends the Thimphu Outcome, the 'Kathmandu Road Map', setting a goal for annual $PM_{2.5}$ concentrations by 2035, and securing supports from development and funding partners for their initiatives (ICIMOD 2024). South Asian countries are also members of the Climate and Clean Air Coalition (CCAC) that pledged to reduce the short-lived climate pollutants (SLCPs). In South Asia, country-specific initiatives are more prominent. In India, the Graded Response Action Plan (GRAP) targets air pollution in Delhi by restricting construction, regulating vehicles, and promoting public transportation (Central Pollution Control Board 2017). In Bangladesh and Pakistan, the National Action Plan for Reducing SLCPs was established with key actions across various sectors like waste management, industries, agriculture, and transportation (Climate and Clean Air Coalition 2023a, b).

5.4.9 Recommendation for Policy Improvement

Immediate and effective interventions against air pollution in SA and SEA are crucial to mitigate the significant public health burden of air pollution among children. Firstly, all governments in both regions play a vital part in mitigating the impact of poor air quality on health by implementing strong policies around air quality monitoring. The implementation of stricter air quality and emission standards targeting both fire-sourced and non-fire-sourced $PM_{2.5}$ should be prioritised due to its strong link with increased ARI risk in children. Furthermore, there is a need for enhanced monitoring and enforcement of air quality not only in the urban areas, but also in vulnerable regions, particularly in the low socioeconomic and rural areas. This initiative will provide timely public health alerts and guide adaptive responses to the affected population. Furthermore, all governments should improve current emissions regulation and prioritise clean energy alternatives. Harmful emissions can be reduced by improving waste management and public transports and utilising flexible grid systems for sustainable energy use (UNICEF South Asia 2017).

The recommended strategies target both the air pollution sources and addressing risk factors among vulnerable populations, with the aim of preventing air pollution-related respiratory health problems in children. Moreover, these strategies emphasised equity by prioritising interventions to rural and low-income communities which will address the critical gap as identified in current studies. More importantly, there is a need to strengthen political will and enhance coordination to find a solution to improve air quality across borders. Compared to the ASEAN strong cooperation in combating transboundary haze, a comparable regional cooperation structure is not yet in place in South Asia. The current situation is further challenged towards bridging countries in Southeast and South Asia countries due to unstable political situations in certain parts of the region (East Asia Forum 2024; UN News 2025).

Secondly, the role of international organisations in safeguarding the respiratory health of children would be required to push the agenda in the regions. In particular, UNICEF advocated for cleaner air for children by supporting strategies to reduce air pollution and safeguard their health (UNICEF South Asia 2017). Given the significant public health burden of air pollution-related respiratory health among children in SEA and South Asia, stronger policy implementation and public health efforts should be introduced with collaborations with all countries in the region. Thirdly, the role of various community stakeholders as the first line of defence should be further emphasised. Communities should also be educated regarding the risks of air pollution on children and actions to minimise exposure, such as improving air filtration in schools and households, cleaner cooking methods, and better indoor ventilation. Moreover, greater efforts are needed in promoting behavioural changes, such as planning and avoiding outdoor activities during peak hours where emission levels are high, as well as minimising the indoor sources of air pollution. South Asia and SEA are rapidly growing economies facing rising air pollution challenges that significantly impact public health, especially among vulnerable populations like children. Future collaborative efforts between both regions could be an opportunity to address transboundary air pollution and protect the health of their most at-risk communities.

5.4.10 Study Limitations

The literature search was conducted on two databases Ovid MEDLINE and PubMed, which may potentially narrow the scope of search and reduce the number of relevant studies extracted in the literature review. The inclusion criteria for this literature review restrict the search to studies published only in English, which additionally limits the number of studies during the search. The risk of bias in this study was not formally assessed using a risk appraisal tool.

5.5 Conclusion

Increasing evidence has shown a significant association between ambient air pollution and ARI risks among children under five years old in SA and SEA. This association is contributed by a variety of factors, namely exposure to varying concentrations of air pollutants, SES, and health vulnerabilities, prominently seen among the populations in these regions. Despite the alarming evidence, there is still insufficient research focusing on addressing the impact of air pollution on children's respiratory health particularly among lower SES groups within these regions, further emphasising the need for immediate action and investigations to better improve our understanding of this association. Longitudinal studies could be utilised as it can provide valuable information of longer-term effects of air pollution on respiratory health among children. Future research should additionally look into potential interventions that could help to mitigate the risk and improve health outcomes.

To this end, collaborative efforts at local, national, and international levels are essential in implementing comprehensive policies and strategies aimed at improving air quality, improving public health strategies through health promotions and delivering targeted healthcare interventions, with the goal of protecting children's health and preserving their right to clean air. In conclusion, safeguarding children's respiratory health in SA and SEA requires collective and sustained effort that would be impossible without comprehensive research, effective policy implementation, and cooperation involving the community, government and nations, with the commitment to work towards securing a healthier future for the younger generation.

Author Contribution WPC: Data management, database search, analysis, investigation, scientific support and manuscript drafting, reviewing, and proofreading. AA: investigation, scientific support, validation, manuscript writing, reviewing, and proofreading. MSMN: conceptualisation, scientific support, validation, manuscript reviewing, and proofreading. TTS: conceptualisation, scientific support, validation, manuscript reviewing and proofreading. HJ: conceptualisation, validation, investigation, scientific support, supervision and manuscript drafting, reviewing, and proofreading.

Conflict of Interest The authors declare no competing interests.

Role of Funding None.

Data Availability This study extracted data based on published journal articles. For further inquiries, please contact the corresponding author.

Appendix: Search Strategy

Search strategy for PubMed:

Search 1:

(((((("Air Pollution"[Mesh]) OR "Air Pollution" OR "Particulate Matter"[Mesh]) OR "Particulate Matter" OR "Nitrogen Dioxide"[Mesh]) OR "Nitrogen Dioxide" OR "Sulfur Dioxide"[Mesh]) OR "Sulfur Dioxide" OR "Ozone"[Mesh]) OR "Ozone" OR PM2.5 OR

PM10) AND ("Child, Preschool"[Mesh] OR "Child" OR "Infant"[Mesh] OR "Infant") AND ("Respiratory Tract Infections"[Mesh] OR "Respiratory Tract Infections") AND Meta Analysis

46 results

Search 2:

((((("Air Pollution"[Mesh]) OR "Air Pollution" OR "Particulate Matter"[Mesh]) OR "Particulate Matter" OR "Nitrogen Dioxide"[Mesh]) OR "Nitrogen Dioxide" OR "Sulfur Dioxide"[Mesh]) OR "Sulfur Dioxide" OR "Ozone"[Mesh]) OR "Ozone" OR PM2.5 OR PM10) AND ("Child, Preschool"[Mesh] OR "Child" OR "Infant"[Mesh] OR "Infant") AND ("Respiratory Tract Infections"[Mesh] OR "Respiratory Tract Infections") AND ("South Asia" OR "Southeast Asia" OR "India")

79 results

Search strategy for Ovid MEDLINE:

Search 1:

1	Air pollution.mp. or Air Pollution/	73,861
2	particulate matter.mp. or Particulate Matter/	41,081
3	sulfur dioxide.mp. or Sulfur Dioxide/	7937
4	carbon monoxide.mp. or Carbon Monoxide/	36,144
5	nitrogen dioxide.mp. or Nitrogen Dioxide/	9853
6	Ozone/or ozone.mp	25,122
7	PM10.mp	8982
8	“PM2.5”.mp	18,467
9	1 or 2 or 3 or 4 or 5 or 6 or 7 or 8	151,274
10	infant.mp. or Infant/	1,339,388
11	Child, Preschool/ or child.mp	2,412,249
12	respiratory infection.mp. or Respiratory Tract Infections/	51,666
13	10 or 11	3,028,454
14	9 and 12 and 13	486
15	limit 14 to ((“all infant (birth to 23 months)” or “newborn infant (birth to 1 month)” or “infant (1 to 23 months)” or “preschool child (2 to 5 years)”) and (meta analysis or “systematic review”))	7

Search 2:

1	Air pollution.mp. or Air Pollution/	73,861
2	particulate matter.mp. or Particulate Matter/	41,081
3	sulfur dioxide.mp. or Sulfur Dioxide/	7937
4	carbon monoxide.mp. or Carbon Monoxide/	36,144
5	nitrogen dioxide.mp. or Nitrogen Dioxide/	9853

(continued)

(continued)

6	Ozone/ or ozone.mp	25,122
7	PM10.mp	8982
8	"PM2.5".mp	18,467
9	1 or 2 or 3 or 4 or 5 or 6 or 7 or 8	151,274
10	infant.mp. or Infant/	1,339,388
11	Child, Preschool/ or child.mp	2,412,249
12	respiratory infection.mp. or Respiratory Tract Infections/	51,666
13	10 or 11	3,028,454
14	South Asia.mp. or Asia, Southern/	5096
15	Southeast asia.mp. or Asia, Southeastern/	17,846
16	14 or 15	22,460
17	9 and 12 and 13 and 16	7

References

Abdul Rahman S et al (2017) A case crossover analysis of primary air pollutants association on acute respiratory infection (ARI) among children in urban region of Klang valley, Malaysia. Ann Trop Med Public Health 10(1):44. Available at: https://doi.org/10.4103/ATMPH.ATMPH_75_17

Abudureyimu K et al (2023) Exposure to fine particulate matter and acute upper- and lower-respiratory tract infections (AURI and ALRI) in children under five years of age in India. Arch Environ Occup Health 78(1):1–6. Available at: https://doi.org/10.1080/19338244.2022.2047584

Adhikary M, Mal P, Saikia N (2024) Exploring the link between particulate matter pollution and acute respiratory infection risk in children using generalized estimating equations analysis: a robust statistical approach. Environ Health: Glob Access Sci Source 23(1):12. Available at: https://doi.org/10.1186/s12940-024-01049-3

Amnuaylojaroen T, Kaewkanchanawong P, Panpeng P (2023) Distribution and meteorological control of $PM_{2.5}$ and its effect on visibility in Northern Thailand. Atmosphere 14(3):538. Available at: https://doi.org/10.3390/atmos14030538

Anderson TD (1979) The demographic conundrum in South Asia. Curr Hist 76(446):150–184

Anwar A et al (2021) Impact of air pollution ($PM_{2.5}$) on child mortality: evidence from sixteen Asian countries. Int J Environ Res Public Health 18(12):6375. Available at: https://doi.org/10.3390/ijerph18126375

Aouizerats B et al (2015) Importance of transboundary transport of biomass burning emissions to regional air quality in Southeast Asia during a high fire event. Atmos Chem Phys 15(1):363–373. Available at: https://doi.org/10.5194/acp-15-363-2015

Arcus-Arth A, Blaisdell RJ (2007) Statistical distributions of daily breathing rates for narrow age groups of infants and children. Risk Anal 27(1):97–110. Available at: https://doi.org/10.1111/j.1539-6924.2006.00862.x

ASEAN Agreement on Transboundary Haze Pollution—ASEAN Haze Portal (2015). Available at: https://hazeportal.asean.org/action/asean-agreement-on-transboundary-haze-pollution/. Accessed 30 November 2024

Assessment, U.E.N.C. for E (2009) Repeated episodes of exposure to ozone alters postnatal development of distal conducting airways in infant rhesus monkeys. Available at: https://hero.epa.gov/hero/index.cfm/reference/details/reference_id/12284. Accessed 9 December 2024

Balasubramani K et al (2022) Spatial epidemiology of acute respiratory infections in children under 5 years and associated risk factors in India: district-level analysis of health, household, and environmental datasets. Front Pub Health 10:906248. Available at: https://doi.org/10.3389/fpubh.2022.906248

Bates JT et al (2019) Review of acellular assays of ambient particulate matter oxidative potential: methods and relationships with composition, sources, and health effects. Environ Sci Technol 53(8):4003–4019. Available at: https://doi.org/10.1021/acs.est.8b03430

Bateson TF, Schwartz J (2008) Children's response to air pollutants. J Toxicol Environ Health, Part A 71(3):238–243. Available at: https://doi.org/10.1080/15287390701598234

Bennett WD, Zeman KL Jarabek AM (2003) Nasal contribution to breathing with exercise: effect of race and gender. J Appl Physiol 95(2):497–503. Available at: https://doi.org/10.1152/japplphysiol.00718.2002

Berman S (1991) Epidemiology of acute respiratory infections in children of developing countries. Clin Infect Dis 13(Supplement_6):S454–S462. Available at: https://doi.org/10.1093/clinids/13.Supplement_6.S454

Bolan S et al (2024) Impacts of climate change on the fate of contaminants through extreme weather events. Sci Total Environ 909:168388. Available at: https://doi.org/10.1016/j.scitotenv.2023.168388

Braun-Fahrländer C et al (2012) Air pollution and respiratory symptoms in preschool children. Am Rev Respir Dis [Preprint]. Available at: https://doi.org/10.1164/ajrccm/145.1.42

Checa J, Aran JM (2020) Reactive oxygen species: drivers of physiological and pathological processes. J Inflamm Res 13:1057–1073. Available at: https://doi.org/10.2147/JIR.S275595

Cherian T et al (1990) Bronchiolitis in tropical South India. Am J Dis Child 144(9):1026–1030. Available at: https://doi.org/10.1001/archpedi.1990.02150330086028

Children in the World by Country 2024 (2024). Available at: https://worldpopulationreview.com/country-rankings/children-in-the-world-by-country. Accessed 6 November 2024

Clean Air, Healthy Children: An Agenda for Action | UNICEF (2024). Available at: https://www.unicef.org/reports/clean-air-healthy-children-agenda-action. Accessed 6 November 2024

Clean the air for children | UNICEF (2016). Available at: https://www.unicef.org/reports/clean-air-children. Accessed 6 November 2024

Cohen Hubal EA et al (2000) Children's exposure assessment: a review of factors influencing Children's exposure, and the data available to characterize and assess that exposure. Environ Health Perspect 108(6):475–486. Available at: https://doi.org/10.1289/ehp.108-1638158

CPCB | Central Pollution Control Board (2017). Available at: https://cpcb.nic.in/graded-response-action-plan-for-delhi-ncr/. Accessed 2 December 2024

Danger in the air | UNICEF South Asia (2017). Available at: https://www.unicef.org/rosa/reports/danger-air. Accessed 30 November 2024

Darrow LA et al (2014) Air pollution and acute respiratory infections among children 0–4 years of age: an 18-year time-series study. Am J Epidemiol 180(10):968–977. Available at: https://doi.org/10.1093/aje/kwu234

Foresti R, Bani-Hani MG, Motterlini R (2008) Use of carbon monoxide as a therapeutic agent: promises and challenges. Intensive Care Med 34(4):649–658. Available at: https://doi.org/10.1007/s00134-008-1011-1

Forouzanfar MH et al (2015) Global, regional, and national comparative risk assessment of 79 behavioural, environmental and occupational, and metabolic risks or clusters of risks in 188 countries, 1990–2013: a systematic analysis for the Global Burden of Disease Study 2013. The Lancet 386(10010):2287–2323. Available at: https://doi.org/10.1016/S0140-6736(15)00128-2

Friedrich MJ (2017) UNICEF reports on the impact of air pollution on children. JAMA 317(3):246. Available at: https://doi.org/10.1001/jama.2016.19034

George PE et al (2024) Impact of ambient air pollution and socio-environmental factors on the health of children younger than 5 years in India: a population-based analysis. Lancet Reg Health SE Asia 20. Available at: https://doi.org/10.1016/j.lansea.2023.100328

Ginsberg GL, Perkovich Foos B, Firestone MP (2005) Review and analysis of inhalation dosimetry methods for application to children's risk assessment. J Toxicol Environ Health, Part A 68(8):573–615. Available at: https://doi.org/10.1080/15287390590921793

Groves AM et al (2012) Prolonged injury and altered lung function after ozone inhalation in mice with chronic lung inflammation. Am J Respir Cell Mol Biol 47(6):776–783. Available at: https://doi.org/10.1165/rcmb.2011-0433OC

Gu Y, Fang T, Yim SHL (2024) Source emission contributions to particulate matter and ozone, and their health impacts in Southeast Asia. Environ Int 186:108578. Available at: https://doi.org/10.1016/j.envint.2024.108578

Guterres welcomes India-Pakistan ceasefire | UN News (2025). Available at: https://news.un.org/en/story/2025/05/1163116. Accessed 12 May 2025

Hajat A, Hsia C, O'Neill MS (2015) Socioeconomic disparities and air pollution exposure: a global review. Curr Environ Health Rep 2(4):440. Available at: https://doi.org/10.1007/s40572-015-0069-5

HEI Collaborative Working Group on Air Pollution, Poverty, and Health in Ho Chi Minh City et al (2012) Effects of short-term exposure to air pollution on hospital admissions of young children for acute lower respiratory infections in Ho Chi Minh City, Vietnam. Res Rep (Health Effects Institute), (169):5–72; discussion 73–83

Ibrahim MF et al (2021) Association between ambient air pollution and childhood respiratory diseases in low- and middle-income Asian countries: a systematic review. Atmos Environ 256:118422. Available at: https://doi.org/10.1016/j.atmosenv.2021.118422

Ibrahim MF et al (2022) Children's exposure to air pollution in a natural gas industrial area and their risk of hospital admission for respiratory diseases. Environ Res 210:112966. Available at: https://doi.org/10.1016/j.envres.2022.112966

Integrating Southeast and South Asia through regional institutions | East Asia Forum (2024) 20 August. Available at: https://eastasiaforum.org/2024/08/20/integrating-southeast-and-south-asia-through-regional-institutions/. Accessed 12 May 2025

Jabbar SA et al (2022) Air quality, pollution and sustainability trends in South Asia: a population-based study. Int J Environ Res Public Health 19(12):7534. Available at: https://doi.org/10.3390/ijerph19127534

Kadiiska MB et al (1997) In vivo evidence of free radical formation in the rat lung after exposure to an emission source air pollution particle. Chem Res Toxicol 10(10):1104–1108. Available at: https://doi.org/10.1021/tx970049r

Kaspersen KA et al (2024) Exposure to air pollution and risk of respiratory tract infections in the adult Danish population—a nationwide study. Clin Microbiol Infect 30(1):122–129. Available at: https://doi.org/10.1016/j.cmi.2023.10.013

Kelly FJ, Fussell JC (2015) Linking ambient particulate matter pollution effects with oxidative biology and immune responses. Ann New York Acad Sci 1340(1):84–94. Available at: https://doi.org/10.1111/nyas.12720

Khwaja MA et al (2012) Overview. Sustainable Development Policy Institute, pp 2–10. Available at: https://www.jstor.org/stable/resrep00582.3. Accessed 7 November 2024

Kim K-N et al (2020) Effects of short-term fine particulate matter exposure on acute respiratory infection in children. Int J Hyg Environ Health 229:113571. Available at: https://doi.org/10.1016/j.ijheh.2020.113571

Lee MH et al (2023) Air quality, meteorological variability and pediatric respiratory syncytial virus infections in Singapore. Sci Rep 13(1):1001. Available at: https://doi.org/10.1038/s41598-022-26184-0

Li J et al (2023) Landscape fire smoke enhances the association between fine particulate matter exposure and acute respiratory infection among children under 5 years of age: findings of a case-crossover study for 48 low- and middle-income countries. Environ Int 171:107665. Available at: https://doi.org/10.1016/j.envint.2022.107665

Li S et al (2023) Rapid increase in tropospheric ozone over Southeast Asia attributed to changes in precursor emission source regions and sectors. Atmos Environ 304:119776. Available at: https://doi.org/10.1016/j.atmosenv.2023.119776

Luong LMT et al (2017) The association between particulate air pollution and respiratory admissions among young children in Hanoi, Vietnam. Sci Total Environ 578:249–255. Available at: https://doi.org/10.1016/j.scitotenv.2016.08.012

Luong LMT et al (2019) Impact of ambient air pollution and wheeze-associated disorders in children in Southeast Asia: a systematic review and meta-analysis. Rev Environ Health 34(2):125–139. Available at: https://doi.org/10.1515/reveh-2018-0079

Luong LTM et al (2020) Particulate air pollution in Ho Chi Minh city and risk of hospital admission for acute lower respiratory infection (ALRI) among young children. Environ Poll 257:113424. Available at: https://doi.org/10.1016/j.envpol.2019.113424

Ma H et al (2012) Sulfur dioxide attenuates LPS-induced acute lung injury via enhancing polymorphonuclear neutrophil apoptosis. Acta Pharmacol Sinica 33(8):983–990. Available at: https://doi.org/10.1038/aps.2012.70

Manager J (2024) The relationship between ozone and temperature in greater Kuala Lumpur, Malaysia. Aerosol Air Qual Res 24(10):240072. Available at: https://doi.org/10.4209/aaqr.240072

Mehta S et al (2013) Ambient particulate air pollution and acute lower respiratory infections: a systematic review and implications for estimating the global burden of disease. Air Qual Atmos Health 6(1):69–83. Available at: https://doi.org/10.1007/s11869-011-0146-3

Miller MR (2020) Oxidative stress and the cardiovascular effects of air pollution. Free Radical Biol Med 151:69–87. Available at: https://doi.org/10.1016/j.freeradbiomed.2020.01.004

Mittal M et al (2014) Reactive oxygen species in inflammation and tissue injury. Antioxid Redox Signal 20(7):1126–1167. Available at: https://doi.org/10.1089/ars.2012.5149

Morgan RL et al (2018) Identifying the PECO: a framework for formulating good questions to explore the association of environmental and other exposures with health outcomes. Environ Int 121(Pt 1):1027–1031. Available at: https://doi.org/10.1016/j.envint.2018.07.015

Million babies under the age of 1 breathe toxic air, majority live in South Asia—UNICEF (2017). Available at: https://www.unicef.org/press-releases/babies-breathe-toxic-air-south-asia. Accessed 7 November 2024

Nathan AM et al (2014) Clinical risk factors for life-threatening lower respiratory tract infections in children: a retrospective study in an Urban City in Malaysia. PLoS ONE 9(10):e111162. Available at: https://doi.org/10.1371/journal.pone.0111162

National Action Plan for Reducing Short-Lived Climate Pollutants | Climate & Clean Air Coalition (2023a). Available at: https://www.ccacoalition.org/policy-database/national-action-plan-reducing-short-lived-climate-pollutants. Accessed 2 December 2024

National Clean Air Policy (NCAP) | Climate & Clean Air Coalition (2023b). Available at: https://www.ccacoalition.org/policy-database/national-clean-air-policy-ncap. Accessed 2 December 2024

Nhung NTT et al (2017) Short-term association between ambient air pollution and pneumonia in children: a systematic review and meta-analysis of time-series and case-crossover studies. Environ Pollut 230:1000–1008. Available at: https://doi.org/10.1016/j.envpol.2017.07.063

Nhung NTT et al (2018) Acute effects of ambient air pollution on lower respiratory infections in Hanoi children: an eight-year time series study. Environ Int 110:139–148. Available at: https://doi.org/10.1016/j.envint.2017.10.024

Nhung NTT et al (2019) Association of ambient air pollution with lengths of hospital stay for Hanoi children with acute lower-respiratory infection, 2007–2016. Environ Pollut 247:752–762. https://doi.org/10.1016/j.envpol.2019.01.115

Nikasinovic L, Momas I, Seta N (2003) Nasal epithelial and inflammatory response to ozone exposure: a review of laboratory-based studies published since 1985. J Toxicol Environ Health, Part B [Preprint]. Available at: https://doi.org/10.1080/10937400306477

Odo DB et al (2022) Ambient air pollution and acute respiratory infection in children aged under 5 years living in 35 developing countries. Environ Int 159:107019. Available at: https://doi.org/10.1016/j.envint.2021.107019

Peltola H (1993) Respiratory infections in the young child. Int J Antimicrob Agents 3:S113–S117. https://doi.org/10.1016/0924-8579(93)90043-5

Petit PC et al (2017) The pathophysiology of nitrogen dioxide during inhaled nitric oxide therapy. ASAIO J 63(1):7. Available at: https://doi.org/10.1097/MAT.0000000000000425

Population of South-Eastern Asia (2024)—Worldometer (2024). Available at: https://www.worldometers.info/world-population/south-eastern-asia-population/. Accessed 21 December 2024

Roadmap on ASEAN Cooperation Towards Transboundary Haze Pollution Control with Means of Implementation—ASEAN Haze Portal (2018). Available at: https://hazeportal.asean.org/publications/roadmap-on-asean-cooperation-towards-transboundary-haze-pollution-control-with-means-of-implementation/. Accessed 13 April 2025

Seah A et al (2023) The influence of air quality and meteorological variations on influenza A and B virus infections in a paediatric population in Singapore. Environ Res 216:114453. Available at: https://doi.org/10.1016/j.envres.2022.114453

Shapiro ED (1998) Epidemiology of acute respiratory infections. Semin Pediatr Infect Dis 9(1):31–36. Available at: https://doi.org/10.1016/S1045-1870(98)80048-4

Sherris AR et al (2021) Associations between ambient fine particulate matter and child respiratory infection: the role of particulate matter source composition in Dhaka, Bangladesh. Environ Pollut 290:118073. Available at: https://doi.org/10.1016/j.envpol.2021.118073

Shukla JB et al (2008) Effect of rain on removal of a gaseous pollutant and two different particulate matters from the atmosphere of a city. Math Comput Modell 48(5):832–844. Available at: https://doi.org/10.1016/j.mcm.2007.10.016

Sigmund E et al (2007) Physical activity patterns of kindergarten children in comparison to teenagers and young adults. Eur J Public Health 17(6):646–651. Available at: https://doi.org/10.1093/eurpub/ckm033

State of Global Air Report 2024 | State of Global Air (2024). Available at: https://www.stateofglobalair.org/resources/report/state-global-air-report-2024. Accessed 21 December 2024

Striving for Clean Air: Air Pollution and Public Health in South Asia (2023) World Bank. Available at: https://www.worldbank.org/en/region/sar/publication/striving-for-clean-air. Accessed 7 December 2024

Suryadhi MAH et al (2020) Nitrogen dioxide and acute respiratory tract infections in children in Indonesia. Arch Environ Occup Health 75(5):274–280. Available at: https://doi.org/10.1080/19338244.2019.1631245

The cost of inaction: tackling air pollution in the ASEAN region (2023) IIASA—International Institute for Applied Systems Analysis. Available at: https://iiasa.ac.at/policy-briefs/oct-2023/cost-of-inaction-tackling-air-pollution-in-asean-region. Accessed 30 November 2024

The Second Roadmap for ASEAN Cooperation on Transboundary Haze Pollution Control with Means of Implementation—ASEAN Haze Portal (2024). Available at: https://hazeportal.asean.org/publications/the-second-roadmap-for-asean-cooperation-on-transboundary-haze-pollution-control-with-means-of-implementation/. Accessed 30 November 2024

The Thimphu Outcome Towards Clean Air in IGP-HF | HimalDoc (2024). Available at: https://lib.icimod.org/record/36533. Accessed 2 December 2024

Thongphunchung K et al (2021) Short-term effects of particulate matter on outpatient department visits for respiratory diseases among children in Bangkok Metropolitan Region: a case-crossover study. Air Qual Atmos Health 14(11):1785–1795. Available at: https://doi.org/10.1007/s11869-021-01053-3

Tian L et al (2013) Ambient carbon monoxide associated with reduced risk of hospital admissions for respiratory tract infections. Am J Respir Crit Care Med 188(10):1240–1245. Available at: https://doi.org/10.1164/rccm.201304-0676OC

Tian X et al (2021) Effects of rain and snow on the air quality index, $PM_{2.5}$ levels, and dry deposition flux of PCDD/Fs. Aerosol Air Qual Res 21(8):210158. Available at: https://doi.org/10.4209/aaqr.210158

Tiwari I et al (2022) Climate change impacts on the health of South Asian children and women subpopulations—a scoping review. Heliyon 8(10):e10811. Available at: https://doi.org/10.1016/j.heliyon.2022.e10811

Types of pollutants (2021). Available at: https://www.who.int/teams/environment-climate-change-and-health/air-quality-and-health/health-impacts/types-of-pollutants. Accessed 1 February 2024

What are the WHO Air quality guidelines? (2021). Available at: https://www.who.int/news-room/feature-stories/detail/what-are-the-who-air-quality-guidelines. Accessed 13 April 2025

WHO and UNICEF launch free online course to address children's environmental health (2024). Available at: https://www.who.int/news/item/13-03-2024-who-and-unicef-launch-free-online-course--to-address-children-s-environmental-health. Accessed 6 Nov 2024

Williams KM, Franzi LM, Last JA (2013) Cell-specific oxidative stress and cytotoxicity after wildfire coarse particulate matter instillation into mouse lung. Toxicol Appl Pharmacol 266(1):48–55. Available at: https://doi.org/10.1016/j.taap.2012.10.017

Wolfe MK et al (2020) Impact of school location on children's air pollution exposure. J Urban Aff 43(8). https://doi.org/10.1080/07352166.2020.1734013

World's Most Polluted Countries in 2023—$PM_{2.5}$ Ranking | IQAir (2023). Available at: https://www.iqair.com/us/world-most-polluted-countries. Accessed 6 Nov 2024

Yang W, Omaye ST (2009) Air pollutants, oxidative stress and human health. Mutat Res Genet Toxicol Environ Mutagen 674(1):45–54. Available at: https://doi.org/10.1016/j.mrgentox.2008.10.005

Yeung W-JJ (2022) Trends in population and socioeconomic development in Southeast Asia. In: Yeung WJJ (ed) Demographic and family transition in Southeast Asia. Springer International Publishing, Cham, pp 17–31. Available at: https://doi.org/10.1007/978-3-030-85679-3_2

Yu W et al (2023) Global estimates of daily ambient fine particulate matter concentrations and unequal spatiotemporal distribution of population exposure: a machine learning modelling study. Lancet Planet Health 7(3):e209–e218. Available at: https://doi.org/10.1016/S2542-5196(23)00008-6

Zhao C et al (2020) Respiratory exposure to $PM_{2.5}$ soluble extract disrupts mucosal barrier function and promotes the development of experimental asthma. Sci Total Environ 730:139145. Available at: https://doi.org/10.1016/j.scitotenv.2020.139145

Ziou M et al (2022) Outdoor particulate matter exposure and upper respiratory tract infections in children and adolescents: a systematic review and meta-analysis. Environ Res 210:112969. Available at: https://doi.org/10.1016/j.envres.2022.112969

2023 IQAir World Air Quality Report (2023). Available at: https://www.iqair.com/newsroom/waqr-2023-pr. Accessed 7 November 2024

Chapter 6
The Effects of Climate Change on Vector-Borne Illnesses in South and South-East Asian Regions: A Narrative Review

Akash Chakraborty, Shreya Mitra, Soheli Saha, Sujan Mandal, and Suchismita Roy

Abstract Climate change is dramatically influencing the transmission dynamics, geographic distribution and seasonality of Vector-Borne Diseases (VBDs) and seriously threatening public health. This chapter examines the interrelationship between climate factors and VBDs like malaria, dengue, Lyme disease and chikungunya, which receive socially disproportionate distribution in tropical-subtropical regions. The main aim of the study is to determine the mechanism through which climate change affects the ecology of vectors and the spread of diseases, while the secondary objective is to integrate health system strategies for climate-sensitive VBDs. The review methodology application requires an extensive synthesis of recent research works, epidemiological case studies and climate impact models. Findings demonstrate that an increase in temperature, with changing rainfall and humidity patterns, expedites increased breeding of vectors, shortens pathogen incubation periods and enables vectors to colonise new habitats, hence leading to increased disease burdens. Socioeconomic inequities and poor healthcare infrastructure keep the vulnerable at greater risk. This chapter posits that measures to incorporate climate adaptation strategies into public health planning alongside initiatives to improve research and development for innovative vector control technologies, as well as trans-regional coordination, will be fundamental in dealing with emerging outbreaks. Within South Asia, interventions that are locally tailored and adapted are of critical importance to reduce vulnerabilities and foster the construction of climate-resilient health systems.

Keywords Vector-borne diseases · Climate change · Disease transmission · South-East Asia · Healthcare challenges

A. Chakraborty · S. Mitra · S. Saha · S. Mandal (✉) · S. Roy
School of Environment and Disaster Management, Ramakrishna Mission Vivekananda Educational and Research Institute, Kolkata, West Bengal, India
e-mail: sujan.m@env.rkmvu.ac.in; sujan.geog@gmail.com

S. Roy
e-mail: suchismita.r@env.rkmvu.ac.in; suchismita.geog@gmail.com

P. Ghosh et al. (eds.), *Climate Crises and Public Health Challenges*, Disaster Risk Reduction, https://doi.org/10.1007/978-981-95-9709-3_6

6.1 Introduction

According to the World Health Organisation (WHO), more than one-half of the people in the world are at risk of Vector-Borne Diseases (VBDs), which are responsible for one-sixth of the global disease and disability. Human diseases caused by bacteria, viruses, and parasites that are spread by vectors are known as VBDs (Müller et al. 2019). Diseases include malaria, dengue, schistosomiasis, human African trypanosomiasis, leishmaniasis, Chagas disease, yellow fever, Japanese encephalitis, and onchocerciasis are some examples of these diseases (Table 6.2 in appendix). VBDs affect over one billion people annually and kill over a million. VBDs alone lead to more than 700,000 deaths annually, accounting for over 17% of all infectious disease deaths. It can be parasitic, bacterial or viral. VBDs are inherently sensitive to weather and climate and to their extreme occurrences. Nonetheless, the debate and uncertainty over the effects of climate change on them centres on how much the weather and climate are driving their occurrence and abundance, versus human efforts to control them, and the vectors that spread them (Lafferty 2009). It also has greater socioeconomic effects, exacerbating health inequities and responding as a brake to socioeconomic development. Many of these VBDs are co-endemic, and it is estimated that more than half the world's population live in areas where more VBDs are present (Wilson et al. 2020). The poorest populations bear the greatest burden of climate-sensitive diseases. More than a billion people are infected each year, and more than one million people die from VBDs such as malaria, dengue, schistosomiasis, leishmaniasis, Chagas disease and African trypanosomiasis (Caminade et al. 2019). In addition to their associated mortality, many VBDs, including lymphatic filariasis and onchocerciasis, can also be highly disabling. Suffering adds to a far larger overall burden of disease. These diseases are important human global causes of morbidity and mortality and have passed primarily to humans, which were once zoonotically transmitted from Anopheles and Aedes spp. in tropical and subtropical areas (Ogden 2017).

Climate change has altered the habitats and behaviours of vectors, hosts and pathogens, leading to the spread of diseases through feeding on diseased hosts and then to humans or animals (Table 6.3 in appendix). There are two types of transmission processes for VBDs: mechanical transmission and biological transmission (Zhang et al. 2008). Mechanical transmission involves the organism being transferred to contaminated mouthparts or other body parts, while biological transmission requires the pathogen to undergo development in the vector's body (Gubler 2009). Cyclodevelopmental transmission involves the pathogen developing but not multiplying within the vector, such as lymphatic filariasis. Propagative transmission involves disease-causing organisms multiplying within the vector's body, such as malaria. Cyclopropagative transmission involves the pathogen both developing and multiplying within the vector, like Malaria. Vertical transmission involves the transmission of pathogens and infectious agents directly into their offspring, such as the Dengue Virus (Ramalho-Ortigao and Gubler 2020). Direct transmission occurs when a vector directly transmits the pathogen to a host without intermediate steps or further development. This cycle of transmission is crucial for the spread of diseases and is influenced by climate change. In warmer climates, vectors (organisms that spread diseases) can inhabit ever-higher elevations and latitudes, and extended seasons of transmission due to warmer and wetter conditions further spread the infectious disease (Agyekum et al. 2021). Rainfall has an exceptionally strong bearing, that is, in the case of vector-borne pathogens with aquatic larval stages (like mosquitoes), but also humidity on diseases such as aquatic larvae have, such as Filaria, and also on those widely considered that they do not, like ticks or sandflies. Aquatic ecosystems are becoming more vulnerable to climate-induced changes, impacting vector populations. Contamination of water sources with organic waste also supports vector breeding (Resh 2009). Deforestation for agricultural expansion can lead to closer contact with human populations, further accelerating the spread of diseases (Elias et al. 2022). Changes in climate and weather conditions also have several more indirect influences, through broader impacts on the natural environment and on human systems, such as drought or wet weather that influence water storage, land-use and irrigation practices and migration, which then influences vector ecology, and human risk of infection (Mills et al. 2010).

Many researchers and think tanks like WHO regional studies, Gubler et al. (2009), have pointed out the effect of climatic variables in VBDs, mainly in the developed regions, but fewer studies around Asian regions and also fewer studies around the challenges around healthcare systems and vector R&D, like less funding. Studies like Zhang et al. (2008) have focused on the public health challenges and their adaptation, but fewer studies around the community-based adaptation and weak regional and cross-border cooperation. So, the main questions are: how do climate changes influence transmission dynamics of major VBDs? What real-world case studies illustrate the impact of climate change on VBDs across different regions, particularly in South and Southeast Asia? How have health systems adapted or responded to the challenges posed by climate-sensitive VBDs in these regions? And what strategies can be developed or enhanced to improve public health resilience against future outbreaks of VBDs driven by climate change? Anthropogenic climate change interacts over the long term with natural variability, affecting the transmission of VBDs over both shorter (annual) and longer (decadal) time scales, with different effects at differing times and in different places (Caminade et al. 2019). These influences can reinforce or compete with each other, such as in locations where the effects on temperature from the warming of the El Nino phase are added to background warming or where the El Nino effects appear to have at least temporarily suppressed the longer term upward temperature trends seen over the past decades or so. The objectives of this study are to (i) assess the impacts of climate change on the biology, distribution and transmission of VBDs; (ii) showcase case studies highlighting issues occurring in the real world across many regions; (iii) integrate health system strategies for climate-sensitive VBDs and (iv) propose strategies for increasing public health resilience against future climate change–induced outbreaks. The study, thereby, seeks to address critical knowledge gaps pertaining to the literature on climate-sensitive public health planning and contribute towards policy development for effective vector control and disease prevention strategies in the changing world.

6.2 Methodology

This study uses a narrative review to integrate the current knowledge on the relationship between climate change and VBDs in South and Southeast Asia. This study aims to assess the impacts of climate change on the biology, distribution and transmission of VBDs with case studies that highlight issues occurring across many regions. To conduct this review, a comprehensive search was done in multiple scientific databases like Google Scholar, ScienceDirect and PubMed with key terms such as 'Vector-borne diseases AND climate change', 'Climate change AND vector-borne diseases in South-east Asia' and 'Public health AND vector control AND adaptation'. The study has to address climate-related factors influencing VBDs, and studies on Southeast Asia with the public health response are mainly considered. Studies with non-climatic factors, like only socioeconomic or genetic factors, are discarded.

6.3 Result

6.3.1 Key Climate Change Impacts on Vector-Borne Diseases

Risk from VBDs is naturally sensitive to changes in weather and climate (Caminade et al. 2019). Many arthropod vectors are found to directly depend on several impacts of changes in the weather and climate, including the direct effects of low and high temperatures on vector mortality rates, the effects of the changes in precipitation and humidity on vector activity and host results, the effects of temperature on rates of vector development from one life stage to the next and the effects of rainfall variability on the availability of the breeding environment for insect vectors (Caminade et al. 2019).

6.3.1.1 Effects of Rising Temperature

Mounting temperatures, due to continuing global warming, are changing the ecology and transmission of VBDs in South and Southeast Asia. Temperature is an important factor for the growth of vector populations and the success of pathogen transmission (Hunter 2003). A secondary effect of rising ambient temperatures is the geographical expansion of the distribution of vectors like Aedes and Anopheles mosquitoes—a trend already observed in the tropical and subtropical parts of Asia—into cooler areas at higher altitudes, which widens the spatial scope of diseases such as dengue, chikungunya and malaria. According to Gubler et al. (2001), temperature affects the risk of VBD transmission through several mechanisms, as it affects vector survival, the rate at which the vector develops, biting rate and the time one pathogen requires to develop in a vector (extrinsic incubation period). These temperature-mediated effects on vector biology are particularly pertinent in monsoon climatic zones of

South and Southeast Asia, where increased temperatures accentuate seasonal transmission windows. The evidence indicates that vector-pathogen interactions in this area are also associated with more general climatic anomalies, including El Nino. For example, epidemics of malaria in South Asia have been strongly linked to El Niño periods, the result of a synergy between increased temperatures and anomalous rainfall changes, which produce optimal conditions for mosquito breeding (Bouma and van der Kaay 1996; Bouma et al. 1996). These weather patterns exacerbate the effect of sustained warming and increase the susceptibility of hot spots, including the populous lowland plains and peri-urban settings. Increments of temperature in certain areas of South and Southeast Asia have also been associated with a greater risk of viral encephalitides such as St. Louis Encephalitis (SLE). Although the viruses have traditionally been restricted by the 20 °C June isotherm in the vast majority of warmer-weathered years, their vectors have spread northward (Hess et al. 1963; Shope 1980), and a history of the disease shows that outbreaks accompany extended heat waves >30 °C (Murray et al. 1986; Monath and Tsai 1987).

6.3.1.2 Effect of Changing Nature of Precipitation and Humidity

Large-scale shifts in precipitation and humidity regimes are now being seen as important drivers for VBD transmission in the regions of South and Southeast Asia in the wake of climate change. These alterations, such as increased rainfall variability, extended dry periods and unseasonal monsoonal oscillations, have a direct impact on vector ecology and pathogen dynamics. Gubler et al. (2001) were aware of multiple pathways through which changed precipitation regimes (a direct consequence of more general climate change processes) might influence the risk of VBDs in these areas. Both the ups and downs of precipitation patterns associated with climate change can form or destroy habitats for disease vectors. Diminished river flows and impounded waters can support sustainable breeding of mosquitoes, especially in urban and semi-urban settings in areas of low rainfall. Conversely, heavy precipitation can create temporary standing water pools that can become abundant habitat for vectors like Culex and Aedes mosquitoes, which are known vectors of diseases like West Nile virus and dengue. Among the climate-driven variation in moisture and thus in rainfall, the direct impact of climate on the promotion of vegetation, whatever it may be, can widen suitable habitats where the vertebrate host populations may increase, making the risk of zoonotic transmission higher. However, too much rainfall and attendant flooding can periodically flood away breeding sites, leading to a temporary reduction in vector and host population densities. Nonetheless, such flooding events often displace vertebrate hosts and human populations, forcing closer interactions that facilitate the transmission of pathogens. In addition, long-term reductions in precipitation, which is another characteristic of regional climate change, have been related to the emergence of SLE, because a switch of a vector mosquito, *Culex pipiens*, to breeding in urban drainage due to reduced surface water has been associated with the outbreaks of the disease in some cases (Mitchell et al. 1980). On the other hand, heavy rainfall events such as those experienced in Florida have provoked SLE outbreaks via enhanced blood feeding by *Culex nigripalpus* (Day and Curtis 1989), suggesting a comparable model in Southeast Asian littoral zones.

6.3.1.3 Effects of Changing Wet and Dry Conditions

The interaction of the variability of wetness and dryness under climate change has become an important determinant of the transmission dynamics of VBD in South and Southeast Asia. The impact of both drought and excessive wetness in transforming the vector ecology, types of breeding and disease epidemics in the region as more extreme and frequent weather patterns become the norm. Longer period of dryness or drought—frequently associated with climate variability—affects multiple components of the vector life cycle, such as survival, population dynamics, distribution and vector-host pathogen interactions (Brown et al. 2014). In South and Southeast Asia, diminished rainfall in dry seasons can lead to the shrinkage of ponds or rivers and expansion of more shallow and still aquatic habitats of water, which are ideal for mosquito breeding (Medlock et al. 2011). These hydrological changes commonly benefit species such as *Aedes aegypti* and *Anopheles* spp., which breed in temporary and man-made containers frequently used in peri-urban environments. These vector populations are mediated primarily through ecological systems such as wetlands. 'Permanent wetlands' are with water year-round and can harbour a predator fauna of invertebrates, amphibians and fish that impact on aquatic stages of mosquitoes (Chase et al. 2003; Medlock et al. 2008, 2011). But, like the Arizona wetland, all of them—even those we think of as permanent wetlands—are being pushed by climate change to a state of abnormal dryness. In the process, predator–prey relationships are disrupted and a decrease in natural mosquito control occurs. When these wetlands are rewetted, mosquitoes frequently recolonise them rapidly, as without predators, and there can be a population explosion as larvae develop (Brown et al. 2014). In contrast, ephemeral wetlands that dry seasonally are adapted to maintain a dynamic balance between mosquitoes and predators. Nevertheless, it is also likely that the erratic shifts in dry and wet cycles under climate change could disrupt these systems, allowing the rise of vectors into new or previously uncolonised sites. In addition, there is a growing understanding of the role of climate-change-induced seasonal irregularities in the epidemiology of VBDs. Disparities in rainfall and temperature patterns have also been associated with variations in the occurrence of malaria in the highland and lowland surfaces in South Asia (Githeko et al. 2000). Evidence from Kenya, though beyond the Asian setting, is by now familiar: climate variability accounts for as much as 26% of the anomalies in highland malaria incidences (Githeko et al. 2000). Similar dynamics are likely occurring in the hill tracts of NE India, Nepal and N Myanmar, where upland ecosystems are experiencing increasing vulnerability to vector establishment.

6.3.2 *Case Studies*

Vector-borne infectious diseases, especially those spread by mosquitoes, represent a significant threat to populations across South and Southeast Asia (Servadio et al. 2018). The region experienced bursts here and there in the first decades of the twenty-first century, particularly with Chikungunya and Dengue outbreaks (WHO 2014). These diseases are also thought to be more endemic in the region (Medlock and Leach 2015). However, with a changing global climate, the dynamics of how variations in ambient temperatures and precipitation levels will affect mosquito populations are critical to predicting the effect this will have on risks of mosquito-borne disease outbreaks (Parham et al. 2015). This study seeks to utilise a large dataset to assess how the risk of outbreaks of mosquito-borne infectious disease correlates with the highest monthly average temperature and precipitation, respectively, for each year in South/Southeast Asia (Khampapongpane et al. 2014).

Chikungunya and dengue epidemics are common in the countries of South and Southeast Asia (Khampapongpane et al. 2014). Similarly, the last large dengue outbreaks that occurred in Laos were in 2010 and 2013 (Khampapongpane et al. 2014). During data collection periods, the incidences of dengue in Thailand were high in 2001, 2002 and 2010 (Limkittikul et al. 2014). Over these three years, the incidence was 0.5 times higher when compared to the average throughout the ten years (Limkittikul et al. 2014). From 2000–2001, there were >800,000 cases of dengue in Laos (Limkittikul et al. 2014). There was a chikungunya outbreak in India from 2005 to 2008, Sri Lanka in 2006, Malaysia in 2006 and 2008 and Thailand between 2008 and 2009 (Pulmanausahakul et al. 2011). Mosquito-borne infectious diseases, their principal vectors, causative agents, and the number of observed outbreaks in the study region between 1980 and 2009 are presented by Servadio et al. 2018.

6.3.2.1 Dengue

Climatic factors exert a decisive influence on transmission of the dengue virus, particularly temperature, rainfall and humidity (Liu-Helmersson et al. 2014). An important factor regarding the transmission dynamics of dengue by *Ae. aegypti* is the diurnal temperature range (Lambrechts et al. 2011). Thermodynamic predictions suggest that increases in DTR contribute to DENV transmission at low mean temperatures (<18 °C), and the reverse occurs at mean temperatures >18 °C (Lambrechts et al. 2011). Indeed, mosquitoes subjected to larger DTRs at 26 °C displayed shorter longevity and increased susceptibility to infection (Lambrechts et al. 2011). According to Carrington and co-workers, small DTRs at high mean temperatures (30 °C) did not influence the vector competence, whereas large DTRs at lower temperatures (20 °C) raised the proportion of infected mosquitoes able to disseminate the infection by 60% (Carrington et al. 2013). The high temperature enhances the development of *Aedes aegypti* mosquitoes and increases viral replication and the biting frequency, both of which contribute to increased transmission (Liu-Helmersson et al.

2014). Precipitation serves as a habitat for aquatic stages of the mosquito life cycle and just as much influences the distribution of the vector (Morin et al. 2013). Effects of precipitation and evaporation on available water sources may bring regulation to size, population and behaviour for Aedes (Ebi and Nealon 2016). High humidity also increases the chances of mosquito survival and activity, thus aiding transmission (Morin et al. 2013). Mosquito habitat modification through changing climatic conditions has greatly modified the geographical incidence of dengue, extending into temperate zones that were considered unaffected—such as higher altitudes and latitudes (Messina et al. 2015). In addition, urban heat islands and unplanned urbanisation due to global warming offer an enabling environment for mosquito proliferation in the peri-urban areas (Campbell et al. 2015). Increased risk of dengue outbreaks has also been attributed to extreme weather events brought about by climate change, like floods and droughts (Lowe et al. 2018). Floods spawn new breeding grounds and disrupt sanitation infrastructure to increase exposure (Lowe et al. 2018), while droughts encourage water storage in the household in wide open containers, acting as a great invitation for mosquito breeding (Semenza and Suk 2018).

In addition to the effects of the weather and climate, its socioeconomic- and public health-related determinants also explain how Aedes and dengue are found in space and how they are transmitted (Ebi and Nealon, 2016). Intensification of farming, damming, irrigation, unplanned urbanisation and increased migration, travel and trade are examples of changes that occur in the natural environment and affect distribution, e.g. by increasing the possibilities for breeding sites or the density of susceptible persons (Ebi and Nealon 2016). Complex and often given temporal variations, nonlinear feedbacks, these interactions occur with climatic, socioeconomic and other eligibility factors (Ebi and Nealon 2016). Poor quality housing in urban areas, poor provisioning of safe water and improved sanitation and limited access to waste management, among many non-climatic factors that would probably be expected to increase rather than reduce the impacts of climate change, depending on the specific socioeconomic context, would typically be found (Campbell et al. 2015).

Dengue Fever (DF) and its severe manifestation, Dengue Haemorrhagic Fever (DHF), are re-emerging arboviral infections of substantial public health concern, and it has become widespread in all the tropical and subtropical nations across the globe (Bhatia et al. 2013). Dengue is a viral disease transmitted by the *Aedes aegypti* mosquito and, less commonly, by *Aedes albopictus* (Urmi et al. 2023). This infection is caused by the dengue virus (DENV), an RNA virus belonging to the genus Flavivirus in the family Flaviviridae, which has four serotypes DENV1, DENV2, DENV3 and DENV4 (Urmi et al. 2023).

South Asian countries are high in population, and the climate and environment are conducive to mosquito breeding, which is also one of the reasons, dengue is more common in this part of the world than in other places (Rahman et al. 2022). These countries also typically lack the resources and infrastructure to successfully control mosquito populations and prevent outbreaks of dengue, which are needed for an effective response (Das et al. 2020).

In Southeast Asian countries, where all serotypes (DENV-1–4) are co-circulating, DF is usually considered to be a disease of early childhood, and clinical DF in adults is unusual (Bhatia et al. 2013). Most cases of DHF/DSS reported in these areas are children aged 2–15 years (Libraty et al. 2002). Because they possess immunity against primary infection, typical older and younger inhabitants are usually immune and do not get DHF (Libraty et al. 2002). There is increasing evidence supporting this age shift, along with the increased incidence of dengue in older age groups in Singapore (linear shift) (Nimmannitya 2002; Kalayanarooj and Nimmannitya 2004; Chareonsook et al. 1999). Among adults and very young infants (1 and 2 months), an increasing number of DHF/DSS cases have also been reported in Thailand (Bhatia et al. 2013).

In Nepal, during the first-ever outbreak in 2010 (vestal soil), the bulk of the cases occurred between the ages of 16 and 45 years (Sedhain et al. 2012) and, in the inaugural DHF irruption in Bangladesh, the age group of 18–33 years was the most affected (Rahman et al. 2002). Sri Lanka, with a chronological overview, demonstrates that the modal age mathematical group affected by dengue has shifted from <15 years of age to 15–34 years of age (MoH, Sri Lanka) (Gupta et al. 2006). In India, a legendary film manufacturer passed out of DHF with multi-organ unsuccessful person at the age of 80 years and, sure enough long-time group was significantly impacted in the last major outbreak in Delhi, India (Gupta et al. 2006).

Dengue's incidence, especially its epidemics, has been coupled with the rainy season and the El-Nino phenomenon (Halstead 2007). Although several studies have been published, the non-existence of compelling data or models that underpin this hypothesis is limited to small countries (Halstead 2007). An analysis in Thailand observed that climate factors also affect the transmission cycle of DHF, but the relative importance of these factors differed with geographical areas (Promprou et al. 2005). In contrast to macro-level climatic conditions, *Ae. aegypti* is a hygrophilic (humidity-loving) species that is subjected to microclimatic impacts to which it has adapted, according to molecular ecology research on the species. But it takes advantage of every opportunity given in peri-domestic areas this rainy season when the temperature goes up and humidity rises (Bhatia et al. 2013). Additional studies and evidence are necessary to explore vector behaviour (Bhatia et al. 2013).

6.3.2.2 Chikungunya

For Southeast Asia, temperature, humidity and rainfall were the three most important climatic factors affecting transmission (Chadsuthi et al. 2016). Increased ambient temperature leads to faster Aedes mosquito development and to a decrease in the Extrinsic Incubation Period (EIP) of Chikungunya virus, both of which facilitate transmission (Chadsuthi et al. 2016). Breeding is enhanced and stagnant water sources are abundant (Chadsuthi et al. 2016) due to high humidity and an increase in rainfall. The outcome was similar in the re-emergence of Chikungunya in Thailand in 2008–2009, where the rainfall and temperature had the greatest effects on the spatio-temporal distribution of the cases, while climatic factors changed the fertility and transmission efficiency of this mosquito (Chadsuthi et al. 2016). The role of climate change in determining the geography of Chikungunya outbreaks in Southeast Asia has been enormous. Aedes mosquitoes are more suitable habitat created through higher temperatures and variable precipitation, allowing the Chikungunya virus to be transmitted beyond its former range (Meason and Paterson 2014). Such cities include, for example, Manila in the tropics, where mean temperatures are high, and seasonality is low, which are zones of very high epidemic suitability for Chikungunya (Mordecai et al. 2017). Besides, the 2015–2016 El Nino contributed to the anomalously dry conditions and warm land surface temperatures in Southeast Asia, leading to amplification of mosquitoes and eliciting epidemics (NASA SVS 2019). Southeast Asia is hit hard by the significant effects of Chikungunya outbreaks that are fueled by climactic factors such as extreme weather events, floods and droughts. Inundation frequently leads to waterlogging situations and development of large volumes of stagnant water, suitable for mosquito breeding (Meason and Paterson 2014). On the other hand, there is a greater hold on water by households so that, in droughts, containers are more often unintended habitats for mosquitoes' larvae (Meason and Paterson 2014). For example, the 2024 Southeast Asia heat wave, exacerbated by climate change and El Niño, resulted in unprecedented temperature and drought conditions, breaking previous national records in several countries, including Malaysia and Laos, which could be as much affected as possible by the impacts on host characteristics that influence mosquito breeding and transmission.

Chikungunya is a highly pathogenic public health problem in India that affects billions. In the year 1963 (Bhatia and Narain 2009), chikungunya was identified in Calcutta. This was followed by further outbreaks and isolated cases in 1964–65 and 1973 (Maharashtra only) from Tamil Nadu, Andhra Pradesh and Maharashtra. In the year 1973, the virus disappeared from India, and in the year of 2005, no case was reported. This virus re-emerged and was epidemic in 2005, almost cosmopolitanizing apart from mainland India many islands in Indian Ocean (Yergolkar et al. 2006; Lahariya and Pradhan 2006). Meanwhile, confirmed cases were reported in the states of India, Andhra Pradesh, Karnataka, Maharashtra, Tamil Nadu, Madhya Pradesh and Gujarat during this epidemic period. In India, the attack rate is 4–45% and approximately 1.3 million people are affected (Lahariya and Pradhan 2006). Chikungunya virus of the African genotype is the causative agent of the epidemics and is vectored by the Aedes vectors (Yergolkar et al. 2006). In India, >3.7 million cases of chikungunya were reported from >18 states/union territories during 2010 (Ministry of Health and Family Welfare 2010).

Chikungunya, a viral disease, was first isolated in 1952 in Africa following an outbreak at the Makonde Plateau in the Mozambique–Tanzania border area. The term Chikungunya is taken from the Makonde word which means that which bends up due to the stooped postures of those affected (Hossain and Akter 2019). The word translates into Swahili as the disease of the walker that bends (Hossain and Akter 2019). If its presence is confirmed, the four most common symptoms are fever, ardent joint pain, muscle pain, headache, nausea, fatigue and rash (Schwartz and Albert 2010; Josseran et al. 2006). *Aedes albopictus* (Asian Tiger mosquitoes) and *Aedes aegypti* (Schwartz and Albert 2010; Josseran et al. 2006) are the two types of mosquitoes that transmit the Virus among humans. In Bangladesh, the key outbreak of Chikungunya was in 2008. A local health official reported another outbreak of fever and potentially fatal arthralgia within this period, and in Dohar Sub-district, Dhaka district, the Breteau index was 35 per 100, and 89% of dissected mosquitoes were *Aedes albopictus* (Khatun et al. 2015).

The records argue that CHIKV has steadily thrived in its range in the Asia–Pacific over the concluding 70 years (Hammon et al. 1960). India (174 criminal records, 46%), Thailand (35 records, 9%) and Malaysia (26 discs, 6.81%) have covered CHIKV circulation since the virus was first of all reported to have got to SEA (South-East Asia) before 1960 (Hammon et al. 1960). The SEA area accounted for 67.02% of records and the CHIKV epidemic were covered only from India, Thailand and Sri Lanka (Wimalasiri-Yapa et al. 2019). The WPR syndromic surveillance track record was seen individually from the above, as they report multiple countries in the same report (Wimalasiri-Yapa et al. 2019). Before 2000, most records go steady from the period 1961–1970 (Wimalasiri-Yapa et al. 2019). The two 10 following the year 2000 accounted for nearly 83% of platter discovery (Wimalasiri-Yapa et al. 2019). From the 1950s to 2017, CHIKV eruptions and/or epidemics frequently occurred in India, Indonesia, Thailand and the Philippines (Wimalasiri-Yapa et al. 2019). A retrospective analysis of human blood serum signals that CHIKV was circulating in the SEA region in 1954 (Presti et al. 2014). The first epidemic state infection of CHIKV with dengue virus was detected in Thailand in 1958 (Presti et al. 2014). India reported the beginning CHIKV outbreak in 1963 (Carey 1971). Yet, a disease with similar clinical features was first marked in 1779 in Indonesia, although several infectious factors can lead to similar symptoms (Carey 1971). Among the WPR republics, Singapore reported the first CHIKV activity in 1960, while Cambodia reported the initial disease cluster infected with CHIKV in 1961 (Bedekar and Pavri 1969).

Future predictions from the study focus on how climate change will influence the geographical suitability for Chikungunya virus transmission under two scenarios: RCP 4.5 (moderate emission) and RCP 8.5 (high emission) (Tjaden et al. 2017). Under both RCP scenarios, climatic suitability for Chikungunya transmission is externalised to expand globally, though RCP 8.5 predict subtle and more intense areas of suitability (Tjaden et al. 2017). Climatic suitability will rise critically everywhere in the country, affecting formerly unaffected zones too (Tjaden et al. 2017). Central parts will experience reduced suitability, while in the southern areas, highly significant suitability will persist (Tjaden et al. 2017). This region will certainly persist with considerable levels of susceptibility due to a significantly high universe population, which will constrain the overall risks in North Australia (Tjaden et al. 2017).

6.3.2.3 Malaria

Despite being one of the most serious of the world's almost significant VBD (Martens 1995), very few infectious diseases have as much influence on the social and economic development of a nation (Martens 1995). Today, the epidemiological range of malaria is subordinated to the tropics and subtropics, but, before the Second World War, malaria was a contagious disease in much of the rest of the world (Martens 1995). The range for the right temperature development of majority of Anopheles vector species of malaria is between 20° Coulomb and 30 °C, corresponding with the general values (Patz 1998). On the other hand, a minimum of 15° Blow is necessary for transmission of *P. vivax*, while transmission by *P. falciparum* requires a minimum of 19°100 (Patz 1998). The window for transmission in relation to the temperature range should be extended over time for the completion of the sporogony (Patz 1998). Survival of malaria has also found to vary according to time of year, wet versus dry season (Russell et al. 1963).

Many ways in which malaria is transmitted have been impacted by climate change because of changes in temperature, humidity and rainfall (Caminade et al. 2014). For example, increasing temperatures will expand the geographic range of Anopheles mosquitoes, the primary vectors of malaria, towards higher altitudes and latitudes that were previously unsuitable for their survival (Caminade et al. 2014). Heavy rainfall and flooding create new breeding sites, while drought may bring human populations and mosquitoes closer together around limited water sources (Ryan et al. 2019). Warmer conditions also accelerate the life cycle of the malaria parasite (Plasmodium spp.) in the mosquito, which could lead to increased transmission rates (Siraj et al. 2014). This is also proven by the fact that variability in climate and not just warming has been linked with outbreaks of malaria, thus making control and predictions difficult (Paaijmans et al. 2009). Therefore, climate change and extreme weather events are going to modernise risk concerning malaria globally, with particular emphasis on populations in a given area with frail health systems.

The climate in Asian parts is very equatorial and brilliant (Githeko 2009). A region that has suffered from malaria for centuries, but it is limited by insecticide and drug resistance (Githeko 2009). Other human pathogenic diseases, such as malaria, are also prominent in the region that has a rich diversity of malaria vectors, depending upon its equally rich ecological diversity (Githeko 2009). Malaria plume will shift to some highland areas where people were once free from malaria transmission (Githeko 2009). Malaria is endemic in all regions of India except above an altitude of 1,800 m and along some coastal regions (Sharma 1996). India is the largest member of the British Commonwealth nations (Githeko 2009). It accounted for 10.2 million malaria cases (Githeko 2009). In 2006, they were the major malaria species, such as Plasmodium falciparum has two (Githeko 2009). VACT, however, in *P. vivax* endemic countries (Githeko 2009). According to Githeko (2009), *An. culicifacies*, *An. dirus*, *An. fluviatilis*, *An. minimus*, *An. philippinensis*, and An. There are two main vectors of malaria, Stephensi are the main vector. Githeko (2009) noted that between 2004 and 2007, government and external financing dedicated to malaria interventions doubled, but the prevalence of *P. falciparum* was little affected. In 2007, there was a slight increase (Githeko 2009). Plasmodium falciparum thrives better in warmer temperate climates. The suitability for malaria transmission is predicted to increase in the North Eastern India regions (Uttar Pradesh, Bihar, Karnataka, Orissa, Rajasthan, Madhya Pradesh and Pondicherry) as a result of climate change (WHO 2008). Background, India has a long history of using indoor residual spraying of

DDT for malaria control (WHO 2008). This lasted until 1985 when the number of cases of malaria reduced from 70 to 2 million (WHO 2008), making the initiative a very successful one. By this time, massive insecticidal resistance had emerged (WHO 2008).

Malaysia and other neighbouring countries have a hot and humid tropical climate that favours malaria transmission (Vythilingam et al. 2008). The country has one of the oldest malaria control programmes and has succeeded in reducing malaria transmission significantly and reporting only 5,569 cases in 2006 (Vythilingam et al. 2008). The country intends to completely eradicate this disease by 2050. There have been recent unpublished reports of new malaria in the highlands of Malaysia, which would be in line with the prevalence of high altitude (Githeko 2009). The nation should step up surveillance in all areas that, due to Climate Change, will become suitable for transmission (Githeko 2009). As history shows, previous successful malaria programmes have experienced reversals of malaria trends; thus, there is much to learn from this and understand that malaria eradication will not be simple (Githeko 2009). Malaria may be reintroduced into the country by immigrants once successfully controlled (Githeko 2009).

Climate change significantly impacts the dynamics of malaria in South and Southeast Asia by altering the geographical and temporal distribution of Anopheles mosquitoes and malaria transmission (Khormi and Kumar 2016). While regions such as southern and southeastern China may see increased climatic suitability for malaria due to reduced cold stress, other areas like India, Indonesia and parts of Indochina might experience decreased suitability due to heat stress (Khormi and Kumar 2016). These changes emphasise the need for targeted, climate-informed malaria control and management strategies. By integrating climatic projections with vector distribution models, health systems can prioritise resources and interventions effectively to mitigate malaria's spread in these evolving hotspots (Khormi and Kumar 2016).

6.3.2.4 Zika

The Zika virus (ZIKV) is an arbovirus predominantly transmitted by *Aedes aegypti* and *Aedes albopictus* mosquitoes, and it has become a global health concern due to recent epidemics in Latin America in 2015–2016 (Chouin-Carneiro et al. 2016). Outside of South Asia, the virus has yet to produce large-scale outbreaks; sporadic cases have been reported, including in India's Gujarat, Rajasthan and Tamil Nadu states (Gupta et al. 2023). Climatic factors, mainly temperature, precipitation and relative humidity, have a significant impact on the transmission of ZIKV (Huber et al. 2018). With the increase in temperature in areas, the range of Aedes mosquito during these years may have been expanded into previously cooler areas, and as a result, high temperature has provided breeding conditions for mosquitoes in the shorter mosquito reproduction cycle that has driven more rapid replication of the extrinsic incubation period of the virus, increasing the risk of RRV transmission (Bellone and Failloux 2020). Concomitantly, an altered pattern of precipitation (recurrent heavy rainfall and droughts) results in the stagnation of water in urban and peri-urban areas, generating the most favourable breeding sites, such as during monsoons (Ahmad et al. 2025). Higher humidity also results in longer mosquito life spans and an increased frequency of biting events, which collectively work to enhance transmission risk (Brown et al. 2023).

Larger urbanisation patterns throughout South Asia in megacities, such as Delhi, Dhaka or Karachi, increased the ZIKV risk due to the urban heat island effect, unplanned water storage and inadequate waste disposal, which promote vector proliferation (Islam et al. 2024). According to climate model projections, with the continued rise in global temperatures, more than 1.3 billion people worldwide would be at risk of the disease by 2050 under the current climate conditions, with a major shift occurring in South Asia (Akhtar 2016). And yet there is little public health preparedness in the region. Although some entomological surveillance is currently in place, it is rarely coupled with the climatic forecast or the ZIKV diagnostic tools (El-kenawy et al. 2024). Therefore, there is an urgent need for climate-informed early warning systems, improved local diagnostic capacity and sustained vector control and community engagement adapted to the urban context. In the absence of adaptation, the region is expected to become more and more fragile to Zika outbreaks given future climate conditions.

6.3.2.5 Leptospirosis

Leptospirosis is an understudied but emerging zoonotic disease in South Asia, which is caused by pathogenic Leptospira and is transmitted mainly through contact with water or soil contaminated with the urine of infected animals, particularly rodents (Choudhary et al. 2023). Unlike Zika, which is spread by infected mosquitoes, leptospirosis is waterborne; its outbreaks are linked to environmental and climatic conditions, particularly rain and flooding (Baharom et al. 2024). Monsoonal large rainfall events leading to urban flooding in cities such as Mumbai, Chennai and Dhaka make for perfect breeding grounds for bacterial persistence and human exposure (Sathyendranath et al. 2020). The survival of Leptospira in water and moist soil is more typical for warmer temperatures and higher humidity, which prevails in the tropical and subtropical climate of South Asia, enhancing the spread of the infection (Bradley and Lockaby 2023). Epidemics that recur after the monsoon season, particularly in squatter settlements and flood-affected areas having poor drainage and sanitation, provide supportive evidence for the environmental vulnerability of the disease. For example, annual bursts in cases in Mumbai following heavy rain disproportionately impact slum-dwelling populations, who are more likely to be exposed to floodwaters and have poor sanitation infrastructure. Likewise, after large outbreaks following extreme weather events (cyclones and extended flooding), Sri Lanka has declared PT as a notifiable disease and engaged in the development of early warning mechanisms that are linked to meteorological data (Athukorala et al. 2024).

With climate change worsening and evidence showing plans for increased intensity and frequency of rainfall events, it can be presumed that the occurrence of leptospirosis is going to increase in the region (Lau et al. 2010). Rodent population dynamics, also an important feature of leptospirosis transmission, are predicted to change according to ecological shifts following these climatic changes, adding further complexity to the control of the disease (Convertino et al. 2021). Nevertheless, leptospirosis continues to be underdiagnosed and underreported owing to poor diagnostic facilities, low clinical suspicion and inadequate surveillance (Sykes et al. 2022). The task of dealing with this challenge calls for a comprehensive One Health approach, which brings together environmental control, better sanitation, flood-resilient infrastructure and public health surveillance systems with climatic risk modelling. Community education about prevention, including not coming into contact with flood waters and wearing protective clothing, is also essential in decreasing exposure. Finally, leptospirosis is illustrative of the complex interaction between climate, urbanisation public health, requiring forward-thinking, climate-proofed governance across South Asia's health systems.

6.3.3 Integrated Health System Strategies for Climate-Sensitive VBDs

6.3.3.1 Preparedness and Surveillance

Preparedness and surveillance are important, especially now, as climate change increases the problem of VBD incidence. Ecosystems altered by climatic change are invaded by new vectors, such as mosquitoes and ticks, into places that were once non-endemic, thus increasing the risk of transmission. Hence, these new challenges call for a proactive approach to healthcare systems (Wilson et al. 2020). Preparedness allows healthcare personnel to build skills and provide resources for managing diseases that could potentially emerge in areas that have never experienced them before (Martens et al. 1995). Surveillance systems are equally important; they give real-time information regarding the presence of vectors, evidence of disease incidence and new emerging patterns. Such an infrastructure for good surveillance must integrate geo-meteorological and epidemiological data and could draw from advancements in technologies such as remote sensing and artificial intelligence (Martens et al. 1995). For both Preparedness and Surveillance, the main component is the Resource allocation (Boussalis et al. 2012). In vulnerable areas, healthcare services do not have enough infrastructure to meet the rising demands created by VBDs in poor regions. Investments in diagnostic tools, vaccines and innovative ways of vector control will make a difference in these problems. The excessive use of insecticides that are causing resistance in the vectors also needs alternative methods, such as biological control or genetically engineered vectors (Lorenz et al. 2017). VBDs are transboundary, as in environments where vectors and pathogens are easily cross-border. Among the various measures and processes that would facilitate such cooperation are international frameworks and agreements, so that there can be a harmonised response to transnational threats (Silva et al. 2019). Moreover, cooperation in research would lead to new tools and strategies, for example, vaccines developed specifically against an emerging pathogen, which can increase preparedness. Such surveillance activities, if embraced by local populations, would even augment data collection and empower the community towards taking ownership of prevention measures (Bardosh et al. 2017). Local leaders and health workers who understand the culture and environment can also close the gap between public health initiatives from popular practices, thus helping to build trust and participation. Interventions if ever a public health incident occurs (Bardosh et al. 2017). Such community involvement would ultimately come into practically all plans of successful preparedness.

6.3.3.2 Vector Control and Prevention

Vector control has historically been used to manage and prevent VBDs for over a century. When applied comprehensively and consistently, this strategy still has no equal, particularly for diseases without treatment or vaccination (Wilson et al. 2020). The history of vector control also offers several important lessons regarding complacency of action, the need for political will and the requirement for sustained investment and innovation (Wilson et al. 2020). The eradication of *Aedes aegypti* in 16 countries of South America was achieved through well-executed control programmes (Seketeli et al. 2002). Instead, the reduction of these efforts led to the resurgence of the vector, resulting in the outbreak of many viruses. Nearly 91% of malaria resurgence events were linked to weakened vector control measures between the 1930s and the 2000s (Cohen et al. 2012). These are a few examples that underline the threats posed by reduced investments and boldly call for continued vigilance and funding for vector control endeavours.

Robust coordination at both the national and worldwide levels would help integrate vector control into healthcare systems (Boatin 2008). Weather extremes are among the varied challenges that require preparations for Vector-Borne Diseases (VBDs) in the face of climate change, with an integrated health system that includes five interconnected components of community involvement, surveillance, resource allocation, vector control and preparedness (Fig. 6.1). Community engagement is primary, as local involvement in prevention and control interventions significantly increases the effect of interventions. Educated and empowered communities are expected to be more likely to eliminate mosquito breeding sites, report disease symptoms promptly and seek adaptive mechanisms against climate fluctuations. Surveillance is as crucial as ever and forms the basis of early warning systems. Meteorology-based infectious disease surveillance combines meteorology data with epidemic records to predict outbreaks and to support targeted actions. Technological enhancements (remote sensing, GIS and AI-based applications) now track vector distribution and disease spread in (near) real-time. The allocation of resources remains central in determining the functioning of the system (investment in diagnostics, in vaccines, in training qualified personnel, as well as in infrastructure). Crucially, investment in R&D and innovation—in vector control technologies such as Wolbachia-infected mosquitoes or gene-editing approaches, for example—must be fast-tracked to combat the changing risks posed by the climate crisis. Vector control proper should also evolve towards integrated and sustainable approaches with the combination of chemical, biological and environmental methods. Bed nets and indoor residual spraying alone will not suffice; vector control must also include urban sanitation and community-based environmental management. Finally, readiness means that health systems can be proactive as well as reactive. Such diverse avenues include climate-informed public health planning, scenario-based outbreak projections and frontline staff capacity development. Taken together, these elements make for a robust and adaptable health infrastructure that can stem the rising challenge of VBDs in an era of worsening climate change, particularly in vulnerable areas such as South and Southeast Asia.

Integrated programmes that signal this benefit include India's initiative for the reduction of Visceral Leishmaniasis (VL) and the Initiative for Elimination in southern Africa, which has clearly shown the importance of cross-border partnerships (Wilson et al. 2020). The joint efforts utilise the sharing of resources, expertise and best practices for an overall response to disease outbreaks. Successful political advocacy and funding by government, NGOs and the private sector will strongly push this success even further; indeed, the commitment must continue to remain (Liese et al. 2014). However, it must be admitted that there is still little funding for vector control as compared to other forms of disease control, such as drugs and vaccine development (Cures 2015). Malaria vector control received only 3% of the whole amount designated for research and development in 2014, and other VBDs, such as lymphatic filariasis, accounted for amounts of funding that came to no more than a tiny fraction

Fig. 6.1 Key components of an integrated healthcare system for managing VBDs in the context of climate change (Author's own creation)

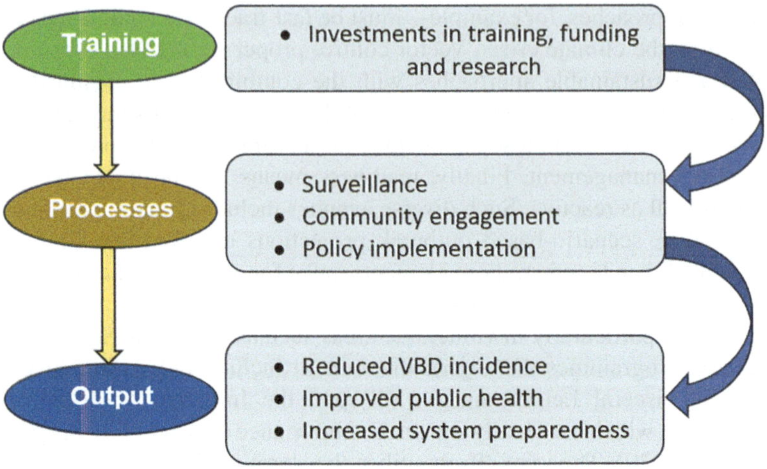

Fig. 6.2 Framework depicting the input–process–output model for mitigating climate-driven VBDs (Author's own creation)

of this total (Wilson et al. 2020). Such disparity hinders both the designing of innovative tools, and they are scaling up in already existing interventions. Funding for vector control must thus be prioritised within healthcare systems in training public health entomologists and technicians and establishing surveillance and monitoring systems (Fig. 6.2). Environmental management and community participation are an integral part of vector control that can be incorporated into the health system (Keiser et al. 2005). Past efforts, as in the example of improved housing against malaria and environmental modification for breeding site reduction, can be considered the success of environmental approaches that have been well-engineered into the health sector (Utzinger et al. 2001). Community-level strategies, which actively involve leveraging local populations in surveillance and intervention, have proven effective against dengue, malaria and Human African Trypanosomiasis (HAT) (Mutero et al. 2015). Healthcare systems are required to formulate flexible, evidence-based recommendations that blend insecticide with non-insecticide interventions specific to venue conditions. This is the theme of the Global Vector Control Response (GVCR): intersectoral collaboration, increased surveillance, community mobilisation and other tools and technologies. Adopting these into the GVCR framework could empower healthcare systems for transformational, flexible adaptation to the challenges of now and the future.

6.3.3.3 Training and Resources

One of the primary ways of enhancing the capacity of health systems against VBDs is training and resource allocation. Such investment in both human development and the availability of critical tools and infrastructure for effective disease prevention and control assures deepening technical know-how among health professionals (Conrad 2023). It is unfortunately determined that the development of comprehensive training programmes for public health workers, such as entomologists and vector control technicians, has not occurred (Mnzava et al. 2014). These individuals would have been provided specialised training for vector species identification, ecological behaviour study and control measures for specific local contexts (Wilson et al. 2020). Without a well-prepared workforce, all interventions, including the spraying of chemicals, environmental management and even those newer innovations like gene-drive mosquitoes, become meaningless. For example, during the 1950s–1960s, South America initially succeeded in eradicating *Aedes aegypti* and saw disease outbreaks again due to a lack of sustained training and oversight (Soper 1963). Resources tools are also important inputs for operationalising strategies for vector control (Table 6.1). For example, Insecticide-Treated Nets (ITNs) and Indoor Residual Spraying (IRS) have had significant effects on the incidence of malaria improvement, depending on sustained funding and logistical support, as described by Wilson et al. (2020). Likewise, the use of state-of-the-art tools such as spatial repellents and Wolbachia-infected mosquitoes will require funding for research and development and eventual implementation at scale (Achee et al. 2012). Lastly, it further establishes the sustainability of vector control measures in community participation and non-health sectors through training programmes. Community-based activities such as those highlighted have shown that local involvement can be effective in the prevention of diseases (Wilson et al. 2020). Involvement in the training of farmers, urban planners and other stakeholders in healthcare will broaden the reach of their initiatives by making them more effective.

6.3.4 Challenges and Future Considerations

6.3.4.1 Climate Adaptation in Healthcare

It demands a multi-disciplinary approach. One of these highly significant issues is the nonlinear and uncertain nature of the climate change impacts on VBDs. Bardosh et al. (2017) pointed out that climatic, social and ecological interactions create a pattern of different spatial–temporal dynamics that are hard to model and predict accurately. This uncertainty makes it a daunting task to design targeted and timely interventions while very likely keeping vulnerable populations at high risk. Increased temperatures and changed precipitation patterns can permit vectors such as mosquitoes and ticks to invade previously unaffected areas, often beyond the reach of poorly equipped healthcare systems. Another great challenge is faced in the form of addressing the socioeconomic and governance barriers to adaptation efforts. A paper by Bardosh et al. (2017) revealed how those in marginalised groups, particularly from low- and middle-income countries, are disadvantaged due to poor access to healthcare services and infrastructure, both financial and human, aggravating such inequities.

Table 6.1 Strategic initiatives and technological instruments to enhance VBDs management in the context of climate change

Aspect	Actions discussed	Tools can be used	References
Preparedness and surveillance	Build infrastructure for real-time data collection Integrate geo-meteorological and epidemiological data	Remote sensing AI-driven surveillance systems	Negev et al. (2015)
Resource allocation	Invest in diagnostic tools and vaccines Develop alternative methods to control vector resistance	Biological control Genetically engineered vectors	Wilcox et al. (2019)
Cross-border cooperation	Facilitate international agreements Promote research collaborations	WHO frameworks Vaccines for emerging pathogens	Duval (2022)
Community involvement	Engage local leaders Empower communities to take ownership of prevention measures	Public health initiatives Tailored to local culture	Bardosh et al. (2017)
Vector control	Implement sustainable environmental management Combine insecticide and non-insecticide methods	Improved housing designs Environmental modification	Wilson et al. (2020)

For example, the chronic underfunding of health systems reduces the capacity for proactive surveillance, monitoring and the implementation of adaptive strategies (Martens et al. 1995). Improving social determinants of health, including housing and sanitation, can reduce the habitats for vectors and improve outcomes. Priority should be given to scalable, adaptable-focused interventions alongside a consistent funding stream and international collaboration (Wilson et al. 2020). Investigation into the highly complex issues associated with VBDs within climate change will inform guiding policy and practice so that health systems are prepared to meet the challenges as yet unknown.

One successful model is Sri Lanka's Malaria Elimination Programme, which has demonstrated the effectiveness of GIS-based surveillance, targeted indoor residual spraying and community-based vector control in supporting disease elimination in tropical settings, which was certified by WHO in 2016 (WHO 2024). Also, the National Centre for VBDs Control (NCVBDC) of India has piloted integrated disease surveillance using climate forecast and mobile reporting tools in endemic zones. Also, Bangladesh's Dengue Early Warning System, which integrates climatic information with entomological indices, has enhanced preparedness during the monsoon seasons (Islam and Hu 2024). Globally, the Pan American Health Organisation (PAHO) has pioneered cross-country vector control training and outbreak simulation for a regional collaboration model. South Asia provides frameworks for strengthening regional data-sharing platforms, standardising entomological surveillance and coordinating emergency response; such Soft Regional Cooperation is offered by existing templates such as the South Asian Association for Regional Cooperation (SAARC) and the Bay of Bengal Initiative for Multi-Sectoral Technical and Economic Cooperation (BIMSTEC). Using these models and frameworks, South Asian states can build resilient health systems responding proactively, technologically and collaboratively against climate-impacted diseases.

6.3.4.2 Research and Development

The main challenge lies in insufficient funding for research and development (R&D) in vector control (Cures 2015). While substantial investments pour into vaccines and pharmaceuticals, development and funding for vector control tools remain poorly supported. For instance, in 2014, a mere 3% of the total research and development investment on malaria went to vector control (Wilson et al. 2020). This major disparity prevents exploring innovative interventions such as genetic modification of vectors or alternative non-insecticidal methods, which become more important as insecticides lose efficacy with rising resistance (Ranson and Lissenden 2016). A paper by Bardosh et al. (2017) shows the challenges in bringing scientific advancement into practice in low-income situations. The vulnerable populations in these areas face systemic inequities such as limited access to health infrastructure and ineffective governance, which act as barriers to implementing research findings. So, future perspectives must be built on the ground of transdisciplinary collaborations. The inclusion of local knowledge and participation of the community in research and development processes can make interventions more relevant and thus more acceptable to the communities (Bardosh et al. 2017). It speaks to the call for community-based adaptation strategies, which centre participatory decision-making and capacity-building at grassroots levels (Abramowitz et al. 2015). Thus, investments in Health and Demographic Surveillance Systems (HDSS) and real-time data analysis are key to effective context-specific adaptations (Geubbels et al. 2015). Innovative approaches that should be prioritised include: Genetic modification of vectors—beings like gene-drive mosquitoes—that will suppress the vector population or block pathogen transmission. (Wang et al. 2022); Biocontrol method using Wolbachia bacteria to inhibit viruses' replication inside mosquitoes. (Iturbe-Ormaetxe et al. 2011); Research and development in next-generation insecticides that overcome resistance, (Van den Berg et al. 2021) and Development in predictive modelling based on climatic and environmental data to anticipate outbreaks like those Kaur et al. (2022) have worked on.

6.3.4.3 Global Collaboration

Resource inequality and variation in technical capacity across regions are among the challenges faced by countries today. Developing nations are generally the most affected by VBDs, and they face crippling deficits in their surveillance infrastructure, trained personnel and access to intervention tools (Islam and Winkel 2017). Collaborative frameworks can pool resources and create a sharing network for diagnostic technologies, vector control strategies and vaccinations (Semenza and Paz 2021). One good example of international work is the Global Vector Control Response (GVCR), which emphasises the need for cross-border partnerships in scaled-up initiatives in integrated vector management. Shared data platforms and real-time information exchange can also enhance preparedness and allow prompt initiation of control measures. The Elimination 8 collaboration programme in Southern Africa is one such example of proving how multinational efforts can complement each other in managing cross-border malaria transmission, as highlighted in the uploaded documentation (Wilson et al. 2020). So, future needs also encourage intersectoral collaboration that includes governments, NGOs, research institutions and communities to foster joint investments in research and development for developing new tools for vector control, such as genetically modified vectors or new classes of insecticides (Wilson et al. 2020). This can contribute towards strengthening the collective ability to combat VBDs. The main role of global collaboration is enhancing resource sharing and collective action by establishing regional centres of excellence demonstrating attribute on vector control and climate health research in South Asia like Korea disease control and prevention agency; through establishing shared surveillance platforms to collect real-time data on vector distribution, climate anomalies and outbreak patterns like described by Institute of Medicine (US); creating support coordination across borders, for instance, concerning cross-border vector control programmes such as the 'Elimination 8' model in Africa (Wilson et al. 2020); strengthening globally applicable international legal frameworks (like revised International Health Regulations), as demand data sharing and joint responses (IHR 2005) and joint training programmes for these entomologists, epidemiologists and climate scientists across borders. An example of collaboration can be seen in the International Collaboration of Medical Research.

6.3.4.4 Challenges Faced by Health Systems in South and Southeast Asia

Health systems in South and Southeast Asia are under growing pressure from the confluence of climate change and Vector-Borne Diseases (VBDs). Changes in climate due to increased temperature, precipitation and extreme weather events have extended the distribution and period of activity of insects such as Aedes, Anopheles and other mosquitoes. These alterations are coinciding with accelerated urbanisation, substandard housing, inefficient water management and a greater population density that can further promote vector breeding and human exposure. The health systems in much of this region are weak and fragmented, with little capability to conduct disease surveillance, early detection of outbreaks and vector control. Chronic underinvestment, poor governance and unequal access to healthcare—particularly in rural and peri-urban areas—contribute to the continued inability to meet new health challenges promptly. Furthermore, the unpredictable nature of disease dynamics, driven by climate change, makes it very difficult for us to plan preventative interventions, leading to reactive health systems rather than anticipatory ones.

6.3.4.5 Existing Capacities and Opportunities for Response

However, in spite of these limitations, several countries in South-East Asia and in South Asia have shown encouraging capability in responding to climate-sensitive VBDs. Digital tools are also being used to monitor and contain disease outbreaks as they unfold; for example, India's National Centre for Vector Borne Disease Control (NCVBDC) has tested the use of an integrated disease surveillance system that includes climate forecasting technologies and mobile applications to predict and prepare for outbreaks, as have cluster tools like Bangladesh's Dengue Early Warning System. The malaria elimination programme of Sri Lanka is a successful example of GIS-based surveillance, focal vector control and community-based interventions. Although they demonstrate the possibilities of adaptive strategies, the implementation of such initiatives is for the most part, not feasible due to financial and institutional barriers. Elsewhere, regional cooperative frameworks including SAARC and BIMSTEC remain largely underutilised for data sharing, aligning surveillance standards and synchronising cross-border responses. Expanding such capabilities calls for investment in public health infrastructure, training of entomologists and frontline workers and integration of climate intelligence into health planning. Capacity building in these areas will be essential to support the development of resilient health systems able to take a proactive approach in managing the emerging risks to climate-change-induced VBDs.

6.4 Conclusion and Way Forward

Concerns over VBDs are further expanding with changes in conditions introduced by climate change. There is a dire need for cohesive and sustainable strategies to combat viruses and diseases transmitted by vectors. It's all about putting together training, resource allocation, global mutuality and community empowerment as the building blocks for effective health systems to tackle such diseases. In this study, it is highlighted that the vector-control histories have allayed fears about complacency and poor investment. Whenever a programme is scaled down, diseases that had been previously controlled begin to emerge with greater force. Continued investment is required not only in vector control instruments like insecticides, bed nets and so on, but also for training public health personnel to deal with extensive ecological dynamics and changing resistance patterns. Such features of entomologists and health personnel are fundamental to the local adaptation of control options that render interventions effective and sustainable. Global cooperation is an important factor through which diseases across borders can be controlled. The effects of climate change on VBDs will also increase with every passing year, and the increased risks need urgent multidimensional responses from public health and environmental sectors. So, strengthening climate-sensitive surveillance systems is important across South Asia to facilitate early detection of changes in vector habitats and outbreaks. Similarly, training public health professionals to enhance community awareness and promote environmental management practices that reduce possible vector breeding sites must be among the local capacity-building investments made. Research and development need to focus on creating new technologies as a matter of priority, including genetic modification of vectors, Wolbachia-based biologic interventions, and next-generation insecticides to deal with increasing insecticide resistance and newly emerging pathogen threats. However, solutions must be country-specific rather than a one-size-fits-all approach since ecological, socioeconomic and cultural conditions vary among South Asian nations, with great disparities in urban density, rural vulnerability and climatic patterns. Regional cooperation will also be paramount: the backbone of an efficient response will hinge on cross-border data sharing, standardised disease control strategies and collaborative financial initiatives. Besides, strengthening the links between climate adaptation strategies for VBDs control and other sectors, such as urbanisation, agricultural practices and water management, will enhance resilience at local and regional levels. Only coordinated, innovative and context-sensitive strategies will be sustainable in the decades to come for South Asia to handle the escalation of the burden caused by VBDs and protect its vulnerable populations due to climate change impacts.

Acknowledgements The authors would like to acknowledge the Pro-Chancellor, Swami Atmapriyanandaji, and the Assistant Administrative head of the Narendrapur Campus, Swami Shivapurnanandaji of the Ramakrishna Mission Vivekananda Educational and Research Institute (RKMVERI), for their invaluable support to this research. We also express our gratitude to Dr P.G. Dhar Chakrabarti for his continuous support.

Declaration of Conflicting Interests: The authors declared no potential conflicts of interest with respect to the research, authorship and/or publication of this article.

Funding: The authors received no financial support for the research, authorship and/or publication of this article.

Contributions: All the authors contributed equally to the preparation of this manuscript.

Appendix

See Tables 6.2 and 6.3.

Table 6.2 Vector-borne diseases and their symptoms

Disease	Pathogen	Vector	Symptom	References
Malaria	*Plasmodium spp.*	Anopheles Mosquito	Fever, chills, general feeling of discomfort, headache, nausea and vomiting, diarrhoea, abdominal pain, muscle or joint pain, fatigue, rapid breathing, rapid heart rate, cough. Sometimes people can experience shivering and chills, followed by a high fever, then sweating and a return to normal temperature	Lindsay and Birley (1996)
Dengue	Dengue virus (DEN-1, DEN-2, DEN-3 and DEN-4)	Aedes mosquito	Fever, bone pain, severe myalgias and rash. Potentially fatal, but this is rare if well managed	Ebi and Nealon (2016)
Chikungunya	*Chikungunya virus*	Aedes mosquito	Fever, shaking chills, myalgias, pharyngitis, mobilising small joints arthralgias, rash like maculopapular rash on the trunk and limbs	Tjaden et al. (2017)
Tick-borne fever	*Rickettsia rickettsii*	Fleas, lice, mites or ticks	Rash on the extremities, abdominal pain, fever, myalgias, eschar	Bouchard et al. (2019)
Zika virus infection	*Zika virus (ZIKV)* Family: Flaviviridae	Aedes albopictus mosquito	Fever, myalgias, arthralgias, rash, headache, conjunctivitis	Marinho et al. (2022)
Typhus(mite-borne)	*Orientia tsutsugamushi*	Rat flea, human body louse, squirrel flea	Headache, fever, myalgia, headache, rash	Kwa (2008)
West Nile virus infection	West Nile virus (WNV)	Culex, Aedes mosquitoes	Nausea, vomiting, diarrhoea, brief rash, fever, headache	Paz (2015)

Table 6.3 Different vector-borne diseases and their potential effects

Disease and pathogen	Vectors and their transmission	Climatic influencers	Example of how climate change influences	References
Plague (*Yersinia pestis*)	Flea, Spread by Zoonosis. like Rats	Certain rainfall patterns cause the rodent and flea populations to increase	Warming in African highlands may reduce transmission as temperatures exceed flea survival thresholds	Githeko et al. (2000)
West Nile Virus: Family *Flaviviridae*	Culex Mosquitoes: Spread by Zoonosis	Changes in weather and land-use changes affecting mosquito life dynamics	An expansion can be seen an expansion in northern Europe with local transmission in Germany after warm weather. Peaks in 23°–25 °C	Githeko et al. (2000)
Malaria (*Plasmodium falciparum* and *Plasmodium vivax*)	Anopheles mosquitos	Diversity in water bodies in response to temperature, rainfall and humidity governs the practical risk of malaria-affected areas otherwise, spacious-temporally	With the rise of 0.2 °C every decade for three decades, places in the Ethiopian highlands would have exposed non-immune populations to malaria	Siraj et al. (2014)
Lyme Disease (*Borrelia spirochete*)	Ixodes tick: Zoonosis-like rats	Linked to the warming atmosphere and effects on tick populations and their hosts. It may play some role in some areas of drought	Expansion in Canadian and Arctic regions. Increased incidence in Northeast USA due to protracted summers and the recovery or revival of deer populations	Githeko et al. (2000)

(continued)

Table 6.3 (continued)

Disease and pathogen	Vectors and their transmission	Climatic influencers	Example of how climate change influences	References
Human Sleeping Sickness (*Trypanosoma brucei,* Cronic: gambiense, Acute: rhodesiense)	Tsetse flies (glossina): Direct transmission	Tsetse flies' populations decline during hot, dry seasonal phases, with increased warming trends worsening the situation	The vector may die out in regions that are already near the upper thermal limits of survival	Papagni et al. (2023)
River Blindness (*Onchocerca volvulus*)	Blackfly, Genus: Simulium	Breeding sites in river systems are influenced by intertropical convergence zones and monsoon rains	Altered patterns in intertropical convergence zones may shift savannah black flies within regions, thereby altering patterns of disease distribution	Githeko et al. (2000)
Zika virus disease *Aedes aegypti, albopictus*	Mostly direct transmission with some zoonotic components	Diversity in water bodies in response to altering rains, temperature and humidity leads to the risk	Warming temperatures could expose more than 1.3 billion new people to Zika virus risk by 2050	Ryan et al. (2021)
Dengue Disease *Aedes aegypti*	Direct transmission	Mosquitoes are more active at certain temperatures, and the difference between the average temperature of the environment and the host temperature affects their behaviour	Dengue cases have surged in other endemic areas, such as Vietnam, where the number of cases in 2022 was four times the number reported in 2021	Piovezan-Borges et al. (2022)

References

Abramowitz SA, McLean KE, McKune SL, Bardosh KL, Fallah M, Monger J, Omidian PA (2015) Community-centered responses to Ebola in urban Liberia: the view from below. PLoS Negl Trop Dis 9(4):e0003706

Achee NL, Bangs MJ, Farlow R, Killeen GF, Lindsay S, Logan JG, Grieco JP (2012) Spatial repellents: from discovery and development to evidence-based validation. Malar J 11:1–9

Agyekum TP, Botwe PK, Arko-Mensah J, Issah I, Acquah AA, Hogarh JN, Fobil JN (2021) A systematic review of the effects of temperature on Anopheles mosquito development and survival: implications for malaria control in a future warmer climate. Int J Environ Res Public Health 18(14):7255

Ahmad A, Chester K, Sathua SK, Das A, Tyagi BK, Sarkar M (2025) Shifting epidemiological dynamics of mosquito-borne diseases and challenges for their elimination in India under the impact of climate change and global warming

Akhtar R (ed) (2016) Climate change and human health scenario in south and southeast Asia. Springer

Athukorala N, Wickramasinghe S, Yatawara L, Weerakoon K (2024) A systematic review of zoonotic disease prevalence in Sri Lanka (2000–2022). Vector-Borne Zoonotic Dis 24(8):453–472

Baharom M, Ahmad N, Hod R, Ja'afar MH, Arsad FS, Tangang F, Osman Y (2024) Environmental and occupational factors associated with leptospirosis: a systematic review. Heliyon 10(1)

Bardosh KL, Ryan SJ, Ebi K, Welburn S, Singer B (2017) Addressing vulnerability, building resilience: community-based adaptation to vector-borne diseases in the context of global change. Infect Dis Poverty 6:1–21

Bedekar SD, Pavri KM (1969) Studies with Chikungunya virus. II. Serological survey of humans and animals in India

Bellone R, Failloux AB (2020) The role of temperature in shaping mosquito-borne viruses transmission. Front Microbiol 11:584846

Bhatia R, Narain JP (2009) Re-emerging chikungunya fever: some lessons from Asia. Trop Med Int Health 14(8):940–946

Bhatia R, Dash AP, Sunyoto T (2013) Changing epidemiology of dengue in South-East Asia. WHO S East Asia J Publ Health 2(1):23–27

Boatin B (2008) The onchocerciasis control programme in West Africa (OCP). Ann Trop Med Parasitol 102(sup1):13–17

Bouchard C, Dibernardo A, Koffi J, Wood H, Leighton PA, Lindsay LR (2019) N Increased risk of tick-borne diseases with climate and environmental changes. Can Commun Dis Rep 45(4):83

Bouma MJ, van der Kaay HJ (1996) The EI Niño Southern oscillation and the historic malaria epidemics on the Indian subcontinent and Sri Lanka: an early warning system for future epidemics? Trop Med Int Health 1(1):86–96

Bouma MJ, Dye C, Van der Kaay HJ (1996) Falciparum malaria and climate change in the Northwest frontier province (Or Pakistan. Am J Trop Med Hyg 55(2):131–137

Boussalis C, Nelson HT, Swaminathan S (2012) Towards comprehensive malaria planning: the effect of government capacity, health policy, and land use variables on malaria incidence in India. Soc Sci Med 75(7):1213–1221

Bradley EA, Lockaby G (2023) Leptospirosis and the environment: a review and future directions. Pathogens 12(9):1167

Brown JJ, Pascual M, Wimberly MC, Johnson LR, Murdock CC (2023) Humidity-the overlooked variable in the thermal biology of mosquito-borne disease. Ecol Lett 26(7):1029–1049

Brown L, Medlock J, Murray V (2014) Impact of drought on vector-borne diseases–how does one manage the risk? PubHealth 128(1):29–37

Caminade C, Kovats S, Rocklov J, Tompkins AM, Morse AP, Colón-González FJ, Lloyd SJ (2014) Impact of climate change on global malaria distribution. Proc Natl Acad Sci 111(9):3286–3291

Caminade C, McIntyre KM, Jones AE (2019) Impact of recent and future climate change on vector-borne diseases. Ann N Y Acad Sci 1436(1):157–173

Campbell LP, Luther C, Moo-Llanes D, Ramsey JM, Danis-Lozano R, Peterson AT (2015) Climate change influences on global distributions of dengue and chikungunya virus vectors. Phil Trans Roy Soc B: Biol Sci 370(1665):20140135

Carey DE (1971) Chikungunya and dengue: a case of mistaken identity? J Hist Med Allied Sci 26(3):243–262

Carrington LB, Armijos MV, Lambrechts L, Scott TW (2013) Fluctuations at a low mean temperature accelerate dengue virus transmission by Aedes aegypti. PLoS Negl Trop Dis 7(4):e2190
Chadsuthi S, Iamsirithaworn S, Triampo W, Cummings DA (2016) The impact of rainfall and temperature on the spatial progression of cases during the chikungunya re-emergence in Thailand in 2008–2009. Trans R Soc Trop Med Hyg 110(2):125–133
Chareonsook O, Foy HM, Teeraratkul A, Silarug N (1999) Changing epidemiology of dengue hemorrhagic fever in Thailand. Epidemiol Infect 122(1):161–166
Chase JM, Knight TM (2003) Drought-induced mosquito outbreaks in wetlands. Ecol Lett 6(11):1017–1024
Choudhary S, Choudhary RK, Kumar M, Singh S, Malik YS (2023) Epidemiological status of leptospirosis in India. J Pure Appl Microbiol 17(4)
Chouin-Carneiro T, Vega-Rua A, Vazeille M, Yebakima A, Girod R, Goindin D, Failloux AB (2016) Differential susceptibilities of Aedes aegypti and Aedes albopictus from the Americas to Zika virus. PLoS Negl Trop Dis 10(3):e0004543
Cohen JM, Smith DL, Cotter C, Ward A, Yamey G, Sabot OJ, Moonen B (2012) Malaria resurgence: a systematic review and assessment of its causes. Malar J 11:1–17
Conrad K (2023) The era of climate change medicine—challenges to health care systems. Ochsner J 23(1):7
Convertino M, Reddy A, Liu Y, Munoz-Zanzi C (2021) Eco-epidemiological scaling of leptospirosis: vulnerability mapping and early warning forecasts. Sci Total Environ 799:149102
Cures P (2015) Neglected disease research and development: the Ebola effect. G-FINDER Rep 2015:69
Das R, Emon MPZ, Shanu SA, Akter D, Islam MR (2020) A haemophilic dengue patient with pleural effusion and earache. Cureus 12(8)
Day JF, Curtis GA (1989) Influence of rainfall on Culex nigripalpus (Diptera: Culicidae) blood-feeding behavior in Indian River County, Florida. Ann Entomol Soc Am 82(1):32–37
Duval L (2022) Climate change, vector-borne diseases, and migration. Handbook of labor, human resources and population economics. Springer International Publishing, Cham, pp 1–15
Ebi KL, Nealon J (2016) Dengue in a changing climate. Environ Res 151:115–123
Elias E, Savoy HM, Swanson DA, Cohnstaedt LW, Peters DP, Derner JD, Rodriguez L (2022) Landscape dynamics of a vector-borne disease in the western US: how vector–habitat relationships inform disease hotspots. Ecosphere 13(11):e4267
El-kenawy ESM, Eid MM, Abualigah LJMOR (2024) Machine learning in public health forecasting and monitoring the zika virus. Metaheuristic Optim Rev 1(2):1–11
Geubbels E, Amri S, Levira F, Schellenberg J, Masanja H, Nathan R (2015) Health & demographic surveillance system profile: the Ifakara rural and urban health and demographic surveillance system (Ifakara HDSS). Int J Epidemiol 44(3):848–861
Githeko AK (2009) Malaria and climate change: special feature. In: Commonwealth health ministers' update 2009. Pro-Brook Publishing, GB, Commonwealth Secretariat
Githeko AK, Lindsay SW, Confalonieri UE, Patz JA (2000) Climate change and vector-borne diseases: a regional analysis. Bull World Health Organ 78(9):1136–1147
Gubler D (2009) Vector-borne diseases. Revue scientifique et technique (International Office of Epizootics) 28(2):583–588
Gubler DJ, Reiter P, Ebi KL, Yap W, Nasci R, Patz JA (2001) Climate variability and change in the United States: potential impacts on vector-and rodent-borne diseases. Environ Health Perspect 109(suppl 2):223–233
Gupta E, Dar L, Kapoor G, Broor S (2006) The changing epidemiology of dengue in Delhi, India. Virol J 3:1–5
Gupta N, Kodan P, Baruah K, Soneja M, Biswas A (2023) Zika virus in India: past, present and future. QJM: Int J Med 116(8):644–649
Halstead, 2007Halstead SB (2007) Dengue. The Lancet 370(9599):1644–1652

Hammon WM, Rundnick A, Sather GE (1960) Viruses associated with epidemic hemorrhagic fevers of the Philippines and Thailand. Science 131(3407):1102–1103
Hess AD, Cherubin CE, LaMotte LC (1963) Relation of temperature to activity of western and St. Louis encephalitis viruses
Home | Ministry of Health and Family Welfare | GOI. (2010). https://mohfw.gov.in/
Hossain MS, Akter S (2019) Analyzing the repercussions of climate change on the outbreak of chikungunya in Bangladesh. J Earth Sci Geotech Eng 9(1):15–31
Huber JH, Childs ML, Caldwell JM, Mordecai EA (2018) Seasonal temperature variation influences climate suitability for dengue, chikungunya, and Zika transmission. PLoS Negl Trop Dis 12(5):e0006451
Hunter PR (2003) Climate change and waterborne and vector-borne disease. J Appl Microbiol 94(s1):37–46
Institute of Medicine (US) Forum on Microbial Threats (2007) Global infectious disease surveillance and detection: assessing the challenges—finding solutions, workshop summary. National Academies Press (US), Washington (DC). 2, Surveillance networks. Available from: https://www.ncbi.nlm.nih.gov/books/NBK52873/
International Health Regulations (2005) (IHR). Avilable from https://www.who.int/health-topics/international-health-regulations#tab=tab_1
Islam J, Hu W (2024) Heatwaves and dengue outbreak in Bangladesh after the pandemic—an urgent call for climate-driven early warning systems. Clin Infect Dis 78(4):1075–1076
Islam J, Asif MH, Rahman S, Hasan M (2024) Exploring mosquito hazards in Bangladesh: challenges and sustainable solutions. IUBAT Rev 7(2):1–29
Islam N, Winkel J (2017) Climate change and social inequality
Iturbe-Ormaetxe I, Walker T, O'Neill SL (2011) Wolbachia and the biological control of mosquito-borne disease. EMBO Rep 12(6):508–518
Josseran L, Paquet C, Zehgnoun A, Caillere N, Le Tertre A, Solet JL, Ledrans M (2006) Chikungunya disease outbreak, Reunion island. Emerg Infect Dis 12(12):1994
Kalayanarooj S, Nimmannitya S (2004) Guidelines for dengue hemorrhagic fever case management. WHO Collaborating Centre for Case Management of Dengue/DHF/DSS, Bangkok, Thailand, p 458
Kaur I, Sandhu AK, Kumar Y (2022) Artificial intelligence techniques for predictive modeling of vector-borne diseases and its pathogens: a systematic review. Arch Comput Methods Eng 29(6):3741–3771
Keiser J, Singer BH, Utzinger J (2005) Reducing the burden of malaria in different eco-epidemiological settings with environmental management: a systematic review. Lancet Infect Dis 5(11):695–708
Khampapongpane B, Lewis HC, Ketmayoon P, Phonekeo D, Somoulay V, Khamsing A, Bryant JE (2014) National dengue surveillance in the Lao People's Democratic Republic, 2006–2012: epidemiological and laboratory findings. W Pac Surveill Response J: WPSAR 5(1):7
Khatun S, Chakraborty A, Rahman M, Nasreen Banu N, Rahman MM, Hasan SM, Gurley ES (2015) An outbreak of chikungunya in rural Bangladesh, 2011. PLoS Negl Trop Dis 9(7):e0003907
Khormi HM, Kumar L (2016) Future malaria spatial pattern based on the potential global warming impact in South and Southeast Asia. Geospatial Health 11(3)
Kwa BH (2008) Environmental change, development and vectorborne disease: Malaysia's experience with filariasis, scrub typhus and dengue. Environ Dev Sustain 10:209–217
Lafferty KD (2009) The ecology of climate change and infectious diseases. Ecology 90(4):888–900
Lahariya C, Pradhan SK (2006) Emergence of chikungunya virus in Indian subcontinent after 32 years: a review. J Vector Borne Dis 43(4):151
Lambrechts L, Paaijmans KP, Fansiri T, Carrington LB, Kramer LD, Thomas MB, Scott TW (2011) Impact of daily temperature fluctuations on dengue virus transmission by Aedes aegypti. Proc Natl Acad Sci 108(18):7460–7465
Lau CL, Smythe LD, Craig SB, Weinstein P (2010) Climate change, flooding, urbanisation and leptospirosis: fuelling the fire? Trans R Soc Trop Med Hyg 104(10):631–638

Libraty DH, Young PR, Pickering D, Endy TP, Kalayanarooj S, Green S, Rothman AL (2002) High circulating levels of the dengue virus nonstructural protein NS1 early in dengue illness correlate with the development of dengue hemorrhagic fever. J Infect Dis 186(8):1165–1168

Liese BH, Houghton N, Teplitskaya L (2014) Development assistance for neglected tropical diseases: progress since 2009. Int Health 6(3):162–171

Limkittikul K, Brett J, L'Azou M (2014) Epidemiological trends of dengue disease in Thailand (2000–2011): a systematic literature review. PLoS Negl Trop Dis 8(11):e3241

Lindsay SW, Birley MH (1996) Climate change and malaria transmission. Ann Trop Med Parasitol 90(5):573–588

Liu-Helmersson J, Stenlund H, Wilder-Smith A, Rocklöv J (2014) Vectorial capacity of Aedes aegypti: effects of temperature and implications for global dengue epidemic potential. PLoS ONE 9(3):e89783

Lorenz C, Azevedo TS, Virginio F, Aguiar BS, Chiaravalloti-Neto F, Suesdek L (2017) Impact of environmental factors on neglected emerging arboviral diseases. PLoS Negl Trop Dis 11(9):e0005959

Lowe R et al (2018) The development of an early warning system for climate-sensitive disease risk in Brazil. Stat Med 37(16):2753–2764. https://doi.org/10.1002/sim.7674

Marinho RDSS, Duro RLS, Mota MTDO, Hunter J, Diaz RS, Kawakubo FS, Komninakis SV (2022) Environmental changes and the impact on the human infections by dengue, chikungunya and Zika viruses in northern Brazil, 2010–2019. Int J Environ Res Public Health 19(19):12665

Martens WJM (1995) Climate change and malaria: exploring the risks. Med Confl Surviv 11(4):202–213

Martens WJM, Jetten TH, Rotmans J, Niessen LW (1995) Climate change and vector-borne diseases: a global modelling perspective. Glob Environ Chang 5(3):195–209

Meason B, Paterson R (2014) Chikungunya, climate change, and human rights. Health Hum Rts J 16:105

Medlock JM, Leach SA (2015) Effect of climate change on vector-borne disease risk in the UK. Lancet Infect Dis 15(6):721–730

Medlock JM, Snow KR (2008) Natural predators and parasites of British mosquitoes—a review. Eur Mosq Bull 25(1):1–11

Medlock JM, Vaux AG (2011) Assessing the possible implications of wetland expansion and management on mosquitoes in Britain. Eur Mosq Bull 29:38–65

Messina JP, Brady OJ, Pigott DM, Golding N, Kraemer MU, Scott TW, Hay SI (2015) The many projected futures of dengue. Nat Rev Microbiol 13(4):230–239

Mills JN, Gage KL, Khan AS (2010) Potential influence of climate change on vector-borne and zoonotic diseases: a review and proposed research plan. Environ Health Perspect 118(11):1507–1514

Mitchell CJ, Francy DB, Monath TP (1980) Arthropod vectors. American Public Health Association, St. Louis encephalitis. Washington, pp 313–379

Mnzava AP, Macdonald MB, Knox TB, Temu EA, Shiff CJ (2014) Malaria vector control at a crossroads: public health entomology and the drive to elimination. Trans R Soc Trop Med Hyg 108(9):550–554

Monath TP, Tsai TF (1987) St. Louis encephalitis: lessons from the last decade. Am J Trop Med Hyg 37(3 Suppl):40S–59S

Mordecai EA, Cohen JM, Evans MV, Gudapati P, Johnson LR, Lippi CA, Weikel DP (2017) Detecting the impact of temperature on transmission of Zika, dengue, and chikungunya using mechanistic models. PLoS Negl Trop Dis 11(4):e0005568

Morin CW, Comrie AC, Ernst K (2013) Climate and dengue transmission: evidence and implications. Environ Health Perspect 121(11–12):1264–1272

Müller R, Reuss F, Kendrovski V, Montag D (2019) Vector-borne diseases. Biodivers Health Face Clim Change 67–90

Murray RA, Habel LA, Mackey KJ, Wallace HG, Peck BA, Mora SJ, Emmons RW (1986) Epidemiological aspects of the 1984 St. Louis encephalitis epidemic in southern California

Mutero CM, Mbogo C, Mwangangi J, Imbahale S, Kibe L, Orindi B, Mukabana WR (2015) An assessment of participatory integrated vector management for malaria control in Kenya. Environ Health Perspect 123(11):1145–1151
NASA Scientific Visualization Studio. (2019) ENSO teleconnections in South East Asia for the period of 2015–2016. https://svs.gsfc.nasa.gov/4697
Negev M, Paz S, Clermont A, Pri-Or NG, Shalom U, Yeger T, Green MS (2015) Impacts of climate change on vector borne diseases in the Mediterranean Basin—implications for preparedness and adaptation policy. Int J Environ Res Public Health 12(6):6745–6770
Nimmannitya S (2002) Dengue haemorrhagic fever: current issues and future research. Asian-Ocean J Paediatr Child Health 1:1–21
Ogden NH (2017) Climate change and vector-borne diseases of public health significance. FEMS Microbiol Lett 364(19):fnx186
Paaijmans KP, Read AF, Thomas MB (2009) Understanding the link between malaria risk and climate. Proc Natl Acad Sci 106(33):13844–13849
Papagni R, Novara R, Minardi ML, Frallonardo L, Panico GG, Pallara E, Nicastri E (2023) Human African trypanosomiasis (sleeping sickness): current knowledge and future challenges. Front Trop Dis 4:1087003
Parham PE, Waldock J, Christophides GK, Hemming D, Agusto F, Evans KJ, Michael E (2015) Climate, environmental and socio-economic change: weighing up the balance in vector-borne disease transmission. Philos Trans Roy Soc B: Biol Sci 370(1665):20130551
Patz JA (1998) Predicting key malaria transmission factors, biting and entomological inoculation rates, using modelled soil moisture in Kenya. Trop Med Int Health 3(10):818–827
Paz S (2015) Climate change impacts on West Nile virus transmission in a global context. Phil Trans Roy Soc B: Biol Sci 370(1665):20130561
Piovezan-Borges AC, Valente-Neto F, Urbieta GL, Laurence SG, de Oliveira Roque F (2022) Global trends in research on the effects of climate change on Aedes aegypti: international collaboration has increased, but some critical countries lag behind. Parasit Vectors 15(1):346
Presti AL, Lai A, Cella E, Zehender G, Ciccozzi M (2014) Chikungunya virus, epidemiology, clinics and phylogenesis: a review. Asian Pac J Trop Med 7(12):925–932
Promprou S, Jaroensutasinee M, Jaroensutasinee K (2005) Climatic factors affecting dengue haemorrhagic fever incidence in southern Thailand
Pulmanausahakul R, Roytrakul S, Auewarakul P, Smith DR (2011) Chikungunya in Southeast Asia: understanding the emergence and finding solutions. Int J Infect Dis 15(10):e671–e676
Rahman FI, Ether SA, Islam MR (2022) Upsurge of dengue prevalence during the third wave of COVID-19 pandemic in Bangladesh: pouring gasoline to fire. Clin Pathol 15:2632010X221076068
Rahman M, Rahman K, Siddque AK, Shoma S, Kamal AHM, Ali KS, Breiman RF (2002) First outbreak of dengue hemorrhagic fever, Bangladesh. Emerg Infect Dis 8(7):738
Ramalho-Ortigao M, Gubler DJ (2020) Human diseases associated with vectors (arthropods in disease transmission). In: Hunter's tropical medicine and emerging infectious diseases. Elsevier, pp 1063–1069
Ranson H, Lissenden N (2016) Insecticide resistance in African Anopheles mosquitoes: a worsening situation that needs urgent action to maintain malaria control. Trends Parasitol 32(3):187–196
Resh VH (2009) Vector-borne diseases of freshwater habitats. Lake Ecosyst Ecol Glob Perspectt 403
Russell PF, West LS, Manwell RD, Macdonald G (1963) Practical malariology
Ryan SJ, Carlson CJ, Mordecai EA, Johnson LR (2019) Global expansion and redistribution of Aedes-borne virus transmission risk with climate change. PLoS Negl Trop Dis 13(3):e0007213
Ryan SJ, Carlson CJ, Tesla B, Bonds MH, Ngonghala CN, Mordecai EA, Murdock CC (2021) Warming temperatures could expose more than 1.3 billion new people to Zika virus risk by 2050. Glob Change Biol 27(1):84–93

Sathyendranath S, Abdulaziz A, Menon N, George G, Evers-King H, Kulk G, Platt T (2020) Building capacity and resilience against diseases transmitted via water under climate perturbations and extreme weather stress. In: Space capacity building in the XXI century, pp 281–298

Schwartz O, Albert ML (2010) Biology and pathogenesis of chikungunya virus. Nat Rev Microbiol 8(7):491–500

Sedhain A, Adhikari S, Bhattarai GR, Regmi S, Subedee LR, Chaudhary SK, Shrestha B (2012) A clinicoradiological and laboratory analysis of dengue cases during an outbreak in central Nepal in 2010. Dengue Bull 36(12):134–148

Seketeli A, Adeoye G, Eyamba A, Nnoruka E, Drameh P, Amazigo UV, Dadzie KY (2002) The achievements and challenges of the African Programme for Onchocerciasis Control (APOC). Ann Trop Med Parasitol 96(sup1):S15–S28

Semenza JC, Paz S (2021) Climate change and infectious disease in Europe: impact, projection and adaptation. Lancet Reg Health Eur 9

Semenza JC, Suk JE (2018) Vector-borne diseases and climate change: a European perspective. FEMS Microbiol Lett 365(2), fnx244

Servadio JL, Rosenthal SR, Carlson L, Bauer C (2018) Climate patterns and mosquito-borne disease outbreaks in South and Southeast Asia. J Infect Public Health 11(4):566–571

Sharma VP (1996) Re-emergence of malaria in India. Indian J Med Res 103:26–45

Shope RE (1980) Arbovirus-related encephalitis. Yale J Biol Med 53(1):93

Silva ASG, Matos ACD, da Cunha MACR, Rehfeld IS, Galinari GCF, Marcelino SAC, Costa EA (2019) West Nile virus associated with equid encephalitis in Brazil, 2018. Transbound Emerg Dis 66(1):445–453

Siraj AS, Santos-Vega M, Bouma MJ, Yadeta D, Carrascal DR, Pascual M (2014) Altitudinal changes in malaria incidence in highlands of Ethiopia and Colombia. Science 343(6175):1154–1158

Soper FL (1963) The elimination of urban yellow fever in the Americas through the eradication of Aedes aegypti. Am J Publ Health Nations Health 53(1):7–16

Sykes JE, Reagan KL, Nally JE, Galloway RL, Haake DA (2022) Role of diagnostics in epidemiology, management, surveillance, and control of leptospirosis. Pathogens 11(4):395

Tjaden NB, Suk JE, Fischer D, Thomas SM, Beierkuhnlein C, Semenza JC (2017) Modelling the effects of global climate change on Chikungunya transmission in the 21st century. Sci Rep 7(1):3813

Urmi TJ, Al Mosharrafa R, Hossain MJ, Rahman MS, Kadir MF, Islam MR (2023) Frequent outbreaks of dengue fever in South Asian countries—a correspondence analyzing causative factors and ways to avert. Health Sci Rep 6(10):e1598

Utzinger J, Tozan Y, Singer BH (2001) Efficacy and cost-effectiveness of environmental management for malaria control. Trop Med Int Health 6(9):677–687

van den Berg H, da Silva Bezerra HS, Al-Eryani S, Chanda E, Nagpal BN, Knox TB, Yadav RS (2021) Recent trends in global insecticide use for disease vector control and potential implications for resistance management. Sci Rep 11(1):23867

Vythilingam I, NoorAzian YM, Huat TC, Jiram AI, Yusri YM, Azahari AH, LokmanHakim S (2008) Plasmodium knowlesi in humans, macaques and mosquitoes in peninsular Malaysia. Parasit Vectors 1:1–10

Wang GH, Du J, Chu CY, Madhav M, Hughes GL, Champer J (2022) Symbionts and gene drive: two strategies to combat vector-borne disease. Trends Genet 38(7):708–723

WHO (2024) Elimination of malaria and prevention of re-establishment in Sri Lanka. source https://www.who.int/publications/i/item/9789240087026

Wilcox BA, Echaubard P, de Garine-Wichatitsky M, Ramirez B (2019) Vector-borne disease and climate change adaptation in African dryland social-ecological systems. Infect Dis Poverty 8:1–12

Wilson AL, Courtenay O, Kelly-Hope LA, Scott TW, Takken W, Torr SJ, Lindsay SW (2020) The importance of vector control for the control and elimination of vector-borne diseases. PLoS Negl Trop Dis 14(1):e0007831

Wimalasiri-Yapa BR, Stassen L, Huang X, Hafner LM, Hu W, Devine GJ, Frentiu FD (2019) Chikungunya virus in Asia–Pacific: a systematic review. Emerg Microbes Infect 8(1):70–79

World Health Organization (2008) World malaria report: 2008. World Health Organization, Geneva

World Health Organization (2014) A global brief on vector-borne diseases (No. WHO/DCO/WHD/2014.1). World Health Organization

Yergolkar PN, Tandale BV, Arankalle VA, Sathe PS, AB S, Gandhe SS, Mishra AC (2006) Chikungunya outbreaks caused by African genotype, India. Emerg Infect Dis 12(10):1580

Zhang Y, Bi P, Hiller JE (2008) Climate change and the transmission of vector-borne diseases: a review. Asia Pac J Publ Health 20(1):64–76

Chapter 7
Navigating Climate Trauma: Mental Health Resilience and Adaptation Strategies in Fiji's Climate-Vulnerable Communities

Mumtaz Alam and Kunal Avishek Gounder

Abstract This chapter elaborates on the interconnection between climate change and mental health trauma in vulnerable communities in Fiji. With an in-depth study of existing literature and policy frameworks, the chapter reveals how the impacts of climate change, like rising sea levels, extreme weather events, forced relocation and livelihood instability, impact the psychological health of the Fijian communities. Indigenous Fijians are connected to their ancestral land, which is a cultural drawback in the relocation of communities; this eventually leads to mental health trauma. Youths are also victims of mental health trauma caused by climate change, as it poses a future insecurity. This chapter analyses the current strategies in place in Fiji to adapt and ensure resilience to climate mental health trauma. Fiji's response, such as Mental Health Decree 2010, National Adaptation Plan, and Climate Relocation of Communities Trust Fund, are some of the strategies currently in place. The findings highlight that there is a need for an upgrade in the health care facilities with more services in mental health and implementation of community consultations to better understand the various challenges faced at the community level; this will allow to make better strategies and decisions. This chapter contributes to the crucial topic of climate-related mental health.

Keywords Community resilience · Eco-anxiety · Cultural adaptation · Disaster mental health

---

M. Alam (✉)
Department of Social Sciences, College of Humanities and Education, Fiji National University, Nasinu, Fiji
e-mail: mumtaz.alam@fnu.ac.fj

K. A. Gounder
College of Humanities and Education, Fiji National University, Nasinu, Fiji

P. Ghosh et al. (eds.), *Climate Crises and Public Health Challenges*, Disaster Risk Reduction, https://doi.org/10.1007/978-981-95-9709-3_7

7.1 Introduction

Climate change has emerged as one of the global challenges that has extended its influence from environment to human psychological health. To date, many research and experiments have been conducted to better understand the causes and impacts of climate change, but least has been done to understand the consequence of climate change on human mental health. This is an alarming concern as the impacts of climate change is getting adverse, posing severe risk to mental health trauma as least support has been provided to frontline population. The correlation of climate change and mental health is at early stage, but a critical one, hence it demands a more research and humanitarian attention.

Charlson et al. (2021) studied 120 studies from 2001 to 2020 on climate change and mental health and found that events associated with climate change leads to psychological distress, extensive mental health, increase in psychiatric admission, increase mortality rate with mental illness and hypered suicide rates. Cianconi et al. (2020) also adds that climate change affects a large number of populations in different ways, mental health is one of them—some mental health trauma maybe common while others may differ based on the type of climatic impacts they experience. The study between climate and mental health has introduced new terminologies such as "ecoanxiety, ecoguilt, ecopsychology, ecological grief, solastalgia, biospheric concern, etc." (Cianconi et al. 2020). According to Lawrence et al. (2021), an estimate of a billion people worldwide suffers from mental health disorders. Yet, the budget allocation in the public budget and health care system is underrepresented, leading to a further rise in mental health problems (WHO 2018). Hence, mental health is given the least priority in the public health sector. Climate change is expected to worsen the mental health scenario as climate change impacts such as rising sea levels, increase in temperature, extreme weather events, and loss of livelihood and culture directly affect humans. IPCC (2022) further highlights that mental health issues may exacerbate in vulnerable population groups such as children, the elderly and those with pre-existing health conditions (Pörtner et al. 2022).

7.1.1 Literature Review and Research Gap

Various interconnected sources can cause mental health challenges due to climate change. The most adverse mental health victims are communities that face the direct impact of climate change, such as tropical cyclones, flooding, rising sea levels etc. These kinds of events can pose two types of psychological impacts: acute psychological responses and eco-anxiety. Acute psychological response, also known as Acute Stress Response (ASR) or Acute Stress Disorder (ASD), is a short-term psychological response to a terrifying or traumatic experience that can have an impact for days or weeks (Friedman 2015). In relation to extreme weather events, individuals may feel intense fear, helplessness and traumatic memories during and aftermath of extreme weather events like tropical cyclones. Due to acute psychological response, individuals may also experience nightmares and flashbacks for days or weeks after the event. Post-traumatic stress disorder (PTSD) is also common as individuals may have symptoms of hypervigilance, emotional numbness and challenges in maintaining daily functions due to life-threatening extreme weather events.

According to the American Psychological Association and ecoAmerica, eco-anxiety, also called climate anxiety, is a "chronic fear of environmental doom" (Mental Health Commission of Canada 2023). This apprehension may arise from firsthand encounters with severe weather phenomena and changes in climate, such as frequent flooding, tropical cyclones and drought. Indigenous people, marginalized communities and youth are vulnerable to climate anxiety. As the Mental Health Commission of Canada (2023) highlights, younger generations are susceptible to eco-anxiety as they express sentiments of despair, anxiety for their future and profound apprehension about possible environmental degradation, whereby Indigenous people endure significant psychological effects when changes in the environment jeopardize cultural traditions, traditional knowledge and hereditary links to specific areas of land.

The effects of climate change on mental health are classified into four primary routes. The first route pertains to the direct effects of severe weather events on mental health, including Post-Traumatic Stress Disorder (PTSD), sadness, and anxiety. The second route pertains to the direct effects of elevated and rising temperatures on mental health, including alterations in mood, aggressive conduct, criminal activity, and substance dependence. Elevated temperatures adversely impact the mental condition of those with pre-existing mental health disorders and lead to a rise in psychiatric hospital admissions. The third route pertains to the suffering induced by persistent and expected climatic and environmental changes, including solastalgia, eco-anxiety, climate anxiety, and eco-paralysis. Children and adolescents are especially susceptible to this kind of suffering. The fourth route pertains to the indirect effects at the communal level, which may undermine community cohesiveness, disturb the feeling of belonging, and adversely influence overall well-being due to losses and damages to natural ecosystems or cultural assets. These effects may incite socio-economic instability, resulting in violent conflicts over scarce resources, displacement, and forced migration. The aggregate effect of personal mental health challenges linked to climate change may lead to problems at the community level, especially in areas with a significant number of vulnerable persons.

The global response to tackling mental health issues has become an international concern as the risk is exacerbated due to climate change. International organizations like the United Nations (UN) and the World Health Organization (WHO) have progressively categorized climate-related mental health as a multifaceted public health imperative, transcending conventional environmental discourse to formulate holistic strategies that combine psychological well-being with climate adaptation initiatives. As WHO states, "There are gaps in understanding the impact of climate change on mental health and psychosocial well-being, but current knowledge is sufficient to act!" (WHO 2022).

During the Stockholm +50 conference in 2022, WHO introduced Mental Health and Climate Change: Policy Brief, it corresponds with a recent publication by the

Intergovernmental Panel on Climate Change (IPCC), which indicated that climate change increasingly endangers mental health and psychosocial well-being, encompassing emotional distress, anxiety, depression, grief, and suicidal tendencies (World Health Organization 2022a, b). The policy brief has emphasized the significant threats to mental health and well-being associated with climate change. The psychological effects of climate change are unevenly allocated, with some demographics disproportionately impacted by variables such as socioeconomic level, gender, and age. A 2021 WHO study including 95 countries revealed that just 9 had included mental health and psychosocial assistance into their national health and climate change strategies. The recent WHO policy brief advocates five critical strategies for governments to mitigate the mental health effects of climate change: incorporating climate factors into mental health initiatives, merging mental health assistance with climate action, leveraging global commitments, establishing community-based strategies to diminish vulnerabilities, and addressing the significant funding deficit for mental health and psychosocial support (World Health Organization 2022a, b).

According to Dr Diarmid Campbell-Lendrum, WHO climate lead, and an IPCC lead author, "WHO's Member States have made it very clear mental health is a priority for them. We are working closely with countries to protect people's physical and mental health from climate threats" (World Health Organization 2022a, b). At COP28 in 2023, 148 countries have endorsed COP28 Declaration on Climate and Health (UNDP 2024). This declaration commits countries to integrate health into their climate policies and climate factors in their health policies. One of the commitments under the declaration is "Promoting a comprehensive response to address the impacts of climate change on health, including for example, mental health and psychosocial wellbeing…" (COP28 UAE Declaration on Climate and Health 2023). This ensures that all 148 countries that have endorsed the declaration must act to response to mental health challenges due to climate change. Usually, the countries allocate an insufficient amount of budget for the health sector, but under this declaration, countries are encouraged to increase the investment in climate and health through the national financial budget and other donor partners.

Fiji is one of the Pacific Island Countries (PIC) facing detrimental impacts of climate change. Although Fiji is among the countries with the lowest carbon emissions—accounting for 0.006% of global emissions—it endures significant repercussions, including economic difficulties (Syngellakis et al. 2018). Like other Small Island Developing States, Fiji is now experiencing elevated sea levels, increased temperatures, coastal erosion, and a heightened frequency and intensity of natural catastrophes. Flooding is the most prevalent natural disaster in Fiji, resulting from heightened precipitation due to inadequate mitigation strategies. In 2012, the Fijian government addressed climate change by implementing the "National Climate Change Policy," which provided a framework for project creation and implementation throughout the country (McNamara and Des Combes 2015). In its capacity as President of COP23, Fiji vigorously supports the Pacific region's stance against climate change on an international level (Corbett et al. 2018). Fiji has several challenges to economic development and climate change mitigation due to its status as a Small Island Developing State with constrained budgetary and resource limitations. The

Fijian government has identified 830 communities requiring relocation due to climate change. It is important to immediately relocate 48 settlements (McMichael and Katonivualiku 2020). Fiji often experiences intense tropical cyclones from November to April due to its tropical position. The Pacific Ocean significantly contributes to the formation of tropical cyclones by providing enough warm ocean water that attracts them. Periodic droughts are a consequence of Fiji's tropical environment. The South Pacific Convergence Zone is mostly the site of El Niño episodes, characterized by the warming of ocean surface waters. Consequently, Fiji has hot, arid conditions from December to February. El Niño also contributed to the most severe drought in Fiji's history, which transpired from 1997 to 1998 (Agrawala et al. 2003). Trenberth and Hoar (1997) assert that the proliferation of El Niño weather patterns is attributable to or associated with global warming, which is directly caused by greenhouse gas emissions. The tropical eastern Pacific Ocean is experiencing a rise in climate associated with El Niño, as shown by Trenberth and Hoar's climate model. Chand et al. (2017) utilize the climate model from the CMIPS experiments to project alterations in ENSO-driven tropical cyclones by the close of the twenty-first century, predicting a 40% increase in the frequency of tropical cyclones during future El Niño events relative to current El Niño events, alongside a decrease during future La Niña events, particularly affecting a cluster of small island nations such as Fiji, Vanuatu, and the Marshall Islands (Chand et al. 2020). Their results have significant implications for the tiny, fragile island governments in the South Pacific concerning climate change mitigation and adaptation programs.

Fiji is particularly vulnerable to devastating weather events, such as bushfires, floods, and rising sea levels. These modifications pose significant risks to people reliant on traditional fishing and farming practices for sustenance and trade. Climate change jeopardizes the mental health and wellness of these individuals. Recent research by Lykins and Cosh (n.d) involving rural Fijians in Fiji indicated that environmental changes, including seasonal fluctuations, increased temperatures, and rising sea levels, have undermined traditional livelihoods and cultural practices. The research highlighted the reduction of conventional fishing methods, particularly the yatule fish, which was traditionally caught using nets (Lykins and Cosh, n.d.). This has generated concerns and sadness for the legacy for future generations. The idea of relocation intensified grieving, as numerous individuals expressed reluctance and considerable anticipated loss due to their profound ties to their ancestral locations. The swiftly changing ecosystem of Fiji resembles that of the Indigenous Inuit communities in the Circumpolar North, encompassing Alaska, northern Canada, and Greenland. The rapid reduction of sea ice significantly affects traditional cultural behaviours, leading to sorrow, anxiety, and mental health issues. Individuals and cultures with profound place-based attachments are particularly susceptible to the mental health repercussions of climate change, sometimes referred to as "eco-grief."

7.1.2 Objective of the Study

This study seeks to thoroughly investigate the intricate relationship between climate change and mental health trauma in Fiji's climate-vulnerable populations through a narrative review of existing literature and policy frameworks. The main aim is to examine the effects of climate-induced environmental changes, such as rising sea levels, extreme weather events, forced relocation, and disruption of livelihoods, on the psychological well-being of Fijian populations, focusing specifically on Indigenous communities and youth.

Secondly, the study aims to identify and analyze the primary pathways through which climate change influences mental health in the Pacific Island context, examining both direct effects, such as acute stress responses from extreme weather events and indirect repercussions, including eco-anxiety, solastalgia, and cultural grief linked to environmental degradation. An essential goal is to analyze the distinct cultural aspects of climate-induced mental health trauma among Indigenous Fijians, with particular emphasis on the spiritual and ancestral ties to land represented by the term "vanua" and the impact of forced relocation on these vital cultural connections.

Thirdly, the investigation seeks to assess the efficacy of existing mental health resilience initiatives and adaptation measures in Fiji, including the Mental Health Decree 2010, the National Adaptation Plan, and the Climate Relocation of Communities Trust Fund. The project aims to discover common patterns and culturally specific responses to climate-induced mental health difficulties through comparative analysis with similar climate-vulnerable populations, especially Indigenous Arctic communities.

Lastly, the study aims to evaluate the sufficiency of current mental health infrastructure and support systems in managing climate-related psychological trauma, pinpointing deficiencies in service delivery, especially in rural and coastal regions most affected by climate change. The research seeks to offer evidence-based suggestions for improving mental health resilience and formulating culturally acceptable therapeutic options that combine ancient healing practices with contemporary psychiatric methods.

7.1.3 Significance of Study

This research tackles a significant knowledge deficiency at the convergence of climate science and mental health psychology, providing vital insights into the nascent discipline of climate psychology, especially in relation to Small Island Developing States. This study's significance spans various dimensions, providing theoretical, practical, and policy-relevant insights into climate-induced mental health difficulties in at-risk Pacific communities.

This study enhances the theoretical comprehension of how environmental deterioration manifests as quantifiable psychological anguish in indigenous populations. This research elucidates the precise mechanisms by which climate stressors impact mental health in Fiji, thereby enhancing the overarching theoretical framework of eco-psychology and environmental grieving studies. The study examines the cultural aspects of climate trauma, especially the disruption of "vanua" connections, revealing how Western-centric mental health frameworks may insufficiently address culturally-specific trauma responses, thus aiding in the creation of more inclusive and culturally-sensitive psychological assessment instruments.

The study establishes a basis for formulating targeted preventative and treatment methods by identifying the most susceptible populations and the unique processes of climate-related psychological stress. A thorough examination of current policy frameworks and their efficacy provides critical insights for enhancing mental health service delivery in climate-vulnerable areas, especially concerning the amalgamation of indigenous healing methods with contemporary psychiatric care. This work is highly significant for national and international climate adaptation policy planning. The findings can guide the formulation of climate policies that clearly include mental health issues, ensuring that adaptation methods address both physical and psychological aspects of climate vulnerability. The study's analysis of deficiencies in existing mental health infrastructure substantiates the need for enhanced resource allocation and international assistance for climate-related mental health services in Pacific Island nations.

As climate change progresses globally, the mental health repercussions identified in this research are expected to exacerbate and extend to other at-risk populations worldwide. The study offers a proactive framework for other island nations to formulate anticipatory mental health strategies prior to the escalation of climate-related psychological stress into crisis levels.

This research enhances the global climate justice discourse by emphasizing the unequal psychological burden experienced by populations that contribute less to global greenhouse gas emissions. The study highlights cultural trauma and spiritual separation, introducing a human rights perspective to climate change discourse and pushing for the acknowledgment of psychological distress as a valid element in climate damage and loss evaluations.

This study's methodological relevance resides in its methodical synthesis of varied evidence sources, including peer-reviewed academic literature, policy papers, and community-based reports. This thorough methodology offers a framework for future investigations into climate psychology, illustrating the effective integration of scientific evidence with cultural knowledge and policy analysis to produce practical insights for at-risk groups.

7.2 Materials and Methods

This chapter utilizes a comprehensive literature review methodology to investigate the correlation between climate change and mental health trauma in Fiji's climate-vulnerable groups. The systematic method adheres to recognized protocols for comprehensive evidence synthesis, including rigorous search tactics, clear inclusion criteria, and organized data analysis to elucidate climate-induced mental health concerns in the Pacific area.

The literature search technique was executed across various academic databases to guarantee thorough coverage of pertinent publications. The systematic search employed PubMed/MEDLINE for peer-reviewed medical and public health literature, Scopus for multidisciplinary academic content in environmental science, psychology, and social sciences, and Web of Science for high-impact research in climate science and mental health. Supplementary sources encompassed Google Scholar for grey literature and reports, PsycINFO for specialized psychological and behavioral health research, the Climate Change Knowledge Portal (World Bank) for climate-specific policy documents, the WHO Global Health Observatory for mental health policy frameworks, and the Pacific Disaster Center for region-specific climate impact studies. The search approach utilized a blend of regulated vocabulary and free-text terms through Boolean operators. The principal search string employed is "((“climate change” OR “global warming” OR “environmental change”) AND (“mental health” OR “psychological trauma” OR “eco-anxiety” OR “solastalgia” OR “PTSD”) AND (“Fiji” OR “Pacific Islands” OR “Small Island Developing States” OR “SIDS”))”. Secondary search terms encompassed climate-related phrases such as “sea level rise,” “tropical cyclones,” “extreme weather,” and “environmental degradation”; mental health concepts including “climate psychology,” “environmental grief,” “disaster mental health,” and “psychological resilience”; as well as geographic references like “Oceania,” “Melanesia,” and “Pacific Island communities.” Boolean operators AND, OR, and NOT were methodically employed to enhance search outcomes, while MeSH terms were utilized where appropriate, including “Climate Change,” “Mental Health,” “Stress Disorders, Post-Traumatic,” and “Island Populations.”

The research defined explicit inclusion and exclusion criteria to guarantee methodical selection of pertinent material. The inclusion criteria comprised peer-reviewed articles published from 2010 to 2024, studies addressing the impacts of climate change on mental health within Pacific Island contexts, research investigating the psychological effects of environmental change on indigenous and vulnerable populations, policy documents and reports from WHO, UN agencies, and Fijian government sources, comparative studies from analogous climate-vulnerable regions such as Arctic communities and small island states, and literature published in the English language. The exclusion criteria removed studies that concentrated exclusively on physical health effects without considering mental health aspects, research conducted outside of Pacific Island or similar climate-vulnerable environments, opinion articles and editorials lacking empirical support, literature published prior to 2010 unless deemed seminal, non-English publications due to resource constraints, and studies characterized by inadequate methodological rigor or ambiguous results. The literature selection adhered to a methodical four-stage procedure, commencing with extensive database searches that produced preliminary results, followed by independent title and abstract screening by researchers to identify potentially pertinent papers. A comprehensive analysis followed, entailing a meticulous evaluation of chosen papers based on inclusion and exclusion criteria, culminating in reference mining by scrutinizing reference lists from the included research to uncover further pertinent sources. A uniform screening form was utilized to maintain consistency in the selection process, with inter-rater reliability measures implemented as numerous reviewers evaluated the same materials. The data extraction and synthesis methodology employed a systematic framework to document study attributes, encompassing author, year, methodology, sample size, geographic focus, population demographics, examined climate stressors such as cyclones, sea-level rise, and temperature fluctuations, mental health outcomes assessed including anxiety, depression, PTSD, and resilience factors, policy interventions and community adaptation strategies, as well as principal findings and recommendations. The study utilized thematic analysis as the principal synthesis methodology, employing inductive coding to discern recurring themes within the literature, comparative analysis to juxtapose findings between Fijian communities and analogous populations such as Inuit communities and other Small Island Developing States, a policy evaluation framework for the critical assessment of existing interventions based on established policy analysis criteria, and gap analysis to pinpoint deficiencies in research and policy implementation. Studies were assessed utilizing modified quality evaluation instruments suitable for the varied types of literature contained, taking into account methodological rigor, sample representativeness, and applicability to the Pacific Island environment. This systematic approach guarantees a thorough and methodologically rigorous analysis of climate-induced mental health issues, establishing a solid basis for evidence-based recommendations for community and policy interventions in Fiji.

7.3 Results and Discussion

7.3.1 Main Causes of Climate-Related Mental Challenges

7.3.1.1 Loss of Ancestral Land/ Forced Relocation

The Pacific Island communities are experiencing significant changes in cultural practices and identity, driven by modifications in local environments that are essential to their communal identity and culture, and are at increased risk for mental health disorders due to climate-induced disasters (Leal Filho et al. 2023). The challenge to relocation in Pacific worldviews arises from the interdependence of land and people. In Fiji, "*vanua*" and in Tuvalu, "fenua" signify both "land" and "people," indicating that land is an extension of the people and vice versa (Bindi 2023). Land transcends mere physicality, including social, relational, cosmic, and spiritual components (Yee et al. 2022). Individuals have the obligation of becoming custodians of ancestral spirits and guardians of holy locations, necessitating a physical presence to establish profound spiritual, cultural, and ancestral linkages (Baleinakorodawa and Boege 2024). Burial grounds are the primary impediment to relocation in the Pacific, since individuals are reluctant to forsake the resting places of their ancestors. Village residences serve as loci of personal, ancestral, intergenerational, and spiritual affiliation, rendering culture, identity, and location as pivotal elements of a stable livelihood of paramount significance. Consequently, recognizing culture, identity, and location as fundamental components of a stable livelihood is essential (Neef et al. 2018). Adverse impacts of climate change, such as rising sea levels, coastal erosion and environmental degradation, are methodically severing these cultural connections, compelling communities to undergo forcible relocation and resulting in unparalleled mental health challenges. Communities that abandon their ancestral lands endure significant losses, including cultural continuity, the transfer of traditional knowledge, and intergenerational social frameworks. This displacement results in communal mourning, disturbance of ecological knowledge systems, loss of community cohesiveness, identity crises, and heightened susceptibility to mental health illnesses such as depression, anxiety, and post-traumatic stress (Rajendran and Binu 2024).

7.3.1.2 Lack of Support

The issue of climate change is exacerbated by insufficient support and poor mental health infrastructure due to limited policies and financial support. The lack of extensive psychological support services in the rural and coastal areas intensifies these issues as they are the front liners to the impacts posed by climate change. Cultural relocation exacerbates unaddressed climate trauma, since Indigenous Fijian groups possess profound spiritual and cultural attachments to their environments. As Sands (2021) elaborates that underscores the detrimental effects of insufficient local participation in relocation choices, the significance of customary land in Fijian culture, and the potential for spiritual disconnection and land disputes arising from top-down methodologies.

Climate trauma significantly impacts the youth of Fiji, who endure environmental instability and stress without adequate mental health care, education, and community engagement. As Veitata (2023) highlights those youths and women are key players during climatic events, taking most of the roles in the community, including construction, farming and evacuating elderly. The healthcare system in Fiji is deficient, and women in climate-vulnerable regions endure disproportionate psychological stress (Dodd et al. 2021). The deficiency of comprehensive data collection and research sustains unresolved mental health concerns (Chang 2011). A thorough plan is required, prioritizing community-centered mental health support, culturally sensitive psychological therapies, local mental health infrastructure improvement, economic stability efforts, and adaptable educational and community support programs. Addressing climate-related mental health concerns as human rights problems requires immediate, comprehensive, and compassionate answers.

7.3.1.3 Extreme Weather Events

Extreme weather occurrences in Fiji have emerged as a substantial psychological trauma affecting community resilience and mental health. Cyclones, floods, and extended droughts have emerged as significant triggers for psychological trauma, impacting both physical infrastructure and psychological well-being (Manning and Clayton 2018; Cianconi et al. 2020). For instance, Cyclone Winston of 2016 exemplifies the psychological ramifications of severe weather catastrophes, acute survival stress, extended dislocation, and deep sorrow for lost communities and cultural traditions (Sattler et al. 2018). Severe weather events activate complex psychological processes (Cianconi et al. 2020). Initial trauma occurs during the incident, characterized by intense fear, helplessness, and a significant awareness of mortality (World Innovation Summit for Health (WISH) 2022). Survivors frequently exhibit persistent memories, heightened startle responses, and significant anxiety that surpasses the immediate threat (Boris 2023). Frequent exposure to severe weather creates a specific form of environmental trauma, leading to a sustained state of psychological hypervigilance in children raised under these conditions (Perry and Polland 1998; Roe et al. 2024).

The psychological effects at the community level are significant. Extreme weather events disrupt conventional social support systems, leading to the fragmentation of extended family networks because of relocation and economic instability (Trombley et al. 2017). Gender dynamics introduce complexity to weather-related psychological stress, as women frequently bear a disproportionate share of the psychological impacts. Limited access to mental health services leads to the neglect of complex emotional experiences, potentially resulting in intergenerational psychological challenges (Wainberg et al. 2017; O'Neill et al. 2018). Severe weather events elicit physiological reactions that can have lasting neurological effects, potentially altering brain chemistry and stress response systems (Cianconi et al. 2020). Agricultural communities experience notable psychological challenges resulting from changing environmental conditions that disrupt traditional subsistence farming practices. Young Fijians constitute an essential demographic for understanding weather-related psychological trauma and developing psychological frameworks in the context of persistent environmental uncertainty.

7.3.1.4 Economic Instability and Livelihood Insecurity

Economic instability and insecurity are significant factors contributing to psychological distress among climate-vulnerable populations in Fiji. These challenges extend beyond immediate financial concerns, affecting not only income generation but also cultural identity and social status. Climate change has disrupted traditional livelihoods in coastal communities, leading to reduced fish populations, ongoing anxiety, depression, and shifts in occupational roles (Kapoor et al. 2021; Andrews et al. 2021). Agricultural communities face considerable psychological stress due to changes in rainfall patterns, increased frequency of severe weather events, and rising temperatures, resulting in persistent uncertainty regarding crop yields and financial instability (Nie et al. 2024). The tourism sector is increasingly vulnerable to severe weather events, coastal erosion, and environmental degradation, resulting in persistent concerns among employees regarding job security and future financial stability (Weatherdon et al. 2016; Shiiba et al. 2023). Women in these communities frequently face disproportionate economic and psychological challenges, managing family finances while adapting to unstable environmental conditions. Limited access to alternative revenue sources intensifies these mental health challenges. Adolescents experience unique psychological challenges linked to economic instability, contending with diminished job prospects and unpredictable financial futures (Li et al. 2023). This leads to reduced educational motivation and increased risk-taking behaviours. The breakdown of traditional economic support systems intensifies these challenges, as extended family networks are strained by climate-related economic pressures experienced by communities. The accumulation of debt intensifies psychological distress, as individuals often incur substantial debt in efforts to adapt to shifting environmental conditions (Hallegatte et al. 2015). It is essential to employ integrated methods that recognize the interconnectedness of economic and psychological resilience to address these challenges (Berkes and Ross 2013). Community-focused economic adaptation initiatives, coupled with mental health support, present potential pathways for advancement, acknowledging traditional knowledge systems while promoting the development of climate-resilient livelihood alternatives. Understanding and addressing the complex interconnections between financial stability, environmental change, and mental health is essential for promoting resilient communities in the face of ongoing climate challenges.

7.3.2 Mental Health Resilience Measures in Fiji

7.3.2.1 Upgrade of Health Facilities in Fiji

Fiji is developing strategies for mental health resilience by integrating infrastructure improvements and legal frameworks to address the rising psychological challenges associated with climate change. Fiji's initiatives on enhancing healthcare infrastructure and enacting extensive mental health laws reflect an acknowledgement of the fundamental relationship between susceptibility and mental health. Enhancing health facilities across Fiji is a crucial advancement to foster mental health resilience. In Fiji, St. Giles Hospital is the principal mental health institution for patients with mental health issues (Ministry of Health and Services- Fiji n.d.). The institution has been upgraded in the past based on the support of the government and other donors. But due to the demand for more beds in the institution to cater for the increasing number of mental health-associated patients, the government has planned to further upgrade to meet the demand for mental health cases in Fiji (Korobiau 2024). The enhancement of health facilities across Fiji is a crucial advancement in fostering mental health resilience. St. Giles Hospital, the principal mental health institution, has seen significant remodeling to cater to the particular requirements of climate-impacted populations (Ministry of Health and Services- Fiji n.d.). Regional health facilities have been systematically improved to provide mental health care, while access remains restricted in isolated regions (Ministry of Health and Services- Fiji n.d.).

St Giles Psychiatric Hospital is scheduled for substantial upgrades to transform into the regional mental health facility for the South Pacific, as stated by Senior Medical Superintendent Dr. Balram Pandit in Suva (Korobiau 2024). The renovation seeks to strengthen the facility's services to accommodate rising demand, including enhanced rehabilitation services for drug misuse patients and assistance for those facing mental distress. A new facility for the "worried well" will be constructed (Korobiau 2024). The project intends to integrate a children's ward, now situated in the Colonial War Memorial Hospital, to optimize services and reduce the burden on healthcare staff. The program aims to enhance the quality of mental health treatment in the area (Korobiau 2024). In 2023, Fiji government stressed out the urgency of investing more in the mental health and prioritize St Giles funding due to intensified admissions, and government's involvement in supporting the institution is crucial, recognizing the intricacies of the issues encountered by those seeking sanctuary and treatment at the nexus of mental health (Cava 2023).

In 2022, a newly refurbished female ward at St Giles Psychiatric Hospital was opened and named "Louis Maraiah Ward" (Radinibaravi 2022). The new ward cost $1 million for the upgrade and can accommodate additional 13 patients to its previous capacity of 132 inpatients (Radinibaravi 2022). Notwithstanding these advancements, considerable infrastructure obstacles remain. Numerous rural populations persist in experiencing restricted access to fundamental mental health treatments, while current institutions contend with resource constraints and personnel deficiencies. The gap in mental health assistance between urban and rural areas underscores the need for a fairer allocation of resources and services.

7.3.2.2 Fiji's Mental Health Degree 2010

The Mental Health Decree 2010 of Fiji constitutes a thorough framework for the administration of mental health, including in-patient treatment, mental health promotion, prevention, community-based care, and human rights concerns (Chang 2016). The Decree, ratified by the Fijian government on 15 October 2010, seeks to provide a framework for the prevention, treatment, care, and management of mental diseases in alignment with international principles, norms, and agreements (Fiji Government 2010). It consists of 11 components, each including many portions. Part 1 addresses operational definitions, principles, aims, obligations of mental health practitioners, and safeguards against prejudice (Fiji Government 2010, Chang 2016). The aims include ensuring, regulating, and coordinating access to mental healthcare, promoting community-based services, and integrating mental healthcare into general healthcare systems (Chang 2016). Part 2 permits the development of mental health facilities (MHFs) in general public hospitals, correctional institutions, health centers, nursing stations, and the private sector. Sections 7.3 and 7.4 describe the criteria for both voluntary and involuntary evaluation and admission to a Mental Health Facility. The Decree permits a 72-h evaluation period, with involuntary assessment criteria including the individual's mental disease, refusal or inability to undergo voluntary assessment, or the need for treatment (Chang 2016).

Part 5 of the Mental Health Decree 2010 in Fiji establishes a thorough framework for the treatment and care of individuals with mental illness, emphasizing human rights concerns and the provision of timely, suitable, and high-quality treatment (Chang 2016). It underscores the significance of honoring patients' rights, including access to interpreters, confidentiality, and a declaration elucidating their rights as inpatients. The Decree mandates periodic medication evaluations and permits prior instructions for prospective inpatient treatment. Part 6 addresses community treatment, inpatient treatment orders, and the administration of electroconvulsive therapy (ECT), as well as emergency and elective surgical operations (Chang 2016). Patients must possess the ability to offer informed consent; if they are unable to do so, the case must be evaluated by the Mental Health Review Board (MHRB). Part 7 delineates the protocols for the admission of minors to mental health institutions (MHFs), including their segregation from the adult demographic, ongoing education, and consistent access to familial connections. The Decree permits the examination of the mental health of those in detention and incarcerated offenders. Part 9 discusses the establishment, structure, procedures, jurisdictions, and functions of two key review bodies: the MHRB and the Board of Mental Health Visitors (BMHV). However, the lack of comprehensive information and training regarding the functions of these entities has hindered their implementation. The Mental Health Decree 2010 provides a comprehensive framework for mental health services in Fiji, highlighting the importance of human rights and ensuring the delivery of timely, appropriate, and high-quality care (Chang 2016).

Pandit et al. (2024) note that the Mental Health Act of 2010 has been modernized and reformed, leading to the decentralization of healthcare services and a shift towards recovery-oriented treatment. The current mental health framework includes St. Giles Hospital in Suva, featuring a Community Mental Health Team, Stress Management Wards, a Telemedicine unit, and Primary Health Centres. Efforts are underway to enhance decentralization, establish specialist clinics, develop Mind Farm, and convert St. Giles Hospital into a Regional Institute of Mental Health Care and Research (Pandit et al. 2024).

7.3.3 Adaptation Strategies in Fiji's Climate-Vulnerable Communities

Fiji's climate adaptation strategy reflects a sophisticated understanding of the various challenges faced by vulnerable populations. The nation has developed a holistic strategy that combines national policy frameworks, community engagement, and financial support mechanisms to address the physical and psychological impacts of climate change. The government of Fiji recognizes the psychological risks linked to climate change and collaborates with stakeholders such as Empower Pacific to tackle these challenges effectively.

7.3.3.1 National Adaptation Plan (NAP)

The National Adaptation Plan (NAP) functions as Fiji's primary policy framework for climate resilience, highlighting the psychological aspects of climate adaptation (Government of the Republic of Fiji 2018). The text acknowledges the psychological impacts of climate change, emphasizing the need for mental health support, and aims to establish a framework for addressing climate-related issues. The NAP seeks to enhance mental health resilience and adaptation strategies in climate-vulnerable areas by integrating mental health considerations into comprehensive climate adaptation planning through community-based initiatives and the fortification of mental health infrastructure.

Improving mental health services within the healthcare system ensures accessibility, especially in remote and vulnerable areas. The NAP underscores the importance of capacity-building initiatives aimed at educating mental health professionals, first responders, and community leaders to address psychological challenges stemming from climate-related stresses (Government of the Republic of Fiji 2018). Marginalized populations, such as children, women, the elderly, and individuals with impairments, experience intensified mental health challenges; therefore, targeted interventions are necessary.

Enhancing social cohesiveness and resilience within communities is crucial, as strong social connections and collective preparedness can mitigate the emotional effects of climate-related disasters. Community-building initiatives and culturally relevant practices can promote emotional healing and harmony. The integration of mental health adaptation within disaster risk reduction frameworks, early warning systems, pre-disaster mental health education, and post-disaster counseling services mitigate the emotional effects of climatic disasters. Collaborative strategies, oversight and assessment frameworks, and partnerships with foreign benefactors, NGOs, and mental health organizations can secure funding and resources for these initiatives. Advocacy and awareness efforts are essential for reducing stigma related to mental health and encouraging individuals affected to seek assistance.

7.3.3.2 Proper Guidelines and Community Consultation

Community consultation is fundamental to Fiji's adaptation strategy, ensuring that local perspectives shape resilience initiatives. Communities participate in structured discourse sessions to collaboratively develop and implement adaptation solutions tailored to their specific needs and cultural values. The discussions involve various stakeholders, such as community elders, youth representatives, women's organizations, and mental health professionals. This participatory method has demonstrated significant effectiveness in identifying psychological support needs and developing culturally appropriate solutions for communities affected by climate change. Regular consultation meetings facilitate continuous input and adjustment of adaptation strategies, ensuring their relevance and effectiveness.

The Fiji Government recognizes the implications of relocating affected communities to new areas. In 2019, the Displacement Guidelines were introduced in relation to climate change and disaster management. The Displacement Guidelines seek to mitigate vulnerabilities linked to displacement and explore sustainable ways to avoid and lessen the factors contributing to relocation in impacted communities in Fiji (Ministry of Economy et al. 2019). It offers guidelines for both state and non-state players in addressing displacement, acknowledging that it is largely not a state-led process (Ministry of Economy et al. 2019). It is consistent with the stipulations of Fiji's 5-Year and 20-Year National Development Plans, Fiji's National Adaptation Plan, the National Climate Change Policy, and other pertinent national, regional, and international frameworks. Fiji's Displacement Guidelines are integral to sustainable development; hence their application will expedite Fiji's advancement toward achieving the Sustainable Development Goals (Ministry of Economy et al. 2019).

Fiji's Displacement Guidelines outline that displacement should be a "People Centred Approach" and necessary support should be provided to the displaced communities, such as capacity building, to ensure that displacement has the least possible impact on the people. Transparency and consultation with communities are the key pillars to better understand the challenges and plan strategies for the resolution and displacement of future communities. The participation and consultation mechanisms of the Paris Agreement, grounded in a human rights-based framework, seek to improve transparency in initiatives and assistance related to climate change and displacement. Article 13 of the Transparency Concept recognizes the diverse capabilities and collective experiences of the Parties (Ministry of Economy et al. 2019). The framework aims to clarify the assistance given and received by specific Parties, while providing a comprehensive overview of total financial support, thereby enabling global stocktaking in line with Article 14 (Ministry of Economy et al. 2019).

7.3.3.3 Climate Relocation of Communities Trust Fund

The Fiji Government is dedicated to protecting and assisting communities and groups negatively impacted by climate change disruptions. The effects encompass sea level rise, inundation, coastal degradation, and catastrophic events such as cyclones and landslides. In 2018, the government established Planned Relocation Guidelines and initiated the Climate Relocation of Communities Trust Fund to support relocation efforts in Fiji. The Climate Change Act of 2021 established a legislative framework to facilitate relocation operations, mandating the Minister Responsible for Climate Change to ensure the orderly, courteous, and dignified evacuation of vulnerable populations (Fiji Government 2023). The Act mandated the Minister to establish Standard Operating Procedures for organized relocation. The Cabinet authorized the Standard Operating Procedures for Planned Relocation in Fiji (SOP) in March 2023 and officially launched it on 18 April 2023 (Fiji Government 2023).

The CROC Trust Fund is financed by the Fiji Government, which allocates 3% of the money generated from VAT on designated services, plastic levy, superyacht charge, and income tax (Fiji Government 2023). The government is engaged in fundraising for this program and intends to obtain more international and bilateral donations to augment the fund progressively.

7.3.3.4 Professional Counseling Support

Psychological health remains a sensitive topic in the Pacific as individuals are not very comfortable talking about it and are afraid to seek assistance even if they face a psychological health issue. Professional counseling support is a need to allow individuals receive the need they require and comfortable to approach. In Pacific and Fiji, Empower Pacific plays a crucial role in providing free and confidential support, with 24 h counseling helpline (Australian Volunteers International 2023). The trained professionals of Empower Pacific respond to natural disaster victims around the country and the Pacific. In August 2023, Empower Pacific trained 18 counsellors with the support from Australian Volunteers International (AVI) in "Disaster Mental Health Counselling and Working with Trauma- Interventions that Foster Resilience" (Australian Volunteers International 2023). The training improved the capacity of Empower Pacific and counsellors to provide better services to victims suffering from mental health distress and trauma post-disaster.

In 2022, the Empower Pacific Labasa Branch facilitated Psychological First Aid (PFA) training for 20 community leaders in Naqaravutu Village, Tikina of Tunuloa in Fiji (Empower Pacific 2022). The organization was contacted by Keisha Maria Foundation, which observed a lack of support and knowledge regarding mental health and well-being in the rural village. The training focused on the principle of "Look, Listen & Link" and assisted the volunteers in identifying individuals in need of support, assessing the situation, working with community police officers, and gathering resources and support services. Empower Pacific, in collaboration with the Australian Humanitarian Partnership, has conducted Psychological First Aid (PFA) training in five communities in Fiji (Empower Pacific 2022). Following a recent flood in December 2024, Empower Pacific is organizing trauma counseling and psychological first aid for affected communities and individuals in evacuation shelters. To provide families with Infant/Baby Packs and Dignity Packs, they have sent out two teams of trained psychological first aid volunteers, social workers, and qualified senior counselors (Kumar 2024). They seek to alleviate acute trauma and avoid detrimental effects over time.

7.4 Recommendation

7.4.1 *Governance*

The 2010 Mental Health Decree in Fiji requires revision to tackle mental health issues arising from climate change. This entails mandating mental health impact evaluations for climate adaptation schemes, with a focus on community relocation activities. Definitive methods for delivering psychological care must be created throughout the duration of climate-related catastrophes and relocations. The National Adaptation Plan should incorporate specific mental health indicators and objectives to ensure the integration of mental health considerations in all climate resilience initiatives. Integrating traditional knowledge systems, particularly the Fijian concept of "vanua," into mental health frameworks is crucial for the creation of culturally appropriate therapies.

Resource allocation is a crucial domain requiring prompt attention. The government should establish a designated mental health funding allocation within the Climate Relocation of Communities Trust Fund. This would support initiatives including the training and deployment of mental health professionals in climate-vulnerable areas, the creation of community-based mental health centers, and the development of mobile mental health units for remote populations. Additionally, resources should be allocated for research initiatives aimed at improving understanding of the mental health impacts associated with climate change and for developing digital mental health infrastructure to support telehealth services.

Collaborations with specialized organizations such as the WHO, UNICEF, and UNESCO are essential for strengthening international partnerships and developing targeted climate-mental health initiatives. Effective execution requires collaboration among governmental bodies, healthcare practitioners, community officials, and international partners. Regular assessment and adjustment of these interventions will ensure their effectiveness in reducing climate-related mental health stress in Fiji.

7.4.2 *Practical Initiatives*

Practical activities serve as essential grassroots interventions to address climate-related mental health challenges in Fiji. These encompass community intervention initiatives designed to improve comprehension of mental trauma and its relationship with climate change. These programs must incorporate monthly seminars, support group meetings, and educational initiatives to normalize discussions surrounding mental health. Involvement of local leaders, religious authorities, and community elders as advocates for mental health is essential to reduce stigma and encourage open discussions regarding psychological well-being.

Capacity-building initiatives should concentrate on strengthening local mental health support systems. This involves training community health professionals, educators, and religious leaders in essential mental health first aid and trauma support. It is essential to establish peer support networks within communities to develop sustainable support systems that can operate independently of external influences. Emphasis should be placed on training youth leaders who can effectively bridge generational divides and improve mental health awareness among younger community members.

Addressing cultural barriers in mental health discussions necessitates a sophisticated and culturally informed approach. Traditional healing methods and beliefs should be acknowledged and integrated into mental health support systems. Community discussions should analyze the incorporation of traditional coping strategies with modern mental health care practices. Public awareness initiatives should utilize local languages and cultural references to improve the accessibility and relatability of mental health discussions.

Consistent evaluation and assessment of mental health in at-risk groups are essential for gauging the efficacy of treatments and pinpointing areas that require further assistance. Success indicators must include heightened community involvement in mental health initiatives, less stigma around mental health discourse, better accessibility to support resources, and strengthened community resilience against climate-related adversities.

7.5 Conclusion

This research considerably advances the subject of climate psychology by evidencing quantifiable connections between environmental deterioration and mental health effects in Small Island Developing States, using Fiji as a pivotal case study. The study addresses a significant deficiency in region-specific climate mental health literature by clearly linking climatic stressors to psychological effects in at-risk Pacific populations, so offering an evidence-based basis for future interventions.

The findings indicate that Fiji's holistic strategy for tackling climate-induced mental health issues, utilizing integrated policy frameworks, stakeholder collaboration, and community activities, serves as a commendable model for other Small Island Developing States confronting analogous risks. The study reveals substantial limitations in the effectiveness of present interventions, attributed to resource restrictions such as limited funding, a shortage of specialists, and poor infrastructure in isolated coastal regions. These shortcomings underscore the pressing necessity for augmented foreign assistance and focused capacity-building initiatives.

This research significantly contributes by demonstrating the deep cultural aspects of climate-induced mental health stress among Indigenous Fijians. The research illustrates that land degradation transcends mere physical loss, incorporating spiritual and cultural anguish, as land underpins indigenous identity and ancient customs. This discovery enhances theoretical comprehension of climate psychology by demonstrating the inadequacy of Western-centric mental health frameworks in addressing culturally specific trauma reactions. The study highlights that successful interventions should amalgamate ancient healing practices with contemporary psychological methodologies, establishing culturally sensitive therapeutic frameworks that honor indigenous knowledge systems.

This study's consequences reach well beyond Fiji, providing essential information for the wider Pacific area and other vulnerable island nations worldwide. As climate change progresses, the recorded mental health effects are expected to escalate and harm more vulnerable populations, highlighting the necessity for culturally sensitive, community-oriented solutions. The findings offer other SIDS a framework for developing proactive mental health initiatives to tackle climate-related psychological issues prior to reaching crisis levels.

This research presents numerous significant opportunities for future inquiry. Creating standardized assessment instruments to evaluate the effects of climate change on mental health across various cultural contexts is crucial for facilitating credible comparative research and evidence-based policy formulation. Longitudinal studies investigating the enduring psychological impacts of climate adaptation and migration tactics might yield significant insights into resilience-building mechanisms. Moreover, a comprehensive assessment of both conventional and contemporary mental treatment methods is essential for determining the most effective therapeutic procedures.

The study significantly influences global climate policy, promoting the integration of mental health factors in climate adaptation strategies and international financial frameworks. The research indicates that the costs of climate change include not just damage to physical infrastructure but also considerable psychological distress, which has been generally neglected in economic evaluations of climate effects.

This study identifies climate-related mental health as a critical global concern for prompt action from researchers, politicians, and international development organizations. As climate impacts escalate, the mental health issues recorded in Fiji are expected to increase among at-risk groups globally, necessitating the creation of effective, culturally sensitive interventions for global health security and climate justice.

Funding: The research was not supported by any funding.

Potential conflicts of interest: On behalf of all authors, the corresponding author states that there is no conflict of interest.

Informed Consent: Not Applicable

Authors Contribution: All authors have contributed equally to this work.

Data availability statement: Not Applicable

References

Agrawala S, Ota T, Risbey J, Hagenstad M, Smith J, van Aalst M, Koshy K, Prasad B (2003) Development and climate change in Fiji: focus on coastal mangroves. Paris, France: OECD, pp 1–56

Andrews N, Bennett NJ, Le Billon P, Green SJ, Cisneros-Montemayor AM, Amongin S, Gray NJ, Sumaila UR (2021) Oil, fisheries and coastal communities: A review of impacts on the environment, livelihoods, space and governance. Energy Res Soc Sci 75:102009

Australian Volunteers International (2023) Health: At the front line of climate change. AVI. https://www.avi.org.au/supporting-mental-health-on-the-front-line-of-climate-change/. Accessed 30 Dec 2024

Baleinakorodawa P, Boege V (2024) Climate change-induced community relocation in Fiji: challenges and ways forward

Berkes F, Ross H (2013) Community resilience: Toward an integrated approach. Soc Nat Res 26(1):5–20

Bindi C (2023) Where the most beautiful clouds pass: place attachment, voluntary immobility, and environmental risks: the role of Vanua, Fenua, and Watan Shirin (Doctoral dissertation, European University Institute)

Boris V (2023) Analysing the effects of trauma on psychological well-being. https://www.jneuropsychiatry.org/peer-review/analysing-the-effects-of-trauma-on-psychological-wellbeing-17418.html

Cava L (2023) Government to prioritize St Giles with increased funding. Fiji Broadcasting Corporation. https://www.fbcnews.com.fj/news/health/government-to-prioritize-st-giles-with-increased-funding/. Accessed 23 Dec 2024

Chand SS, Tory KJ, Ye H, Walsh KJE (2017) Projected increase in El Niño-driven tropical cyclone frequency in the Pacific. Nat Clim Chang 7:123–127

Chand SS, Dowdy A, Bell S, Tory K (2020) A review of South Pacific tropical cyclones: impacts of natural climate variability and climate change. Climate change and impacts in the Pacific, pp 251–273

Chang O (2016) Fiji's mental health decree 2010. Bjpsych Int 13(2):41–43

Chang O (2011) Mental health care in Fiji

Charlson F, Ali S, Benmarhnia T, Pearl M, Massazza A, Augustinavicius J, Scott JG (2021) Climate change and mental health: a scoping review. Int J Environ Res Public Health 18(9):4486

Cianconi P, Betrò S, Janiri L (2020) The impact of climate change on mental health: a systematic descriptive review. Front Psych 11:490206

European Climate and Health Observatory (2022) Climate change impacts on mental health in Europe: an overview of evidence. In: European climate and health observatory. https://climate-adapt.eea.europa.eu/en/observatory/evidence/health-effects/mental-health-effects

COP28 UAE Declaration on climate and health (2023) WHO. https://www.who.int/publications/m/item/cop28-uae-declaration-on-climate-and-health. Accessed 4 Dec 2024

Corbett J, Yi-Chong X, Weller P (2018) Climate change and the active participation of small states in international organisations. Table 107(1):103–105

Dodd R, Shanthosh J, Lung T et al (2021) Gender, health and ageing in Fiji: a mixed methods analysis. Int J Equity Health 20:205. https://doi.org/10.1186/s12939-021-01529-9

Empower Pacific (2022) Post-disaster response enhanced by proper training—empower Pacific. https://www.empowerpacific.com/blog-impact/2022/5/2/post-disaster-response-enhanced-by-proper-training. Accessed 30 Dec 2024

Fiji Government (2010) Mental health decree 2010, Fiji

Fiji Government (2023) FIJi's climate relocation of Communities trust fund. INFORMATION BRIEF 1: An introduction to the climate relocation of Communities trust fund for communities and the public. In: Ministry of Health and Services Fiji (Decree No.54 of 2010). Government of Fiji. https://fijiclimatechangeportal.gov.fj/wp-content/uploads/2023/05/CROCTF-Information-Brief-1.pdf. Accessed 30 Dec 2024

Friedman MJ (2015) Posttraumatic and acute stress disorders. Springer

Government of the Republic of Fiji (2018) Republic of Fiji national adaptation plan [Report]. https://www4.unfccc.int/sites/NAPC/Documents/Parties/National%20Adaptation%20Plan_Fiji.pdf

Hallegatte S, Bangalore M, Bonzanigo L, Fay M, Kane T, Narloch U, Rozenberg J, Treguer D, Vogt-Schilb A (2015) 26 Poverty and climate change: natural disasters, agricultural impacts and health shocks. Towards a workable and effective climate regime, vol 369

Kapoor A, Alcayna T, de Boer T, Gleason K (2021) Climate change impacts on health and livelihoods: Fiji Assessment

Korobiau S (2024) St Giles hospital to get a facelift https://fijisun.com.fj/2024/01/09/st-giles-hospital-to-get-a-facelift/#:~:text=St%20Giles%20Psychiatric%20Hospital%20will,the%20revelation%20in%20Suva%20yesterday. Accessed 23 Dec 2024

Kumar NA (2024) Empower Pacific provides psychological first aid. Fiji Broadcasting Corporation. https://www.fbcnews.com.fj/news/empower-pacific-provides-psychological-first-aid/. Accessed 30 Dec 2024

Lawrance E, Thompson R, Fontana G, Jennings N (2021) The impact of climate change on mental health and emotional wellbeing: current evidence and implications for policy and practice. Grantham Institute briefing paper, vol 36, pp 1–36

Leal Filho W, Krishnapillai M, Minhas A, Ali S, Nagle Alverio G, Hendy Ahmed MS, Naidu R, Prasad RR, Bhullar N, Sharifi A, Nagy GJ (2023) Climate change, extreme events and mental health in the Pacific region. Int J Clim Chang Strat Manag 15(1):20–40

Li L, Serido J, Vosylis R, Sorgente A, Lep Ž, Zhang Y, Fonseca G, Crespo C, Relvas AP, Zupančič M, Lanz M (2023) Employment disruption and wellbeing among young adults: a cross-national study of perceived impact of the COVID-19 lockdown. J Happiness Stud 24(3):991–1012

Lykins A, Cosh S (n.d.) We don't know what tomorrow will bring: how climate change is affecting Fijians mental health. The conversation. https://theconversation.com/we-dont-know-what-tomorrow-will-bring-how-climate-change-is-affecting-fijians-mental-health-219506. Accessed 6 Dec 2024

Manning C, Clayton S (2018) Threats to mental health and wellbeing associated with climate change. In: Psychology and climate change. Academic Press, pp 217–244

McMichael C, Katonivualiku M (2020) Thick temporalities of planned relocation in Fiji. Geoforum 108:286–294

McNamara KE, Des Combes HJ (2015) Planning for community relocations due to climate change in Fiji. Int J Disaster Risk Sci 6:315–319

Mental Health Commission of Canada (2023) Understanding and coping with eco-anxiety—mental health commission of Canada. https://mentalhealthcommission.ca/resource/understanding-and-coping-with-eco-anxiety/. Accessed 3 Dec 2024

Ministry of Economy, Deutsche Gesellschaft fur Internationale Zusammenarbeit GmbH, & German Federal Ministry for Economic Cooperation and Development (2019) Displacement guidelines. In: Displacement guidelines [Report]. https://fijiclimatechangeportal.gov.fj/wp-content/uploads/2022/01/Displacement-Guidelines-Jan-2020.pdf

Ministry of Health and Services- Fiji (n.d.) Mental Health—MHMS FIJI. https://www.health.gov.fj/mental-health/#:~:text=St%20Giles%20Hospital%20is%20for,also%20undertakes%20outreach%20throughout%20Fiji. Accessed 23 Dec 2024

Neef A, Benge L, Boruff B, Pauli N, Weber E, Varea R (2018) Climate adaptation strategies in Fiji: the role of social norms and cultural values. World Dev 107:125–137

Nie P, Zhao K, Ma D, Liu H, Amin S, Yasin I (2024) Global climate change, mental health, and socio-economic stressors: toward sustainable interventions across regions. Sustainability 16(19):8693

O'Neill L, Fraser T, Kitchenham A, McDonald V (2018) Hidden burdens: a review of intergenerational, historical and complex trauma, implications for indigenous families. J Child Adolesc Trauma 11:173–186

Pandit B, Evans M, Gill N (2024) Fiji: Mental health care and law in an Island Nation. In: Mental health and human rights: the challenges of the United Nations convention on the rights of persons with disabilities to mental health care. Cham: Springer International Publishing, pp 115–120

Perry BD, Pollard R (1998) Homeostasis, stress, trauma, and adaptation: a neurodevelopmental view of childhood trauma. Child Adolesc Psychiatr Clin 7(1):33–51

Pörtner HO, Roberts DC, Poloczanska ES, Mintenbeck K, Tignor M, Alegría A, Craig M, Langsdorf S, Löschke S, Möller V, Okem A (2022). IPCC, 2022: summary for policymakers

Radinibaravi (2022) Hospital female ward refurbished, The Fiji Times, 9 September 2022

Rajendran RK, Binu S (2024) Navigating the intersection of global climate change and mental health. In: The climate change crisis and its impact on mental health. IGI Global, pp 120–131

Roe EA, Khov E, Kim H, Demelo B, Zavitz S, Cao-Lei L (2024) The impact of climate change-related disasters on mental health and epigenetics: a narrative review. Front Epigenetics Epigenomics 2:1464276

Sands B (2021) Climate change displacement in Fiji: Legal landscapes in the face of rising tides. J S Pac Law 22:140–163

Sattler DN, Whippy A, Graham JM, Johnson J (2018) A psychological model of climate change adaptation: Influence of resource loss, posttraumatic growth, norms, and risk perception following Cyclone Winston in Fiji. In: Climate change impacts and adaptation strategies for coastal communities, pp 427–443. https://doi.org/10.1007/978-3-319-70703-7_22

Shiiba N, Singh P, Charan D, Raj K, Stuart J, Pratap A, Maekawa M (2023) Climate change and coastal resiliency of Suva, Fiji: a holistic approach for measuring climate risk using the climate and ocean risk vulnerability index (CORVI). Mitig Adapt Strat Glob Change 28(2):9

Syngellakis K, Anastasia O, Schulz A, Sagar A, Kang JS, Johnston P, Raturi A, Prasad R, Wairui M, Mani F (2018) Fiji low emission development strategy 2018–2050. Ministry of economy, Government of the republic of Fiji: Fiji

Trenberth KE, Hoar TJ (1997) El Niño and climate change. Geophys Res Lett 24(23):3057–3060

Trombley J, Chalupka S, Anderko L (2017) Climate change and mental health. AJN Am J Nurs 117(4):44–52

UNDP (2024) The climate crisis is a health crisis—here's why. UNDP climate promise. https://climatepromise.undp.org/news-and-stories/climate-crisis-health-crisis-heres-why. Accessed 4 Dec 2024

Veitata SN (2023) A study of the potential for post-disaster resilience in indigenous Fijian communities

Wainberg ML, Scorza P, Shultz JM, Helpman L, Mootz JJ, Johnson KA, Neria Y, Bradford JM, Oquendo MA, Arbuckle MR (2017) Challenges and opportunities in global mental health: a research-to-practice perspective. Curr Psychiatry Rep 19:1–10

Weatherdon LV, Magnan AK, Rogers AD, Sumaila UR, Cheung WW (2016) Observed and projected impacts of climate change on marine fisheries, aquaculture, coastal tourism, and human health: an update. Front Mar Sci 3:48

WHO (2018) Mental Health ATLAS 2017. https://www.who.int/publications/i/item/9789241514019. Accessed 3 Dec 2024

World Health Organization (2022) Mental health and climate change: policy brief. https://www.who.int/publications/i/item/9789240045125. Accessed 4 Dec 2024

World Health Organization (2022) Why mental health is a priority for action on climate change. World Health Organization. https://www.who.int/news/item/03-06-2022-why-mental-health-is-a-priority-for-action-on-climate-change. Accessed 4 Dec 2024
World Innovation Summit for Health (WISH) (2022) Our children face “pretraumatic stress” from worries about climate change. WISH. https://wish.org.qa/our-children-face-pretraumatic-stress-from-worries-about-climate-change/. Accessed 20 Dec 2024
Yee M, McNamara KE, Piggott-McKellar AE, McMichael C (2022) The role of Vanua in climate-related voluntary immobility in Fiji. Front Clim 4:1034765

Chapter 8
Impacts of Climate Change on Nutritional Health and Food Insecurity in South and Southeast Asia

Nitish Karn, Atin Kumar, Sharad Sachan, and Mohamed A. E. AbdelRahman

Abstract Climate change has emerged as an alarming global crisis with consequences for both natural ecosystems and human societies. This chapter discusses the effect of climate change on food security and nutritional health, specifically in South and Southeast Asia, and the fact that rising temperatures and precipitation are some of the worst impacts on the environment. The consequences of such climate change are adverse to crop yields, reducing supplies of nutrient-dense foods, and widening existing inequalities, which disproportionately affect low-income and vulnerable populations. The aim is to study the adaptation in farming using climate-smart agriculture, biofortification, and precision farming to increase the resilience and nutritional quality of food systems. The chapter (methodologically) synthesizes recent research findings, socioeconomic analysis, and policy framework of the link between climate variability, food access, and nutritional outcomes. Results also reveal that climate change decreases dietary diversity, and micronutrient deficiencies and increases food insecurity among the most marginalized. The contributions of this chapter are that they conclude for climatic impacts to be mitigated, innovative, sustainable agricultural practices should be adopted that are integrated with indigenous knowledge, and also that there is a need for international cooperation.

N. Karn · A. Kumar (✉)
School of Agriculture, Uttaranchal University, Dehradun 248007, Uttarakhand, India
e-mail: atinchaudhary0019@gmail.com

N. Karn
e-mail: nitishkarnak47@gmail.com

S. Sachan
Department of Agricultural Economics and Extension, School of Agriculture, Lovely Professional University, Phagwara 144411, Punjab, India
e-mail: sharad.19461@lpu.co.in

M. A. E. AbdelRahman
Division of Environmental Studies and Land Use, National Authority for Remote Sensing and Space Sciences (NARSS), Cairo 1564, Egypt
e-mail: maekaoud@yahoo.com

P. Ghosh et al. (eds.), *Climate Crises and Public Health Challenges*, Disaster Risk Reduction, https://doi.org/10.1007/978-981-95-9709-3_8

Keywords Climate change · Food insecurity · Nutritional health · South East Asia

8.1 Introduction

Climate change is best described as a shift or variability of the measures of climate features to a new spatial/temporal distribution. This is evident from a change in the distribution of statistical properties of the climate system around the mean over roughly 30 years. Climate change includes fluctuations in atmospheric carbon dioxide levels, changes in world temperature and rainfall, and impacts on the levels of sea and sea water salts, crop, soil, nitrogen, plant species, and diseases affecting crops. The main contributory factor to climate change is the emission of greenhouse gases, and the climatic factors affected include temperature, relative humidity, and rainfall, and ultraviolet thereby causing variation in climate. The IPCC and other authors employed global climate models to estimate average global warming of 1.5–5.88 °C and an average of global precipitation of 5–15% by the end of the century. Weather hazards are the abrupt changes in weather variables, and the magnitude of these weather variables causes geographical disasters that consist of the unexpected and localized occurrences of extreme hydrological, geophysical, meteorological, or climatological character (NASA 2023).

The average global temperature has risen by 1.1 degrees Celsius above the average temperature that prevailed during the pre-industrial period. This warming is associated with enhanced concentration of carbon dioxide (CO_2), methane (CH_4%), and nitrous oxide (N_2O%) in the atmosphere. The change makes the greenhouse atmosphere trap more heat, which contributes to global warming. For instance, it is important to note that the period of 2010 was identified to have been the warmest decade, and the years 2020 together with the year 2016 were noted to have been the warmest years experienced on the globe. (NOAA 2023). An increase in average temperature and patterns of rainfall increases infection rates of diseases that are spread by vectors, such as malaria and dengue fever. Recognizing heat stress as a major public health challenge, especially in densely populated regions of the world (WHO 2023). Air pollution caused by climatic factors also worsens respiratory and cardiovascular diseases. In the same way, cryospheric processes of the Polar ice sheets, both in Antarctica and Greenland, show vast increases in the melting rates and hence sea level. Global sea level has risen by 15–25 cm during the twentieth century since 1900, with higher rates observed for the last few decades (IPCC 2019). Coastal areas face higher risks of flooding, land erosion, and habitat destruction, directly threatening over 680 million people living in low-lying coastal zones (UNEP 2022). Also, the frequency and intensity of extreme weather events, such as hurricanes, droughts, heat waves, and floods, have increased due to climate change. For instance, the 2023 heatwave in Europe broke several records, affecting human health and agricultural productivity (WMO 2023). Similarly, prolonged droughts in Sub-Saharan Africa have exacerbated food insecurity for millions (FAO 2022). Climate change affects agriculture through changes in growth duration, reduced availability of water, and increases in pests and diseases. The future estimated crop output in the world may be reduced by 10–25% by the year 2050 if things remain like this. Wheat, rice, and maize–based crops all around the world can be severely affected by shifts in temperature and rainfall intensity (World Bank 2022).

In essence, health and food supply as dimensions of the question of sustainable development are interrelated concepts. Nutritional health deals with the status of nutrients essential to health and the utilization of these nutrients for health enhancement, whereas food security involves adequacy, accessibility, and stability in the provision of safe and nutritious food for everyone. Climate change, population increase, and increased socioeconomic differences have exerted increased pressure on the food systems that compromise both nutrition and food security globally (FAO 2006).

In one way or the other, climate change impacts food production, quantity, and quality. In light of this, rising temperatures, changing rainfall patterns, and frequent occurrences of extreme weather conditions degrade agricultural production. For instance, predicted yields include: Losses of population for wheat and maize crops by 6 and 7.4%, respectively, for every degree centigrade of global temperature increases (Challinor et al. 2014). Reduced crop diversity further limits dietary diversity, exacerbating malnutrition risks (Myers et al. 2017). Similarly, the interruptions of food supply chains affected by extreme climate impacts raise food prices and put a lot of pressure on the most affected groups of people. It has been estimated that climate-related shocks raise the average food prices by 10–20%, imposing all the more imposing impacts on the poorer sections of the population (Hertel et al. 2010). This economic impact reduces the ability to feed on healthy foods, which further worsens the occurrence of malnutrition. Volatility in climate conditions reduces the reliability of food chains, leading to regular deficits and food market instability. Such a disaster in the Russian Federation, for example, the unprecedented heat waves in the summer of 2010 affected the yield and disrupted many wheat supplies, which consequently forced high prices, affecting food security in countries that rely on imports. Such disruptions increase food insecurity risk (Lagi et al. 2011).

Although the link between climate and nutrition is growing, many research gaps regarding climate and nutrition remain unexplored. A major area is the understanding of the nutritional effects of elevated levels of atmospheric CO_2. The results of studies have shown that the quality of staple crops such as rice and wheat decreased with increasing CO_2, but the extent to which this occurs is not known for all food groups and agroecosystems. For example, it remains to be agreed what relationships exist between increased CO_2 and soil nutrient availability, and how this affects the plant nutrient uptake (Smith et al. 2021). Regarding dietary diversity, another missed aspect is related to how climate change impacts this. Despite the growing evidence of the effects of climate variability on crop yields and quality, there is scant research on the impact of climate variability on fruit, vegetable, and legume availability, affordability, and accessibility, all sources of nutrient-dense foods. Additionally, little is known about the socio-economic mediators of change in food consumption patterns at different climatic scenarios (Fanzo et al. 2020a, b). More importantly, the vast majority of current studies concern global or regional trends, at the expense of population-specific nutritional impact. However, vulnerable groups, such as women, children, and poor communities in developing countries, are more susceptible to nutritional risks associated with climate variability, but little has been devoted to tailored studies on these demographics (Rosenzweig et al. 2020). A broader, open gap is in the integration of climate and nutrition policies. To the uncertainty of how far climate change mitigation and adaptation strategies are consistent with or not with nutrition-sensitive targets. As climate response mechanisms start being embedded with nutrition outcomes, further research will be necessary to help elaborate policy frameworks (Haddad et al. 2018). Finally, technological advances like biofortification and nanotechnology can be used to increase food nutrients in the climate stress, however, there is little evidence on their sustainability, effectiveness, and adaptability across different socio-economic and environmental contexts in the long term. These research deficits underscore the need for interdisciplinary approaches that draw on the latest climate science, nutrition, and agronomic research as well as new policies to help produce resilient and equitable food systems in light of climate change. This study seeks to explore how climate-smart agricultural strategies can be effectively implemented to enhance food security and nutritional outcomes among vulnerable populations affected by climate change in South and Southeast Asia.

The primary objectives of this study are to assess the impact of climate change on food insecurity and nutritional health in South and Southeast Asia, and analyze the effectiveness of climate-smart agricultural practices in enhancing resilience and improving nutritional quality. It also aims to identify policy and technological opportunities for strengthening adaptation strategies among vulnerable populations.

8.2 Methodology

A literature review was conducted to learn from existing knowledge of climate change, food security, and nutritional health at the intersection with South and Southeast Asian contexts, using an overall comprehensive scope. Therefore, the major academic databases and search engines like PubMed, Scopus, Web of Science, Google Scholar, ScienceDirect, and SpringerLink were used for literature search. Other than peer-reviewed journals, grey literature and official reports from reputable bodies such as FAO, WHO, IPCC, World Bank, and ICAR were also examined for a more holistic point of view.

The keywords and search terms used to identify relevant studies included combinations such as "climate change" and "food security," "nutritional health" or "micronutrient deficiency," "climate-smart agriculture" or "biofortification" or "precision farming," along with geographical identifiers like "South Asia" or "Southeast Asia," and terms related to crop nutrition such as "elevated CO_2" and "crop nutrition." Search filters were applied to limit results to articles published in English between 2005 and 2024. For this narrative review the inclusion criteria encompassed peer-reviewed articles, institutional reports, or policy briefs focusing on climate-nutrition linkages; studies conducted in or relevant to South and Southeast Asia; and the articles discussing climate adaptation strategies in agriculture with nutritional outcomes. The exclusion criteria comprised non-English publications; studies focusing exclusively on high-income countries; and editorials or opinion pieces lacking empirical evidence.

8.3 Result and Discussion

8.3.1 Impact of Climate Change on Nutritional Health and Food Insecurity

There is numerous case study that explains climate change effects on nutritional health and food insecurity under changing climatic conditions, and they are following:

8.3.1.1 Regional Impacts of Climate Change on Food Security

Climate change affects food security in different areas, and especially in the Limpopo Province of South Africa. Kom (2022) notes that due to rain-fed farming and low adaptive capacity, smallholder farmers are highly susceptible in this area. The study shows that 56% of such farmers have suffered food insecurity lasting for more extended period attributable to climate shocks, including high temperatures and reduced rainfall. These climatic changes have affected crop production and, as a result resulted in less food production and income for farmers. As a result, the farmers have developed other income sources, but these practices have not been effective in eliminating food insecurity. It therefore becomes obvious that there is a need for a sustainable livelihood framework and policies, as well as models that will ensure high production yields in crops to counter any impacts of climate change on food security in the region.

Likewise, in Bangladesh, climate change has impacted food security, especially in the coastal areas, people as Kalapara, Patuakhali, and so on. Exploration carried out by Rahman 2021. Further shows that effects of climate change, such as cyclones and high sea levels, have led to a rise in soil salinity and a decrease in the amount of fertile land, hence low crop production. On the aspect of climate-induced impacts, the study realized that climate change impacts have put pressure on ponds and rainwater, which is used for cooking and drinking, hence a concern for freshwater supplies. Also, livelihood dependencies and income levels have been affected, and food insecurity prevails among the people of the ravaged region. The study recommends the need for coping strategies to negate the impacts of climate change on food security in Bangladesh: This includes the development of saline-tolerant crop varieties and a change in livelihood practices.

8.3.1.2 India's Climate-Resilient Agriculture Program

As a country that is experiencing changes in climate variability, India launched the NICRA, or the National Innovations in Climate Resilient Agriculture, in 2011. The program was based on four pillars of capacity building for the identification of climate resilient crop varieties, enhanced water use efficiency, and integrated farming systems. For instance, one released drought-tolerant rice varieties that would ensure reasonable yields in areas with low water supply. To avoid the impacts of climate change on farming, farmers were empowered the use resource-saving technologies, including zero tillage and effective water use. Farmers in the pilot regions of the program said they experienced much better yields of crops of 20–30% better yields, and better hardiness to climatic shocks. These ventures point to the need to cover food security through enhancing the applicability of scientifically-informed innovations at the place level in the context of climate change (ICAR 2020).

8.3.1.3 Climate-Smart Agriculture (CSA) in Vietnam

This paper assessed how CSA has helped Vietnam to implement and fight food insecurity amidst climate change challenges, even though it has grown in the recent past. Vietnam's agricultural sector is focused on the rice industry, and poor farmers that our country due to influence of sea level rising and extreme weather conditions, resulting in a loss of the ability to produce rice with high productivity, and poverty in rural areas increases. To overcome these challenges, the government, in partnership with international organizations such as the Food and Agriculture Organization (FAO), popularized CSA practices. Others were SRI—the System of Rice Intensification for increased and improved water use efficiency and the planting of salt-tolerant rice in coastal areas. Further, the practice of integrating the company farming systems, such as fish farming, animal farming, and crop farming, enhanced the overall shock resilience to adverse climate conditions. The described CSA practices have resulted in increased rice grain yields by 20–30%, mostly reduced water consumption by 30%, most importantly reduced methane emissions by 25%, and all this was accompanied by improved farmers' income. This initiative reflects how sustainable practices are possible to implement to eliminate food insecurity while meeting the climate change crisis (Nguyen et al. 2020).

8.3.1.4 Vulnerable Populations and Food Access Disparities

In terms of people, the affordability of healthy food is another process incorporated into vulnerability concerning climate change impacts on agriculture. The vulnerable households at the baseline include the low-income earners, the smallholder farmers, the poor and needy who have no means of income, hence less access to food, and the increase in poverty due to COVID-19 restrictions is worsening the problem (FAO 2021a, b). Climate change exacerbates these disparities by reducing harvests on farms and increasing the costs of food, nutrients are thus a preserve of the rich. For example, elevated temperatures and fluctuating rainfall disrupt food production in housing and food-providing families, the neoclassical poor, those leading a subsistence livelihood, as well as exacerbate food insecurity (Castañeda-Navarrete et al. 2020). Because economic privileges consolidate rather than eliminate these differences, the inequalities enhance these disparities. Through widespread technology, developed households and regions can work around climate problems in agriculture, such as flood or drought, by using technologies like irrigation or better seeds, while "the poor cannot" (Hertel et al. 2021). Additionally, gender plays a significant role in food access disparities, as women in many rural areas are often responsible for food production and preparation but have limited control over resources, land ownership, and decision-making (Quisumbing et al. 2020). These problems are also reinforced by geographic discrepancies. In areas that are already prone to climate shocks, that is, sub-Saharan Africa and South Asia, people experience disruptions to the supply of food, causing localized shortages and reliance on imports. This leads to a use of less healthy and nutrient-rich processed foods to feed the bodies, a situation which poses a threat to the health of vulnerable persons (Pingali et al. 2021). These issues call for appropriate socioeconomic policies that focus on funding for rural development, technologies, equipment, and structures to farmer, and proactively assisting suppressed social segments. As for better target groups' conditions, there are conditional cash transfers, crop insurance for farmers, community-based resilience programs and others.

8.3.1.5 Economic Challenges in Ensuring Food Security

The availability and access dimensions of food security are closely related to the economic issues that occur as regards food security. The affordability of food can be determined by the income levels, food prices, and formation of economic policies. Shortage in production due to climate change factors, such as high cost of irrigation water, fertilizers, and pest control, leads to high food prices that are relatively high for the poor. In almost all the developing countries, a large percentage of income generated at the household level is spent on purchasing foods and other necessities, hence any slight adjustment of the price of foods puts the affected population, especially the households, at a high risk of being food insecure and experiencing malnutrition (FAO 2020; HLPE 2021).

Besides, smallholder farmers and other rural dwellers seem to be locked out from accessing credit facilities, markets for their produce, and the necessary infrastructure required to adopt climate-smart agriculture practices and machinery. This inequality not only will slow down economic growth but also will make those groups more susceptible to economic fluctuations, for example, related to extreme weather events or disruptions of the global value chain (World Bank 2020).

Weather-induced fluctuations accompanied by export bans and high global food prices unstable an import-reliant country of food can be significantly food-reliant. Also, economic crises, social aspects such as roles to be played by female farmers, and equality to overcome gender barriers that hinder the access to resources and decision making of women who are farmers, were delayed in achieving sustainable food security (FAO 2021a, b; IFPRI 2019).

To address these challenges, targeted interventions are needed to support smallholder farmers, improve access to financial and technological resources, and strengthen social protection systems. By addressing the economic barriers to food security, policymakers can ensure more resilient and equitable food systems that meet the needs of all populations.

8.3.2 Strategies for Mitigating the Impact of Climate Change

8.3.2.1 Role of Sustainable Farming Practices in Food Security

Sustainable Agriculture is an effective measure towards attaining food security in the ways it increases the adaptability of the agricultural sector to climate variability and change, lessens impacts on the natural resources base and in the ways, it fosters long-term yield improvements. These practices include the minimal exploitation of natural resources, preservation of species, and minimization of injurious inputs that in one way or the other assist in protecting both the frequency and volume of food output. Some of the best practices include Crop rotation and agroforestry, organic farming, wildlife and weed control through natural eco-friendly methods among others since they enhance soil fertility by reducing the use of chemical fertilizers, and increase on natural enemies for pests and other diseases that affect crops, among others, to enhance healthy ecosystems (Altieri 2020). Conservation tillage, which reduces soil erosion and enhances soil moisture retention, helps maintain the fertility of agricultural land, thus increasing the ability to produce food even under challenging conditions like drought (Lal 2021). Integrated pest management (IPM) also contributes by reducing the reliance on chemical pesticides, which can harm both the environment and human health (Zhao et al. 2021). Furthermore, the use of drought-resistant crop varieties and water-efficient irrigation systems can improve crop yields in regions where water scarcity is becoming a critical issue due to climate change (Chaves et al. 2020). These practices are also useful in mitigating, for instance impacts of climate variation and thus assist in enhancing long-term food security since the vulnerability of the agriculture-based food systems is minimized. Investigations have also revealed that sustainable agriculture means improved yields of 15–20% in some locations, better soil health, production of greenhouse gases, besides greater control of carbon addition in the soil. Implementation of these sustainable practices, hence, can go a long way in attaining food security goals at the global level, especially with the emergence of what with increasing population pressure as well as other environmental challenges (Beddington et al. 2020).

8.3.2.2 Climate-Smart Agriculture and Its Implementation

Climate-smart agriculture (CSA) is an approach that helps to address the problems of climate change while feeding the nation. It is designed to raise agricultural productivity and output, prepare for the worst effects of climate change, and reduce GHG emissions as far as possible. CSA includes management practices and measures combined with innovations that increase farm systems' adaptive ability to sustain and make the best use of the resources. One key aspect of CSA is the promotion of resilient crop varieties, such as drought-tolerant and heat-resistant cultivars, which can sustain yields under adverse climatic conditions (Lipper et al. 2021). The conservation agriculture techniques, such as minimum tillage, crop rotation, and cover crops, make the soil healthier and more water reserves, and are less exposed to climatic risks. Further, technologies such as sensors, UAVs, and satellite farming allow the farmers to regulate water and fertilization, hence reducing wastage while increasing returns (Thornton et al. 2022). The implementation of agroforestry systems, which integrate trees with crops or livestock, provides multiple benefits, such as carbon sequestration, improved biodiversity, and enhanced soil fertility, contributing to long-term sustainability (Mbow et al. 2019). Moreover, appropriate sustainable water control measures such as drip irrigation as well as rainwater collection, are vital to address the water deficit problem, especially in the arid and semi-arid areas. It is apparent from the study that policy support is important in the promotion of CSA in a large way. Governments and organizations should ensure farmers get access to capital, technical skills, and reliable climate-resilient technologies, tools, and infrastructure for the adoption of CSA practices, especially by smallholder farmers. For example, in sub-Saharan Africa, concepts such as climate-smart villages have shown that more can be achieved in terms of production and community capacity (Aggarwal et al. 2020).

8.3.2.3 Resilient Crop Varieties and Genetic Improvements

Durable seed varieties and breeding advances are major findings whose implications are critical to underpinning food security and nutrition across the world, especially about climate variability and population explosion. These developments allow crops to be resilient to biotic and abiotic factors such as drought, salinity, heat, and diseases, considered main challenges as they arise from the changing weather patterns. For instance, through genetic modification, we can grow drought-resistant maize or salt-susceptible rice among other crops in regions that could not support normal crop production (Lobell et al. 2020a, b). Genetic engineering and marker-assisted breeding have brought the development of improved varieties with desired quality characteristics concerning pest and disease resistance and nutrient requirements for growth in the field, lowering the dependence on chemical fertilizers and pesticides for sustainable agriculture (Fischer et al. 2021). Additionally, biofortification through genetic improvement has increased the nutritional value of staple crops like wheat, rice, and maize, addressing malnutrition and improving public health in vulnerable populations (Saltzman et al. 2021). It also plays a key role in arriving at sustainable solutions for agriculture and hence reducing agriculture's environmental impact through sufficient water and input use. In using resilient crops in the agricultural systems, countries achieve the use of reducing the risks occasioned by climate variation and the establishment of food security for the countries in the future.

8.3.2.4 Impact of Climate Change on Nutritional Quality

Both macro and micronutrient concentrations in crops are influenced by climate change, especially the increase in the amount of atmospheric carbon dioxide (CO_2), and pace positioning it as a worldwide nutritional issue. High levels of CO_2 have been proven to raise carbohydrate amounts in crops, starch and sugar levels are boosted, while the concentrations of protein, an essential macronutrient, go down. This effect is felt most in food commodities, where crops such as wheat, rice, and maize are lethal for billions of people as a source of protein (Myers et al. 2014). Similarly, the levels of micronutrients, including iron, zinc, and magnesium, tend to decline under elevated CO_2 conditions, exacerbating the risk of nutrient deficiencies in populations dependent on these crops for their dietary needs (Loladze 2014). Heat stress, in addition to changes in precipitation patterns, affects nutrient uptake by affecting the availability of nutrients in the soil as well as the metabolic activity of the plant. For example, a shortage of moisture decreases the efficiency of nutrient uptake by crops, while isolated downpours may wash out nutrients away from the plant roots (Gupta et al. 2021). On this account, these changes in crop nutrient composition present a major challenge to food security and public health, particularly in regions with low-income diets. Some of the measures include efforts towards developing better-yielding, nutritious crops, Human efforts on soil management, and integrated farming systems.

8.3.2.5 Implications for Dietary Diversity and Human Health

The implications of changes in food access and quality due to alterations of climatic conditions are based on the projections of yield changes and the structure of the food systems that cultivators offer. Climate change pressures that are proportional to heat, erratic rainfall, and enhanced frequency of extreme climate events decrease the production, quality, and quantity of various horticultural crop varieties, including fruits and vegetables, pulse crops, tree-borne nuts and seeds (Myers et al. 2017). As staple crops like wheat, rice, and maize are often prioritized for their caloric value in response to declining arable land, the production of nutrient-rich crops is marginalized, leading to reduced dietary diversity (Fanzo et al. 2020a, b). It can lead to malnutrition through micronutrient deficiency, also known as "hidden hunger," which is a widespread problem in low and middle-income countries affecting the most vulnerable population. In addition, on aspects of the nutrient density of crops, which include protein and minerals, restrictive factors like high levels of CO_2 content threaten to complicate the task of feeding the populace. Loss of dietary diversity is related to the rising trend in non-communicable diseases such as obesity and diabetes because people depend on high-energy, dense foods that are nutrient-poor foods. Meeting these challenges calls for the adoption of climate-responsive agricultural practices and sound policies and practices that encourage the production and consumption of diverse nutrient-dense foods to support human health and wellbeing in the face of climate change (Smith et al. 2021).

8.3.3 Emerging Solutions and Innovations for Strengthening Adaptation Strategies Among Vulnerable Populations

The different emerging solutions and innovations to overcome the ill effects due to the impact of climate change are explained in the points below. They are:

8.3.3.1 Role of Biotechnology in Improving Crop Resilience and Nutrition

This scientific discipline contributes significantly to improving the performance of plant genetic resources, as well as the nutritional value of crops in response to climate change, population, and food security imperatives. Using biotechnologies thus helps in their liberation from difficulties related to climate conditions, including drought, salinity, and extreme temperatures, as well as pests. For instance, the integration of drought-tolerant receptors into Maize and wheat has enhanced productivity under dry conditions as it delivers standard yields in dry areas (Lobell et al. 2020a, b). To the same extent, biotechnology has produced salt-tolerant rice, which can be grown in saline soils attributed to increasing sea level and water-induced salinity (Qadir et al. 2021). With biotechnology, resilience is also boosted, and through biofortification, the nutritional quality of crops is also increased. For example, golden rice, in which the genes were altered to incite production of beta-carotene, contains vitamin A in areas where rice is eaten. In the same way, pro-vitamin A or iron and zinc fortified wheat and legumes fight hidden hunger or micronutrient deficiencies in resource-limited environments. Such methods of gene engineering go further than merely helping to decrease dependence on such other biological additions to diet as supplements, but also assist in attaining better standards of common wellness of the population (Tang et al. 2017; Saltzman et al. 2021).

In addition, biotechnology gives the ability to enhance the protein content, composition, and healthier fatty acids, and thus reduces the incidence of non-communicable diseases such as cardiovascular diseases. In particular, biotechnology does not merely improve food security through the introduction of resilient and nutrient-dense crops into farming systems but also optimizes sustainable farm management in conditions of climate change and population pressures. Further studies that address biotechnological innovations and related policies are crucial for maximizing the utilization of available potential in enhancing a strong window of resilience to future shocks and ensuring adequate capability to feed the growing population (Khosla 2020).

8.3.3.2 Adoption of Agroecological Approaches for Sustainability

Using agroecological techniques is gradually receiving recognition as one of the successes and a major goal of sustainable agriculture since it blends ecological principles in the development of agriculturally productive, stable, and environmentally friendly systems. Agroecology embraces biological diversification, including practices for improving soil structures, fertility, and offering various ecosystem services that reject monoculture options in favor of multiple cropping, intercropping, and trees in crop production. Practices like crop rotation, cover cropping, and minimum tillage give agroecology the characteristics associated with improved soil physical characteristics as well as climate change mitigation and adaptation (Tittonell et al. 2020). It also reduces dependency on synthetic inputs, such as fertilizers and pesticides, which can harm ecosystems and human health (Altieri and Nicholls 2020). Besides, the strategies of agroecological systems also depend on local knowledge and practices of the farmer's innovative strength to change and adapt to the existing environment and social circumstances. For example, livestock operation with cropping increases nutrient turnover and yield, and agroforestry systems increase microclimatic conditions and various wealth. These practices help to increase the levels of adaptation to climate change volatility and food security because they guarantee homestead food security and range diversification (Wezel et al. 2020; Pretty et al. 2018). Studies have shown that agroecological approaches increase long-term productivity while preserving natural resources, making them a viable pathway to sustainable agriculture (Francis et al. 2021).

The adoption of agroecological practices aligns with global sustainability goals, addressing environmental challenges such as soil degradation, water scarcity, and biodiversity loss while promoting social equity and economic viability in farming systems.

8.3.3.3 Use of Nanotechnology in Enhancing Nutrient Availability

Nanotechnology has become a subject matter in agriculture as a tool for improving nutrient uptake and utilization. Nanotechnology preserves nutrients from leaching by directly applying nanoparticles to the plant tissues; nutrients are not lost through volatilization and fixation, thus enhancing the use of fertilizers. These bio-chemical entities, which are controlled-release fertilizers at the nano scale, feed crops and plants nutrients such as nitrogen, phosphorus, and potassium in a flow that complements the crop's growth cycle (Chen et al. 2021), (Subramanian et al. 2015). For example, nano-phosphates can address the issue of low phosphorus availability in soils by improving solubility and plant uptake. Similarly, nano-encapsulation of nitrogen fertilizers reduces volatilization, ensuring higher nitrogen use efficiency and mitigating environmental pollution (Ramesha et al. 2020). It has also been used to increase the bioavailability of micronutrients, including zinc and iron, for the proper growth of plants and human health. For example, nano-based zinc fertilizers enhance the relative availability and absorption by the plant of the zinc content, hence increasing the yield and nutritional value of crops. Similarly, soil conditioners containing nanomaterials improve physical properties and nutrient retention of soil and sustain agriculture (Kah et al. 2018).

Another such revolutionary application is the incorporation of micro sensors for detecting the soil nutrient content and crop status continuously, so that nutrients can be added only when required, and there is no wastage. These innovations not only increase the aspects of nutrient use efficiency but also decrease the impacts of agriculture on the environment. Further research and development in this field can set nanotechnology within farming systems design and bring timely nutrient provision to fruition, at the same time maintaining food security and environmental conservation (Dasgupta et al. 2021).

8.3.4 Future Strategies for Integrating Nutrition-Sensitive Agriculture into Policy Frameworks

There are a few things for which detailed work needs to be done to overcome the effects of climate change shortly. Some of them are:

8.3.4.1 Climate-Resilient Crops

Planting and trialing climate change-tolerant foods that withstand drought, flood, and heat is essential if food security is to be achieved. These crops should ideally have improved nutritional contents that would benefit the health of people. Genetic manipulations and methods such as CRISPR will make it possible to grow crops that are both productive and withstand many hard conditions and are nutritious (FAO 2023; Tadele 2019).

8.3.4.2 Sustainable Agricultural Practices

Climate Change, which is a rising worry on the environment, will be managed through sustainable farming practices such as agroecology, conservation tillage, and integrated pest management (IPM). The practices underline sustainable use of the natural resource by supporting and maintaining diversity among biotic components, soil conservation, and water usage, thus leading to a sustainable food system (Godfray and Garnett 2014).

8.3.4.3 Diversification of Food Systems

The expansion of dietary and production diversification, particularly in the targeting of underutilized fish stocks, local and indigenous food commodities, is crucial towards improving climate risk stability. This approach can also go a long way in raising nutritional status by eating diverse, micronutrient-dense foods (FAO 2021a, b).

8.3.4.4 Nutritional Education and Awareness

As Malnutrition underscores, consumer awareness on the nutritional value of climate-resilient seeds and locally available foods will go a long way in supporting food security, translating to improved nutrition outcomes of communities. By endorsing the consumption of different foods at different seasons and foods produced in different regions, the constantly advertised nutrient-poor, climate-sensitive crops could be minimized (Hoddinott and Haddad 2022).

8.3.4.5 Integrated Climate-Health Policies

Application of climate change mitigation polices together with those issues such as public health and food security, is important. Such policies should include not only ecological farming for sustainable agriculture but also strategies for food and nutrition security and health. Climate-related risks to food security will have to be managed by local and global governance structures with a social justice concern (IPCC 2022; Headey and Ecker 2013).

8.3.4.6 Technology and Data-Driven Solutions

Going forward, there will be usage in precision agriculture, satellite, and AI solutions to forecast climate impacts on food production. This rationale is important in ensuring that sufficient food is produced to feed the population, as well as combat malnutrition through using data in decision-making. This will enhance input use efficiencies, minimize wastage, and subsequently increase yield for the farmers (Rosegrant and Cline 2020).

8.4 Conclusion

This concludes the chapter in a very important step further to understanding the complex interplay between climate change, agriculture, and nutrition in vulnerable regions like South and Southeast Asia. This calls attention to the importance of incorporating, among others, integrated, sustainable, and innovative agricultural practices, including climate-smart agriculture, biofortification, and precision agriculture, which should be able to increase resilience to climate variability, together with better quality of the food systems. The chapter stresses that indigenous knowledge should be integrated in climate change responses to food security and public health, and proposes a complete framework for addressing some of the most pressing issues raised by climate change to food security and public health by bringing in the international cooperation aspect. The present study contributes to existing knowledge by pointing out that interrelations between climate adaptation strategies and nutritional outcomes need to be explicitly made, which requires multi-disciplinary mechanisms that address the contexts of socio-economic factors, technological breakthroughs, and policies. This shows that these measures help to alleviate the impacts on vulnerable populations from both climate stressors and health and nutrition disparities that are exacerbated by climate stressors. The implications of this research are very wide. Through offering valuable guidance for policymakers, researchers as well and practitioners that are working to create durable food systems in an atmosphere of ongoing climate change, it is suitable for its target audience. The study advocates for solutions that are not neglect to be scalable and culturally sensitive; as a result, policies would be designed to advocate for climate adaptation, social equity, and nutritional security. It further recommends the key roles played by innovation, education, and capacity building at the local and global levels to successfully execute adaptive strategies. However, the insights from this chapter confirm the significance of proactive, holistic strategies to defend food security and nutritional health. Such strategies will need to play their part in making the delivery of sustainable development, reducing inequalities, and promoting health and well-being across vulnerable populations in the world, as climate change progresses faster and faster.

8.5 Potential Conflicts of Interest

The authors declare that there is no conflict of interest regarding the publication of this chapter.

8.6 Informed Consent

Not applicable, as this study does not involve human participants or animals requiring informed consent.

8.7 Data Availability Statement

No new data were generated or analyzed in support of this research. All information is derived from previously published sources, which are cited accordingly in the chapter.

Acknowledgements The authors would like to express their sincere gratitude to the School of Agriculture, Uttaranchal University, for providing the necessary resources and support for this work. Special thanks are extended to colleagues and peers for their constructive feedback and encouragement throughout the development of this chapter.

Authors Contribution NK contributed to conceptualization, literature review, writing the original draft. AK supported for supervision, methodology, writing review and editing; SS contributed to data collection, reference management, literature analysis, visualization. MAEA supported critical review, technical inputs, and final approval of manuscript.

Funding This research did not receive any specific grant from funding agencies in the public, commercial, or not-for-profit sectors.

References

Aggarwal PK, Gupta R, Singh A (2020) Scaling out climate-smart villages for enhancing resilience and food security in sub-Saharan Africa. Agric Syst 179:102753. https://doi.org/10.1016/j.agsy.2019.102753

Altieri MA (2020) Agroecology and the search for a sustainable food system. J Sustain Agric 44(3):227–239. https://doi.org/10.1080/10440046.2020.1710564

Altieri MA, Nicholls CI (2020) Agroecology: Principles for sustainable agriculture. Springer. https://doi.org/10.1007/978-3-030-21280-2

Beddington J, Asaduzzaman M, Fernandes S (2020) Food security and sustainable farming: the role of ecological practices in enhancing food production. Food Secur 12(3):553–561. https://doi.org/10.1007/s12571-020-01022-7

Castañeda-Navarrete J, Manoli G, Heller J (2020) Climate shocks and rural household welfare: evidence from developing countries. World Dev 127:104748. https://doi.org/10.1016/j.worlddev.2019.104748

Challinor AJ, Watson J, Kilsby CG, Downy MR (2014) A meta-analysis of crop yield under climate change and adaptation. Nat Clim Chang 4:287–291. https://doi.org/10.1038/nclimate2153

Chaves MM, Oliveira MM, Munné-Bosch S (2020) Drought resistance and water management in agriculture: strategies for food security. Environ Exp Bot 174:104048. https://doi.org/10.1016/j.envexpbot.2020.104048

Chen H, Zhang C, Wang J, Zhang X (2021) Nanotechnology in agriculture: advances in nano-fertilizers for sustainable nutrient management. Environ Sci Technol 55(4):2519–2530. https://doi.org/10.1021/acs.est.0c06745

Dasgupta N, Sinha S, Prasad K (2021) Nano-sensors for nutrient management in precision agriculture. Nano Today 38:100976. https://doi.org/10.1016/j.nantod.2020.100976

Food and Agriculture Organization (FAO) (2021) Diversification and resilience in food systems. FAO Policy Brief.

Fanzo J, Davis C, Yaktine AL (2020a) The importance of dietary diversity for human health and nutrition in a changing climate. Nutr Rev 78(1):1–8. https://doi.org/10.1093/nutrit/nuz069

Fanzo J, Davis C, McLaren R, Choufani J (2020b) Climate change, dietary diversity, and global nutrition challenges. Glob Food Sec 26:100411. https://doi.org/10.1016/j.gfs.2020.100411

FAO (2006) Food security: policy brief. FAO, Rome

FAO (2019) The state of food security and nutrition in the world 2019. Food and Agriculture Organization of the United Nations. http://www.fao.org/3/ca5162en/ca5162en.pdf

FAO (2020) The state of food security and nutrition in the world 2020. Food and Agriculture Organization of the United Nations. http://www.fao.org/publications/sofi/2020/en/

FAO (2021) The impact of trade restrictions on global food security. Food and Agriculture Organization of the United Nations. https://www.fao.org/3/cb6673en/cb6673en.pdf

FAO (2021) The state of food security and nutrition in the world 2021. Food and Agriculture Organization of the United Nations. http://www.fao.org/publications/sofi/2021/en/

FAO (2022) Climate change and food security. Food and Agriculture Organization of the United Nations

Fischer RA, Byerlee D, Edmeades GO (2021) Advances in breeding for sustainable agriculture under climate change. Field Crop Res 261:107952. https://doi.org/10.1016/j.fcr.2021.107952

Food and Agriculture Organization (FAO) (2023) Climate-resilient crops for food security. FAO Report.

Francis CA, Lieblein G, Soto G (2021) Agroecology: a global movement for food security and environmental resilience. Adv Agron 163:1–45. https://doi.org/10.1016/bs.agron.2020.12.001

Godfray HCJ, Garnett T (2014) Food security and sustainable intensification. Philos Trans R Soc B: Biol Sci 369(1639):20120273. https://doi.org/10.1098/rstb.2012.0273

Gupta N, Garg S, Singh P (2021) Impact of climate change on soil nutrient availability and its implications for crop nutrition. Agron J 113(2):951–963. https://doi.org/10.1002/agj2.20698

Haddad L, Waage J, Webb P (2018) Integrating climate adaptation and nutrition for sustainable development. Climate Dev 10(3):189–202. https://doi.org/10.1080/17565529.2017.1372269

Headey D, Ecker O (2013) The political economy of food security. Annu Rev Resour Econ 5(1):1–31. https://doi.org/10.1146/annurev-resource-091912-151818

Hertel TW, Burke MB, Lobell DB (2010) The poverty implications of climate-induced crop yield changes by 2030. Glob Environ Chang 20(4):577–585. https://doi.org/10.1016/j.gloenvcha.2010.07.001

Hertel TW, Baldos ULC, Fujimori S (2021) Economic dimensions of climate change and food security. Nat Sustain 4(5):389–397. https://doi.org/10.1038/s41893-021-00674-9

HLPE (2021) Impacts of climate change on food security and nutrition. High Level Panel of Experts on Food Security and Nutrition. http://www.fao.org/3/cb4784en/cb4784en.pdf

Hoddinott J, Haddad L (2022) The impact of nutrition interventions on food security and nutrition. World Bank Research Papers

ICAR (2020) National innovations in climate resilient agriculture: achievements and impacts. Indian Council of Agricultural Research. http://www.icar.org.in/files/icar-publications/NICRA/NICRA%20Impact%20and%20Achievements%20Report%202020.pdf

IFPRI (2019) Gender inequality, food security, and sustainable development: an intersectional analysis. International Food Policy Research Institute. https://doi.org/10.2499/9780896293487

Intergovernmental Panel on Climate Change (IPCC) (2022) Climate change and land. IPCC Special Report

IPCC (2019) Special report on the ocean and cryosphere in a changing climate. Intergovernmental Panel on Climate Change
Kah M, Hirst M, Laird M (2018) Zinc oxide nanoparticles as nano-fertilizers: Implications for soil and crop nutrition. J Agric Food Chem 66(20):5132–5143. https://doi.org/10.1021/acs.jafc.8b01956
Khosla P (2020) Advances in biotechnology for health-promoting crop profiles. Trends Food Sci Technol 102:23–34. https://doi.org/10.1016/j.tifs.2020.06.015
Kom M (2022) Climate change, food security, and farmers' livelihoods: a case study in Limpopo province, South Africa.
Lagi M, Bertrand KZ, Bar-Yam Y (2011) The food crises and political instability in North Africa and the Middle East. SSRN Electron J. https://doi.org/10.2139/ssrn.1909605
Lal R (2021) Conservation tillage and its role in sustainable food production systems. Soil Tillage Res 208:104784. https://doi.org/10.1016/j.still.2020.104784
Lipper L, McCarthy N, Zilberman D, Cavatassi R (2021) Climate-smart agriculture: building resilience to climate change. Agric Food Secur 10(1):1–12. https://doi.org/10.1186/s40066-021-00312-5
Lobell DB, Schlenker W, Costa-Roberts J (2020a) Climate-resilient crops: advances in biotechnology for stress tolerance. Nat Clim Chang 10(4):373–383. https://doi.org/10.1038/s41558-020-0703-7
Lobell DB, Schlenker W, Costa-Roberts J (2020b) Genetic improvements in crop resilience to climate stress. Nat Clim Chang 10(6):457–463. https://doi.org/10.1038/s41558-020-0836-x
Loladze I (2014) Hidden shift of the ionome of plants exposed to elevated CO_2 depletes minerals at the base of human nutrition. eLife 3:e02245. https://doi.org/10.7554/eLife.02245
Mbow C, Rosenzweig C, Barioni LG (2019) Agroforestry as a resilient strategy for climate-smart agriculture. Curr Opin Environ Sustain 41:48–54. https://doi.org/10.1016/j.cosust.2019.01.003
Myers SS, Smith MR, Guth S, Golden CD, Vaitla B, Mueller ND (2014) Increasing CO_2 threatens human nutrition. Nature 510(7503):139–142. https://doi.org/10.1038/nature13179
Myers SS, Frumkin H, Gaffigan R, Margo L (2017) Climate change and global food systems: potential impacts on food security and undernutrition. Annu Rev Public Health 38:365–377. https://doi.org/10.1146/annurev-publhealth-031816-044327
NASA (2023) Global climate change: vital signs of the planet. National Aeronautics and Space Administration
Nguyen VT, Tran QT, Le NH, Hoang HT (2020) Scaling up climate-smart rice production in Vietnam: Lessons from the Mekong Delta. Clim Chang Agric 15(2):121–134. https://doi.org/10.1007/s13353-020-00304-3
NOAA (2023) Climate change and coral reefs. National Oceanic and Atmospheric Administration
Pingali P, Rosegrant M (2021) Transforming food systems for a rising world population: addressing access and equity challenges. Annu Rev Resour Econ 13:41–58. https://doi.org/10.1146/annurev-resource-110819-010820
Pretty J, Benton TG, Giller KE (2018) Sustainable intensification in agricultural systems. Agr Ecosyst Environ 225:1–10. https://doi.org/10.1016/j.agee.2016.04.022
Qadir M, Rengel Z, Murtaza G (2021) Developing salt-tolerant crops for sustainable agriculture. Nat Sustain 4(1):14–23. https://doi.org/10.1038/s41893-020-00651-3
Quisumbing AR, Meinzen-Dick R, Behrman JA (2020) Gender and food security: closing the gap. Glob Food Sec 26:100409. https://doi.org/10.1016/j.gfs.2020.100409
Rahman A (2021) Climate change and its impact on food security in bangladesh: a case study on Kalapara, Patuakhali
Ramesha P, Manohar SS, Sudhakar M (2020) Nanotechnology applications in improving nitrogen use efficiency of fertilizers. Agronomy 10(6):845. https://doi.org/10.3390/agronomy10060845
Rosegrant MW, Cline SA (2020) Global food security in the face of climate change: the role of technology. Food Policy 94:101930. https://doi.org/10.1016/j.foodpol.2020.101930
Rosenzweig C, Hillel D, Ortiz-Bobea A, Elliott J (2020) Food security and nutrition under a changing climate. Nature Sustain 3(11):1021–1028. https://doi.org/10.1038/s41893-020-00643-1

Saltzman A, Birol E, Kumar S (2021) Biofortification: addressing micronutrient malnutrition through biotechnology. Glob Food Sec 28:100499. https://doi.org/10.1016/j.gfs.2021.100499
Smith MR, Myers SS, Golden CD (2021) Elevated CO_2 and nutrient content of crops: current understanding and future challenges. Lancet Planet Health 5(1):e6–e14. https://doi.org/10.1016/S2542-5196(20)30201-8
Subramanian KS, Radhakrishnan R, Kumar V (2015) Nano-fertilizers for balanced crop nutrition: a review. J Plant Nutr 38(5):723–737. https://doi.org/10.1080/01904167.2015.1013975
Tadele Z (2019) Climate change and food security: implications for African agriculture. Afr J Agric Res 14(15):3811–3820. https://doi.org/10.5897/AJAR2019.12345
Tang G, Hussain SH, Savaiano DA (2017) Golden rice is an effective source of vitamin A. Am J Clin Nutr 105(6):1549–1557. https://doi.org/10.3945/ajcn.116.148287
Thornton PK, Ericksen PJ, Herrero M (2022) Transformative approaches to climate-smart agriculture. Food Policy 105:101896. https://doi.org/10.1016/j.foodpol.2022.101896
Tittonell P, Vanlauwe B, Giller KE (2020) Agroecology for sustainable intensification of agriculture in the global South. Nature Sustainability 3(8):616–624. https://doi.org/10.1038/s41893-020-0534-2
UNEP (2022) Rising seas: the impacts of climate change. United Nations Environment Programme.
Wezel A, Bellon S, Dore T, Francis CA, Vandenbroucke P (2020) The development of agroecology: from practices to science and policies. Agron Sustain Dev 40(1):1–18. https://doi.org/10.1007/s13593-020-0626-3
WHO (2023) Climate change and health. World Health Organization
WMO (2023) Global weather extremes. World Meteorological Organization
World Bank (2020) Agriculture and food security: building resilience to climate change. World Bank Group. https://www.worldbank.org/en/topic/agrifood
World Bank (2022) Climate change and agricultural impacts. World Bank
Zhao D, Cheng X, Yang Y, Zhang Z (2021) Integrated pest management and food security: sustainable approaches to pest control in agriculture. Agr Ecosyst Environ 319:107542. https://doi.org/10.1016/j.agee.2021.107542

Part II
Intervention of Climate Change on Healthcare Infrastructure and Healthcare Practice

Chapter 9
Climate Change-Inflicted Occupational Health Risk of Fisherfolk of Kerala, India

K. Sree Lakshmi, K. H. Anagha, Ajay Goyal, and Roopam Shukla

Abstract The intensity and frequency of climate change driven extreme weather events and natural hazards are on the rise often affecting human health directly and indirectly. According to the International Labor Organization, the fishing industry accounts for 7% of all laborer fatalities. This chapter mainly focuses on the fishing communities of Kerala with the objective of characterizing climate change driven occupational health risk as per the Intergovernmental Panel on Climate Change (IPCC) Fifth Assessed Report (AR5) and Sixth Assessment Report (AR6) framework. According to the IPCC AR5 and AR6, risk is a function of hazard, vulnerability, and exposure. We reviewed the literature and conducted interviews to synthesize information based on these frameworks in the context of the occupational health risks of Kerala fisherfolks. The study focuses on reviewing such impacts from repositories, literature and real-time interviews with the impacted communities. Subsequently, the chapter elaborates on the study's findings and their relevance in resilience measures, policy changes, and other risk reduction measures. The study finds that climate change is evident in Kerala and is driving hazards, often affecting health directly and indirectly. The sociocultural aspects of fisherfolks add to the vulnerability thereby increasing the risk, while government schemes and initiatives act as an efficient mechanism to reduce the risks. Interaction with residents of Valiyathura points to the various health issues they face and their perception of changing climate. This chapter thus synthesizes the critical adversities and vulnerabilities the fisherfolks of Kerala encounter due to climate change driven occupational health hazards and the ways of navigating through the adversities.

Keywords Climate change · Occupational health hazard · Fisherfolks · Kerala · Disaster · Risk

K. Sree Lakshmi (✉) · A. Goyal
National Institute of Mental Health and Neuroscience (NIMHANS), Bangalore, Karnataka, India
e-mail: kizhakootlakshmi.official@gmail.com

K. H. Anagha · R. Shukla
Indian Institute of Technology, Roorkee, Uttarakhand, India
e-mail: roopam.shukla@dm.iitr.ac.in

P. Ghosh et al. (eds.), *Climate Crises and Public Health Challenges*, Disaster Risk Reduction, https://doi.org/10.1007/978-981-95-9709-3_9

9.1 Introduction

The intensity and frequency of climate-change driven extreme weather events and natural hazards are on the rise, causing unrest in society, the economy and the environment (Singh et al. 2014; Ghosh et al. 2012; Goswami et al. 2006). Evidence supports projection of increasing sea surface temperature, sea level rise, ocean acidification, intense storms, ocean current and rainfall anomalies (Shaffril et al. 2011) which could lead to multifaceted repercussions on ecosystems and society (Weatherdon et al. 2016). Climate change and adverse weather conditions have both direct and indirect negative impacts on the health of society (McMichael 2001). Direct effects include increased mortality and morbidity from heat waves, floods, and droughts. Indirect impacts will arise from infectious disease patterns, food security, and water quality changes. WHO reports that 3% of diarrheal, 3% of malaria, and 3.8% of dengue fever-related deaths in the year 2004 worldwide are attributable to climate change, and mortality due to climate change accounted for 0.1 million (0.2%) deaths globally. Studies also indicate that cardiac deaths could increase due to climate change (Shaffril et al. 2011). In the context of ocean systems, human exposure to health impacts is probable to increase under climate change (Burge et al. 2014). These health impacts will interact with existing socio-economic determinants, further exacerbating risk. According to the World Health Organisation (WHO) and the International Labour Organisation (ILO), almost 2 million fatalities are reported under occupational hazards only in the year 2016, out of which 7% is within the fishing industry. These numbers indicate that the fishing industry falls into one of the high-risk occupation categories. Fishing is characterized by numerous risks (Matheson et al. 2001;GBD 2015; WHO; Frantzedkou et al. 2012; Udolisa et al. 2013) including capsizing of boats, long working hours in risky environment, working with heavy machinery, risks of toppling or fall hazards due to slippery surfaces, lack of safety equipment and harness for the handling of heavy loads, fish attacks in the sea and poisoning from jellyfish, sting rays, snakes, insects etc. Further, there is susceptibility to physical harm from extreme weather, scorching sun, strong winds, extreme cold, rough seas, salt water and noise, which may result in various adverse health conditions like physical injuries, skin damage, respiratory issues, auditory disabilities and in numerous cases fatalities (Mandal et al. 2017; WHO 2021). Acute and chronic mental health disorders are also an aftermath of exposure to hazardous environments and conditions of the sea. These environmental conditions could be aggravated by climate change, hence affecting the lives and livelihood of fisherfolks (Shaffril et al. 2011).

Climate change is also driving differential impacts resulting in the increased vulnerability of specific individuals or groups like mothers and children, the ageing population, and certain occupational groups. Areas with substandard healthcare structures, mostly in developing countries, will struggle to survive this situation without appropriate planning and timely intervention (WHO 2023). Hence, reducing the risk posed by adverse climatic conditions on the physical and mental health of the fisherfolks community and shifting focus to providing technological improvements, framing better policies, improving healthcare systems and ensuring financial stability are vital for improving the plight of fisherfolks (Chand and Murthy 2008). Though susceptibility to health impacts of climate change has been much discussed in academia, susceptibility among workers has less academic attention (Kjellstrom et al. 2009a, 2009b; Hanna et al. 2011; Bourbonnais et al. 2013; Adam-Poupart et al. 2013; Applebaum et al. 2016). While there is acknowledgement of dangers and high mortality associated with fishing, there has been little research in this field (Matheson et al. 2001). Additionally, there exists limited scholarship on the convergence of climate change inflicted occupational health hazards especially in the fisheries sector. There is also a gap in comprehensively deconstructing climate change inflicted occupational health risk of Kerala fisherfolk. We tried to bridge these gaps using Intergovernmental Panel on Climate Change Fifth and Sixth Assessed Report (IPCC AR5 & AR6) framework providing a comprehensive picture of climate change driven occupational health hazards.

Through this study, we address the question of how climate change lead to occupational health risk of fisherfolks. Our research questions were:

(i) How climate change is affecting Kerala and what are its evident manifestations especially in the coastal systems?
(ii) What is the state of affairs of climate driven hazards in the Kerala coast?
(iii) How exposed and vulnerable are the population of Kerala coast to coastal hazards?
(iv) What are the existing governance mechanism to reduce the health risk?
(v) What are the major occupational health risk of fisherfolks in the context of climate change and climate change driven hazards?
(vi) What are the perceptions and evidence of climate change and occupational health risks taking the case of Valiyathura?
(vii) What are the recommendations to reduce the occupational health risk inflicted by climate change?

The objective of the study is to characterize the climate change driven occupational health risk associated with the fisherfolk as per the IPCC AR5 and AR6 framework. Through this study, we could comprehensively draw a picture of state of affairs of climate-driven occupational health risk of Kerala fisherfolks.

9.2 Methodology

In this study, we followed IPCC AR5 and AR6 frameworks to conceptualize the occupational health risk of fisherfolks due to climate change. According to the IPCC AR5 and AR6, risk is a function of hazard, vulnerability, and exposure where; Hazard is "A hazard is the potentially harmful event or condition caused by the climate system or natural processes." (AR6; UNDDR 2009). Vulnerability is "The propensity or predisposition to be adversely affected" (IPCC AR6). Exposure is "elements at risk"

(IPCC 2014, 2022). We reviewed articles, reports, newspapers to synthesize information based on the IPCC AR5 and AR6 framework regarding the occupational health risks of Kerala fisherfolks and demystified the elements of occupational health risk of fisherfolks due to climate change into the components of risks–hazard, exposure, and vulnerability.

To provide more insights at a grassroots level, a case study was conducted in the coastal community of Valiyathura, Trivandrum, incorporating health statistics, and qualitative insights from community interactions and interviews.

9.3 Brief Overview of Kerala

Kerala, a southern state of India stretches geographically across 14 districts with a land area of 38,863 sq. km. Of the 14 districts, 9 are coastal, with Arabian Sea in the west. Forty-four rivers flow through Kerala, with 41 west-flowing and 3 east-flowing. The coastline of Kerala is diverse with its lagoons, estuaries, splits, rocky headlands, sandy beaches, and cliffs (Sheela Nair et al. 2018). Kerala experiences a tropical climate with three distinct seasons. The pre-monsoon season starts from February to May followed by the southwest monsoon season from June to September and completing the cycle with the post-monsoon season from October to January (Sheela Nair et al. 2018). Kerala is one of India's highly populated states, with about 50% of its population living in the coastal districts (AP et al. 2014). The coastal regions of Kerala face natural hazards like coastal erosion, cyclones and flooding which are further exacerbated by increasing sea levels and climate changes (RKDP 2018). The 2018 flood and the Ockhi cyclone 2017 perilously impacted Kerala's coastal resources (CPDAC report 2017; RKDP 2018) (Fig. 9.1).

9.4 Result and Discussion

9.4.1 Climate Change Dynamics in India and Kerala

The SST (Sea Surface Temperature), over the past few years, has witnessed increase around the Indian seas (Vivekanandan 2013) and a rise in the sea level that is projected to rise further (Unnikrishnan and Shankar 2007). The dangerous effect of changes of the sea puts the fisherfolk in and around the coastal area extremely at risk to natural hazards like flooding, cyclones and erosion.

Kerala, a maritime state of peninsular India, receives abundant rainfall with spatiotemporal variations. Study shows a 44% SST anomaly during 2001– 2005 in Kerala (Vivekanandan et al. 2009). The rainfall data of the past century shows a decrease (Guhathakurta and Rajeevan, 2008; Krishnakumar et al. 2009; Nair et al. 2014; Nikhil et al. 2012; Simon et al. 2004; Soman et al. 1988), while two years

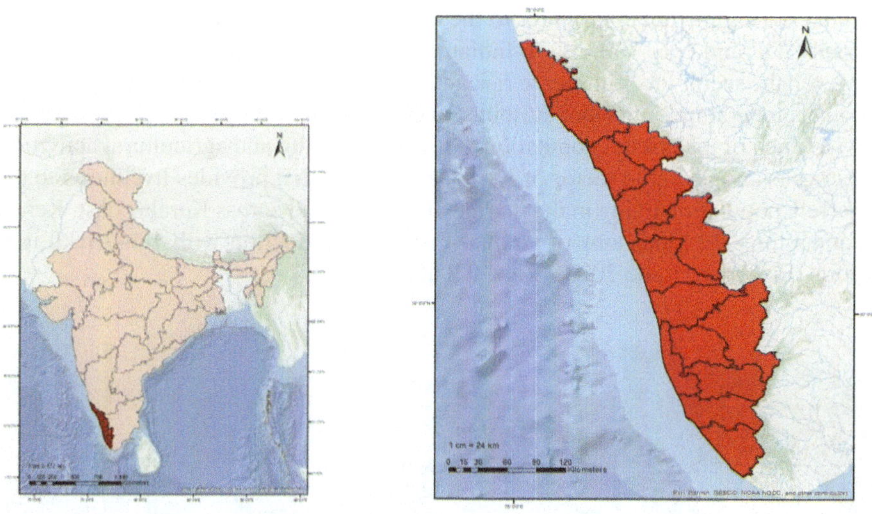

Fig. 9.1 **a** Map of India showing Kerala. **b** Map of Kerala

(2018 and 2019) had heavy rainfall intensity for shorter duration during July and August, causing extensive devastation. These extreme events will likely be driven by anomalous atmospheric conditions exacerbated by climate variability and climate change (Surendran et al. 2020).

Cyclonic disasters aggravate the pressure on coastal communities by putting the fishers at sea at high risk of accidents and even fatalities. This could adversely influence the livelihoods of all stakeholders in the fisheries sector and dependent business chains. Severe cyclonic and storms are anticipated to increase in the Arabian Sea due to the rising surface temperatures of the seas and recent anomalies in the behavioral trend of the monsoon winds (Murakami et al. 2017). Cyclone Ockhi, which can be a prime example of climate change-induced calamities took the lives of over 350 people in the regions of southern Tamil Nadu and Kerala, thus adding to the fear, distress and increased uncertainties of the fisherfolk about the changing weather and climatic conditions. To add to the environmental issues, the fisherfolks in general are socioeconomically marginalized, and have limited accessibility to good healthcare, educational and other public systems placing them at a high-risk climate impacts (Salim et al. 2014).

9.4.2 Fisheries Sector

India has 11,098 km of coastline shared with the Indian Ocean, the Arabian Sea and the Bay of Bengal. Coastal region of India is home to 20% of the nation's population, with fisheries sector as the main livelihood of millions. The fisheries sector of India is a significant contributor to the country's economy with a share of 1.07% of total GDP and over 300 billion Indian rupees towards export earnings annually. However, the sector faces immense risk and is struggling due to the recent increase in oceanic temperatures and other attributes of global climate changes (FRAD, CMFRI, 2017). Much of the coastal population depends on fishing and agricultural activities. Fisheries is an inevitable sector of Kerala's economy that provides livelihood to the fisherfolk population living in the marine villages spread across Kerala coast. Kerala accounts for 8.27 Lakh tons of Annual Fisheries production with 1.447 lakh tons exports (Fisheries Guide 2023) (Figs. 9.2 and 9.3).

Fig. 9.2 Women involved in fish-selling activities, Trivandrum district

Fig. 9.3 Fish-selling activities, Ernakulam district

9.4.3 Hazard Profile of Kerala Coast

Coastal systems, often exposed to multiple hazards like cyclones, tsunamis, floods, storm surges, sea level rise and erosion, have been identified to be at the forefront of climate-induced risks (Islam et al. 2016). Coastal areas host a huge population and support many human activities like fishing, agriculture, tourism, and industry. The aggregated impacts of environmental hazards and anthropogenic stressors amplify the threats to coastal systems and communities living there (Balica et al. 2012; Mahendra et al. 2011). A study by Mohapatra 2015 classifies districts prone to cyclones, among which 7 coastal districts of Kerala are classified as moderately prone. Kerala has witnessed cyclones like Ockhi, mainly affecting Thiruvananthapuram, Kozhikode and Ernakulam in 2017 and Tauktae in 2021 (Public EMDAT 2024). The 2004 Indian Ocean tsunami devastated some parts of the Kerala coast, affecting Kerala's Kollam, Alappuzha, and Ernakulam districts (Kurian et al. 2006). Kurian et al. (2009) categorized the coastal flooding phenomena in the Indian context based on the causative factors: Synoptic scale flooding—due to storm surge were driven by large-scale weather systems; Mesoscale flooding which is a high amplitude sea-level oscillation that occurs quite frequently; Remote forcing—"kallakkadal," distinguished with its frequency and local impact which is mostly witnessed in Kerala. A study by Muis et al. (2016) classifies Kerala to be at high risk due to coastal flooding.

A storm surge refers to the temporary rise in sea level along a coastline, typically caused by meteorological and/or oceanographic factors, leading to rapid flooding of coastal areas (Krishna et al. 2023). According to a study by Balachandran et al. (2024), 684 such events were recorded along the Kerala coast between 2012 and 2023. Erosion is one of the other significant coastal hazards along the Kerala coast (Selvan et al. 2020).

9.4.4 Exposure

An estimated population of 10,60,078 fisherfolks lives in Kerala, which includes 4,22,465 men, 3,74,226 women and 2,63,387 children (Fisheries Guide 2023). Across Kerala and India, the fisherfolk community is heterogeneous: seafaring fisherfolk work long hours in the sea, devoid of any facilities except for essential supplies, while inland fisherfolk spend their night fishing for their livelihood in their boats. Women in the fisherfolk community are involved in fish processing and selling activities (Samantray and Pathak 1998).

9.4.5 *Vulnerability of the Fisherfolk*

Fisherfolk, accounting for ~3.5% of Kerala's population, faces marginalization in various realms despite Kerala's achievements in social sector development and social security schemes (Sheeja et al. 2023). The Human Development Index value for Kerala's marine fisherfolk was 9% lower than that of the state's general population in 2016 (Salim 2024). The fisherfolk of Kerala still follow their traditional occupation with low occupational mobility, often depriving them of the minimum standard of living (Rajan, 2002; Shyjan, 2011). Deprivations, vulnerabilities, insecurities, and inequalities at individual, household and community levels have resulted in poverty, unemployment, illiteracy, and health issues among the fisherfolks (Devika 2014).

The health condition of the fisherfolk is compounded by income inequality and financial issues, lack of nutritious food, and basic facilities like sanitation and drinking water (Vijayan et al. 1997; Asawa et al. 2014). The patriarchal nature of the country and Kerala extends to fisherfolks as well with traditional gender roles which position husband or son as family head, while women remain passive and submissive in household decision-making (Devika 2014), despite several initiatives for women and women's groups (Shruthi et al. 2016).

9.4.6 *Adaptive Capacity Through Government Schemes*

A study by Sheeja et al. (2023) listed 23 major social security schemes in Kerala for fisherfolks, which are as follows:

The Group Insurance scheme for all active fishermen aims to provide insurance coverage to active fishermen for death while fishing. The Old Age Pension and Widow Pension scheme substantially benefits beneficiaries. The Marine Fishing Implements scheme includes modernizing crafts and insurance coverage for fishing implements. Theeramythri and Micro Enterprises scheme, address the livelihood needs of the fishing community. The scheme for the management of Fishery Harbour helps with dredging works and maintenance of harbour amenities. The scheme for the modernization of fish markets and value addition was initiated to set up fish marketing areas with a sophisticated and clean environment for handling fish products and related activities. The scheme for Investment in New Fishing Harbours and Landing Centres aims at investigation work, model studies and Environmental Impact Assessment (EIA). The Saving Cum Relief Scheme focuses on thrift habits and provides aid to registered fisherfolk. The Fisherman Welfare Fund Assisted Housing Scheme gives housing to underprivileged fisherfolks. The Extension and Training scheme encourages aquaculture. The Scheme for Coastal Social Infrastructure provides financial aid for developmental activities, emphasizing setting up dispensaries and guide lights. It is highly relevant as the current dispensaries do not provide the necessary healthcare requirements for the fishermen. The Fishing Equipment Insurance scheme provides insurance for conventional fishing equipment. Most schemes provide short-term or immediate financial aid without addressing the long-term requirements and perspectives. The lack of necessary measures in social sectors constrains human development in coastal areas, leading to inequalities. The accessibility and attainability of the schemes aggravate the inherent drawbacks of the existing schemes.

9.4.7 Occupational Health Risk of Fisherfolk Due to Climate Change

The health impacts of fisherfolks are acute but understudied. Occupational asthma, musculoskeletal disorders, lip cancer, eye damage and irritations, allergic reactions to skin, Injury by allergic responses to marine life and hearing loss are some of the significant health concerns of fisherfolks. Global studies on fisherfolk indicate skin and respiratory problems are the most prevalent health issues (ILO 1999). Skin diseases are common due to fish handling and exposure to extreme temperatures. The health issue of skin and eye damage from sunburn is very probable at sea due to the unobstructed sunlight, resulting in burns and an increased chance of skin cancer. It may also cause sunstroke. Extreme weather conditions like scorching sun, strong winds, extreme cold, rough seas could adversely affect the health and even lead to death (Mandal et al. 2017; WHO 2021). One hundred forty-three were dead or missing during Ockhi. (Ockhi Parliamentary Standing Committee Full Report 2018). Acute and chronic mental health disorders are also an aftermath of exposure to hazardous environments and conditions of the sea. Exposure to low temperature causes a drop in blood circulation in muscle tissues, affecting their ability to function. In contrast, long-term exposure to low temperatures can worsen rheumatic conditions and result in Raynaud's syndrome, which could lead to permanent cold hands and feet with blue coloration and constant aches (Nair and Rajan 2007).

Environmental aspects encompassing climate change, pollution, nutritional changes, and chemical exposure, have been recognized to impact thyroid function and health (Street et al. 2024). According to (Rowe and Talley 2024), exposure to extreme cold waves was associated with increased TSH levels, while exposure to heat was associated with decreased TSH. The thyroid gland plays an important role in the physiological processes. According to Dhanabhakyam (2023), Fisherfolk also find maintaining a work-life balance challenging because of lengthy working hours and irregular schedules. Lack of appropriate government aid, limited access to markets, health problems, and the shortage of social security aggravate the issues. These health issues fisherfolks are exposed to could be exacerbated by climate change (Burge et al. 2014).

Table 9.1 Major health issues in Valiyathutra

Major health issues in Valiyathura	Percentage of population affected (%)
Tobacco use	16.3
Alcohol use	16.7
Self-reported hypertension	25.9
Self-reported diabetes	20.4
Cardiac disorders	5.5
Stroke	1.0
Renal disorders	1.4
Thyroid disorders	5.5
Cancer	0.7
Injury	1.0
Tuberculosis	0.6

9.4.8 *Valiyathura, Trivandrum—A Case Study*

Valiyathura, a coastal ward in Trivandrum district with a marginalized socioeconomic setting, has a significant part of the adult population involved in fishing and associated works. A household survey by the Kerala Institute for Local Administration (KILA) a decade back reported a total population of 4847, with ~2350 females, which is considered an adverse sex ratio. The average household size is 5.4 members per household, compared to 4.4 among non-fishery households. Fishery households are predominantly from the Christian faith, followed by Islam and Hinduism (Table 9.1) (Shaffia et al. 2016).

9.4.9 *Reports from Interaction*

Everyone we talked to unanimously stated about the increasing temperature, especially in April and May of 2024. When asked about climate change, Robert Panipilla, founder of a coastal NGO, Friends of Marine Life stated,

> *"My ancestors were fisherfolks, and I recall them going seas without shirts on, and now it's unimaginable to go to seas without jackets".*

He stated that fisherfolks generally go fishing in the evenings and return in the morning before it gets hot. Recently, the climate has changed, and it gets very cold at night. Those who could not afford jackets wear plastic sheets on them, stated Babu, a fisherman who has been fishing for the past 38 years. He further stated many suffer health issues due to cold weather and the increasing instances of storms at sea. Robert Panipilla also stated the sea has eroded several lanes of houses, and Vinitha, a fisherfolk, claimed that the waters have encroached landward and pointed at the previous shoreline location, confirming the rising sea level triggered by climate change. According to the data provided by TSSC (Trivandrum Social Service Society), many deaths have happened in Muthalappozhi from 2000–2024. "Pozhi" in the local language refers to an estuary. Leeja stated that fisherfolks travel through the estuaries to sea during rough seasons. Among the 78 deaths, 16 were recorded till 2014, while the rest were recent. News reports suggest that the unscientific construction of breakwaters compounded by worsening of weather conditions as the main reason for the deaths. Sand mounds are created by rough sea waves striking breakwater and river sediment influx into the seas. Boats strike on these sand mounds, causing the death of fisherfolks.

Babu talked about the various health issues they face. He said his friends had been taken backshore after encountering cold-related health issues. He is a heart patient and was once taken back to shore after experiencing heart issues while fishing on the seas. Another issue of concern is mental health. Fisherfolks face psychological health issues as a direct and indirect aftermath of climate change.

In general, the whole community is stricken with a newly developed fear of the sea and the environment after the Ockhi cyclone that took a devastating toll on the lives and livelihood of the fisherfolks as stated by Babu. He further stated that fisherfolks are also scared of lightning and that there is no shelter from lightning. This anxiety and fear can aggravate the already vulnerable population with compromised health.

For the fishermen that go into the deep seas, fishing and trawling work are carried out during the night, leading to irregular sleep and disrupted daily routines, continually affecting their health and well-being. As far as women are concerned, the pressure of being part of a patriarchal society weighs heavily on them in addition to the domestic tensions at home.

During our interaction, Vinitha, a fish vendor, talked about the health issues women face, such as headaches, back pain, skin issues, and allergies. Men catch fish during the night and return by morning, while women buy it from them during auctions and sell it in the fish market. Hence, women are more exposed to the scorching heat of the sun. They also spend a considerable amount of time sitting still in the heat. "Many won't even drink water," says Vinitha. There is also a lack of pure drinking water, with more than 50% of families not having access to piped water supply in fishing villages of Kerala. This could lead to kidney-related issues. Vinitha also talked about skin-related problems that are prevalent due to exposure to the sun. From the interaction itself it is very evident that they consider all these problems are integral and unavoidable at any circumstances hence they are still suffering.

9.5 Recommendations

Drawing from the insights gathered through field interactions and literature review, this section presents a comprehensive set of recommendations. These recommendations are designed to address the intersecting challenges of climate change, occupational health hazards, and socio-economic stressors. By integrating advanced technologies, culturally rooted communication strategies, psychosocial support systems, and robust governance mechanisms, the proposed actions aim to empower the community risk reduction. The recommendations are structured across five key domains: technical, socio-cultural, psychological, governance, and financial, ensuring a holistic approach to risk reduction.

9.5.1 Technical and Technological Recommendations

To address ongoing fears of unpredictable events, it is essential to safeguard the livelihoods of fisherfolks. Given the region's high dependence on fishing and its vulnerability to extreme weather, the adoption of advanced technologies and localized early warning systems must be prioritized. Fisherfolks should be encouraged to regularly access meteorological and oceanographic updates through user-friendly platforms that provide real-time alerts about sea conditions. Locally relevant, scientifically validated information such as safe fishing windows and weather advisories should be disseminated through accessible channels like community radio, satellite messaging, and mobile alerts in the local language. Capacity-building programs should focus on equipping fishers with essential safety gear, such as life jackets, and training them in disaster preparedness. Additionally, promoting the use of portable health-monitoring tools like pulse oximeters can help monitor physical well-being during long fishing expeditions, especially in the face of changing sea temperatures and weather patterns.

9.5.2 Socio-Cultural

To ensure effectiveness, communication strategies especially early warnings and safety advisories must be aligned with messages delivered through trusted spiritual leaders and conveyed in the local language, as the fisherfolk community places deep faith in these sources. Promoting diverse livelihood opportunities such as aquaculture, ecotourism, and marine conservation, could strengthen community resilience.

9.5.3 Psychological

Given the incidence of psychological stress there is a need to prioritize psychosocial and mental health support as a core component of risk reduction and public health planning. Efficient mental health services should be integrated into existing healthcare infrastructure, including Community Health Centres (CHCs), Family Health Centres (FHCs), and Primary Health Centres (PHCs). Programs like Tele-MANAS, a Government of India initiative, offer free, 24/7 multilingual mental health support could be adapted for coastal communities. Strengthening social support networks and ensuring access to mental health care will improve individual well-being and enhance the overall resilience of the communities.

9.5.4 Governance

The policy and governance mechanisms must take a proactive and integrated approach to address the complex health and livelihood challenges faced by the fisherfolk with a focus on efficient schemes, incentives, and welfare initiatives that directly target the well-being of coastal communities. National and state-level frameworks such as the National Action Plan on Climate Change (NAPCC) and the State Action Plan on Climate Change (SAPCC) must incorporate the needs of vulnerable fishing populations. Given the existence of mental health issues, policies must prioritize psychosocial interventions. Strengthening the institutional capacity of FHCs, PHCs, CHCs, and ASHAs is also essential. Stakeholders including government agencies, NGOs, and local organizations should conduct community-specific sensitization campaigns to raise awareness about health risks and available services. Regular assessment studies must be undertaken to ensure that interventions are evidence-based and tailored to the evolving needs of the community. Emphasis should also be placed on enforcing the Occupational Safety, Health and Working Conditions Code, ensuring minimum safety standards and routine health check-ups for fisherfolk. Furthermore, investment in research and development of climate-resilient fishing practices is critical to safeguarding both health and livelihoods. To alleviate the economic stress, the government should actively promote alternative livelihoods and diversification programs, enabling fisherfolk especially youth and women to pursue sustainable and dignified employment options. There needs to be impetus and focus on ensuring the well-being of women in the aspect of health, finance and agency.

9.5.5 Financial

To strengthen the financial resilience of fisherfolk, a greater emphasis is needed for insurance schemes such as the Government of India's Pradhan Mantri Matsya Sampada Yojana (PMMSY). However, awareness and accessibility remain limited in many coastal communities. To address the economic vulnerabilities such as loss of income due to extreme weather events, damaged equipment, and health-related disruptions, parametric insurance models should be introduced. These models offer pre-agreed payouts based on measurable triggers like storm intensity or fishing bans, ensuring timely compensation without lengthy claims processes. Such financial instruments can provide a critical safety net for fisherfolk, reducing the cascading effects of income loss on mental health, education, and long-term livelihood security.

9.5.6 Limitations and Strengths

The strength of this study is that we have done different approaches of reviewing existing literature and gaining insights from on field interactions. We could also structure occupational health risk as per IPCC AR5 risk framework thus comprehensively giving a picture about climate change driven occupational health risk. Often studies treat climate impacts or health risks in isolation, this work integrates both dimensions to present a holistic risk profile The limitation of this study is that we could not extensively broaden the fieldwork and gather more field insights. The possible bias from interaction is another drawback. We couldn't consult health officials in gaining more insights.

9.6 Conclusion

This study throws light on the risk the fisherfolk community is exposed to, in the face of climate change and occupational health hazards. By employing different approaches and grounding the analysis in the IPCC AR5 and AR6 risk frameworks, the research offers a nuanced understanding of how environmental, socio-economic, and psychosocial stressors influence the well-being of fisherfolks.

Often studies treat climate impacts or health risks in isolation, this work integrates both dimensions to present a holistic risk profile. The inclusion of community voices through interviews and field observations adds depth and authenticity to the findings, making this study not only academically relevant but also socially grounded.

This work advances the field by bridging the gap between climate adaptation policy and grassroots health realities, offering actionable insights for targeted interventions. It calls for a reorientation of existing governance mechanisms to explicitly include fisherfolk communities in their adaptation agendas. It also highlights the need for integrated health systems, culturally sensitive mental health support, and livelihood diversification as essential components of risk reduction and resilience building.

The implications of this study extend beyond Valiyathura, as coastal communities across the global south face similar threats, the findings and recommendations presented here can inform broader frameworks for climate-resilient public health, inclusive policy design, and community-based adaptation strategies. Future research could expand on this work by incorporating longitudinal health data, evaluating the effectiveness of implemented interventions, and exploring scalable models for other vulnerable coastal regions. Thus, this study lays a foundational pathway toward a more equitable, sustainable, and health-secure future for fisherfolk communities navigating the frontlines of climate change.

References

Adam-Poupart A, Labrèche F, Smargiassi A et al (2013) Climate change and occupational health and safety in a temperate climate: potential impacts and research priorities in Quebec. Canada Ind Health 51(1):68–78

AP S, Menon N, Sankaranarayanan KC (2014) An analysis of various coastal issues in Kerala. Int J Sci Res Educ 2:1993–2001

Applebaum KM, Graham J, Gray GM et al (2016) An overview of occupational risks from climate change. Curr Envir Health Rpt 3:13–22. https://doi.org/10.1007/s40572-016-0081-4

Asawa K, Pujara P, Tak M, Nagarajappa R, Aapaliya P, Bhanushali N et al (2014) Oral health status of fishermen and non-fishermen community of Kutch district, Gujarat, India: a comparative study. Int Marit Health 65(1):1–6

Balachandran S, Joseph S, Sreelakshmi M (2024) An evidence-based study of spatiotemporal trends and hotspots of sea surge events in Kerala. Regional studies in marine science, India, p 103949

Balica S, Wright NG, Van der meulen F (2012) A flood vulnerability index for coastal cities and its use in assessing climate change impacts. Nat Hazards 64:73–105

Bourbonnais R, Zayed J, Lévesque M, Busque MA, Duguay P, Truchon G (2013) Identification of workers exposed concomitantly to heat stress and chemicals. Ind Health 51(1):25–33

Burge CA, Mark Eakin C, Friedman CS, Froelich B, Hershberger PK, Hofmann EE et al (2014) Climate change influences on marine infectious diseases: implications for management and society. Annu Rev Mar Sci 6:249–277. https://doi.org/10.1146/annurev-marine-010213-135029

Chand PK, Murthy P (2008) Climate change and mental health. RegNal Health Forum 12(1):43–48

CPDAC (2017) Report of sub-committee of coastal protection and development advisory committee (CPDAC) on coastal data collection, compilation and publication. New Delhi, India: CPDAC Secretariat, CWC

Decent working conditions, safety and social protection–work in Fishing Convention No. 188 and Recommendation No. 199 (2007)

Department of Fisheries, Government of Kerala (2023) Fisheries guide 2023. https://fisheries.kerala.gov.in/en/fisheries-guide-2023.

Devika J (2014) Land, politics, work and home-life at Adimalathura: towards a local history. Centre for development studies, p 64

Dhanabhakyam M (2023) A systematic literature review on the quality of work life of fishermen in Kerala. J Surv Fish Sci 10(4):3525–3528

Global Burden Disease (GBD) Mortality and Causes of Death Collaborators (2015) Global, regional, and national life expectancy, all-cause mortality, and cause-specific mortality for 249 causes of death. 1980–2015: a systematic analysis for the global burden of disease study 2015. Lancet 388:1459–1544

Doddamani A, Ballala ABK, Madhyastha SP, Kamath A, Kulkarni MM (2021) A cross-sectional study to identify the determinants of non-communicable diseases among fishermen in Southern India. BMC Public Health 21(1):414

FRAD, CMFRI (2017) Marine fish landings in India 2016. Technical Report, Central Marine Fisheries Research Institute, Kochi, India

Frantzeskou E, Kastania AN, Riza E, Jensen OC, Linos A (2012) Risk factors for fishermen's health and safety in Greece. Int Marit Health 63:155–161

Ghosh S, Das D, Kao SC, Ganguly AR (2012) Lack of uniform trends but increasing spatial variability in observed Indian rainfall extremes. Nat Clim Chang 2(2):86–91

Goswami BN, Venugopal V, Sengupta D, Madhusoodanan MS, Xavier PK (2006) Increasing trend of extreme rain events over India in a warming environment. Science 314(5804):1442–1445

Government of Kerala (2011) District census handbook. Department of Information and Public Relations, Kerala, Thiruvananthapuram

Government of Kerala (Various issues) Economic review. Kerala State Planning Board, Thiruvananthapuram, Kerala, India

Guhathakurta P, Rajeevan M (2008) Trends in the rainfall pattern over India. Int J Climatol 28(11):1453–1469

Hanna EG, Kjellstrom T, Bennett C, Dear K (2011) Climate change and rising heat: population health implications for working people in Australia. Asia Pac J Public Health 23(2 Suppl):14S – 26

https://sdma.kerala.gov.in/wp-content/uploads/2020/08/Ockhi-Parliamentary-Standing-Committee-Full-Report-2018.pdf

https://timesofndia.indiatimes.com/city/thiruvananthapuram/145-fshermen-died-at-sea-in-5-years-highest-in-kerala/articleshow/87842439.cms

https://www.ilo.org/sector/Resources/publications/WCMS_161209/lang--en/index.htm

https://www.newindianexpress.com/states/kerala/2021/sep/03/alappuzha-boat-accident-leaves-tharayilkadavu-devastated-2353547.html

https://www.newindianexpress.com/states/kerala/2021/sep/13/kerala-speed-boat-drivers-daredevilry-saves-three-fishermen-from-choppy-sea-2358051.html

https://www.preventionweb.net/files/67193_rebuildkeraladevelopmentprogramme.pdf

IPCC (2001) Climate change 2001: impacts, adaptation, and vulnerability. Cambridge University Press, Cambridge and New York

IPCC (2012) Managing the risks of extreme events and disasters to advance climate change adaptation. Cambridge University Press, Cambridge and New York

IPCC (2014) Climate change 2014: impacts, adaptation, and vulnerability. Cambridge University Press, Cambridge and New York

IPCC (2022) Climate change 2022: impacts, adaptation, and vulnerability. Cambridge University Press, Cambridge and New York

Islam MA, Mitra D, Dewan A, Akhter SH (2016) Coastal multi-hazard vulnerability assessment along the Ganges deltaic coast of Bangladesh–a geospatial approach. Ocean Coast Manag 127:1–15. Jentoft S, Eide A, Bavinck M, Chuenpagdee R, Raakjær J (2011) A better future: prospects for small-scale fishing peoples. In: Jentoft S, Eide A (eds) Poverty mosaics: realities and prospects in small-scale fisheries. Springer, Dordrecht. https://doi.org/10.1007/978-94-007-1582-0_20

Kjellstrom T, Gabrysch S, Lemke B, Dear K (2009a) The 'Hothaps' programme for assessing climate change impacts on occupational health and productivity: an invitation to carry out field studies. Glob Health Action 11:2

Kjellstrom T, Holmer I, Lemke B (2009) Workplace heat stress, health and productivity—an increasing challenge for low and middle-income countries during climate change. Glob Health Action 2:1–6. A global examination of climate change-related heat risk and occupational health in LMI countries

Kurian NP, Baba M, Rajith K, Nirupama N, Murty TS (2006) Analysis of the tsunami of December 26, 2004, on the Kerala Coast of India—part i: amplitudes. Mar Geod 29(4):265–270

Kurian NP, Nirupama N, Baba M, Thomas KV (2009) Coastal flooding due to synoptic scale, meso-scale, and remote forcing. Nat Hazards 48:259–273

Mahendra R, Mohanty P, Bisoyi H, Kumar TS, Nayak S (2011) Assessment and management of coastal multi-hazard vulnerability along the Cuddalore-Villupuram, east coast of India using geospatial techniques. Ocean Coast Manag 54:302–311

Matheson C, Morrison S, Murphy E, Lawrie T, Ritchie L, Bond C (2001) The health of fishermen in the catching sector of the fishing industry: a gap analysis. Occup Med 51(5):305–311. https://doi.org/10.1093/occmed/51.5.305

McMichael AJ (2001) Health consequences of global climate change. J Roy Soc Med 94(3):111–114

Muis S, Verlaan M, Winsemius HC, Aerts JCJH, Ward PJ (2016) A global reanalysis of storm surge and extreme sea levels. Nat Commun 7:1–11. https://doi.org/10.1038/ncomms11969

Murakami H, Vecchi G, Underwood S (2017) Increasing frequency of extremely severe cyclonic storms over the Arabian Sea. Nat Clim Chang 7(12):885–889

Nair A, Joseph KA, Nair KS (2014) Spatio-temporal analysis of rainfall trends over India's maritime state (Kerala) during the last 100 years. Atmos Environ 88:123–132

Nikhil Raj PP, Azeez PA (2012) Trend analysis of rainfall in Bharathapuzha River basin, Kerala, India. Int J Climatol 32(4):533–539

Olaoye OJ, Ojebiyi WG (2021) Occupational health hazards associated with Nigerian fisheries. IntechOpen eBooks. https://doi.org/10.5772/intechopen.9

Oppenheimer M, Glavovic B, Hinkel J, van de Wal R, Magnan AK, Abd-Elgawad A, Cai R, Cifuentes-Jara M, Deconto RM, Ghosh T, Hay J, Isla F, Marzeion B, Meyssignac B, Sebesvari Z (2019) Sea level rise and implications for low lying Islands. Coast Commun Supplement. http://hdl.handle.net/11554/9280. Last accessed 1 Nov 2021

Parliamentary Standing Committee on Home Affairs (2018) Ockhi Parliamentary Standing Committee Full Report 2018. Kerala State Disaster Management Authority

Patz JA, Epstein PR, Burke TA, Balbus JM (1996) Global climate change and emerging infectious diseases. JAMA 275(3):217–223

Pörtner HO, Farrell AP (2008) Ecology: physiology and climate change. Science 322:690–692

Rajan SI (2002) Social security for the unorganized sector in South Asia. Int Soc Secur Rev 55(4):143–156. https://doi.org/10.1111/1468-246X.00143

Ramachandran Nair G, Rajan JB (2007) Kerala fisheries: Exploring data needs for resource management—Kerala Institute for Environment and Development (KIED)

Recognizing the invisible role of women in India's fisheries (2022). https://blogs.dal.ca/openthink/recognizing-the-invisible-role-of-women-in-indias-fisheries/. Accessed 28 June 2022

RKDP Report (2018) Rebuild Kerala development programme: a resilient recovery policy framework and action plan for shaping Kerala's resilient, risk-informed development and recovery from 2018 floods. Government of Kerala initiative

Roshan M (2020) Cyclone Ockhi: disaster risk management and sea safety in the Indian marine fisheries sector

Rowe CW, Talley NJ (2024) Will climate change impact patients with thyroid dysfunction? Clin Thyroid 36:402–405. https://doi.org/10.1089/ct.2024;36.402-405

Salim SS, Kripa V, Zachariah PU et al (2014) Vulnerability assessment of coastal fisher households in Kerala: a climate change perspective

Salim SS, Sathianandan TV, Mohamed KS, Narayankumar R, Athira NR (2024) A fisher development index (FDI) for assessing human development in Kerala, India marine fishers. Environ Dev Sustain. https://doi.org/10.1007/s10668-024-04923-9

Samantray K, Pathak SC (1998) An overview of women in the Indian fisheries sector and suggestions for improvement of their socioeconomic status. In: International symposium on women in asian fisheries. Chiang Mai: ICLARM-The World Fish Centre

Sachin et al APS, Menon N, Sankaranarayanan KC (2014) An analysis of various coastal issues in Kerala. Int J Sci Res Educ 2:1993–2001

Selvan SC, Kankara RS, Prabhu K, Rajan B (2020) Shoreline change along Kerala, south-west coast of India, using geo-spatial techniques and field measurement. Nat Hazards 100(1):17–38

Shaffia M, TNa A, Aa MP et al (2016) Developing a health registry for coastal population: protocol of the Valiyathura population based health registry in Kerala, the southern coast of India. J Health Syst 2(1)

Shaffril HM, Samah BA, Uli J, D'Silva JL (2011) The potential impact of climate change environmental hazards on quality of life of fishermen community in Malaysia. Aust J Basic Appl Sci 5(7):507–515

Sheeja SR, Ajay A (2023) State led social security and inclusion of marine fisherfolk: analyzing the case of Kerala, India. Mar Policy 147:105392

Sheela Nair L, Prasad R, Rafeeque MK, Prakash TN (2018) Coastal morphology and long-term shoreline changes along the southwest coast of India. J Geol Soc India 92:588–595. https://doi.org/10.1007/s12594-018-1072-x

Shyam SS, Geetha R (2013) Empowerment of fisherwomen in Kerala—an assessment. Indian J Fish 60(3):73–80

Shyjan D (2011) Dimensions of deprivation among the marine fisherfolk of Kerala: an empirical examination. Labour Dev 18:57–80. V. V. Giri National Labour Institute, Delhi

Simon A, Mohankumar K (2004) Spatial variability and rainfall characteristics of Kerala. J Earth Syst Sci 113(2):211–221

Singh D, Tsiang M, Rajaratnam B, Diffenbaugh NS (2014) Observed extreme wet and dry spell changes during the South Asian summer monsoon season. Nat Clim Chang 4(6):456–461

Solomon S, Qin D, Manning M et al (2007) The physical science basis. Contribution of working group I to the fourth assessment report of the intergovernmental panel on climate change (IPCC). Cambridge: Cambridge University Press

Soman MK, Krishna KK, Singh N (1988) Decreasing trend in the rainfall of Kerala. Curr Sci 57(7):7–12

Sreelakshmi M, Balachandran S, Abdurazak F et al (2024) From perils to preparedness: decoding coastal hazard risks on Kerala's central coast through the IPCC framework. Nat Hazards 120:8639–8666. https://doi.org/10.1007/s11069-024-06543-y

Sruthi P, Jayalal L, Gopal N (2016) Gender roles in fisheries along the Vembanadestuarine system. Asian Fish Sci 29(Supplement):193–203

Street ME, Shulhai AM, Petraroli M et al (2024) The impact of environmental factors and contaminants on thyroid function and disease from fetal to adult life: current evidence and future directions. Front Endocrinol 15, Article 1429884. https://doi.org/10.3389/fendo.2024.1429884

Surendran DE, Sridhar L, Kumari A et al (2020) Impact of climate change on the heavy rainfall events during June to September over Kerala. Curr Sci 119:545–550. https://doi.org/10.2307/27138890

Swathy Krishna PS, Sheela Nair L, Ramesh M (2023) Coastal flooding in India: an overview. In: Jayaraju N, Sreenivasulu G, Madakka M, Manjulatha M (Eds) Coasts, estuaries and lakes. Springer, Cham. https://doi.org/10.1007/978-3-031-21644-2_2.

The New Indian Express (2021) Alappuzha: boat accident leaves Tharayilkadavudevastated. Accessed 7 Dec 2024

The New Indian Express (2021) Kerala: speed boat driver's daredevilry saves three fishermen from choppy sea. Accessed 7 Dec 2024

The Times of India (2021) 145 Fishermen died at sea in 5 years in Thiruvananthapuram, the highest in Kerala. Accessed 7 Dec 2024

Udolisa REK, Akinyemi AA, Olaoye OJ (2013) Occupational and health hazards in Nigerian coastal artisanal fisheries. J Fish Aquat Sci 8:14–20

UNDDR (2009) Terminology on disaster risk reduction. https://www.undrr.org/publication/2009-unisdr-terminology-disaster-risk-reduction. Accessed 5 June 2023

UNDRR (2007) Sendai framework terminology on disaster risk reduction. Accessed 5 June 2023 https://www.undrr.org/terminology/vulnerability.

Unnikrishnan AS, Shankar D (2007) Are sea level rise trends along the coasts of northern Indian Ocean consistent with global estimates? Glob Planet Change 57:301–307

Vivekanandan E, Rajagopalan M, Pillai NGK (2009) Recent trends in sea surface temperature and its impact on oil sardine. In: Aggarwal PK (ed) Global climate change and Indian agriculture (ICAR, New Delhi), pp 89–92

Vivekanandan E (2013) Climate change: challenging the sustainability of marine fisheries and ecosystems. J Aquatic Biol Fish 11(2):54–67

Vijayan A, Gracy M, Alexander M, Nayak N (1997) Women first: Report of the women in fisheries programme of ICSF in India. International collective in support of fishworkers (ICSF). https://www.icsf.net/wp-content/uploads/1997/09/930.ICSF092.pdf

Weatherdon LV, Magnan AK, Rogers AD, Sumaila UR, Cheung WW (2016) Observed and projected impacts of climate change on marine fisheries, aquaculture, coastal tourism, and human health: an update. Front Mar Sci 3:48

Ziervogel G, New M, Archer van Garderen E, Midgley G, Taylor A, Hamann R et al (2014) Climate change impacts and adaptation in South Africa. Wires Clim Change 5:605–620. https://doi.org/10.1111/gcb.12597

Chapter 10
Climate-Induced Labour Precarity: Occupational Health Challenges for Migrant Labourers in India

Rameswar Mukherjee and Pamela Deb

Abstract The migrant labourers of India play a crucial role in the key sectors such as agriculture, construction, and manufacturing. However, because of hazardous working conditions, lack of effective social security measures, and exclusion from welfare programmes, they are among the most vulnerable segments of the labour force in the country. Their vulnerabilities are worsened by climate change. The frequency and severity of various health issues like heat stress and vector-borne diseases have increased due to rising global temperatures, extreme weather events, and unpredictable rainfall which disproportionately affect the migrant labourers. This study attempts to understand the role of climate change which amplifies the vulnerability of the migrant labourers in India through sectoral analysis. Various cases studies were selected which focused on occupational risks due to the nature of work and climate-change-driven vulnerabilities, drawn from various academic sources such as Google Scholar, Scopus, and Pub Med. The findings of this study urge for climate-resilient labour policies, modification of existing policies, and effective implementation to mitigate these risks.

Keywords Occupational hazard · Heatwaves · Labour intensive industries · Vulnerability · Labour rights

R. Mukherjee
Department of Geography, Samsi College, Kandaran, Malda, India

P. Deb (✉)
Department of Geography, Amity Institute of Social Sciences, Amity University Kolkata, Kolkata, India
e-mail: pamela9.deb@gmail.com

P. Ghosh et al. (eds.), *Climate Crises and Public Health Challenges*, Disaster Risk Reduction, https://doi.org/10.1007/978-981-95-9709-3_10

10.1 Introduction

10.1.1 Backgrounds

Migrant labourers in India make a significant contribution to the country's workforce, especially in urban areas. According to the Census of India (2011), there are approximately 450 million internal migrants of whom 15.6% are rural–urban migrants. Furthermore, among the inter-district (117.9 million) and inter-state migrants (54.4 million), the latter one is engaged with low-paying and insecure jobs, such as manufacturing, construction, hospitality, transportation, and domestic work (Hannan and Hossain 2023; Pareek and Sole 2020). A substantial number of people are migrants in major cities like Delhi and Mumbai (about 43%), primarily from less urbanized states such as Bihar and Uttar Pradesh (Roy et al. 2015). However, the informal sector absorbs nearly 67% of urban migrant workers, offering low wages and job insecurity (NSSO 2012). Despite their indispensable role in the economy, they often work in hazardous conditions with inadequate social security, making them one of the most vulnerable groups in the country's labour force (Srivastava 2020). Additionally, climate change has intensified their vulnerabilities through extreme variation in weather events, such as variability and scarcity of rainfall, increasing number of heatwaves, the spread of vector-borne diseases, and so on.

Rising temperatures, irregular monsoons, and the occurrence of extreme events, including cyclones, floods, and heatwaves, are now commonplace. These changes, especially, have a great impact on the people living in the marginal society. According to the United Nations Call to Action on Extreme Heat release, India has reported over 100 deaths and 40,000 cases of suspected heat stroke as of mid-June 2024 (Hindustan Times 2024). Sharma and Majumdar (2017) reported a significant rise in heatwave intensity, particularly in Northwest and Central India with prolonged drought conditions. Additionally, the recent Open Government Data (OGD) has revealed several new findings such as notable increase in the number of heatwave days in 2024 with significant fluctuations during 2022–24, Odisha experiencing the highest increase in the number of heatwave days followed by Gangetic West Bengal and Bihar, occurrence of heatwaves in the regions where no such events were recorded in the past like the interior part of North Karnataka, and overall expansion of the areas under heatwave in Eastern India. During July–August 2024, extreme monsoon rains caused significant flooding in Delhi, Mumbai, Pune, Guwahati, and many other cities which led to fatalities, displacement, and disruptions to transportation and power supply. All the major cities and metropolises are suffering from various problems related to climate change by affecting the most vulnerable but important city dwellers, i.e., migrant labourers. For example, a city like Bengaluru, where almost 20% of its population is migrant labourers, is facing various problems associated with climate change-driven temperature rise and unpredictable rainfall (Deccan Herald 2024). On the one hand, the workers in manufacturing and outdoor jobs are suffering from heatwaves. Furthermore, unpredictable rainfall and frequent flooding have worsened their living conditions in informal settlements, leading to poor sanitation, waterlogging, and the spreading of vector-borne diseases. According to a report by McKinsey Global Institute, rising temperatures have significantly reduced outdoor working hours, with projections indicating a potential GDP loss of 2.5% to 4.5% annually by 2030 due to heat stress (Woetzel et al. 2020). Additionally, Koll et al. (2019) predict that the increasing frequency of lethal heatwaves will put 160–200 million people at severe risk. Until now the mainstay of the rural economy in India is agriculture, which is climate-dependent. But the climate change-driven unpredictable and variable rainfall and prolonged droughts have disrupted rural economies, which compelled more workers to migrate to urban areas and become involved in precarious urban jobs (Viswanathan and Kumar 2015).

Several important factors define climate change-related vulnerabilities among the migrant workers. For instance, many migrant workers are in an informal category, meaning most of them do not fall under minimum wage regulations or standard work conditions (Srivastava and Sasikumar 2003). A lack of legal protection, health coverage, and job security leads to workers being exploited due to the absence of contracts and being stranded without work (Deshingkar 2017). Furthermore, migrant workers usually have less influence or labour rights. They fight to get higher salaries, working conditions, or benefits without union involvement or collective bargaining agreements (Sharma et al. 2020). This power difference lets companies enforce negative terms without thinking about consequences. Moreover, many migrant workers are ignorant of their legal rights or fear consequences should they voice opinions, adding to their vulnerability and exploitation (Rai 2020). These challenges are made harder for migrant workers by their deep level of reliance on industries vulnerable to climate change such as sectors like farming, construction, fishing, and so on (Keshri and Bhagat 2013). In addition, the lack of access to occupational health treatment that migrant workers have is surely more burdensome than this. Many do not receive regular health check-ups, preventative treatment, or treatment for diseases and injuries, which they receive while working (Borhade et al. 2020). Such limited availability of healthcare services leads to many unresolved health challenges due to financial, geographical, and systematic barriers that are difficult to overcome (Jain and Sharma 2019).

Under this backdrop, the study attempts to explore the occupational hazards that are intensified by climate change, particularly among the migrant labourers of India who migrate due to livelihood loss and food insecurity to the urban areas and again enter into the environment where climate change-related concerns make them more vulnerable, such as vector-borne diseases, heat stress, dehydration, and kidney disorders.

10.1.2 Evidence in Previous Literature and Research Gaps

There is extensive research documented on the vulnerabilities of migrant labourers in India, yet it has conspicuously neglected the conjunctive effects of climate change and occupational health hazards on this population. Earlier inquiries have either mapped migratory flows and labour contexts (Deshingkar 2017; Srivastava 2020) or dissected climate vulnerabilities in urban agglomerations (Sharma and Majumdar 2017; Avashia and Garg 2020). However, these studies have not sufficiently accounted for the synergistic vulnerabilities encountered by migrant labourers who are compelled to navigate conditions of labour precarity while simultaneously confronting the escalating hazards associated with climate change.

A cohort of studies has delineated the perilous contexts faced by migrant labourers, highlighting precarious, informal employment agreements, the void of social safety provisions, and consistent exposure to both mechanical and chemical hazards (Srivastava and Sasikumar 2003; Borhade et al. 2020). In parallel, research addressing climatic perturbations within the Indian subcontinent has predominantly examined economic and environmental consequences; the McKinsey Global Institute, for instance, projects an annual contraction of 2.5–4.5 per cent of gross domestic product by 2030, attributable to heat-related morbidity and loss of productivity (Woetzel et al. 2020).

The existing body of literature reveals several pressing research gaps. To begin with, most investigations have not disaggregated the effects of climatic change on economic sectors that concentrate on migrant labour, particularly construction, manufacturing, agriculture, and informal services. Moreover, while there is general information available about heat-related illnesses and deaths, there has not yet been a full investigation into sector-specific occupational health hazards. Another significant aspect is that the climate-migration connection has not been fully investigated when it comes to the increased risks faced by migrating women, who typically work in jobs that are sensitive to climate change and face many barriers to getting the resources they need. Finally, the studies did not focus on how climate change is acting as a potent factor both in rural livelihood vulnerability and urban employment adversities for these marginal communities who live both in rural and urban areas as a vulnerable group.

10.1.3 Objectives

The study aims to examine how climate change has intensified the occupational health risks of migrant labourers both in rural and urban areas of India. It also seeks to analyse the sector-specific vulnerabilities due to climate change among the migrant labourers through published case studies and research. Moreover, the research attempts to illustrate the existing vulnerabilities faced by migrant labourers, related to unsafe working conditions, gender bias, limited or lack of access to social security, healthcare measures, and welfare programmes. Finally, it recommends climate-resilient labour policies, modification of existing policies, and effective implementation to mitigate these risks.

10.2 Database and Methodology

The study is entirely based on secondary data. It includes peer-reviewed journal articles, government reports, and policy documents. By drawing on case studies and secondary data from scholarly sources such as Google Scholar, Scopus, and PubMed, the study aims to highlight sector-specific occupational health risks and challenges. The literature for this study was collected on climate change, labour vulnerability, and occupational risks and hazards in India. It has adopted a case study-based approach to conceptualize the correlation between labour precarity, climate change, and occupational health risks in India. The case studies were focused on four sectors: agriculture, informal labour, construction, and industry. These case studies were selected based on three criteria: i. Relevance of various sectors, i.e., the sectors that are affected more by climate change and highly labour-intensive; ii. The availability of empirical literature, i.e., the availability of literature that provides insights into climate change and labour precarity; iii. Case studies were chosen from different geographical regions to understand the spatial variation. Based on these case studies, the inherent causes of labour vulnerabilities, workplace risk, and gaps in the government policies were analysed in the light of the political economy perspective, which examines how climate change amplifies the existing labour vulnerabilities in India.

10.3 Results and Discussion

10.3.1 Climate Change and Labour Vulnerability

The extent of climate change impacts is a growing concern for India, primarily due to its geographical location and large population, as well as its heavy dependence on climate-sensitive sectors like agriculture (Revi et al. 2014). Some of these vulnerabilities are extreme droughts, changes in precipitation patterns, etc. Additionally, many of the severe heatwaves in India have changed in intensity and frequency (Mazdiyasni et al. 2017). India witnesses prolonged periods of intense heat, which can be dangerous for people who are more sensitive, like the elderly and those who work outdoors. Extreme heatwaves can significantly increase the death rate, cause discomfort, and lead to dehydration (Pattanaik et al. 2017). The impact on agriculture is huge since heat stress can have adverse effects on the health of livestock and can diminish agricultural productivity (Sharma et al. 2021). In India, the timing of monsoons is equally shifting, which makes agriculture very challenging (Roxy et al. 2015). Monsoons are unpredictable and can bring either floods, droughts, or anything in between, all of which harm agriculture. A drought lowers a crop's yield, damages crops, and creates a water deficit. Insufficient rainfall results in crop destruction, soil erosion, and the loss of fertile land (Mishra et al. 2020). These problems are detrimental to the livelihood of farmers, driving them deeper into poverty and famine.

Without a doubt, the most concerning aspect of this issue is the increased severity of storms. Severe thunderstorms coupled with cyclone activity have become life-threatening, especially for those living in poor socio-economic conditions (Singh et al. 2019a, b). Migrant workers often find themselves with no choice but to live in 'temporary' housing, which puts people in extreme jeopardy during storms. The damage done to a country with these natural disasters leads to the failure of basic civic amenities. Water-stressed areas are on the rise in many parts of India. Due to shifts in rainfall, decreased river flows, and decreased groundwater volume, some regions are becoming drier (Rodell et al. 2018). This trend affects agriculture, the availability of potable water, and countless other industries. Farmers trying to irrigate their land end up lowering the yield of crops, causing their businesses to go into debt. Insufficient supply of water in the urban centres can also greatly affect the quality of life and further complicate the issue of water distribution (Immerzeel et al. 2020).

10.3.2 Climate-Induced Migration Patterns

One of the significant factors for internal migration in India is climate change, especially for small and marginal farmers who are undergoing agricultural distress because of prolonged droughts, erratic rainfall, and declining crop yields (Bhatta et al. 2015; Kumar and Viswanathan 2012). Farmers experiencing intense climate stress are obliged to either migrate alone or with family to places with more favourable environmental conditions or better economic opportunities, which results in distress migration (Deshingkar 2012; Viswanathan and Kumar 2015). On the one hand, the rural labour force is depleted because of the migration, which in turn disrupts the traditional farming operations not only in the region that is left behind but the area that is being settled, while on the other, economic growth becomes stagnant in the region that is abandoned (Kattumuri et al. 2017; Mahapatra et al. 2018).

Moreover, seasonal migration patterns that have normally helped supplement incomes in rural areas during non-harvest periods are severely changing due to weather conditions and agricultural cycle shifts (Deshingkar and Start 2003; Shah and Lerche 2020). Unpredictable weather trends make it impossible for migrants to adhere to certain pathways, resulting in employment volatility alongside difficulty in obtaining social and familial connections (Choithani 2017; Mishra 2016). In addition, the heightened unsustainability of rural livelihoods propelled urbanization in the form of a higher concentration of economically active people moving to cities with the hope of gaining consistent work opportunities (Denis and Zérah 2017; Kundu 2018).

Nonetheless, these new migrants are creating an overwhelming burden on city infrastructure and services, causing congestion, unemployment, and inadequate living conditions, especially within the peripheral and dilapidated urban areas (Agarwal and Taneja 2019; Revi 2008). As a reaction to these problems, new corridors of migration are emerging due to climate migrants looking for safe places to settle economically and socially, motivated by potential employment, resource availability, and pre-established social structures (Bhagat 2017; Chandrasekhar and Sharma 2014). The growing proportion of migration resulting from climatic changes emphasizes the importance of formulating policies to deal with the increasing climate vulnerability of migrants as well as implementing sustainable climate change adaptation approaches such as climate-smart agriculture, rural job creation, and urban development (Kattumuri et al. 2017; Mukherji 2018).

10.3.3 Labour Precarity and Occupational Health Vulnerabilities Among Migrant Workers

Migration driven by climate change has worsened labour instability and exposure to risks among migrant workers who operate in the informal and climate-sensitive industries such as fishing, agriculture, and construction (Rao et al. 2020; Srivastava and Jha 2016). Casual labourers or migrants are employed under highly informal working arrangements and therefore do not have contracts or legal protection (NCEUS 2009; Srivastava 2020). In a system where migrants are completely unsupported, job losses, exploitation, dangerous working environments, and financial instability are chronic issues that become the norm over time (Breman 2013; Mosse et al. 2002).

Employment for such workers is often dispersed with a distribution that is irregular and unpredictable. In construction and agriculture for example, there are seasonal limits to the work which means intervention is made to suppress the ability to earn money and make plans (Breman 2016; Pattenden 2018). In addition, lacking the ability to undertake collective bargaining means that these employees cannot negotiate for basic rights such as minimum wage, proper working conditions, and job security which permits and encourages employers to abuse the workers without supervision (Lerche 2010; Sharma 2006).

Many climate migrants continue to be reliant on industries that are directly influenced by climate changes, in which their livelihood is interrupted by catastrophic events such as erratic monsoons, and extreme heatwaves and storms (Kattumuri et al. 2017; Srivastava and Kumar 2018). For example, productivity in the agricultural sector is diminished owing to continuous droughts, while the construction sector is stalled due to high temperatures which generates jobless economic suffering (Breman 2016; Kumar et al. 2020a, b). Moreover, migrant workers are exposed to dangerous levels of occupational disease such as heat stress and other forms of high-temperature illnesses (Ayyappan et al. 2009; Lundgren et al. 2013). Workers engaged in construction or agricultural activities are particularly vulnerable to heat-induced exhaustion, cramps, dehydration, or even deadly heat strokes unless treated in a timely manner (Dash and Kjellstrom 2011; Venugopal et al. 2016).

Without shaded rest spaces, cooling amenities, and access to clean drinking water, many migrant workers experience extreme exhaustion and dehydration, leading to an eventual decline in productivity and a weakened immune system (Sett and Sahu 2014; Venugopal et al. 2021a, b). Moreover, seasonal phenomena such as rainstorms greatly increase workplace risk by making floors slippery, which creates the potential for mechanical malfunctions, falls, and equipment mishandling (Nag et al. 2009; Srinivasan et al. 2016). While farming employees contend with perilous conditions and machinery breakdowns, construction employees are more likely to experience falls from elevated positions (Patel and Joshi 2018; Singh et al. 2019a, b).

The lack of adequate construction facilities in workplaces such as urban slums, alongside air pollution, exposes construction workers and slum-dwelling migrants to higher risks of developing respiratory issues like asthma and Chronic Obstructive Pulmonary Disease (COPD), to name a few (Kulkarni et al. 2016; Thomas et al. 2018). Lung function is affected by the inhalation of fine particulate matter and industry-specific pollutants over prolonged periods (Balakrishnan et al. 2019; Guttikunda and Jawahar, 2018). Addressing the demand of health needs of these communities is challenging, as sociocultural factors, like financial limitations, obstructions in geography, and insufficient knowledge about medical intervention, delay aid and lead to chronic conditions, as well as expectancy gap for migrant workers (Mohan et al. 2016; Parasuraman et al. 2018). Financial crisis is so severe that these migrant labourers are unable to go for regular health check-up. Most of them prefer the Over-the-Counter medications (OTC) where they purchase medicine without prescription than meeting a doctor due to high Out-of-Pocket-Expenditure (OOPE) (Bhoi et al. 2022; Banerjee and Goswami 2019).

10.3.4 Psychological Impacts of Climate-Induced Labour Migration

Climate-related labour migration gives rise to intricate psychological vulnerabilities, as workers from climate-vulnerable regions face a combination of climate anxiety, displacement trauma, and occupational precarity (Farbotko et al. 2022; Heinz and Brandt 2024). These kinds of effects disproportionately impact already marginalized migrant groups, who are coping with the grief associated with family separation from their native land. Hence, the mental health concerns of climate migrants reflect the problems observable among the displaced populations. Relocation tends to impose upon individuals multiple psychological burdens which demand holistic understanding and targeted solutions (Bhugra et al. 2013; Fazel et al. 2019). The psychological vulnerabilities of climate migrants are comparable to those of refugee populations, who suffer from severe mental health consequences because of being forcibly displaced and having long-term uncertainty about their legal status (Fazel et al. 2019). According to Heinz and Brandt (2024), there is a 0.9% increase in mental health-related morbidity for every degree Celsius. An increase in temperature occurs. These findings are consistent with broader evidence that links extreme temperatures and climate disasters to higher rates of anxiety, affective disorders, and suicide.

The financial strains associated with climate migration may jeopardize the flexibility of mobility plans. Although remittances are a vital source of support for households that are vulnerable to climate change, workers experience significant stress due to the pressure to maintain income flows while managing personal mental health crises, particularly when relationship breakdowns or family emergencies occur abroad (Farbotko et al. 2022). Workers still have very limited access to culturally competent mental health services, and when they are in extreme distress, they turn to unofficial diaspora networks rather than formal support networks. The lack of access to mental health services shows bigger problems with the mental health of migrants, as different pressures faced during migration cause emotional pain and require major changes in policies to help (Bhugra et al. 2013; Frounfelker and Rousseau 2018).

10.3.5 Sector-Specific Vulnerabilities: Case Studies

Occupational hazards induced by climate change affect various sectors differently, depending on the nature of work and the geographic location. The labourers in the construction sector suffer from extremely high temperatures and unpredictable rainfall. The lack of shade, proper rest, and hydration makes the scenario worse. The illnesses related to heat stress are very common during the summer season, while during the rainy season, the slippery surface heightens the risk of accidents. Similarly, the agricultural labourers and industrial labourers, also suffer the same problem, especially from heat. The agricultural labourers apart from heat stress, also suffer from respiratory diseases due to the increasing use of pesticides, which is further due to climate-induced pest outbreaks. Furthermore, due to changing sowing and harvesting patterns based on changes in rainfall variability, low productivity also forces them to adjust to irregular work patterns. In the industrial sector, the labourers are compelled to work on indoor activities in the poorly ventilated rooms, exposed to hazardous chemicals during the extremely hot season. It increases the risk of skin and respiratory diseases. The following sector-specific case studies clearly contextualize these issues.

10.3.5.1 Case Study I: Agricultural Sector

The first case study shows the impact of heat on the health and work productivity of the workers in the rice fields of Murshidabad during the hottest months of the year. The economic condition of the labourers is very poor, which compelled them to work during extreme heat conditions with mere wages of Rs 120–140/day. The total number of workers was 124, with an age group ranging 18–45 years. Out of that, 48 workers underwent detailed assessments, including heart rate monitoring during and after work. The work productivity was measured as the number of rice bundles collected hourly. It showed how the work productivity rate decreased by 5% for every 1 degree rise of Wet Bulb Global Temperature (WBGT). Consequently, people suffer exhaustion, body pain, slower recovery of heart rates, and cardiovascular strain. This study also revealed that the number of days when the WBGT reaches more than 29 degrees at midday has increased from 120 to 160 days in 1980 and 2011, respectively, which signifies the phenomenon of climate change (Sahu et al. 2013).

10.3.5.2 Case Study II: Informal and Urban Labour Sector

The informal sector workforce in Indian cities involves diverse economic activities such as construction, factory work, street vending, and home-based labour. This case study is based on the informal workers in the Indore city of Madhya Pradesh who are in a vulnerable condition due to several reasons, e.g., injury of factory workers due to usage of unsafe machinery, minimum or no support from employer for carrying health expenditure due to injury at the workplace, and so on. Additionally, the gender-biased treatment is also noticed, such as the assignment of physically demanding roles to women, which leads to musculoskeletal pain and backaches, and gender pay differences, where women get low-paying jobs compared to men. In this already helpless situation, the problems related to climate change, such as heatwaves, flooding, and water scarcity, make their lives miserable. For instance, due to unavailability of toilets, female workers consume less water, which leads to dehydration; the factory workers who operate near heat-emitting machinery must bear extreme heat stress. This process also amplifies when they go home, where cooking on biomass stoves in poorly ventilated single-room dwellings exacerbates the situation. Since the responsibility of fetching water is typically assigned to female household members, they have to stand in queues, which takes a considerable amount of time. As a result, they either miss workdays and incur wage losses or be penalized due to late arrival. Similarly, the flood also impacts various ways, such as the damage of informal settlements, waterlogging, economic loss, and vector-borne diseases. This study revealed the gendered impact of water scarcity among the informal working group (Sverdlik et al. 2024).

10.3.5.3 Case Study III: Construction Sector

In India, after agriculture, the construction sector employs a substantial portion of the unorganized workforce. This case study focuses on Bangalore's construction sector, a sector known for its high health risks and associated vulnerabilities. The workers here suffer from a high rate of morbidity due to precarious working conditions. Different types of work include masonry, scaffolding, and material handling. They often encounter musculoskeletal issues due to heavy lifting and postural strain. Additionally, respiratory issues due to exposure to paint, dust, poor air quality, gastrointestinal problems due to poor sanitation, and unsafe drinking water are also common among them. Workplace injury is another major concern due to unsafe working conditions. As the working class belongs to a destitute economic class with no or limited primary education level, it's not possible to understand the functionality of social security measures or the policies for them. Their already precarious situation is amplified by climate change, which is increasing the air temperature, pollution, reduced rainfall, and related health risks associated with it, such as heat stress, fatigue, water scarcity, unsafe drinking water, and poor sanitation (Nirmala et al. 2019).

10.3.5.4 Case Study IV: Manufacturing and Industrial Sector

This case study (Venugopal et al. 2020) considered several sectors of the informal workforce who are the victims of climate-induced occupational health risks. This study was conducted on 1,480 workers across agriculture, construction, salt pans, brick manufacturing, auto parts, foundries, garments, and steel industries. This study also showed how increasing Wet Bulb Globe Temperature (WBGT) measurements (ranging from 21.2 °C to 41.7 °C) experienced heat levels above safe thresholds. In all three sectors, the most common health issue is dehydration, which further leads to kidney-related problems. Such as in the agricultural field, 73.1% of workers suffered from dehydration, in the salt pans due to intense heat and reflective surfaces, 59.1% of workers reported kidney-related issues, and in the steel industry, the problem of kidney stones was also reported by 33% of respondents. Apart from this excessive sweating, fatigue, headaches, nausea, and fainting are very common (84%) among all. From the construction sector, 55.8% of workers and from the agricultural sector, 45.2% of workers reported urogenital issues, while urine specific gravity was reported from salt pan workers in the highest share (27.8%).

10.3.6 Climate Change and Migrant Labour Trap: Rural Distress to Urban Precarity

The migrant labourers in the urban areas of India are intertwined in a tenacious cycle of rural distress and urban adversities, which is intensified by climate change, one of the important factors of labour vulnerability. Although rural–urban migration is not a new phenomenon in a developing country like India, climate change has played a significant role in moulding the migration scenario for these labourers in the urban areas. Most of them come from an unstable rural economic base such as agriculture, fishing, and forestry which is dependent on climatic parameters and their fluctuations. However, recent climate changes, including irregular rainfall, rising number of heat-wave days, and prolonged drought conditions, have rendered agriculture unviable. The farmers, especially the marginal and landless, are compelled to migrate to the urban areas for better livelihood opportunities (Fig. 10.1). But they are absorbed in the informal economy, mostly engaged in construction, manufacturing, domestic help, or the solid waste management sector. With low wages, risky and hazardous working sites, lack of social security measures and local safeguards, and limited healthcare facilities, they are vulnerable to exploitation. Additionally, their living conditions are also very harsh. They generally live in the slum areas with extremely high population density, poor sanitation, and lack of safe drinking water, often located near fragile areas such as near the industries, beside the railway lines, and near the garbage dumps, which are prone to flooding and air and water pollution. This social and economic segregation from the mainstream urban dwellers, distance from family, and economic insecurity led to stress, anxiety, and emotional breakdown among them. After all these layers of vulnerability, climate change makes their existence more pathetic. The impact of increasing temperatures, prolonged droughts, an increasing number of heatwaves, and urban flooding has a greater effect on them compared to that on settled people there. Urban life offers no escape from the changing climate. In fact, cities amplify the effects of climate change. The phenomenon of urban heat islands makes summer temperatures unbearable, and erratic rainfall often leads to flash floods and waterlogging. For these workers, already living in substandard housing and working in open or poorly ventilated environments, exposure to these conditions takes a serious toll on their physical and mental health. In essence, migrant labourers remain trapped between failing rural livelihoods and unsustainable urban environments while carrying the invisible weight of psychological trauma. For both the source and the destinations, climate change has aggravated their vulnerability.

10.3.7 Major Findings

The case studies show how climate change, poor living conditions, and prevailing economic inequality impact the lives of workers in different sectors within India. These stories go beyond concerns of heat stress or risks at the workplace; they reveal structural weaknesses embedded within the labour hierarchy, social exclusion, and policy neglect (Ranjan 2016; Srivastava and Jha 2016). In agriculture, construction, or informal employment, the environmental and occupational challenges are borne mainly by the lowest strata of society—the migrant and women workers and other informal economy participants (Rao et al. 2020; Shah and Lerche 2020; Williams et al. 2022).

These studies also highlight the worrying trend of risk normalization in labour-intensive industries. Agricultural workers do not stop working in the field under the scorching sun, not because they are unaware of the risks, but because their need for income overshadows the possibility of incurring health complications (Venugopal et al. 2021a, b; Dutta et al. 2020). Likewise, construction and factory workers are exposed to dangerous working conditions with little to no protective measures, as the fear of job loss and reliance on meagre wages ensures that the bare minimum is demanded (Kreshpaj et al. 2022; Singh et al. 2019a, b). The informal nature of the work means that suffering due to climate and health shocks is inevitable, as there is no support from employers or protection from government policies (Jain et al. 2021; Srivastava 2020).

Vulnerabilities with gendered differences further deepen these problems. Women working in the informal sectors of cities like Indore not only perform strenuous work in dangerous environments but also shoulder the weight of unpaid reproductive work (Rao et al. 2019; Kesar and Bhattacharya 2020). Their economic contributions are

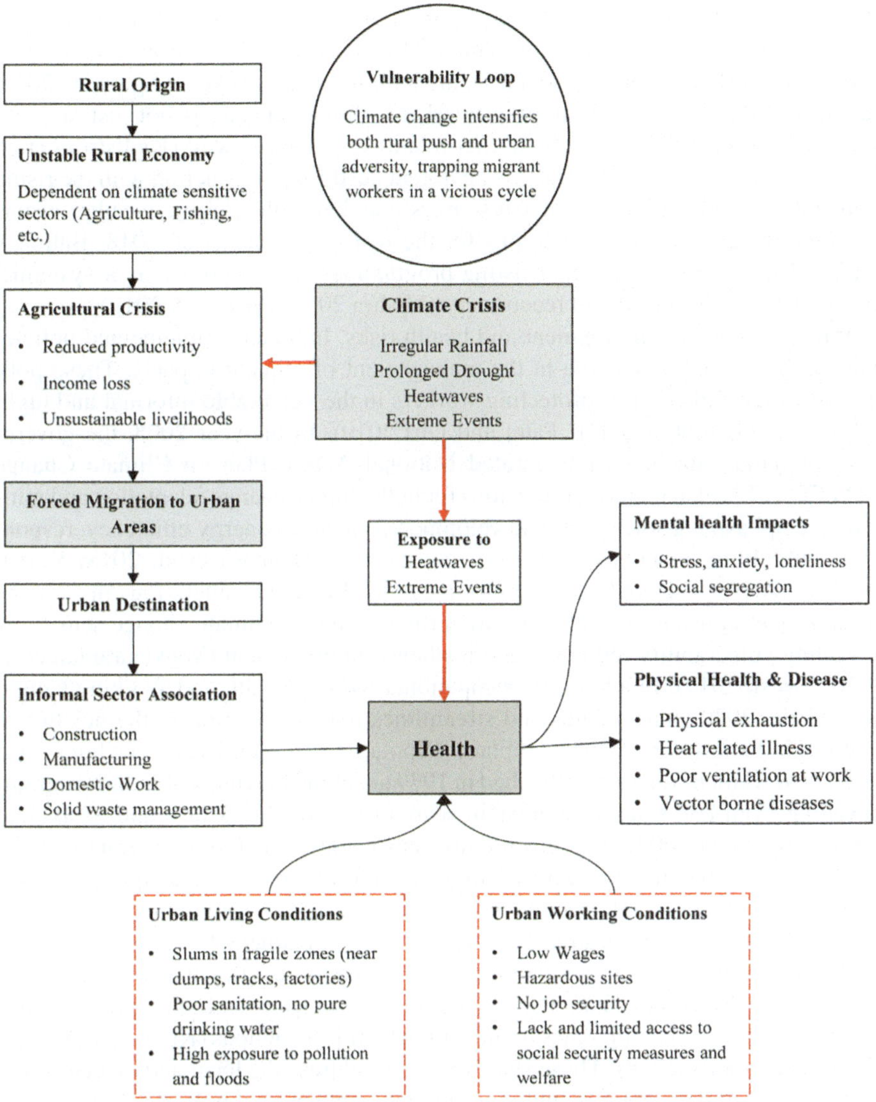

Fig. 10.1 Climate change and migrant labour trap: rural distress to urban precarity

undervalued as they suffer from greater occupational risks due to their inability to access basic sanitation, water facilities, and healthcare (Bhan et al. 2020). Within agricultural contexts, more frequent heatwaves coupled with erratic rainfall patterns lower productivity, which worsens food scarcity and intensifies rural distress (Geethalakshmi et al. 2017; Surie and Prasanna 2022). Besides personal health concerns, these examples demonstrate another form of governance failure by not providing adequate protection on labour and climate change adaptation (Garg et al. 2018;

Mikulewicz 2020). These workers live with no protection because there is no legislation providing basic safety requirements for the workplace, no enforcement of laws at the workplace, and no social security for informal workers (Agarwala 2019; Satpathy, 2018). Among the migrant workers, climate change is not just an environmental problem; it is a problem that relates to labour as well (Kjellstrom et al. 2016; Maharjan et al. 2020). There is an urgent need for policy action with the rising temperatures and shrinking water resources coupled with extreme weather events changing the employment conditions for the worse (Parsons et al. 2018; Bathiany et al. 2018). Unfortunately, the existing programmes are not proactive or systemic; they look more dispersed and reactionary (Mishra 2017; Patil et al. 2019).

With reference to employment and health risks, India has implemented policies that can be termed actionable in the management of climate impacts. These policies, however, fall short of protecting workers in the vulnerable informal and high-risk sectors (Jain et al. 2021; Patel and Giri 2019). In the year 2008, the government of India launched an Integrated National Action Plan on Climate Change (NAPCC). This plan includes provisions for both climate change adaptation and mitigation, emphasizing solar energy development, enhanced energy efficiency, responsible agricultural practices, and water conservation (Dubash et al. 2018; Kumar and Naik 2019). The NAPCC seeks to integrate climate considerations in as many areas of operations as possible with the primary goal of climate change sensitivity reframing vulnerability reduction and resilience improvement (Venkataraman et al. 2016; Jogesh 2019). In addition, Occupational Safety, Health and Working Conditions Code, 2020, consolidates and streamlines policies relating to the health and safety of Indian workers in the workplace (Papola 2020; Sarkar 2021). The Inter-State Migrant Workmen Act was established in 1979 to defend the rights of Indian migrants travelling from one state to another in search of work (Srivastava and Sasikumar 2003; Deshingkar 2009). This law requires businesses that employ migrant workers to register with the government and stipulates minimum wages, adequate working conditions, and reasonable accommodation requirements (Sengupta and Jha 2020; Roy 2018). Besides, it assures fair and unbiased treatment for migrant workers while preventing the exploitation of their provisional status (Rajan and Mishra 2020; Pattenden 2018). Besides national provisions, several states in India have legislated to address work and health issues found within their jurisdictions (Kumar and Managi 2020; Chakraborty 2018). These state strategies address the local and diverse needs of every region in the country and thus, complement national policies (Nathan et al. 2018; Roy 2019).

Even with the policies that address the impacts of climate change on occupational health and labour, there is still considerable difficulty in implementation. One main challenge is the lack of enforcement capacity, meaning that there are regulatory bodies that have insufficient personnel, skills, or resources to carry out proper monitoring (Pal and Dutta 2020; Roychowdhury et al. 2020). As a result, businesses can flout regulations without suffering serious consequences (Bhatt et al. 2018; Janardhanan et al. 2020). Fragmentation of authority into several institutions with overlapping national and state jurisdictions makes the situation worse, leading to a lack of understanding by employers on what the compliance obligations are (Sharma et al. 2019;

Lahiri-Dutt 2019). This leads to a lack of clarity among many employers with respect to what compliance needs are (Kumar et al. 2018; Mehrotra 2019). Coordination between the various governmental agencies charged with responsibility for labour, health, and climate policies is poor, and such agencies operating in isolation and with little communication, invariably leads to duplication of activities and gaps in service provision (Bhushan et al. 2020; Sharma and Dube 2019). In addition, monitoring compliance with labour and health regulations is expensive in terms of financial, personnel, and equipment resources which are not sufficient now (Basole 2021). In the absence of adequate spending for mobile inspection units, digital track-and-trace systems, and reporting infrastructures, there can be little meaningful enforcement of regulations, which is what sets the stage for regulations to be useless instead of effective (Das and Mishra 2020; Purohit and Purohit 2018).

Additionally, there is a limited awareness among workers and employers about occupational health risks. Many migrant workers know nothing about the possible risks associated with their jobs or the steps they should follow to keep themselves safe (Srivastava and Sutradhar 2016; Tripathy and Dash 2018). In addition, many healthcare professionals lack the necessary skills to identify and manage occupational health problems (Gulati et al. 2019; Tiwari et al. 2019). This discrepancy especially causes problems in rural or underdeveloped areas where migrant workers are somewhat common (Paliwal et al. 2020; Choudhury and Sindhi 2020). Furthermore, accurate reporting of occupational health issues can help one have the correct knowledge of their degree and breadth (Karan and Sakthivel 2016; Agrawal and Kalia 2020). Unfortunately, many events still go unreported or are not adequately documented, depleting the body of accurate data (Shyam 2019; Rajalakshmi and Mohan 2020). This shortcoming lowers the capacity to create meaningful policies and activities. Especially in industrial accidents or extreme storms, emergency response systems typically fail to handle health issues faced by migrant workers (Viswanathan et al. 2018; Nori et al. 2018). Limited resources, lack of coordination, and inadequate action can dramatically worsen the effects of crises on worker health.

Several policies have been introduced in India to address the issues of labour vulnerability. For instance, Mahatma Gandhi National Rural Employment Guarantee Act (MGNREGA), the Inter-State Migrant Workmen Act (GoI 1979), the National Disaster Management Plan (NDMA 2019), One Nation One Ration Card (ONORC) scheme, and the National and State Action Plans on Climate Change (NAPCC and SAPCCs). Recently in 2021, after exposing the vulnerability of the informal workforce during urban–rural exodus due to Covid-19, Government introduced e-Shram portal which is a national database for unorganized workforce (Press Information Bureau 2023). Additionally, ESIC was also formed to monitor the Employee's State Insurance (ESI) Scheme for the organized workforce. But all these schemes and initiatives are unable to meet the grievances of the migrant labours. These schemes may be successful in paper but not in reality. For instance, the MGNREGA promises to give 100 days' job guarantee for the people in the rural areas. The success in ground reality is yet to be verified (Mallik and Paltasingh 2024; Sukhtankar 2012). Although Government attempted to make a national database e-Shram, till now the Government does not have the complete information of the informal workforce as most of the people are not aware about its know-how and benefits. Migrant labourers, engaged in circular or seasonal patterns of employment, frequently fall through administrative gaps, lacking access to health care, accident protection, pensions, and other entitlements due to barriers such as portability issues, lack of documentation, and inadequate awareness (Deshingkar and Akter 2009).

To mitigate such risks, the government ought to broaden the scope of social security provisions so that entitlements accrue to individuals regardless of their geographic location, the specific nature of their occupation, or their migration status. Care should be taken about the portability of the welfare schemes such as ration and healthcare facilities. The schemes like ONORC should be effectively implemented across the states just like across the districts (Tumbe and Jha 2024). The occupational health and safety standards should be upgraded and incorporate the climate-related risk prevention, provision of shade, drinking water, work break, and flexible working hours. Additionally, climate-resilient infrastructure should be provided in the informal settlements to prevent from heat stress, waterlogging, vector-borne diseases. The urban and disaster management planners must go with sustainable urban development to include all strata of the society for getting the benefits of urban infrastructure and provision essential services like emergency relief, food, and healthcare services (United Nations Human Settlements Programme 2020). Additionally, policymakers should focus on strengthening the early warning network system, climate surveillance. Eventually all these measures should be aligned with Sustainable Development Goals, for reducing poverty, inequality, maintaining decent work and good health, and sustainability for creating a more inclusive and effective development.

10.4 Conclusion

The intersection of climate change, workforce exploitation, and occupational health hazards poses a complex challenge to the Indian migrant labour force, especially those in the informal economy. In the present study, through the sector-specific analysis and case studies, it has become evident that migrant labourers in agriculture, construction, and urban informal work environments face compounded risks due to extreme climatic conditions and systemic socio-economic inequalities. The prevalence of hazardous working conditions, inadequate social protections, and limited access to healthcare exacerbates their vulnerabilities, with significant implications for their health, productivity, and economic security.

The National Action Plan on Climate Change (NAPCC), the Occupational Safety, Health and Working Conditions Code, 2020, and the Inter-State Migrant Workmen Act provide some attention to worker exploitation and health. However, the fragmented implementation and weak enforcement of these policies result in substantial gaps in worker protections. This underscores the necessity for a more integrated and robust approach to safeguarding worker health and wellbeing.

Addressing these challenges requires targeted interventions that bridge policy and practice. The strengthening of primary healthcare services in labour-intensive areas, instituting a specialized occupational health programme, and upgrading health surveillance systems are imperative for dealing with emerging climate change health threats. Workplace interventions aimed at heat stress mitigation, increased ventilation, provision of personal protective equipment (PPE), and periodic health examinations can greatly reduce health hazards associated with work. Furthermore, robust systems must be put in place to guard workers against industrial mishaps and extreme weather conditions for emergency response.

To facilitate meaningful development, there needs to be improved and stronger enforcement mechanisms and better coordination among government agencies are essential. Male and female migrant workers have multiple vulnerabilities which can be addressed through social security coverage of informal workers and implementation of gender-based policies. There is no doubt that governments, industries, and civil society need to work together if we are to broaden labour protection and include the health of workers in climate adaptation strategies.

Ultimately, a healthy and empowered workforce in India is fundamental for the economic growth, social justice, and ecological wellbeing of the country. As climate change transforms working environments, there is a need to prioritize employee health, safety, and financial stability to create a more balanced and diverse workforce. Millions of labourers, who are the backbone of the nation, will only be protected through coordinated political action, sufficient funding, and continuous evaluation. Hence, policymakers should focus on an inclusive approach that integrates labour rights, environmental governance, and public health.

Acknowledgements Not applicable.

Authors Contribution: Dr. Rameswar Mukherjee and Dr. Pamela Deb jointly conducted the comprehensive literature review, conceptualized the structure of the paper, and prepared the initial draft. Pamela Deb further contributed to the refinement of the conceptual framework, carried out critical revisions, and finalized the manuscript for submission. Both authors reviewed and approved the final version of the manuscript.

Availability of Supporting Data: This study has taken secondary data from data.gov.in website, Open Government Data (OGD) Platform India.

Funding The research was not supported by any funding.

Potential Conflicts of Interest: There are no conflicts of interest.

References

Agarwal S, Taneja S (2019) All slums are not equal: child health conditions among the urban poor. Indian Pediatr 42(3):233–244

Agarwala R (2019) The politics of India's reformed labor model. Curr Hist 118(807):123–129

Agrawal T, Kalia S (2020) Occupational health and safety management in developing countries: a study of India. J Work Health Saf 68(12):582–591

Avashia V, Garg A (2020) Implications of land use transitions and climate change on local flooding in urban areas: an assessment of 42 Indian cities. Land Use Policy 95:104571. https://doi.org/10.1016/j.landusepol.2020.104571

Ayyappan R, Sankar S, Rajkumar P, Balakrishnan K (2009) Work-related heat stress concerns in automotive industries: a case study from Chennai, India. Glob Health Action 2(1):2060

Balakrishnan K, Dey S, Gupta T, Dhaliwal RS, Brauer M, Cohen AJ, Dandona L (2019) The impact of air pollution on deaths, disease burden, and life expectancy across the states of India: the global burden of disease study 2017. Lancet Planet Health 3(1):e26–e39

Banerjee S, Goswami K (2019) The occupational illness of slum dwellers across industries: a case study in West Bengal. Int J Ind Ergon 73:102834. https://doi.org/10.1016/j.ergon.2019.102834

Basole A (2021) Informality and precarity in India's labour markets. Econ Pol Wkly 56(9):45–52

Bathiany S, Dakos V, Scheffer M, Lenton TM (2018) Climate models predict increasing temperature variability in poor countries. Sci Adv 4(5):eaar5809

Bhagat RB (2017) Migration, gender and right to the city: the Indian context. Econ Pol Wkly 52(32):35–40

Bhan G, Surie A, Srivatsa S, Anand R, Rao N (2020) Urban health and packed food: consumption patterns of workers in India's informal economy. J Urban Health 97(4):517–529

Bhatt M, Jain A, Gupta OM (2018) Occupational safety and health in India: a study of enforcement agencies. Saf Sci 113:461–471

Bhatta GD, Aggarwal PK, Poudel S, Belgrave DA (2015) Climate-induced migration in South Asia: migration decisions and the gender dimensions of adverse climatic events. J Rural Community Dev 10(4):1–23

Bhoi SR, Joshi SH, Joshi A (2022) Out-of-Pocket health expenditure among migrant workers in India: a narrative review. Cureus 14(10) https://doi.org/10.7759/cureus.30948

Bhugra D, Gupta S, Bhui K, Singh J (2013) Migration and mental health. Lancet 381(9860):125–135

Bhushan C, Goswami N, Kumarankandath A (2020) State of India's environment: a citizens' report. Econ Pol Wkly 55(1):19–22

Biswas G, Bhattacharya A, Bhattacharya R (2017) Occupational health status of construction workers: a review. Int J Med Sci Public Health 6(4):669–675. https://doi.org/10.5455/ijmsph.2017.0745302112016

Borhade A, Dey S, Tripathi A, Mavalankar D, Ramani KV (2020) Migration and health: a review of policies and initiatives in low and middle-income countries. Lancet Glob Health 8(5):e1296–e1306

Breman J (2016) On pauperism in present and past. Oxford University Press

Breman J (2013) At work in the informal economy of India: a perspective from the bottom up. Oxford University Press

Chakraborty S (2018) Can the labour code on occupational safety, health and working conditions protect Indian workers? Econ Pol Wkly 53(38):19–22

Chandrasekhar S, Sharma A (2014) Internal migration for education and employment among youth in India. Indira Gandhi Institute of Development Research

Choithani C (2017) Understanding the linkages between migration and household food security in India. Geogr Res 556681(2):192–205

Choudhury PR, Sindhi S (2020) Internal migration in India: distress and opportunities, a study of migrant workers in India. Int J Migr Dev 2(1):132–153

Das S, Mishra AJ (2020) Promoting occupational health and safety climate in India. Am J Ind Med 63(8):663–675

Dash SK, Kjellstrom T (2011) Workplace heat stress in the context of rising temperature in India. Curr Sci 101(4):496–503

Denis E, Zérah MH (eds) (2017) Subaltern urbanisation in India: an introduction to the dynamics of ordinary towns. Springer

Deshingkar P (2012) Environmental risk, resilience and migration: implications for natural resource management and agriculture. Environ Res Lett 7(1):015603

Deshingkar P (2017) Towards contextualised, disaggregated and intersectional understandings of migration in India. Asian Popul Stud 13(2):119–135
Deshingkar P, Akter S (2009) Migration and human development in India (Human Development Research Paper No. 13). United Nations Development Programme, Human Development Report Office. https://hdr.undp.org/
Deshingkar P, Start D (2003) Seasonal migration for livelihoods in India: Coping, accumulation and exclusion. Overseas Development Institute
Deshingkar P (2009) Migrant transfers and local development (Working Paper 2009(34)). Overseas Development Institute
Dodman D, Sverdlik A, Agarwal S, Kadungure A, Kothiwal K, Machemedze R, Verma S (2023) Climate change and informal workers: towards an agenda for research and practice. Urban Clim 48:101401. https://doi.org/10.1016/j.uclim.2022.101401
Dubash NK, Khosla R, Rao ND, Bhardwaj A (2018) India's energy and emissions future: an interpretive analysis of model scenarios. Environ Res Lett 13(7):074018
Dutta S, Ghosh S, Gopalakrishnan S, Bijoy CR, Yasmin H (2020) Climate-dependent heat stress in the context of Indian agricultural labor. Int J Biometeorol 64(11):1889–1902
Employees' State Insurance Corporation (n.d.) About us. https://www.esic.gov.in/about-us
Farbotko C, Kitara T, Dun O, Evans C (2022) A climate justice perspective on international labour migration and climate change adaptation among Tuvaluan workers. Oxf Open Clim Chang 2(1):kgac002. https://doi.org/10.1093/oxfclm/kgac002
Fazel M, Reed RV, Panter-Brick C, Stein A (2019) Mental health of refugees and asylum seekers. Lancet 376(9755):1801–1811
Frounfelker RL, Rousseau C (2018) The UCL–lancet commission on migration and health. Lancet 392(10164):2530–2531
Garg T, Jagnani M, Taraz V (2018) Temperature and human capital in India. J Assoc Environ Resour Econ 7(6):1113–1150
Geethalakshmi V, Lakshmanan A, Rajalakshmi D, Jagannathan R, Sridhar G, Ramesh AP, Bhuvaneshwari K, Gurusamy L, Anbhazhagan R, Bhuvaneswari K, Kokilavani S, Kowsalya S (2017) Climate change impact assessment and adaptation strategies to sustain rice production in Cauvery basin of Tamil Nadu. Curr Sci 112(2):256–263
Government of India (1979) The Inter-State Migrant Workmen (Regulation of Employment and Conditions of Service) Act, 1979. https://labour.gov.in/
Government of India (2008) National Action Plan on Climate Change. Prime Minister's Council on Climate Change, Ministry of Environment and Forest
Government of India, Ministry of Labour and Employment (1979) The Inter State Migrant Workmen (Regulation of Employment and Conditions of Service) Act, 1979 (Act No. 30 of 1979, assented June 11, 1979; commenced June 1, 1987). https://www.indiacode.nic.in/bitstream/123456789/13209/1/the_inter-state_migrant_workmen_regulation_of_employment_and_conditions__of_service_act_1979.pdf
Gulati S, Paterson M, Medves J, Luce-Kapler R (2019) Occupational health issues of hospital workers in India: findings from focus group discussions. Work 62(4):609–619
Guttikunda SK, Jawahar P (2018) Atmospheric emissions and pollution from the coal-fired thermal power plants in India. Atmos Environ 92:449–460
Hannan A, Hossain F (2021) COVID-19, migrant workforce and regional health infrastructure in North Bengal. Soc Chang Dev 18(1–2):103–116
Heinz A, Brandt L (2024) Climate change and mental health: direct, indirect, and intersectional effects. Lancet RegNal Health-Eur 43:100969. https://doi.org/10.1016/j.lanepe.2024.100969
Deccan Herald (2024) Migrant workers struggle to survive amid the rubble of climate change. https://www.deccanherald.com/india/karnataka/bengaluru/migrant-workers-struggle-to-survive-amid-the-rubble-of-climate-change-3288246
Immerzeel WW, Lutz AF, Andrade M, Bahl A, Biemans H, Bolch T, Hyde S, Brumby S, Davies BJ, Elmore AC, Emmer A, Feng M, Fernández A, Haritashya U, Kargel JS, Koppes M, Kraaijenbrink

PDA, Kulkarni AV, Mayewski PA, Baillie JEM (2020) Importance and vulnerability of the world's water towers. Nature 577(7790):364–369

Jain VK, Sharma A (2019) Social determinants of health among migrant workers in India. Indian J Community Med 44(2):159–162

Jain P, Sharma A, Khanna P (2021) Effects of climate change on occupational health and productivity in the unorganized sector in India. SAGE Open 11(3):21582440211030228

Janardhanan N, Ikeda E, Ikeda M (2020) Regulatory institutions and policy coordination for climate action in Japan and India. Clim Policy 20(6):697–709

Jogesh A (2019) Inter-ministerial coordination on climate policy in India: Exploring the mechanisms. Environ Policy Gov 29(6):440–450

Karan A, Sakthivel S (2016) Gender and informality in the Indian labour market. Indian J Labour Econ 59(3):359–382

Kattumuri R, Ravindranath D, Esteves T (2017) Local adaptation strategies in semi-arid regions: study of two villages in Karnataka, India. Clim Dev 9(1):36–49

Kesar S, Bhattacharya S (2020) Dualism and structural transformation: the informal manufacturing sector in India. Eur J Dev Res 32(4):1185–1215

Keshri K, Bhagat RB (2012) Temporary and seasonal migration: Regional pattern, characteristics and associated factors. Econ Polit Wkly, 81–88

Keshri K, Bhagat RB (2013) Socioeconomic determinants of temporary labour migration in India. Asian Popul Stud 9(2):175–195

Kjellstrom T, Briggs D, Freyberg C, Lemke B, Otto M, Hyatt O (2016) Heat, human performance, and occupational health: a key issue for the assessment of global climate change impacts. Annu Rev Public Health 37:97–112

Koll R, Krishnan R, Sinha A, Roxy MK (2019) Extreme events and climate adaptation-mitigation linkages in India. Curr Sci 117(2):161–163

Kreshpaj B, Orellana C, Burström B, Davis L, Hemmingsson T, Johansson G, Kjellberg K, Jonsson J, Wegman DH, Bodin T (2022) What is precarious employment? A systematic review of definitions and operationalizations from quantitative and qualitative studies. Scand J Work Environ Health 48(1):79–90

Kulkarni H, Narlawar UW, Sukhsohale ND, Thakre SS, Ughade SN (2016) Prevalence and risk factors for chronic bronchitis among rural women: a community-based study. Indian J Community Health 28(3):225–230

Kumar P, Naik GD (2019) Climate change and agriculture in India: impacts and adaptation strategies. J Environ Biol 40(3):352–358

Kumar A, Singh DP (2021) Heat stroke-related deaths in India: an analysis of natural causes of deaths, associated with the regional heatwave. J Therm Biol 95:102792. https://doi.org/10.1016/j.jtherbio.2020.102792

Kumar S, Sharma J, Jat SR (2018) Occupational health and safety in Indian micro, small and medium enterprises. Econ Pol Wkly 53(41):61–66

Kumar P, Rupa R, Ravinder J, Kattarkandi B, Manjeet S (2020a) Climate change impacts on crop yield, crop water productivity and food security—a review. Prog Nat Sci 29(1):16–28

Kumar S, Sharma J, Sirohi C (2020b) Climate change and Indian agriculture: impact and adaptation strategies. Indian J Agric Sci 90(6):935–939

Kumar KS, Viswanathan B (2012) Weather variability and agriculture: Implications for long and short-term migration in India. Centre for Development Economics, Delhi School of Economics

Kundu A (2018) Urbanisation in India: trends, patterns and policy challenges. In: Subaltern urbanisation in India, pp 53–74. Springer

Lahiri-Dutt K (2019) The coal nation': Histories, ecologies and politics of coal in India. Taylor & Francis

Lerche J (2010) From rural labour to classes of labour: class fragmentation, caste and class struggle at the bottom of the Indian labour hierarchy. In: The comparative political economy of development, pp 90–118. Routledge

Lundgren K, Kuklane K, Gao C, Holmér I (2013) Effects of heat stress on working populations when facing climate change. Ind Health 51(1):3–15
Mahapatra SK, Ratha KC, Mahapatra S (2018) Climate change and migration: evidence from the vulnerable districts of Odisha, India. Int J Environ Sci Dev 9(2):46–52
Maharjan A, de Campos RS, Singh C, Das S, Srinivas A, Bhuiyan MRA, Ishaq S, Umar MA, Dilshad T, Shrestha K, Bhadwal S, Ghosh T, Suckall N, Vincent K (2020) Migration and household adaptation in climate-sensitive hotspots in South Asia. Curr Clim Chang Rep 6(1):1–16
Mallik P, Paltasingh T (2024) MGNREGA and its impact on tribal communities: a policy perspective. Bihar J Public Adm 21(1):16–27
Mazdiyasni O, AghaKouchak A, Davis SJ, Madadgar S, Mehran A, Ragno E, Sadegh M, Sengupta A, Ghosh S, Dhanya CT, Niknejad M (2017) Increasing probability of mortality during Indian heat waves. Sci Adv 3(6):e1700066
Mehrotra S (2019) Informal employment trends in the Indian economy: Persistent informality, but growing positive development (Working Paper 254). ILO Employment Policy Department
Mikulewicz M (2020) The discursive politics of adaptation to climate change. Ann Am Assoc Geogr 110(6):1807–1830
Ministry of Environment, Forest and Climate Change (2008) National action plan on climate change (NAPCC). Government of India. https://moef.gov.in/
Ministry of Environment, Forest and Climate Change (2008) National Action Plan on Climate Change (NAPCC). Government of India. https://moef.gov.in/uploads/2018/04/NAP_E.pdf
Mishra DK (2016) Seasonal migration and unfree labour in globalising India: insights from field surveys in Odisha. Indian J Labour Econ 59(1):25–48
Mishra D (2017) Internal migration and labour market integration: breakthroughs in research and practice. Econ Pol Wkly 52(7):19–22
Mishra V, Thirumalai K, Singh D, Aadhar S (2020) Future exacerbation of hot and dry summer monsoon extremes in India. NPJ Clim Atmos Sci 3(1):1–9
Mohan M, Pant K, Upadhyay P, Upadhyay SK (2016) Living on the edge: vulnerability of migrant workers in India. Indian J Labour Econ 59(1):125–140
Mosse D, Gupta S, Mehta M, Shah V, Rees JF, Team KP (2002) Brokered livelihoods: Debt, labour migration and development in tribal western India. J Dev Stud 38(5):59–88
Mukherji A (2018) Sustainable groundwater management in India needs a water-energy-food nexus approach. Appl Econ Perspect Policy 40(4):644–667
Nag PK, Nag A, Sekhar P, Pandit S (2009) Vulnerability to heat and respiratory distress among workers in different occupations in India. Am J Ind Med 52(1):45–56
Nandi J (2024) India reported 40,000 suspected heat stroke cases this summer. Hindustan Times. https://www.hindustantimes.com/india-news/india-reported-40-000-suspected-heat-stroke-cases-this-summer-101721933015106.html
Nathan D, Tewari I, Sarkar S (2018) Labour markets in India: developments and challenges. J Adv Stud Humlties Soc Sci 7(4):402–424
National Commission for Enterprises in the Unorganised Sector (2009) The challenge of employment in India: an informal economy perspective. Academic Foundation
National Disaster Management Authority (2019) National Disaster Management Plan (NDMP 2019). Government of India. https://ndma.gov.in/sites/default/files/PDF/ndmp-2019.pdf
National Sample Survey Organisation (2012) Employment, unemployment and migration survey, July 2007–June 2008. Ministry of Statistics and Programme Implementation, Government of India
NCEUS (2009) The challenge of employment in India: an informal economy perspective. Government of India. https://ruralindiaonline.org/en/library/resource/the-challenge-of-employment-in-india-an-informal-economy-perspective/
Nirmala CJ, Prasad SD (2019) Occupational hazards and public health concerns of migrant construction workers: an epidemiological study in southern India. Int J Community Med Public Health 6:818–822. https://doi.org/10.18203/2394-6040.ijcmph20190214

Nori M, Taylor M, Sensi A (2018) Agro-pastoralism and climate change: risks, vulnerability and resilience. Sci Total Environ 645:643–654

Pal R, Dutta P (2020) Non-standard employment in India: trend and determinants. Econ Pol Wkly 55(32):61–70

Paliwal A, Singh S, Mohanty S, Sharma C (2020) Climate change and Indian agriculture: impacts on crop productivity. Indian J Agric Sci 90(6):1046–1060

Papola TS (2020) Labour, employment and economic growth in India. Cambridge University Press

Parasuraman G, Unni J, Srivastava R (2018) Informal economy in India: perspectives and policies. Routledge India

Pareek U, Sole NA (2020) Indian migrant workers during the COVID-19 pandemic: perspectives and policy responses of the state. In: Kumar A, Nayak JK, Mishra V (eds) COVID-19: diagnosis and management-part II. Bentham Science Publishers, pp 267–289

Parsons LA, Shindell D, Tigchelaar M, Zhang Y, Spector JT (2018) Heightened risk of heat stress to human health with rising temperature. Environ Res Lett 13(6):064007

Patel SK, Giri B (2019) Climate change, migration and women: analysing construction workers in Odisha. Soc Chang 49(1):97–113

Patel SA, Joshi GJ (2018) An empirical study of risks in construction projects. Int J Eng Manag Res 8(6):45–52

Patil R, Roy S, Kamath R (2019) Climate change and mental health: a study of vulnerability of coastal communities in Karnataka, India. Clim Dev 11(8):679–685

Pattanaik DR, Mohapatra M, Srivastava AK, Kumar A (2017) Heat wave over India during summer 2015: an assessment of real time extended range forecast. Meteorol Atmos Phys 129(4):375–393

Pattenden J (2018) The politics of classes of labour: fragmentation, reproduction zones and collective action in Karnataka, India. J Peasant Stud 45(5–6):1039–1059

Press Information Bureau (2023) Salient features of e-Shram portal. Government of India. https://www.pib.gov.in/PressReleasePage.aspx?PRID=1943207

Purohit P, Purohit K (2018) Renewable energy for sustainable development: key issues and policy priorities. Energ Effi 11(5):1233–1251

Rai P (2020) Seasonal masculinities: seasonal labor migration and masculinities in rural western India. Gend Place Cult 27(2):260–280

Rajalakshmi K, Mohan VR (2020) Occupational health profile of workers employed in the manufacturing sector of India. Natl Med J India 33(1):8–13

Rajan SI, Bhagat RB (2022) Internal migration and the Covid-19 pandemic in India. In Migration and pandemics. Springer, pp 227–241. https://doi.org/10.1007/978-3-030-81210-2_12

Rajan SI, Mishra US (2020) Handbook of internal migration in India. SAGE Publishing India

Ranjan P (2016) Occupational health hazards among migrant construction workers in India. Int J Med Public Health 6(3):115–120

Rao N, Lawson ET, Raditloaneng WN, Solomon D, Angula MN (2019) Gendered vulnerabilities to climate change: insights from the semi-arid regions of Africa and Asia. Climate Dev 11(1):14–26

Rao N, Singh C, Solomon D, Camfield L, Sidiki R, Angula M, Poonacha P, Sidibe A, Lawson ET (2020) Managing risk, changing aspirations and household dynamics: implications for wellbeing and adaptation in semi-arid Africa and India. World Dev 125:104667

Revi A (2008) Climate change risk: an adaptation and mitigation agenda for Indian cities. Environ Urban 20(1):207–229

Revi A, Satterthwaite DE, Aragón-Durand F, Corfee-Morlot J, Kiunsi RBR, Pelling M, Roberts DC, Solecki W (2014) Urban areas. In: Field CB, Barros VR, Dokken DJ, Mach KJ, Mastrandrea MD, Bilir TE, Chatterjee M, Ebi KL, Estrada YO, Genova RC, Girma B, Kissel ES, Levy AN, MacCracken S, Mastrandrea PR, White LL (eds) Climate change 2014: impacts, adaptation, and vulnerability. Part A: global and sectoral aspects. Contribution of Working Group II to the Fifth Assessment Report of the Intergovernmental Panel on Climate Change, pp 535–612. Cambridge University Press

Rodell M, Famiglietti JS, Wiese DN, Reager JT, Beaudoing HK, Landerer FW, Lo MH (2018) Emerging trends in global freshwater availability. Nature 557(7707):651–659

Roxy MK, Ritika K, Terray P, Murtugudde R, Ashok K, Goswami BN (2015) Drying of Indian subcontinent by rapid Indian ocean warming and a weakening land-sea thermal gradient. Nat Commun 6(1):7423

Roy S (2019) Discriminations in urban labor markets and rural-urban migration. World Dev 117:327–343

Roy AK, Singh P, Roy UN (2015) Impact of rural-urban labour migration on education of children: a case study of left behind and accompanied migrant children in India. Space Cult 2(4):17–34. https://doi.org/10.20896/saci.v2i4.74

Roy I (2018) Marginal matters: migrant labor in India. SpringerBriefs in Political Science

Roychowdhury A, Nasim U, Chandola T (2020) Occupational health, safety and well-being in India: status, challenges and opportunities. Prog Disaster Sci 7:100121

Sahu S, Sett M, Kjellstrom T (2013) Heat exposure, cardiovascular stress and work productivity in rice harvesters in India: implications for a climate change future. Ind Health 51(4):424–431

Sarkar R (2019) Urbanisation in India before and after the economic reforms: what does the census data reveal? J Asian Afr Stud 54(8):1213–1226

Satpathy A (2018) Informalisation of work and the fading dream of an egalitarian society. Indian J Labour Econ 61(3):403–414

Sengupta S, Jha MK (2020) Social policy, COVID-19 and impoverished migrants: challenges and prospects in locked down India. Int J Community Soc Dev 2(2):152–172

Sett M, Sahu S (2014) Effects of occupational heat exposure on female brick workers in West Bengal, India. Glob Health Action 7(1):21923

Shah A, Lerche J (2020) Migration and the invisible economies of care: production, social reproduction and seasonal migrant labour in India. Trans Inst Br Geogr 45(4):719–734

Sharma AN (2005) Agrarian relations and socio-economic change in Bihar. Econ Pol Wkly 40(10):960–972

Sharma AN (2006) Flexibility, employment and labour market reforms in India. Econ Pol Wkly 41(21):2078–2085

Sharma R, Dube S (2019) Climate change and its impact on rainfed agriculture in India. Int J Adv Sci Technol 28(2):275–283

Sharma S, Mujumdar P (2017) Increasing frequency and spatial extent of concurrent meteorological droughts and heatwaves in India. Sci Rep 7(1):15582

Sharma K, Gulati A, Jha G (2019) Assessing agricultural policies of India. Econ Pol Wkly 54(52):12–15

Sharma A, Anand A, Roy N (2020) Mapping poverty and migration in relation to climate vulnerability: a case study of Odisha, India. Int J Popul Res 2020:3672342

Sharma S, Mujumdar P, Goswami BN (2021) Multidecadal variations in phytoplankton phenology in the Arabian Sea: linkages to climate variability. Prog Oceanogr 193:102550

Shyam SS (2019) Climate-induced migration and refugees: a case study from the Sundarbans deltaic system. Forced Migr Rev 63:44–46

Singh MS, Ghosh S, Sahana AS (2019a) Future changes in extreme precipitation over the Himalayan region as inferred from CORDEX-South Asia projections. Clim Dyn 52(5):3625–3640

Singh R, Bhattacherjee A, Chau N (2019b) Analyses of injury incidence and duration among construction workers in India. Int J Inj Contr Saf Promot 26(1):62–71

Srinivasan K, Jain P, Sharma S, Singh A (2016) Occupational health and safety in Indian construction industry: a case study. Int J Eng Res Technol 5(1):594–598

Srivastava R (2020) Growing precarity, circular migration, and the lockdown in India. Indian J Labour Econ 63(4):1–8. https://doi.org/10.1007/s41027-020-00260-3

Srivastava R, Jha A (2016) Capital and labour standards in the organised construction industry in India: a study based on fieldwork in the National Capital Region of Delhi. Jawaharlal Nehru University, Centre for the Study of Regional Development

Srivastava R, Kumar S (2018) Internal migration and employment in India: recent trends and patterns. Indian J Labour Econ 61(3):417–441

Srivastava R, Sutradhar R (2016) Labour migration to the construction sector in India and its impact on rural poverty. Indian J Hum Dev 10(1):27–48

Srivastava R, Sasikumar SK (2003) An overview of migration in India, its impacts and key issues. In: Regional conference on migration, development and pro-poor policy choices in Asia, pp 22–24

Sukhtankar S (2012) Corruption and the Mahatma Gandhi National rural employment guarantee act. Ideas for India. https://www.ideasforindia.in/topics/governance/corruption-and-the-mahatma-gandhi-national-rural-employment-guarantee-act.html

Surie A, Prasanna S (2022) Climate change, agrarian distress, and farmer suicides in India. World Dev 151:105773

Sverdlik A, Kothiwal K, Kadungure A, Agarwal S, Machemedze R, Verma S, Loewenson R (2024) Understanding the interplay of occupational, public health, and climate-related risks for informal workers: a new framework with findings from Zimbabwe and India. Soc Sci Med 348:116750. https://doi.org/10.1016/j.socscimed.2024.116750

The inter-state migrant workmen (regulation of employment and conditions of service) act (1979). https://www.indiacode.nic.in/bitstream/123456789/13209/1/the_inter-state_migrant_workmen_regulation_of_employment_and_conditions__of_service_act_1979.pdf

The occupational safety, health and working conditions code (2020) Ministry of Labour and Employment. https://labour.gov.in/sites/default/files/OSH_Gazette.pdf

Thomas BE, Charles N, Watson B, Chandrasekaran V, Kumar RS, Dhanalakshmi A, Swaminathan S (2018) Prevalence of chest symptoms amongst brick kiln migrant workers and care seeking behaviour: a study from South India. J Public Health 40(3):625–633

Tiwari RR, Zodpey SP, Deshpande SG (2019) Occupational health education in India: current status and future needs. J Occup Health 61(3):238–244

Tripathy SN, Dash JK (2018) Migrant labour and issue of social security in India. Sarup Book Publishers

Tumbe C, Jha RK (2024) One nation, one ration, limited interstate traction: migration and PDS portability in India. Urbanisation 9(1):87103. https://doi.org/10.1177/24557471241237098

United Nations Development Programme (2023) Sustainable development goals. https://www.undp.org/sustainable-development-goals

United Nations Human Settlements Programme (2020) World Cities Report 2020: the value of sustainable urbanization. UN-Habitat. https://unhabitat.org/world-cities-report-2020-the-value-of-sustainable-urbanization

Venkataraman C, Ghosh S, Thakur JS (2016) Climate change and its impact on health and livelihoods in India. Econ Pol Wkly 51(8):44–51

Venugopal V, Chinnadurai JS, Lucas RA, Kjellstrom T (2016) Occupational heat stress profiles in selected workplaces in India. Int J Environ Res Public Health 13(1):89

Venugopal V, Latha PK, Shanmugam R, Krishnamoorthy M, Omprashanth R, Lennqvist R, Johnson P (2020) Epidemiological evidence from south Indian working population—the heat exposures and health linkage. J Eposure Sci Environ Epidemiol 31(1):177–186. https://doi.org/10.1038/s41370-020-00261-w

Venugopal V, Chinnadurai JS, Lucas RA, Kjellstrom T (2021a) Heat stress in South India: perceptions of occupational health risks and adaptation behaviors of vulnerable communities. Int J Environ Res Public Health 18(12):6501

Venugopal V, Latha PK, Shanmugam R, Krishnamoorthy M, Rehman T, Balakrishnan K, Lucas RA (2021) Heat stress, its effect and vulnerability on the agricultural farmers of Tamil Nadu, India. Sci Total Environ 757:143761

Viswanathan B, Kumar KSK (2015) Weather, agriculture and rural migration: evidence from state and district level migration in India. Environ Dev Econ 20(4):469–492

Viswanathan PK, Mishra P, Bhattarai M, Iyengar K (2018) Institutional linkages and climate change adaptation: the case of rainfed agriculture in India. Econ Pol Wkly 53(25):29–38

Williams G, Guérin I, Venkatasubramanian G, Kock S (2022) Making social protection work for the informal economy: adaptation of national provisions to protect informal workers during the COVID-19 crisis in India

Woetzel J, Pinner D, Samandari H, Engel H, Krishnan M, Boland B, Powis C (2020) Climate risk and response: Physical hazards and socioeconomic impacts. McKinsey & Company. https://www.mckinsey.com/capabilities/sustainability/our-insights/climate-risk-and-response-physical-hazards-and-socioeconomic-impacts

Chapter 11
Safe, Sustainable, and Ready: Climate-Resilient Healthcare Facilities in South Asia

Mohd Mudassir, Shubham Tandon, Sakshi Bhutani, and Hari Kumar

Abstract Healthcare facilities are vital institutions for providing medical services to the nearby population on a regular basis and during/after disaster emergencies; thus, their safety and functionality remain essential after any climatic events for community resilience. Disruption of healthcare facilities due to disasters associated with climate change increases the vulnerability of surrounding populations (https://www.hospitalesporlasaludambiental.net/wpcontent/uploads/2016/07/Global-Green-and-Healthy-Hospitals-Agenda.pdf). The health sector is a significant contributor to global carbon emissions due to its energy-intensive operations and resource consumption. Healthcare facilities are both significant contributors to carbon emissions and vulnerable to the impacts of climate events; it is essential to integrate both resilience and minimal environmental impact. This paper outlines the key important factors that can keep health facilities operational during and after extreme events while minimizing carbon footprints based on learnings from several hospital safety and resilience projects implemented in parts of South Asia and India. The objective is to build the climate and disaster resilient healthcare facilities to withstand climate-related hazards and promote sustainable and climate-smart practices. A qualitative method was adopted with numerous case studies to highlight the narratives around the topic. The results contribute to high level recommendations and policy-oriented actions for healthcare professionals, decision makers, and health agencies.

Keywords Climate change · Healthcare facility · Safety and functionality · Green measures · Sustainability · South Asia

---

M. Mudassir (✉) · S. Tandon
United Nations Development Programme, New Delhi, India
e-mail: mudassir.afz@gmail.com

S. Tandon
e-mail: Subham.tandon@undp.org

S. Bhutani
Chandigarh, India

H. Kumar
GeoHazards Society, New Delhi, India

P. Ghosh et al. (eds.), *Climate Crises and Public Health Challenges*, Disaster Risk Reduction, https://doi.org/10.1007/978-981-95-9709-3_11

11.1 Introduction

11.1.1 Background of the Study

The world is grappling with a wide range of climate induced health challenges ranging from heat stress, floods, changing disease patterns, water-and-vector borne diseases. These issues are further aggravated by the occurrence of frequent and intense hydro-meteorological events. The impact of climate change is both direct and indirect on human health. Direct impacts can be attributed to extreme weather events like floods or cyclones which can adversely affect health infrastructure, workforce, and patient load. Thus, mortality and morbidity can be more due to such failure of health facilities following such events. Indirect impacts of climate change mainly arise due to environmental mediation, wherein alterations such as increased humidity and variability in precipitation patterns contribute to secondary effects. These include elevated levels of air pollution, as well as heightened risks of food-, water-, and vector-borne diseases. Moreover, shift in temperature pattern may disturb the ecosystem, limit fresh water supply, and exacerbate air pollution and catalyze infectious diseases (WHO Fact Sheet 2018).

The phenomenon of Climate Change not only acts as a risk factor but also triggers the already existing vulnerabilities. It exacerbates pre-existing health vulnerabilities, deteriorates overall population health, and undermines the capacity of health systems to provide continuous, high-quality care (Sherman et al. 2023). When healthcare facilities are exposed to climate-induced hazards, vulnerabilities so produced extend beyond health-related impacts to larger social and economic repercussions. The health impact of hospitals being affected by climate linked hazards include the disruption in medical care being provided to survivors of a disaster event along with disruption in preventive medicine and overall public health response to the affected community. The disruption of health facilities due to extreme events has significant social repercussions, including a decline in community morale. This diminished morale can hinder long-term recovery efforts and negatively influence the community's overall sense of well-being (Choudhary and Kumar 2022).

Health sector plays a key role in protecting and improving human health. Ironically, this sector is also one of the key major contributors to the emissions of greenhouse gases and pollutants which in turn leads to climate change and the global burden of diseases. Hospital functions 24×7 with high consumption of water, energy, generates waste and disposal, requires moving of medical equipment, pharmaceuticals, and personnel to provide healthcare and cater the needs of the community. All these functional activities in hospitals significantly increase their overall carbon footprints. For instance, in India, healthcare emissions are estimated to be equivalent to the annual output of 10 coal-fired power plants or the emissions of over 8 million passenger vehicles (HCWH and Arup 2019).[1] The healthcare sector contributes approximately

[1] HCWH and Arup. (2019). Healthcare's climate footprint report. HWCH. https:// noharm-global.org/sites/default/files/documents-files/5961/ HealthCaresClimateFootprint_092319.pdf.

to 1.5% of India's total emissions, with over 50% stemming from supply chain activities, including production, transportation, packaging, and disposal of medical equipment and pharmaceuticals. Direct emissions from healthcare facilities and vehicles account for 17% of the sector's global footprint (HCWH and Arup 2019).

Climate change is documented today as one of the major threats of the twenty-first century that has the potential to destroy decades of public health gains (Costello et al. 2009). Among this, low- and middle-income countries are most vulnerable to the anticipated impact of climate change on account of their insufficient health infrastructure systems, poor governance, inadequate resources, and prevalent social and economic inequalities. Additionally, the health of a community is also closely linked with the climate of a region as it affects the environmental determinants of health leading to a significant rise in the number of patients with various diseases and other threat of microbial emergency (WHO Fact Sheet 2018).

The South Asian Region in particular is susceptible to numerous climatic hazards indicated from its risk profile. The risk of climate change in South Asia could be substantial and is expected to impact countries differently (Hijioka et al. 2014). The physiography of the regions is such that arid areas are subject to drought conditions, low-lying coastal areas are prone to floods and sea level rise, tropical regions are highly prone to frequent cyclones, and mountainous ranges are vulnerable to rainfall-induced landslides, melting of glaciers, and Glacial Lake Outburst Flood (GLOF). The overall increase in the intensity of rains accelerates flood risk, particularly in South Asia, south-east Asia, and the western Amazon (Eccles et al. 2019). This risk profile of the South Asian region makes it susceptible to almost all climatic hazards that may affect different sectors of economy including health and hospitals.

In view of climate vulnerabilities and risks associated with public health, the term "Climate-smart Health Care Facility (HCF)" is increasingly receiving attention. It is a collective term used for measures that build resilience to climate change while also addressing carbon footprints (Choudhary and Kumar 2022). Climate smart health facilities broadly have two dimensions (a). Resilience and (b). Mitigation (environmentally sustainable). WHO defines climate-resilient and environmentally sustainable health care facilities–*that are capable to anticipate, respond to, cope with, recover from and adapt to climate-related shocks and stress, so as to bring ongoing and sustained health care to their target populations, despite an unstable climate, with access to minimum standards of WASH (and wastes) and energy services, an informed health workforce, and solid infrastructure, and which protects its environment for the benefit of its workers, patients and surrounding communities* (WHO 2020). Beyond safety and operationality, the concept of "climate-smart healthcare" includes strategies to enhance resilience from the adverse effects of climate change alongside addressing greenhouse gas emissions associated with healthcare operations.

Moreover, the term "Resilience" is broadly understood as "*the capacity of a system to absorb disturbance and reorganize while undergoing change so as to still retain essentially the same function, structure, identity and feedbacks*" (Walker et al. 2004). Resilience encompasses both physical and societal systems and can be characterized by four foundational principles, commonly referred to as the "Four Rs." These include: **robustness**, defined as the inherent strength to withstand external shocks without a loss of functionality; **redundancy**, referring to the presence of alternative options or systems that can be utilized under stress; **resourcefulness**, the capacity to identify, mobilize, and deploy necessary resources in response to disruption; and **rapidity**, the ability to restore functionality swiftly following a disturbance (Bruneau 2007). So, the concept of climate resilience as applied to HCF has been centered around "acute care" with "massive dependencies" and "complexities of interconnected systems." Healthcare facilities operate as complex, interdependent networks of systems, wherein the failure of a single subsystem can trigger cascading effects that compromise overall functionality and continuity of care. Consequently, the resilience of a hospital is inherently constrained by the vulnerability of its weakest component. In this context, critical systems are only as resilient as their most fragile link, emphasizing the need for a holistic and integrated approach to resilience planning.

The connections between climate change and public health goes back to the late 90 s. However, climate-smart and disaster resilient HCF is an emerging concept that usually overlaps with other agendas and missions like universal healthcare/one health, disaster risk reduction, strengthening healthcare system and CCA. Climate-related threats to health systems are particularly evident at the level of healthcare facilities. These facilities are frequently not designed to withstand the growing risks associated with climate change, including extreme hydro-meteorological events such as storms, floods, and droughts; elevated temperatures; wildfires; sea-level rise; and shifts in the geographic and temporal distribution of climate-sensitive diseases. These weather extreme events are and will remain vital features of the atmosphere putting the health care delivery system at risk. Although it is recognized that health care facilities, through their operations, are an important source of greenhouse gases, and therefore contributing to climate change, often smaller facilities in less developed countries do not contribute to the problem. It is therefore key to build climate resilience, to address the main four environmental fundamentals to deliver harmless and quality care: a well-informed health personnel, adequate access to WASH (and wastes) services, minimum standards for energy, and a strategically located, and solid health infrastructure. Interventions to build climate smart and resilient hospitals may include four areas namely risk assessments, structural and non-structural safety, emergency and preparedness, and greening measures (WHO 2020).

11.1.2 Research Questions

Given the dual role of the health sector—as both a victim and a contributor to climate change, this paper explores two critical following questions. How can healthcare facilities maintain operational functionality during and after extreme climate events while minimizing their carbon footprints? What key DRR and mitigation strategies can contribute to building climate-resilient and environmentally sustainable healthcare infrastructure?

11.1.3 Conceptual Framework

The conceptual framework for this study outlines an integrated, systems-based approach to enhancing the resilience and sustainability of healthcare facilities, ensuring continuity of medical services and safeguarding public health amid growing climate risks. The following framework components are based on WHO (2020) operational framework for building climate-resilient and environmental sustainability in HCF. **Climate risk mapping**, this includes site specific risk assessment of healthcare facilities to protect them from extreme events and climate stresses. **Resilient infrastructure design**, insists hospitals to comply building norms and ensures smooth operation and maintenance. **Environmentally sustainable** hospitals transit from high- to low-carbon energy sources, manage waste sustainably, and optimize water usages. **Emergency preparedness planning**, having a robust contingency plan and testing them through mock drills and simulation-based exercise can ensure their readiness in disaster situations triggered by climate impacts. **Governance, Policy and Finance** can foster leadership, suitable policy provisions, and required funding mechanism while **Community and System Integration** will enable healthcare facilities having top to bottom approach for strengthening partnership and developing community-based strategies. The development of **economic feasibility and financial models** also requires for bringing transition to make climate-smart healthcare facilities.

11.1.4 Problem Statement

Healthcare facilities face a paradox–being both highly vulnerable to climate change and a significant contributor to it. However, most existing research tend to address disaster risk reduction (DRR) and climate mitigation in isolation, which often lead to lack of integrated frameworks or models that combine DRR and mitigation strategies in healthcare settings, particularly in low- and middle-income countries like those in South Asia and India. The review of literature shows the scarcity of standardized tools to measure and compare the effectiveness of interventions. It highlights the need for robust performance indicators, assessment tools, and benchmarks for developing resilient and green hospitals.

Current research often treats these aspects in isolation, and implementation is further hindered by limited data, inadequate policy support, and insufficient stakeholder engagement. Resilience and sustainability approach require participation from various stakeholders. The existing research studies available lack analysis on participatory approach that can be used to co-design and implement effective interventions.

11.1.5 Research Objectives

The objective of this paper is to analyze resilience and mitigation (greening) measures of Healthcare Facilities (HCFs) in the context of four identified areas citing examples and case studies from South Asia, particularly from India in the context of disaster risks attributed to climate change effects. It also aims to provide actionable insights for enhancing the resilience of healthcare systems while minimizing their carbon footprints. The four key areas include: **(a).** Structural safety of hospital buildings **(b).** Non-structural safety of hospital building components **(c).** Emergency preparedness planning **(d).** Climate change mitigation (greening measures).

11.2 Materials and Methods

The authors have drawn extensively on field-based insights gained through their active involvement in multiple hospital safety assessment projects across South Asia, particularly in various regions of India. The qualitative observations presented in this paper are grounded in first-hand experience from implementing initiatives aimed at enhancing the resilience of healthcare facilities. These initiatives encompassed a wide range of activities, including structural and non-structural risk assessments conducted using the World Health Organization's Hospital Safety Index (HSI), a globally recognized tool.

In addition, hospital safety assessment, the authors have contributed to the development of hospital disaster management plans for healthcare facilities of varying scales, in collaboration with WHO SEARO, UNDP, and several State Governments in India. These efforts adhered to both national and international guidelines. The authors have also been involved in designing, conceptualizing, and implementing environmentally sustainable ("green") measures for hospitals. Throughout these projects, a variety of context-specific risk reduction and mitigation strategies were developed and applied, considering the geographic, environmental, and institutional settings of the respective healthcare facilities. The case studies and examples included in this paper are derived from these implementation experiences.

Furthermore, the authors conducted a comprehensive review of relevant literature, including international publications, academic articles, technical reports, and policy documents, using a systematic approach. The paper is organized thematically, integrating insights from the literature review to offer a narrative analysis and a holistic perspective on the subject.

The study is guided by following hypothesis "Healthcare facilities that integrate climate-smart and disaster-resilient strategies demonstrate higher operational continuity and service delivery capacity during and after extreme events compared to facilities that do not implement such strategies."

11.3 Results

The failure of an HCF to maintain functionality during and after extreme events can be attributed to non-compliance to building rules and regulations that includes by passing local municipal byelaws (such as neglecting minimum flood elevations requirements outlined in flood zonation maps), location of heavy utility equipment such as generators, diesel fuel storage in basement that normally get flooded, inadequate retrofitting safety audit measures in place for existing hospital buildings, not having emergency preparedness protocols, and untrained personnel to manage mass casualty events. The findings of this study underscore multiple dimensions of risk, including the exposure and vulnerability of HCFs to climate-related hazards, the failure of both structural and non-structural components of hospital buildings, deficiencies in disaster preparedness systems, and challenges related to environmental sustainability and greening initiatives.

11.3.1 Climate Risk Assessment/Siting

Healthcare facilities are recognized as critical infrastructure among others like water, transportation, telecommunication, and educational facilities by Sendai Framework for DRR (2015). The Coalition for Disaster Resilient Infrastructure defines critical infrastructure as "*physical structures, facilities, networks, and other assets that provide services that are indispensable to the social and economic functioning of society and are necessary for managing disaster risk*" (CDRI 2023). Thus, HCF is a critical infrastructure, that is, a high-reliability organization whose resilience relies on several factors (or systems) including suitable siting. When risk assessment is carried out for a hospital, it must use a system-based approach. Understanding systematic risk in the context of HCF becomes important due to its inherent dependency on internal as well as external systems that cover safety of all contents of buildings including hospital staff, uninterrupted supplies of water, power and oxygen, and maintenance of clean WASH environment. All these factors are interconnected with each other and thus failure of one component can trigger cascading effects across the entire facility which may have ripple effects. For example, following extreme weather effects, disruption to external water facilities that supply water to a hospital may hinder the functioning of the Central Sterile Supply Department (CSSD) of the hospital leading to cancellation of emergency surgeries even if the buildings remain intact. Similarly, disruption of road network may impede supply chain or inflows of patients to the hospital.

In climate risk assessment process, understanding site selection of an HCF is critically important not for keeping it away from floodplains, landslides/mudslides, or storm surges, but with access to public transportation, use of green energy sources, building with natural ventilation–all these contribute to resilience and mitigation. Health facilities located in multi-hazard prone areas must account for an all-hazard approach during risk assessment considering multiple scenarios such as concurrent hazards, cascading and ripple effects to address a wide range of interconnection of risks (explained below in reference to HCF). For better understanding siting of HCF, the following question must first be answered: What puts healthcare facilities at-risk? This is explained below demonstration on the multi-hazard risk profile of one of the district hospitals in India.

11.3.2 Hospital Siting and Multi-Hazard Risk: A Case Study of the District Hospital in Uttarkashi, India

This case study analyses the siting of a 110-bed district hospital in Uttarkashi, Uttarakhand, India—a region highly susceptible to multiple natural hazards including landslides, floods, and glacial lake outburst floods (GLOFs). The study is based on observational assessment during a comprehensive hospital safety training program that was carried out by GeoHazards International upon request from local authorities. The hospital's proximity to both the Bhagirathi River and the landslide-prone Bhagirathi Parvat illustrates the cascading and compounding risks that threaten the continuity of healthcare services during extreme climate events. The district hospital in Uttarkashi is in an H-shaped building structure situated between two dominant geographical features: Bhagirathi Parvat, a steep hillside located within 150 m of the hospital, known for its landslide and rockfall incidents, particularly during the monsoon season. Bhagirathi River, a major Himalayan River situated less than 300 m from the hospital site, which is prone to seasonal flooding and glacial lake outburst floods (GLOFs). This topographic arrangement places the hospital in a high-risk zone, characterized by compound hazards that can trigger one another and escalate into cascading events (Fig. 11.1).

The region has a documented history of **climate-induced and seismic hazards**, notably the **2012 rainfall-induced landslides** that severely affected local infrastructure. The hospital's location exposes it to several interconnected hazards:

(a) **Landslides and Rockfalls**: Triggered by either intense precipitation or seismic activity, these events threaten structural integrity and access routes.
(b) **Riverine Flooding**: Bhagirathi River's flood risk is exacerbated by glacial melt and monsoonal surges, with the potential to submerge surrounding roads and bridges—critical lifelines for both patient access and supply logistics.
(c) **Glacial Lake Outburst Floods (GLOFs)**: Increasingly common in Himalayan regions due to glacier retreat linked with climate change, GLOFs pose a sudden and destructive flood risk.

The siting of the Uttarkashi district hospital exemplifies the vulnerability of healthcare infrastructure in geophysical unstable locations. In the wake of climate extremes or geological shocks, the hospital's operational continuity is at high risk. The implications include: disruption of health services for surrounding rural populations due to physical isolation, delayed emergency response due to blocked bridges or roads,

Fig. 11.1 Uttarakhand district hospital. *Source* Google Earth Image 2023

and increased mortality and morbidity during disaster events due to lack of access to timely care.

This case underscores the urgent need for risk-informed siting protocols in the design and construction of healthcare facilities, especially in mountainous and hazard-prone regions. To ensure health system resilience, planners and policymakers must incorporate multi-hazard risk assessments and climate projections into hospital siting decisions. The key actions that can be undertaken to build resilience:

(a) Conducting comprehensive geotechnical and hydrological assessments prior to site finalization
(b) Avoiding siting in known hazard-prone zones, especially in proximity to rivers, unstable slopes, or glacial systems
(c) Integrating early warning systems and redundant access routes into facility design
(d) Including climate resilience criteria in national health infrastructure guidelines

A synergistic approach with the integration of adaptation and risk reduction is paramount when it comes to building climate resilience in the healthcare system. It is imperative that planning for siting of a hospital should anticipate change in the climate scenarios over the next 30 years or 50 years. The NDMA Guidelines (2008) on Management of Floods recommends that critical infrastructure such as hospitals be constructed with plinth elevation of important buildings (i.e. hospitals) above the estimated elevation of a flood with a 100-year return period. This underscores the need to revisit and update site planning to reflect evolving climate change scenarios including alteration in the frequency, intensity, and magnitude of all hydro-meteorological hazards and projected uncertainties associated with these extreme events. Moreover, with projected uncertainties associated with these extreme events it is essential to have credible and reliable data and knowledge about projections on disaster risk considering anticipated effects of climate change for mid and end century.

11.3.3 Structural Safety of HCF

The whole built environment of health infrastructure can be broadly divided into two major categories namely, (a) existing buildings (b) new buildings. The existing health infrastructure includes the older hospital buildings, which predate current knowledge and codes and have come up organically with little or no consideration of safety and resilience standards. For existing infrastructure, more resources are to be invested in retrofitting, safety audit, operation and maintenance, and recovery mechanisms (to restore medical services quickly aftermath of extreme events). This can be a step-by-step process involving critically important departments such as Emergency, CSSD, Radiology and Operation Theatre (to name a few) first followed by other units. For development of the new health infrastructure, the fundamental focus should be on in three important aspects, (a) designing of health infrastructure to future risk based on climate change modelling and projections, (b) building health infrastructure that must be low or zero carbon emission (c) the infrastructure should be envisioned as part of a broader, interconnected system, recognizing the interdependencies across sectors such as energy, water, transportation, and communication.

As the structural safety aspects of building design and construction of hospitals especially in climate risk prone areas are governed by building codes and standards, it is important to underline that the national codes and standards are generally prescriptive (recommendatory) in nature. The enforcement by local regulatory authorities is yet highly desirable. The buildings of critical functions like hospitals require strict codes compliance. The National Disaster Management Authority (NDMA) Guidelines prescribe that hospital buildings should be designed beyond national codes and standards (NDMA 2016 p.32).

One common observation in several healthcare facilities in South Asia has been that buildings are designed and constructed such that they are anchored to each other through ramps, staircases, corridors or passageways. The buildings are commonly attached for quick mobility of patients and staff from one department to another. In many cases, the old buildings of the hospital are attached/enhanced with new construction. The attached buildings have shared critical utilities like electricity, water pipelines, oxygen supplies, heating and air conditioning systems which means damage to one building in extreme events can potentially obstruct the supplies of this critical system to other connected units or departments. In the event of disaster emergency, routes can be blocked or hindered making harder for the personnel and patients to evacuate. In the design and construction process, risk in building systems must be accounted for to avoid potential losses and ensuring continuity of care and operational resilience during extreme events.

11.3.4 Non-Structural Safety HCF

The non-structural components are those elements that are attached to the structural parts of the buildings including walls, staircases, architectural elements, contents, utility systems, medical equipment and supplies. They are generally referred as non-structural components or elements as they do not bear the load of the building. Failure or damage to equipment, ceilings, utility distribution network, and conduits can have harmful effects, including injuries, deaths, loss of operation, and economic costs. Despite not contributing to the building's structural integrity, their failure can be just as dangerous and disruptive as structural damage—leading to injuries, fatalities, operational downtime, and significant economic losses. For instance, strong windstorms can often impact rooftop water tanks, roof-mounted ventilation or window air conditioners. For example, a strong wind can cause rooftop water tanks to slide (if not anchored properly) and may interrupt their pipe connections or fall from the roof (if situated close to the edge). It might cause flooding in the building, interrupt water supplies to wards, and may result in downtime losses. Another example of medical oxygen tanks which typically exhibit a height-to-diameter ratio of approximately 2.5, resulting in a form that is markedly elongated relative to their base dimensions often topple in strong wind forces (if the anchored bolts are loose or the support frame is weak to resist strong wind events) causing connections to break cutting off supply of medical gas, and thus could cause leakage that may further lead to fire hazard. In oxygen deprived supply environment, patients on the life support system may not survive. Additionally, the external environment of the hospital is also as vital as safety of internal systems, for example, access to the hospital can also be compromised if trees or electric poles along the access routes fell during cyclone or cyclonic winds.

Thus, the retrofitting measures for non-structural components of hospital buildings are also significant matter of economic concerns given that hospitals house many expensive equipment, and studies have shown that cost of these medical equipment and systems is relatively expensive than the cost of the entire building. Furthermore, the electrical and mechanical systems in these facilities are more expensive than those found in conventional buildings, resulting in higher repair and maintenance costs. Protecting non-structural contents in and around the hospital should be seen from economic benefits. According to WHO (2009), retrofitting non-structural elements typically cost about **1% of a hospital's total value**, yet it can protect up to **90% of the facility's assets**. The National Hospital Safety Guidelines (2016) of Govt of India emphasizes four aspects to ensure the resilience of structural and non-structural components of hospital buildings. These guidelines encompass the following components for new and existing buildings. For new hospital buildings, it focuses on structural design and construction practices, while for existing hospital buildings, the emphasis is on pre-disaster safety measures, retrofitting strategies, and post-disaster needs assessment measures. Moreover, the guidelines prescribe over, and above safety standards as recommended in national standards to ensure safety, design, and retrofitting particularly for critical units of hospital buildings.

11.3.5 Emergency Preparedness Planning

The functionality of a hospital is not merely restricted to its physical integrity or safety of non-structural elements, but it goes beyond covering the readiness of the hospital to deal with disaster situations. Imagine a health facility that can withstand a 500-year flood, but what if its personnel cannot come to work due to unforeseen disaster events? What if the storage of medicines or oxygen get depleted after 72 h (as prescribed by WHO)? So, the disruption of personnel services in the wake of extreme events can limit hospital functionality as much as physical failure to its building. Before ensuring robust emergency preparedness and planning for HCF, let's understand from the below demonstrations he possible three scenarios for which a hospital or its staff needs to be prepared.

Scenario 1: Internal Emergency: During this scenario, the hospital gets affected by an internal event such as fire, explosion, or utility failure. The community is not affected, but the personnel may or may not be affected in this scenario. As this is limited to hospital only, the hospital must be prepared to trigger evacuation of in-patients to safer locations. Such scenario requires high level of preparedness and response measures in place to carry out successful evacuation without disrupting regular inflows of patients.

Scenario 2: External Emergency: In this scenario, healthcare facility is not affected, but the surrounding community gets impacted by any extreme events, for example, a major landslide impacted several people. In the event of a substantial influx of patients requiring urgent medical care, hospital surge capacity may become insufficient to meet demand, potentially compromising timely and effective treatment delivery. The hospital may get overwhelmed and would need mass casualty management actions.

Scenario 3: Combined Emergency: Unlike above two, in this scenario, hospital as well as community are both affected. For example, a major flood event can trigger this situation that can aggravate challenges to hospitals, as they have to keep the hospital functional, and also be ready to address sudden increase in demand. This is a critical situation for the hospital as they have to compromise with limited supplies, loss of essential services like water and sanitation, power and oxygen, and reduction in personnel availability.

Based on the above three possible events, the hospital should work upon developing an emergency contingency plan. The first step to this is to form a hospital disaster management committee involving several departments. The committee is tasked with overseeing and implementing all activities related to disaster management. The plan should also include elements like development of Hospital Incident Command System (HICS) for mass casualty events, developing coordination with nearby hospitals for referral or shifting patients, pre-agreement arrangements for essential supplies, procedures for horizontal and vertical evacuations, triage protocols, marking evacuation sites, and setting up a hospital command and control center. To be self-sufficient, the hospital should have department wise action plans for water supply, electricity, medical gas supply to estimate the need in normal as well as disaster situations and alternatively find ways to store enough supplies in advance.

A skilled and adequately trained health workforce is a fundamental prerequisite for the safe operation of healthcare facilities. Strengthening capacity-building initiatives and implementing effective human resource management strategies are essential for managing climate-related health risks and ensuring the continued functionality of hospitals under stress.

Robust coordination mechanisms are critical to integrating diverse stakeholders, thereby enabling a comprehensive approach to preparedness planning, emergency response implementation, and post-disaster recovery. Furthermore, intersectoral coordination is necessary to establish efficient referral pathways and manage mass casualty events. Such collaboration also enhances the role of hospitals in broader disaster risk reduction efforts at the community level and beyond.

One notable example is from GNRC Hospital Six-Mile Branch, a 100-bedded super specialty tertiary care facility in Guwahati city of India. This initiative is part of a unique effort to integrate local auto owners into hospital preparedness planning. GNRC has a parking stand outside the hospital wherein auto owners park their autos. The hospital has initiated a training program for them to serve the hospital in disaster emergencies. As they are permanent residents around the hospital, familiar with local routes and environment, their roles include assisting the hospital with patient transport, evacuation, and traffic / crowd control during and after any mass casualty events. This is a win–win situation for both the hospital and auto owners by reducing stress on the hospital's resources and enabling them with an income stream. Such a small level and low-cost coordination model provides opportunities for hospitals to leverage planning efforts involving local community. Another case of Rwanda healthcare services highlighted below is the maintenance of healthcare services during the COVID-19 pandemic.

11.3.6 The Interconnections of Risks in HCF

To understand different dynamics and characteristics of risk in the context of healthcare facilities, it is important to underscore how the risk is interconnected across systems, sectors, and society. When an extreme monsoon season causes damage to one system let's say the water system, the disaster will have impact on the environment on which the system relies and the surrounding region. Road network may be affected by landslides that carry medical equipment and supplies to hospitals, community may have to travel long to access hospital services, the electric power grid may also not be operational as the long transmission lines rest on individual towers, any number of which could be toppled by landslides or storms. All these factors can have cascading effects thereby requiring multiple interventions as this is a complex and interconnected system.

The IGMC Hospital in Shimla presents similar level of complexities and challenges. The hospital's vulnerability scenario has been studied in detail. This is based on learnings from a UNDP supported project for developing a Disaster Management Plan for IGMC in 2018. It has also examined the hospital's exposure to compound risks and highlights the urgent need for systemic resilience.

11.3.7 Resilience Under Risk: A Case Study of Indira Gandhi Medical College (IGMC), Shimla

The Indira Gandhi Medical College (IGMC), located in Shimla, Himachal Pradesh, serves as a vital healthcare hub for both urban and rural populations across the state. As a multi-specialty, 800-bed tertiary care institution, IGMC is not only a center of clinical excellence but also a critical node in the region's emergency and disaster response system. However, its geographical location and systemic dependencies render it highly vulnerable to climate-induced extreme events.

Situated atop a hill and surrounded by steep, landslide-prone terrain, IGMC comprises of five interconnected blocks accessible through ramps and staircases. While this location offers strategic oversight, it also exposes the infrastructure to topographic hazards exacerbated by climate variability. Increasing temperature, erratic precipitation, and more frequent landslides—projected under the Himachal Pradesh State Climate Change Action Plan (HPSCCAP 2012)—underscore the need to reassess vulnerability through an integrated risk framework.

Like many healthcare facilities, IGMC operates at the intersection of interdependent systems—water supply, power, transportation, and healthcare service delivery. These systems are tightly dependent and any disruption in one can have cascading effects on the others.

Water supply in Shimla is sourced from riverine and spring-fed systems through a combination of pumping and gravity-based infrastructure. Vulnerabilities such as landslides can severely disrupt supply chains by damaging pumping stations, pipelines, and reservoirs (GHI Report 2020). According to WHO standards, hospitals should maintain three days' worth of water at 300 L per bed per day. For IGMC, this equates to a minimum storage requirement of **720,000 L**. Current storage capacity stands at **554,800 L**, resulting in a critical shortfall of **165,200 L**. Given the hospital's dual role as a treatment center and a hotspot for potential infection spread, this deficit significantly undermines its ability to maintain basic hygiene, sanitation, and infection control protocols during emergencies.

Nearly 90% of Shimla's water supply infrastructure depends **on electricity**, creating a feedback loop between energy and water security. IGMC's dependence on grid electricity also extends to life-saving medical technologies, ventilation systems, diagnostic equipment, and lighting. During climate-induced extreme events such as landslides or cloudbursts, the risk of power outages escalates as transmission lines are in unstable areas. The hospital's current backup power system consists of nine diesel generators with a total fuel reserve of 1,000 L, which would sustain critical operations for **only about 5 h** under full load. This is significantly below the recommended **72-h minimum autonomy period** for health facilities in disaster scenarios. The nearest fuel station, a 20-min drive away, may also be inaccessible due to blocked roads during such events.

The case of IGMC exemplifies the cascading nature of risks in interconnected infrastructure systems. A landslide may not only impede physical access to the hospital but simultaneously disable water supply systems, sever electrical connections, and prevent emergency fuel deliveries—thus paralyzing the hospital's functional capacity during the very time it is most needed.

11.3.7.1 Toward Systemic Resilience

Considering these vulnerabilities, it is imperative for IGMC and health facilities with similar features in South Asia to have the following systemic changes to strengthen resilience:

(a) **Enhance On-site Redundancies**: Increase in-campus water storage to meet WHO standards and expand fuel reserves for generator-based power continuity.
(b) **Promote Decentralized Resource Systems**: Invest in solar energy, rainwater harvesting, and greywater reuse to reduce dependence on external grids and municipal supply lines.
(c) **Conduct Climate Risk Audits**: Regular assessments must identify weak links in interdependent systems and develop adaptation plans.
(d) **Integrate Health Facilities into Urban Resilience Planning**: Health care facilities must be prioritized in climate resilience policies at both local and state levels.

11.3.8 Greening Measures

Health Care Facilities (HCFs) operate on a 24-h cycle of operations thereby leaving behind a trail of carbon emissions. Healthcare Sector is positioned as one of the "large carbon emitting" sectors with its climate footprint of 2.0 gigatons in 2019 alone. As per an estimate by World Bank in 2017, the sector's contribution accounts for an over 5% to the global Greenhouse Gas (GHG) emissions equivalent to 2.6 billion metric tons of CO2 emissions.

In a healthcare facility, 65% of its energy consumption occurs through water heating, lighting, cooling, and ventilation. Moreover, the construction materials and waste generation add to the impact that HCF's have on the environment. Adopting Green designs and concepts into the building designs can reduce its impact on the GHG emissions, increase its energy efficiency, cut down its operational costs, and protect public health from the effects of climate change (NPCCHH 2023).

Furthermore, the World Meteorological Organization (WMO) in a report on *"Worsening Climate Impacts"* has urged for implementation of "Green Hospitals" to protect human lives and the environment to turn to climate action. Green hospitals prioritize both environments along with humans. Green infrastructure enhances environmental sustainability through using environment-based products and practices. Green hospitals provide better indoor conditions for their occupants which makes them very crucial considering the health implications of greenhouse gas emissions and hazardous waste. The greening measures in the hospitals can be majorly categorized into four major groups namely:

(a) Building—walls (insulated), roofing (cool roofing), ventilation (adequate focus during construction stage), and lighting (LED lights)
(b) Energy—transition from complete grid dependence to energy mix to complete renewable energy where feasible
(c) Water—wastewater recycling, treatment and reuse including sewage treatment plant (STP) and where laboratory facilities etc. are available, additional infrastructure for effluent treatment plant (ETPs). Rainwater harvesting and rainwater recharging units can also be adopted as feasible. Water metering should be mandatory for effective monitoring of use.
(d) Waste—effective waste management channels

11.3.9 Building Design and Construction

A green healthcare facility can be defined that can reduce its carbon footprint through its construction design including its operation and maintenance. Appropriate use of building materials in the construction of a hospital is all embodied energy, which is generally defined as the energy required to produce a material. As construction materials, these have a large ecological impact. The soil has to be quarried to make bricks resulting in brick quarries. Concrete and mortar need good quality sand. This results in sand mining with its resultant ecological degradation and effect on water systems. Thus, in the case of building materials, we have to calculate the energy in the production, packaging, and transportation of the material, component, or product.

The green building norms also insist on optimum and efficient utilization of all natural resources and man-made resources, i.e., land, water, air, energy through integrated water, energy and waste management, greening of spaces, carbon footprint reduction, and use of renewables. Thus, promoting and encouraging green building adoption and increased use for healthcare facility is a win–win for governments, service agencies (for water supply, energy supply, waste management), building owners, developers, and in the end for consumers and occupants. With incentives, the following benefits (a) substantial saving of water use, (b) savings in energy use (c) treatment of all wastes in all forms of use (d) reduction in carbon footprints (e) expanding tree and green cover around hospital premise for ideal temperature (f) enhancing overall quality of life of all occupants.

Green building measures often point to a result that must be achieved. The result may be one or more however, in case of green hospitals, the primary objectives are to keep them (a) Operational during and post disaster emergencies (b) Reduce their dependencies on external resources (c) Make them self-sufficient in the face of disaster events (d) Carbon neutrality. Along the same lines, some of the measures that can be adopted based on the green codes, tools, and practices include using environmentally friendly materials, adopting water conservation techniques such as rainwater harvesting, recycling, and treatment, utilizing energy-efficient devices, using nearby facilities to reduce transportation costs. Moreover, designing cool roofs which absorbs less heat through practices such as applying paints on roofs, solar passive provisioning, and thermocouple insulation can significantly lower the indoor temperature, thereby reducing the need for air conditioning (Vellingiri et al., 2020). Additionally, using appropriate building facade will enhance the cooling and lighting. Measures such as using High Efficiency Particulate Matter Filters (HEPA) along with proper ventilation having at least 6 air changes per hour could lead to significant improvements in the air quality. (Gilkeson et al., 2013). The process of retrofitting existing structures or constructing resilient healthcare facilities necessitates comprehensive planning and close coordination among structural engineers and green building specialists. This interdisciplinary collaboration ensures that resilience measures are integrated with sustainable design principles, thereby enhancing both the structural integrity and environmental performance of healthcare infrastructure. State and local governments should recognize the potential benefits of green measures in hospitals and must encourage sustainable, green development through incentives (i.e., reduction in property tax, speedier issuance of NOC, reduction in utilities tariffs etc.).

11.3.10 Energy Conservation

The healthcare sector is the largest consumer of energy for its operationality. It is important to note that without adequate energy, even the most basic medical services cannot be provided. Meanwhile, using fossil fuels to power the HCF's utilities increases the indoor pollution coupled consequently leading to increased fatality (Pozzer et al. 2020). As per the Indian Public Health Standards (IPHS) an estimate 3KW of energy gets consumed per bed daily. Hospitals require uninterrupted energy for the operationality of medical equipment and devices on 24 × 7 basis. Moreover, larger dependencies of power grids result into carbon emissions.

Therefore, using renewable energy and other clean energy sources such as wind and / or solar will not only reduce GHG emissions, but also significantly protect public health from the adverse effects of climate risks including frequent flooding, water scarcity such as vector borne diseases, and heat-related illnesses (Healthcare without Harm). It is important to adopt measures for its conservation and choosing sustainable alternatives. One such initiative was taken by Mary Johnston Hospital (MJH) in Tondo, Manila where the Hospital could save an amount of P 5,374.00 daily or P 1,961,510 annually. The energy thus saved was equivalent to 2,299 tree seedlings grown for 10 years, 165 acres of U.S. Forest in one year.

It involved a comprehensive audit of the hospital's existing lighting systems, planning and logistics for replacement, procurement of mercury-free alternatives (primarily LED technologies), and post-installation monitoring. Stakeholder interviews and environmental performance assessments were conducted to evaluate the outcomes in terms of mercury elimination and carbon emissions reduction complying SDGs (SDG 3, 7, 13).

11.3.10.1 Implementation Highlights

(a) Hundreds of CFLs were safely removed and disposed of according to hazardous waste protocols.
(b) Replacements included high-efficiency LED lighting across administrative, patient care, and support areas.
(c) Training sessions were held for hospital maintenance and operations staff to ensure long-term sustainability of the initiative.

The pilot project achieved several key results:

(a) **100% elimination** of mercury-based lighting from hospital premises.
(b) Estimated **energy savings of 40–60%** depending on location and usage.
(c) Enhanced **lighting quality** in patient care areas, contributing to improved working conditions and safety.
(d) Avoided future **mercury exposure incidents** and reduced hospital liability.
(e) Strengthened institutional commitment to sustainable development and climate action.

Another case study from Afghanistan provides an account on ensuring energy resilience in crisis in Herat, where frequent power outages endangered patients reliant on life-saving medical equipment. In response, Afghanistan's national power utility, Da Afghanistan Breshna Sherkat (DABS), installed solar energy systems in ten hospitals. The initiative generated a total of 75 kW of power, supporting critical healthcare infrastructure such as ventilators, defibrillators, operation theatres, and patient monitoring systems.

11.3.10.2 Key Highlights

(a) Solar panels installed on hospital rooftops with battery backup systems.
(b) Power distributed to operate **ventilators**, **defibrillators**, **ICU equipment**, **operation theatres**, and **critical care units**.
(c) Reduced dependence on unreliable grid electricity and diesel generators.

11.3.10.3 Findings and Analysis

(a) **Power Reliability**: The solar systems provided stable energy, compensating for frequent grid failures.
(b) **Healthcare Continuity**: All ten hospitals maintained uninterrupted use of ventilators and ICU equipment during critical periods.
(c) **Cost-Effectiveness**: Lower operational costs compared to diesel-powered backup systems.
(d) **Environmental Benefit**: Reduced emissions and air pollution at a time when respiratory health was paramount.
(e) **Community Trust**: Public confidence in healthcare services increased due to better facility performance.

11.3.10.4 Water

Water is one of the most essential components of life. However, more than 1 billion persons suffer for want of access to potable drinking water while nearly two million people die from diarrhea annually, with children occupying the major share. Water is thus a critical utility in hospitals as it is used for drinking, dialysis, operation theatres etc.

However, lack of proper water and sanitation infrastructure is one of the biggest challenges in tackling healthcare delivery along with overburdening the population with diseases. As per WHO estimates 94% of diarrheal deaths can be managed through interventions including sanitation and clean water.

The unsafe and unhygienic patient care results in healthcare-associated infections (HCAI) and anti-microbial resistance (AMR) subsequently leading to chronic health conditions, lengthy hospitalization, and high treatment costs (water Aid). Measures such as coming up with simple technologies suitable for small- to medium-scale healthcare facilities, especially in low- and middle-income countries, where resource constraints necessitate cost-effective and context-specific solutions.

Hospitals can serve as critical public health assets by providing access to potable water not only for their own operations but also for surrounding communities. This contributes to the prevention of waterborne diseases and reduces the burden on both medical and natural resources required for disease treatment. However, when water is readily available, hospitals often become significant consumers due to the high demand across various operational functions, including clinical services, sanitation, and facility maintenance. Therefore, to assure a seamless supply of water, appropriate planning and implementation strategies shall be devised to clearly lay down its operation and maintenance policy.

11.3.10.5 Waste Management

The health sector is a significant contributor to overall waste generation, producing substantial volumes of infectious, non-infectious, hazardous, general, and chemical waste. According to the United Nations Environment Programme (UNEP), approximately 25% of biomedical waste is classified as hazardous, posing considerable risks to both human health and the environment if not properly managed.

Moreover, the World Health Organization (WHO), through its core principles has advocated for safe management of healthcare-related waste while the World Health Assembly has called upon greater action on medical waste. This is because medical waste is infectious and toxic, presenting a significant public health threat.

An important concern is that hospitals generate a staggering 5 million tons of waste annually as per estimates by Healthcare without Harm. Moreover, the waste generated is difficult to manage and vast requiring varying levels of treatment and enormous amount of energy, particularly if it gets mixed with non-infectious waste. (Khobragade 2019). For this, it is imperative to adopt measures such as recycling, composting, reusing products, minimizing packaging, reduce, and segregate.

In this regard, reducing the waste on the first hand is a preventive step in reducing the adverse impacts of climate risks. Additionally, segregating the waste generated at source improves overall waste management along with preventing sanitation workers from occupation-related health risks (Datta et al. 2018).

Furthermore, wastewater generated from hospitals is also a significant public health concern as it contains drug resistant pathogens, chemicals, as well as hazardous materials. This needs to be managed properly so as to avoid its adverse impacts.

However, to ensure long term sustainability, the products shall be so designed which can be reused and recycled, i.e., manufacturing the product considering its lifecycle. This can be done by advocating its benefits to drive demand in favor of such products. One such initiative was made by Mary Johnston Hospital (MJH), located in Tondo, Manila by exploring and implanting alternatives to PPE items.

11.3.10.6 Adaptive Waste Management and PPE Reuse During COVID-19: A Case Study of Mary Johnston Hospital, Manila

The COVID-19 pandemic triggered an unprecedented surge in healthcare waste, particularly infectious plastic waste from personal protective equipment (PPE). Mary Johnston Hospital (MJH), located in Tondo, Manila, faced dual challenges of escalating hazardous waste and critical shortages of PPE. In response, the hospital implemented adaptive waste management strategies, including the reprocessing of select PPE items and the exploration of alternatives made from water-resistant microfiber fabrics. MJH also leveraged decision-making frameworks through a COVID-19 bulletin to guide rapid response efforts. These initiatives led to a significant reduction of 10,821 kg of hazardous waste and gross cost savings of Php8,673,350.

11.4 Discussions

11.4.1 Key Findings of the Study

The projected impact of climate change poses a great challenge across sectors. The health industry is uniquely positioned as both a victim and a contributor to climate change. The rising frequency and intensity of hazard events globally aligns with projections outlined in the Intergovernmental Panel on Climate Change (IPCC) Sixth Assessment Report (AR6), which warns that surpassing 1.5 °C of global warming will expose the world to an unprecedented level of climate-related hazards, including extreme weather events, sea-level rise, and ecosystem disruptions (IPCC 2021). If the business-as-usual scenario continues, the health industry could triple emissions between now and 2050.

This paper offers some action points that can be undertaken by hospital administrators, decision makers, government agencies, and other relevant stakeholders associated with the healthcare system. The assessment strategies in healthcare facilities particularly for those located in moderate to high climate risk zones must consider interconnection of risks. The tools for assessment must account system-based thinking to determine vulnerabilities of different sectors linked with HCF.

A key requirement to construct a safe and climate resilient hospital should be that suitable financial models or in-kind incentives should be initiated through larger programs. These programs can mandate health facilities to adopt green norms, conduct annual safety audits, may insist on encouraging voluntary structural maintenance of buildings, and can establish an incentive system. So, these can be part of large-scale social demand by government authorities.

The healthcare industry in several South Asian countries is yet to adopt greening measures on a wider scale. The green building movement was initiated around the world in the 1990s to promote sustainable development norms for buildings, through environmentally friendly and ecologically appropriate initiatives including energy saving and water saving efficiencies. At present, there are 65 Green Building Councils in different countries operating and encouraging green building initiatives. To become resilient and adapt to climate and disaster risks, healthcare facilities need to be prepared for climate threats while minimizing their emissions. Fundamentally, hospitals can take leadership in reducing carbon footprints by increasing their dependency on renewable energy sources and setting examples of safe and functional green buildings. The use of water in sustainable manner and harnessing available technologies to conserve water resources will be useful way toward sustainability. Adopting practices to comply with segregation, storage and treatment policies will help healthcare facilities reduce waste footprints. Transportation activities for supply chain and personnel commuting need adaptation measures to reduce travel footprint. The hospitals need to work collaboratively with other industries and establishments to ensure sustainable buying and procurement for a low-carbon health supply chain.

Several international guidelines provide valuable frameworks for achieving climate resilience in healthcare facilities, including the SMART Hospitals Toolkit

(PAHO 2017), the WHO Hospital Safety Index (HSI), the Framework for Climate-Resilient Health Care Systems (WHO 2015), as well as standards from the US Green Building Council (USGBC), Indian Green Building Council (IGBC), Leadership in Energy and Environmental Design (LEED), Green Rating for Integrated Habitat Assessment (GRIHA), and the California Green Building Code (CGBC). These international guidelines offer comprehensive frameworks and standards for enhancing climate resilience of hospitals that can be adopted.

11.5 Summary of Applications

Area of Application	Guidelines/Standards
Climate adaptation and disaster resilience	SMART Hospital Toolkit, WHO HSI, WHO Framework
Environmental sustainability	LEED, IGBC, GRIHA, CGBC
Energy and water efficiency	LEED, IGBC, SMART Hospitals Toolkit, CGBC
Regulatory compliance	CGBC (California), national adoption of LEED, IGBC, or GRIHA
Health system widening planning	WHO Framework for Climate-Resilient Health Systems

The efforts for building climate resilient health system requires strategies to be led by examples. The first Asian city to develop and implement a Heat Action Plan (HAP) is Ahmedabad from Gujarat State of India. This plan is based on an early warning and early actions. The experiences show compelling evidence that heath action plans can meaningfully reduce life losses due to heat event. The outcome of the plan shows a reduction of 25% mortality (excess from the month of May) in two years (2014–15) since the plan was put in place (Hess et al. 2018) saving hundreds of lives. HAP heat-health research plays a critical role in enhancing community resilience to extreme heat events. These areas of study are essential for the prevention and effective management of heat-related illnesses, particularly in vulnerable populations disproportionately affected by rising temperatures and poor air quality. The other vulnerable cities in South Asia can replicate customized strategies in heat prone regions to reduce burden on health infrastructure and save lives.

Implementing green measures in hospitals can be challenging due to financial constraints, technical complexities, and regulatory hurdles. Transition to climate-smart healthcare infrastructure requires upfront investment. There is also need for economic feasibility and investment models to be developed for cost effectiveness, return on investment for implementing resilience and sustainability systems in vulnerable regions. The growing awareness of environmental issues and the increasing availability of green technologies are slowly driving the adoption of sustainable practices in the healthcare sector.

Community-based approach is also an important step towards climate resilient hospitals. For the implementation of international commitments such as NDCs and NAPs, many countries have adopted community-based models (Fransen et al. 2015; PEER 2009). The partnership between healthcare providers and local community can help reduce the impact of disaster events. The partnership approach promotes both community health and climate resilience.

11.5.1 Limitation and Strength

This paper presents our understanding of healthcare vulnerabilities based on field visits conducted over five years or more combined with first-hand experience. The findings included learnings derived from hospital safety and resilience projects implemented in parts of South Asia. While these insights offer valuable perspectives, some limitations may affect the accuracy and comprehensiveness of the results. For example, some of the examples/case studies provided are not of recent years, it is likely that information/data given may have changed. Additionally, due to the absence of any financial support to write this paper hence, field visits were not conducted, which limits the scope of site-specific vulnerabilities and tailored recommendations. The data used in calculating time and distance of hospitals have been indirectly captured through Google Earth and Google maps that may vary depending on the scale of the extreme events.

Despite all these, the strength of the paper lies in assumptions or hypothetical scenarios given for conducting analysis describe the foreseeable impact of plausible, but hypothetical extreme events on health facilities. These points trigger system-based understanding for the audience to underscore inherent and dynamic vulnerabilities and conduct careful planning. These action points can be adopted by (medical) personnels/administrators, decision makers, to support the development of climate smart and disaster resilient healthcare infrastructure.

11.6 Conclusion

The projected impacts of climate change highlight the urgent need to strengthen the resilience of healthcare facilities in South Asia. These facilities not only serve as critical infrastructure during climate-related emergencies but are also substantial contributors to carbon emissions. Therefore, it is imperative to integrate climate resilience and environmental sustainability into the design, construction, and operation of healthcare systems. Such integration ensures that facilities remain functional during extreme weather events while minimizing their ecological footprint.

South Asia's unique geography, coupled with its vulnerability to a wide range of hazards across different regions, necessitates the revision of the "One Size Fits All" approach. This implies a more tailored strategy, considering the specific geographical characteristics of each region, should be adopted. Such localized planning which includes area-specific building regulations, safety standards, and resilience approaches must be established before embarking on the development of climate- and disaster-resilient healthcare facilities (HCFs).

The evidence is clear that HCF must evolve to become climate-smart, integrating both resilience and mitigation measures to sustain medical services. Across South Asia, several examples highlight both risks posed by climate change and emerging best practices in climate-smart HCF. For example, hospitals in Bangladesh in low-laying costal districts like Khulana and Barisal are often affected by tidal surges and cyclones. The events like Cyclone Sidr in 2007 and Amphan caused massive disruption to hospital services due to power outages. In response, the Government, with support from WHO implemented solar-powered water filtration and backup energy systems in rural healthcare centers. Sri Lanka also piloted green hospitals initiative in partnership with UNDP and WHO. The Kalutara District General Hospital, for instance, has installed rainwater harvesting systems, solar lighting, and improved waste segregation, serving as a model for eco-friendly hospital design in disaster prone areas.

Building resilience in healthcare systems also means continuous training of health workers, ensuring access to WASH and developing redundancies in supply chains, water, and power sources. As South Asia faces a future with increasing climate-related health risks, such investments are essential to protect public health and secure sustainable development.

In conclusion, the path forward demands multi-sectoral collaboration, robust policy frameworks, and localized innovations that prioritize both human and environment health. Climate-resilient and environmentally sustainable healthcare is not only doable—it is imperative.

Acknowledgements This paper presents the findings of hospital safety assessment projects carried out by the authors for organizations such as UNDP, WHO, GeoHazards International, and GeoHazards Society across regions of India and South Asia. We gratefully acknowledge the support of our extended project team and subject-matter experts, whose collaboration enriched these efforts. We also extend our sincere thanks to the administrations of each participating hospital for facilitating safety assessments and granting access to critical data.

Authors' Contribution: All authors have contributed equally to this work.

Data Availability Statement

The datasets used and/or analyzed during this study are available from the corresponding author upon reasonable request.

Funding: This research paper received no external funding.

Potential conflicts of interest: On behalf of all authors, the corresponding author declares that there are no conflicts of interest.

Informed Consent: Informed consent was obtained from all individual participants included in this study.

References

Alotaiby RAM, Krenyácz É (2023) Energy efficiency in healthcare institutions. Soc Econ. https://doi.org/10.1556/204.2023.00013

Alotaiby F et al (2023) Climate-resilient healthcare systems and sustainable hospital infrastructure: a review. Int J Environ Res Pub Health 20(4):3125

Association for Healthcare Providers, India (2015) The standard for green and clean hospitals. https://www.scribd.com/document/357186023/AHPI-Standardfor-Green-Clean-Hospital

Bruneau M et al (2007) A framework to quantitatively assess and enhance the seismic resilience of communities. Earthq Spectra 19(4):733–752

Choudhary Y, Kumar H (2022) Green and climate resilient health infrastructure. Climate change and health sector, healing the world

Coalition for Disaster Resilient Infrastructure (CDRI) (2023) Report on disaster resilient health infrastructure. CDRI

Costello A, Abbas M, Allen A, Ball S, Bell S, Bellamy R et al (2009) Managing the health effects of climate change. The Lancet 373(9676):1693–1733

Datta P et al (2018) Climate change, disaster management and health systems in India. Int J Disaster Risk Reduction 31:691–700

Eccles R, Zhang H, Hamilton D (2019) A review of the effects of climate change on riverine flooding in subtropical and tropical regions. J Water Clim Chang 10. https://doi.org/10.2166/wcc.2019.175

Fransen T, Levin K, Mogelgaard K, Northrop E (2015) National climate plans (INDCs) by the numbers. World Resources Institute. https://www.wri.org/insights/national-climate-plans-indcs-numbers

GeoHazards International (2020) A way forward to increase earthquake and landslide resilience of water supply systems in Shimla, Aizawl, and Similar Hill Cities

Gilkeson CA, Camargo-Valero MA, Pickin LE, Noakes CJ (2013) Measurement of ventilation and airborne infection risk in large naturally ventilated hospital wards. Build Environ 65:35–48. https://doi.org/10.1016/j.buildenv.2013.03.0062

Global Climate and Health Forum (2018) Call to action on climate and health. https://www.globalclimateandhealthforum.org/call-to-action

Global Green and Healthy Hospital Case Studies (2019) A solar energy initiative to reduce cost and carbon emission: Bhagat Chandra Hospital, India. https://www.hospitalesporlasaludambiental.net/wp-content/uploads/2018/05/SolarEnergy-Initiative-India.pdf

Green M et al (2019) Sustainable healthcare systems and climate resilience: emerging global practices. J Clean Prod 230:1342–1351

Healthcare Without Harm and Arup (2019) Healthcare's climate footprint: how the health sector contributes to the global climate crisis and opportunities for action. https://noharm-global.org/sites/default/files/documents/files/5961/HealthCaresClimateFootprint_092319.pdf

Hess JJ et al (2018) Building resilience to climate change: pilot evaluation of the sustainable and climate-resilient health care facilities Toolkit. BMC Publ Health 18:639

Hijioka Y, Lin E, Pereira JJ, Corlett RT, Cui X, Insarov GE, Lasco RD, Lindgren E, Surjan A (2014). Asia. In: Barros VR, Field CB, Dokken DJ, Mastrandrea MD, Mach KJ, Bilir TE, Chatterjee M, Ebi KL, Estrada YO, Genova RC, Girma B, Kissel ES, Levy AN, MacCracken S, Mastrandrea PR, White LL (eds) Climate change 2014: impacts, adaptation, and vulnerability. Part B: regional aspects. Contribution of working group II to the fifth assessment report of the intergovernmental panel on climate change. Cambridge University Press, Cambridge, United Kingdom and New York, NY, USA, pp 1327–1370

Himachal Pradesh State Climate Change Action Plan (HPSCCAP) (2012) Strategy and action plan on climate change. Government of Himachal Pradesh

Intergovernmental Panel on Climate Change (IPCC) (2021) Climate change 2021: the physical science basis. Cambridge University Press

Karliner J, Guenther R (2011) A comprehensive environmental health Agenda for hospitals and health systems around the world, Pg 6, Global Green and Healthy Hospitals, HCWH. https://www.hospitalesporlasaludambiental.net/wpcontent/uploads/2016/07/Global-Green-and-Healthy-Hospitals-Agenda.pdf

Kumagai K et al (2013) Resilient healthcare infrastructure and disaster preparedness in hospitals. Int J Disaster Risk Sci 4(3):123–131

Khobragade SD (2019) Health care waste: avoiding hazards to living and non living environment by efficient management. Fortune J Health Sci 2(2):014–029. http://www.fortunejournals.com/articles/health-care-wasteavoiding-hazards-to-living-and-non-livingnbspenvironment-by-efficientmanagementp.pdf

Kumagai S, Koda S, Miyakita T, Ueno (2013) Guidelines for green and resilient health care facilities. Green and climate resilient healthcare facilities guidelines by NPCCHH, MoHFW

National Disaster Management Authority (NDMA) (2008) National disaster management guidelines: management of biological disasters. Government of India

National Disaster Management Authority (NDMA) (2016) National disaster management plan. Government of India

National Disaster Management Authority, Government of India (2016) National disaster management guidelines: hospital safety. A publication of the National Disaster Management Authority, Government of India. ISBN: 978-93-84792-03-9

National Programme on Climate Change and Human Health (NPCCHH) (2023) Operational guidelines for climate change and human health. Ministry of Health and Family Welfare, Government of India

Pan American Health Organization. (2017) Smart hospitals toolkit. https://www.paho.org/disasters/index.php?option=com_content&view=article&id=1742:smart-hospitals-toolkit&Itemid=1248&lang=en

Partnership for European Environmental Research (PEER) (2009). Europe adapts to climate change: comparing national adaptation strategies. PEER. A Buildings and their impact on the environment: a statistical summary. www.epa.gov/greenbuilding/pubs/gbstats.pdf

Pozzer A, Dominici F, Haines A, Witt C, Münzel T, Lelieveld J (2020) Regional and global contributions of air pollution to risk of death from COVID-19. Cardiovasc Res 116(14):2247–2253. https://doi.org/10.1093/cvr/cvaa288

Sherman JD, MacNeill AJ, Biddinger PD, Ergun O, Salas RN, Eckelman MJ (2023) Sustainable and resilient health care in the face of a changing climate. Ann Rev Publ Health 44:255–277

Sustainable Energy for All (SEforALL) (2020a) Chilling prospects: providing sustainable energy for all. SEforAll. https://www.seforall.org/sites/default/files/SEforALL_CoolingForAll-Report.pdf

United Nations Environment Programme (2007) Buildings and climate change: status, challenges and opportunities

USEPA (2017) Sustainable materials management basics. https://www.epa.gov/smm/sustainable-materials-management-basics#lifecycle

United States Environmental Protection Agency (USEPA) (2017) Energy star guidelines for energy management. U.S. Environmental Protection Agency

Vellingiri S, Dutta P, Singh S, Sathish LM, Pingle S, Brahmbhatt B (2020) Combating climate change-induced heat stress: assessing cool roofs and its impact on the indoor ambient temperature of the households in the Urban Slums of Ahmedabad. Indian J Occup Environ Med 24(1):25–29. https://doi.org/10.4103/ijoem.IJOEM_120_19

WHO Fact Sheet (2018) Climate change and health. https://www.who.int/news-room/fact-sheets/detail/climate-change-and-health

World Health Organization (2009) Hospital safety: texas dodged a bullet. Bull World Health Organ 87:11. http://www.who.int/bulletin/volumes/87/11/09-021109/en/

World Health Organization (WHO) (2015) Operational framework for building climate resilient health systems. World Health Organization

World Health Organization (WHO) (2020) WHO guidance for climate resilient and environmentally sustainable health care facilities. World Health Organization
Walker B, Holling CS, Carpenter SR, Kinzig A (2004) Resilience, adaptability and transformability in social–ecological systems. Ecol Soc 9(2):5

Chapter 12
Gender-Responsive Climate Change Adaptation: Policy Framework and Indigenous Knowledge Systems for Maternal Health Resilience in Meghalaya, Northeast India

S. Elika Assumi and Basil N. D. Diengdoh

Abstract Maternal health in Meghalaya particularly in the Khasi Hills continues to suffer from structural inequities, weak institutional outreach, and increasing climate-related stress. Despite the presence of a matrilineal kinship system and deeply embedded Indigenous medicinal knowledge, traditional maternal health practices remain largely excluded from formal health policy. This study investigates the interface between Khasi Indigenous health systems and state-led biomedical care in the context of climate change. In this context, the study addresses three key research questions: (1) What structural and policy barriers hinder the formal recognition of Indigenous maternal health knowledge? (2) How can women-led institutions such as the *Seng Kynthei* strengthen maternal health outcomes under climate stress? (3) What hybrid models can support pluralistic, climate-adaptive healthcare systems? Using qualitative methods, including literature review, policy analysis, and interviews with traditional healers, women leaders, and health workers, the study maps existing synergies and gaps. Findings indicate that integration of Indigenous practices with formal systems—particularly through culturally anchored women's groups—can enhance maternal resilience. The study concludes by suggesting policy recommendations for inclusive, gender-responsive health systems that leverage traditional knowledge to meet the dual challenges of climate vulnerability and maternal health inequity.

Keywords Meghalaya · Seng Kynthei · Women · Maternal health · Health policy · Indigenous health

S. E. Assumi (✉) · B. N. D. Diengdoh
National Law University Meghalaya, Shillong, Meghalaya, India
e-mail: eliassumi@gmail.com

P. Ghosh et al. (eds.), *Climate Crises and Public Health Challenges*, Disaster Risk Reduction, https://doi.org/10.1007/978-981-95-9709-3_12

12.1 Introduction

12.1.1 The Challenges of Maternal Health in Khasi Hills

The state of Meghalaya is one of eight that collectively comprise Northeast India. Despite matrilineality and its community-oriented sociality, the convergence between the health and well-being of women and children, arguably, remains vulnerable. This is starkly evident among the Khasis of Meghalaya, one of the major tribes of the state, where growing disparities in class, income, and accessibility pose systemic healthcare challenges, especially for the most vulnerable sections of its society. Maternal health indices, for instance, consistently lag behind both the national average and those of neighboring states in the region. These challenges are shaped by the distinctive socio-cultural fabric of the Khasi Hills but also climate change-induced stressors, where the study of the matrilineal social structure can be a useful frame to understand better persistent issues in healthcare, nutrition, and public awareness around these concerns. While the kinship system offers women a central role in lineage and inheritance, it is clear that this does not insulate them from pervasive health inequities, particularly those affecting maternal health outcomes. Notably, the Khasi Hills face significant maternal health challenges, including high rates of teenage pregnancies, lack of maintenance support for single mothers, and widespread illiteracy among women. Additional challenges include inadequate access to healthcare facilities, high maternal mortality rates, and insufficient prenatal and postnatal care services (Bonney 2016; Borkotoky 2016; Khyriemmujat and Basaiawmoit 2016; Lyngdoh 2016, 2017; Laloo 2016; Mawkhroh 2016; Contractor et al. 2018; Keeni et al. 2018; Dutta 2022; Soreide 2022; Shullai et al. 2023; Aborigo et al. 2015).

In recent years, climate change has emerged as a critical stressor that compounds pre-existing maternal health challenges in Meghalaya. The Khasi Hills, with their ecologically fragile landscapes, are experiencing increasingly erratic rainfall, soil degradation, water scarcity, and disruptions in food systems—conditions that disproportionately affect pregnant women and new mothers. Seasonal flooding, landslides, and crop failures restrict physical access to health services. Such climate stressors also undermine household food security leading to nutritional deficiencies that increase the risk of maternal anemia and low birth weight. These environmental changes are particularly severe in remote and forest-dependent villages, where livelihoods and well-being are closely tied to climate-sensitive resources. Capacity has to be built for the public health systems to respond to climate stressors that may compromise prenatal and postnatal care, as the root cause can be traced to the degradation of local ecosystems. In this context, one can postulate that climate change does not act in isolation but intensifies the barriers of geography, poverty, and structural neglect already faced by women in the Khasi Hills (Nelson et al. 2002). Thus, any effort to improve maternal health outcomes in the region must also function as a form of climate change adaptation—addressing both immediate health needs and long-term ecological vulnerabilities.

As of 2024, some districts still lack district hospitals, while the capital, Shillong, continues to hold a disproportionate share of medical facilities (Comptroller and Auditor General 2019). Although institutional births have increased from 51.4% in 2016 to 58% in 2019–20, they remain below national levels (IIPS and ICF 2021). Nutritional deficiencies significantly exacerbate maternal health outcomes. In Meghalaya, 53.3% of pregnant women are anemic, surpassing the national prevalence of 50.4% (Shullai et al. 2023). These deficiencies are closely correlated with elevated rates of low birth weight, which constitute 11.7% of births in the state ("Mobilising collaborative actions to improve nutrition status", 2021). Further compounding these challenges, teenage pregnancies persist at alarmingly high levels, despite initiatives aimed at raising awareness through sex education programs and the implementation of punitive measures, notwithstanding the demerits of a legal system that punishes underage but consensual sexual relations, such as the Protection of Children from Sexual Offences (POCSO) Act. Notably, the POCSO Act is among the few legislative measures translated into the Khasi language to facilitate greater dissemination ("POCSO Act in Khasi hailed", 2024). Notwithstanding, reports indicate that teenage pregnancy rates reach up to 53% in certain areas (Shullai et al. 2023). This cohort, comprising individuals under 18 years of age, is particularly susceptible to adverse outcomes, as teenage mothers frequently lack the physiological and socio-economic preparedness essential for childbirth and child-rearing.

The Khasi Hills region can illustrate, as this paper contends, the possible interplay of socio-cultural factors and health outcomes in the context of climate change in Meghalaya. Cultural and religious taboos surrounding contraception and abortion persist, contributing to a low contraceptive usage rate of 27.4% and far below the national target of 90% (Shullai et al. 2023). This reluctance is compounded by a lack of awareness and the limited involvement of men in family planning, which collectively impede efforts to address high fertility rates and inadequate birth spacing (Pereira and Khatso 2007). A holistic approach including the cognizance of climate change factors that addresses healthcare infrastructure and these kinds of cultural barriers is essential for improving maternal health in the state (Nelson et al. 2002). Given the significant gaps in access to reliable healthcare infrastructure—whether due to systemic inadequacies, distrust in the quality of formal services, or a cultural preference for traditional healing—future interventions must prioritize the integration of Indigenous knowledge systems with evidence-based medical strategies. Such an approach ensures that healthcare responses are not only effective and sustainable but also culturally attuned and locally meaningful.

12.1.2 Indigenous Medicinal Knowledge and Customary Health Practices in the Khasi Hills Districts

The Khasi medical tradition demonstrates profound ecological consciousness. The sacred landscapes, forest groves, and watersheds are not only sources of medicinal plants but are embedded in cultural taboos and stewardship practices that prevent overextraction and environmental degradation. These place-based ethics, often transmitted orally through clan elders and local healers, offer a parallel system of health governance that is climate-responsive by design. Traditional medicinal knowledge in the Khasi Hills districts is tied to cultural identity, ecological stewardship, and community health resilience. Despite this, it still remains marginalized within formal healthcare systems which is a consequence of both colonial epistemologies and contemporary development paradigms. British colonial ethnographies constructed Indigenous health practices through a primitivist lens (Ray 2023), alienating them from institutional legitimacy and shaping a legacy of exclusion that persists in health governance. In recent years, with the emphasis on knowledge systems also veering toward Indigenous knowledge, traditional healers, who have maintained a pivotal role in traditional healthcare for the Khasis, and perpetuating practices through intergenerational transmission (Kharbangar 2017; Lyngdoh 2017; Pde 2020), have gained institutional prominence and recognition, even if this is still a nascent development. Khasi traditional medicine *(dawai Khasi)*—practiced by *nongai dawai kynbat* (herbalists) and *nong khem* (physical therapists)—continues to be a vital, community-embedded practice, passed down through oral transmission, observation, and apprenticeship (Kharbangar 2017; Pde 2020). These healers draw on intimate ecological knowledge and community trust, embodying a holistic understanding of health rooted in the Khasi cosmology that views the earth as *Ka Meiramew*, a nurturing maternal force (Miri 2022). The concept of *sap*, denoting intuitive aptitude for recognizing medicinal plants, for instance, and communal recognition underpins the healer's legitimacy (Albert 2014).

Beyond their role in healthcare, the *nongai dawai* hold a significant position within the traditional social structures of the Khasi community. Their expertise and the trust they command imbue them with a form of socio-political authority that extends into decision-making processes at the community level. This centrality underscores the *nongai dawai* as integral to the cultural and governance fabric of the Khasi society, where health and well-being are deeply interconnected with traditional leadership and community management. Khasi healing practices often transcend conventional biomedical paradigms, rather, coexisting pragmatically with biomedicine: traditional practitioners refer to doctors as *doktor kot* (book doctors) while identifying as *doktor sla* (plant doctors) (Albert and Porter 2015). Their practices, while empirically grounded and distinct from ritual healing (*nongkñia*), reflect culturally specific diagnostic categories such as *lait thied sohpet* (gastric), often incongruent with biomedical classifications. Nonetheless, the absence of reciprocal recognition hampers meaningful integration, even as traditional practitioners refer patients to hospitals for conditions beyond their scope.

The formal recognition of the *nongai dawai* through the "Protection and Promotion of Khasi Traditional Medicine Act" (2011) presents an opportunity not only for their integration within the mainstream healthcare systems of the state but also for fostering a synergistic relationship between climate change, traditional medicine, and grassroots governance structures, notably the *Seng Kynthei*. The *Seng Kynthei* is the association of adult women in a village or *Shnong* in the Khasi Hills districts.

These associations work to improve the lives of women in the village, and also help the *Dorbar Shnong*, or village council and play a crucial role in community welfare, addressing social, economic, and even climate change-related health issues specific to women (Lyngdoh 2020). Operating quasi-formally within Khasi social and cultural society, it often addresses the nuanced needs of women at the local level. Integrating this women's institution into healthcare planning could enhance the identification of specific vulnerabilities and needs, given their intimate connection with the daily lived experiences of Khasi women. The *Seng Kynthei's* "last mile" reach, drawn from their embedded presence in community life, can provide critical insights into health disparities and aid in tailoring interventions to meet the unique requirements of different stakeholders. Here too, the aspect of incentivizing this healthcare-centric role as well as parallel capacity-building has to be considered in order to make this kind of community-based intervention more effective. *Seng Kynthei* members who are also part of the Accredited Social Health Activist (ASHA) network could be further incentivized to act as vital bridges between this women's body and the integrated holistic traditional and state health delivery systems (National Rural Health Mission 2009). This kind of integration would serve to strengthen the healthcare framework and also empower these women to play an essential role in linking traditional practices with state mechanisms especially in shaping climate policies, thus facilitating a more comprehensive approach to community health. With the Khasi medical tradition demonstrating profound ecological consciousness and commitment to biodiversity conservation, the *nongai dawai* function as custodians of medicinal flora, contributing significantly to the preservation of Meghalaya's unique biodiversity. This role has received formal recognition through initiatives such as the Khasi Hills Autonomous District Council's "Protection and Promotion of Khasi Traditional Medicine Act" (2011), which aims to institutionalize and support these practices whilst acknowledging their accessibility, affordability, and contribution to sustainable livelihoods (Albert 2014).

Nonetheless, challenges persist. Khasi medicine is largely undocumented, unlike codified systems such as Ayurveda or Unani, rendering it vulnerable to erosion despite community reliance—87% of rural residents express confidence in traditional healing, and nearly half report recent use (Albert and Porter 2015). Efforts by organizations such as the Institute of Bioresources and Sustainable Development to catalogue medicinal flora (IBSD 2023) remain at early stages. Population growth, ecological degradation, and the scarcity of trained healers further limit scalability. Integration within public health frameworks necessitates addressing cultural barriers and fostering structured collaborations between traditional and biomedical practitioners but also climate adaptation specialists. As a practice the *nongai dawai* regularly refer patients to more mainstream healthcare facilities for conditions beyond their expertise, but reciprocal recognition and formal partnerships remain limited. This asymmetry reflects broader challenges of medical pluralism in India, where non-codified systems like traditional medicine and health practices remain subordinate to state-sanctioned AYUSH institutions, despite their demonstrated local efficacy and cultural resonance (Albert and Porter 2015). Here too the *Seng Kynthei* could play a pivotal role in bridging this gap by facilitating a sustained interface and cooperation between various *nongai dawai* and mainstream biomedical practitioners. Given their established presence in community life, the same *Seng Kynthei* members can act as intermediaries, helping to address cultural barriers and fostering trust between the two systems. A multi-pronged strategy through systematic documentation, institutional support for knowledge preservation, training and certification pathways, and policy frameworks could help recognize and formalize collaborations. Equally crucial is enabling community-based organizations like the *Seng Kynthei* to participate in policy feedback loops, ensuring culturally congruent and climate-responsive maternal and public health interventions. A pluralistic, integrative model—responsive to Khasi ecological, social, and epistemic contexts—can advance a more inclusive and resilient health system. Current health policies and programs in Meghalaya reflect a predominantly biomedical orientation, which inadequately accounts for Indigenous epistemologies, social structures, and ecological knowledge. This disjuncture results in three major gaps: Firstly, the absence of formal mechanisms for documenting and integrating Khasi medicinal knowledge into state healthcare systems; secondly, a lack of policy attention to the role of women's community institutions like the *Seng Kynthei* in advancing maternal health; and thirdly, the limited development of models that incorporate traditional medicine into climate-resilient public health planning.

These gaps are further exacerbated by the rapid pace of ecological change and requires a culturally informed climate and gender-sensitive health interventions. This study thus explores the intersection of Indigenous Khasi maternal health practices with modern biomedical systems and their resilience in the face of growing climate challenges by positing three main research questions: What are the barriers that impede on the formalization of Indigenous knowledge? In the face of climate change, how can Khasi women's groups like the *Seng Kynthei* amplify maternal health outcomes? What replicable models for climate-adaptive as well as pluralistic healthcare systems can shape policy for Meghalaya?

12.1.3 Objectives of the Study

This study investigates the convergence of Indigenous health practices, gendered climate adaptation, and maternal healthcare in Meghalaya's Khasi Hills. The following objectives are framed in response to the study's central research questions and the key objectives are: To identify and critically analyze the structural, epistemological, and policy-related barriers that hinder the formalization and institutional integration of Khasi Indigenous medicinal knowledge within mainstream health governance systems. To evaluate the role and potential of Khasi women's groups particularly the *Seng Kynthei* in improving maternal health outcomes under climate stress emphasizing on their embeddedness in community life, local governance, and informal health mediation. To develop and propose context-specific, replicable models of climate-adaptive and pluralistic healthcare systems that can inform inclusive health policy in Meghalaya by integrating Indigenous knowledge with biomedical approaches.

12.1.4 Significance of the Study

This study contributes both empirical and theoretical insights to ongoing debates in Indigenous health governance, gender-responsive climate policy as well as public health pluralism. It centers Khasi women's perspectives and community-based knowledge systems in reimagining inclusive healthcare frameworks amidst growing ecological uncertainty. The research outputs comprise two primary contributions: First, evidence-based policy recommendations to enhance the integration of traditional and formal health systems; second, theoretical advancement in the fields of medical pluralism, Indigenous health governance, and gender-responsive climate adaptation. These components provide a holistic and culturally attuned understanding of maternal health vulnerabilities and resilience within the Khasi context.

12.2 Research Design

This study examines the complex interplay between the goal of gender-responsive climate adaptation, the embedded Indigenous knowledge systems, and the issues in maternal health in Meghalaya's Khasi Hills. It considers the integration of traditional maternal health practices within formal policy frameworks to enhance community resilience and health outcomes in the context of growing environmental pressures. A summary of the research design is shown in Fig. 1.1 below:

Figure 12.1 outlines the overall research design, which integrates three interrelated domains of indigenous knowledge, formal health policies, and environmental/climate pressures. These domains form the conceptual foundation of the study, providing the analytical lens through which maternal health outcomes in the Khasi Hills are examined. The study utilizes a triangulated qualitative approach, combining policy analysis; qualitative surveys, and interviews with traditional healers, health professionals, and community women; case-based evaluation of maternal health outcomes. The design foregrounds institutional engagement with *Seng Kynthei*, a grassroots women's organization central to Khasi village governance, positioning it as a key mediator of maternal health resilience.

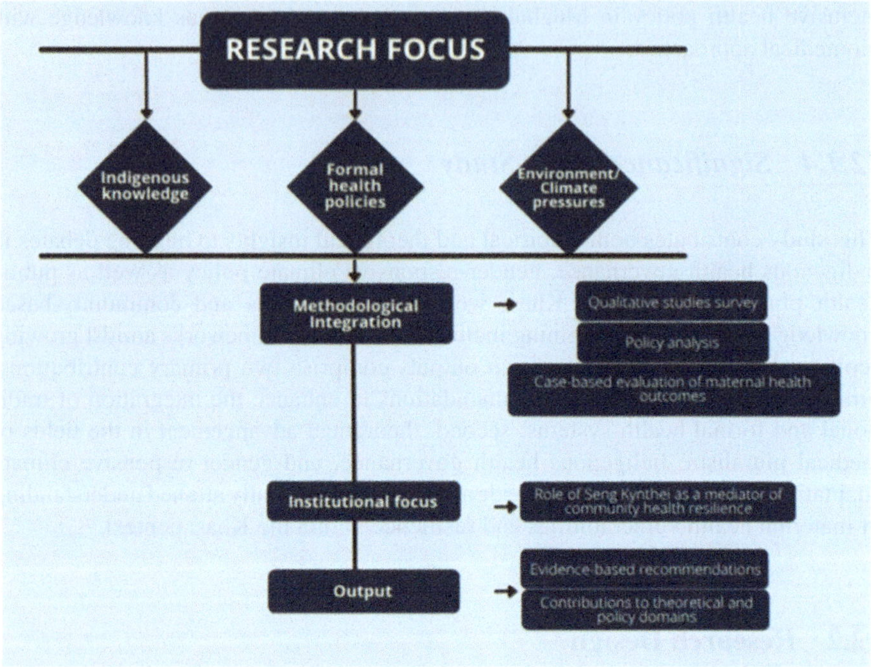

Fig. 12.1 Research design summary

12.3 Literature Review Methodology

12.3.1 Inclusion and Exclusion Criteria

Literature was systematically searched across the following databases and sources: PubMed, Scopus, JSTOR, and Google Scholar as well as published books, unpublished theses, national and state government reports and reports from the National Health Missions.

Searches were conducted between August 2024 and March 2025. To ensure relevance and rigor, the following inclusion and exclusion criteria guided the literature selection:

Inclusion criteria	Exclusion criteria
Peer-reviewed academic publications (2000–2025) focusing Indigenous health systems, maternal health, gender, and climate adaptation, particularly within Indian Northeast	Generalized studies with no regional or socio-cultural specificity

(continued)

(continued)

Inclusion criteria	Exclusion criteria
Government and institutional reports (e.g. NFHS, MoHFW, CAG, Meghalaya health missions) directly assessing maternal health metrics or healthcare infrastructure in Meghalaya	Biomedical literature lacking socio-cultural context
Region-specific ethnographic and historical studies on Khasi society, matriliny, and traditional medicine practices	Opinion pieces or grey literature without empirical grounding
Interdisciplinary studies engaging with medical pluralism, climate resilience, and women's health in tribal or marginalized contexts	Studies not available in English

12.3.2 Screening and Selection Process

All identified literature was exported into a reference management system (Zotero) with the initial screening of titles and abstracts by the primary author. Full texts were subsequently reviewed against the inclusion/exclusion criteria. Where relevance was unclear, a consensus method was applied: articles were re-evaluated in consultation with two independent researchers familiar with the subject domain to ensure consistency and thematic alignment.

12.3.3 Data Extraction and Thematic Synthesis

A thematic analysis approach was adopted for synthesizing literature. Key data points were extracted into an Excel matrix, categorized by thematic relevance (e.g., maternal health, traditional healing, climate pressures); geographical/tribal specificities; methodological quality (e.g., empirical, policy-based, ethnographic); and policy implications. An inductive coding process was applied to identity emergent themes and were iteratively refined to develop a comprehensive understanding of the structural, cultural, and institutional dimensions shaping maternal health and climate resilience in the Khasi Hills.

12.4 Results and Discussion

12.4.1 The Impact of Climate Change on Maternal Health in Meghalaya

The Khasi traditional medicine system (*nongai dawai kynbat*) represents a profound repository of Indigenous knowledge that has served as the cornerstone of healthcare in the Khasi Hills. This extends to the maternal and reproductive interventions of the tribal people. Traditional healing practices are grounded in an intimate understanding of local biodiversity and cultural practices. These traditional approaches have offered culturally resonant and accessible healthcare solutions, particularly in remote areas where mainstream biomedical services remain scarce (Shangpliang 2012). A sophisticated knowledge of local flora for managing various aspects of maternal health, from morning sickness to postpartum recovery is available. These Indigenous practices reflect the interconnectedness between women's health and their natural environment (Contractor et al. 2018; Shangpliang 2012). However, the preservation and practice of this traditional knowledge face many unprecedented challenges in the contemporary context. The advent of modern healthcare and a booming population in the Khasi Hills has catalyzed a significant transformation in maternal healthcare seeking behaviors. This shift echoes trends observed in other regions, such as Ghana, where traditional healers have an increasingly diminished role in obstetric care (Aborigo et al. 2015). The erosion of traditional knowledge has created a void in communities where modern healthcare facilities remain inaccessible or inadequate (Dutta 2022). Climate change has also emerged as another formidable challenge to Indigenous knowledge systems, particularly affecting the availability and sustainability of medicinal plants crucial for traditional maternal care (Awiti 2022). The degradation of local ecosystems due to changing weather patterns and environmental stress has disrupted the natural habitat of various medicinal plants traditionally used in maternal care (CIFOR and CGIAR 2015). This environmental impact presents a dual challenge—not only does it threaten the practical application of traditional medicine, but it also risks the permanent loss of Indigenous knowledge about these botanical resources (Shangpliang 2012; Dutta 2022).

The interplay between modernization and climate change-induced stressors has necessitated adaptive strategies within Indigenous knowledge systems in Meghalaya. Rising temperatures, erratic rainfall patterns, and the depletion of forest resources have begun to directly affect maternal health—both by altering the availability of medicinal plants used in Khasi ethnomedicine and by disrupting traditional food systems essential to maternal nutrition (Albert and Porter 2015; Dutta 2022). Some communities have begun systematically documenting traditional practices as a response recognizing the urgent need to preserve and adapt this knowledge in the face of climate variability (Albert and Porter 2015). Traditional healers are also increasingly collaborating with modern healthcare providers, co-developing hybrid care models that integrate Indigenous wisdom with biomedical approaches to manage climate-sensitive maternal health issues such as anemia, heat stress during pregnancy, and water-borne infections (Aborigo et al. 2015; Nengnong et al. 2024). Nonetheless, the sustainability of these Indigenous systems requires more than documentation—it demands institutional support, legal recognition, and the incorporation of Indigenous knowledge into public health frameworks, especially in climate-vulnerable zones (Soreide 2022; Shangpliang 2012). This integration is particularly critical in remote areas where climatic disruptions have further weakened formal healthcare access, and where traditional practices still serve as frontline maternal care. The formalization and standardization of Indigenous knowledge—while respecting its cultural integrity—are key to ensuring its resilience and relevance in addressing maternal health under conditions of climate stress. As Khasi communities continue to navigate the dual pressures of modernization and environmental change, the resilience embedded in their traditional maternal health systems offers instructive, context-sensitive models for sustainable and culturally attuned healthcare delivery (Diengdoh and Wahlang 2016; Mawlong 2016; Contractor et al. 2018; Patel and Gebreyes 2020).

12.4.2 Climate-Induced Vulnerabilities in the Health Status of Indigenous Mothers

Climate change magnifies existing socio-economic inequalities, disproportionately affecting women due to entrenched structural vulnerabilities. It is increasingly recognized as a public health crisis, with maternal health in Indigenous and rural communities facing compounded risks from environmental stressors (Meherali et al. 2024). Women, particularly in agrarian and Indigenous communities, face increased susceptibility to chronic malnutrition, a condition exacerbated by climate-induced food insecurity. This nutritional stress is especially detrimental during pregnancy and lactation, leading to cascading effects on maternal and child health outcomes (Awiti 2022; Caswell and Jang 2024; CIFOR 2015). Furthermore, the resurgence and geographical spread of vector-borne diseases such as malaria, driven by altered ecological conditions, pose significant health risks, including low birth weight and heightened neonatal mortality (Awiti 2022). Climate-related disasters—floods, droughts, and heatwaves—compound these vulnerabilities. These events not only disrupt healthcare infrastructure and impede access to maternal services but also increase the prevalence of adverse reproductive outcomes such as pre-eclampsia and preterm birth (Awiti 2022). Table 12.1 outlines these vulnerabilities. These impacts underscore the urgent need for targeted, gender-responsive interventions supported by sex-disaggregated data to prioritize women's health in climate adaptation strategies.

In Meghalaya, climatic shifts have been marked by increased rainfall variability, rising temperatures, and a growing frequency of extreme weather events. These environmental changes threaten the delicate ecological balance that communities in the Khasi Hills districts have historically depended on for food, health, and cultural identity. Loss of biodiversity—including the decline of medicinal plants

Table 12.1 Climate risks, maternal health impacts, and indigenous responses in Meghalaya

Climate stressor	Maternal health impact	Indigenous response	Key gap
Food insecurity	Anemia, low birth weight, poor maternal nutrition	Traditional diets and herbal supplements	Not integrated into health schemes (Shangpliang 2012; Albert and Porter 2015; Chaand et al. 2019; Bordoloi 2021; Govt. of Meghalaya 2021; Shullai et al. 2023; Nongkynrih et al. 2024; Singh et al. 2024)
Vector-borne diseases	Miscarriage, neonatal mortality	Use of medicinal plants	Limited formal recognition (Albert and Porter 2015; Govt. of Meghalaya 2021; Shullai et al. 2023; Nengnong et al. 2024; Nongkynrih et al. 2024)
Extreme weather	Disrupted care, preterm births	Reliance on *Seng Kynthei*, local healers	Lack of gender-sensitive emergency plans (Marbaniang and Singh 2022; Nengnong et al. 2024; Nongkyrih et al. 2024; Shullai et al. 2023; Nongsiej and Dutta 2024)
Biodiversity loss (medicinal plants)	Decline in traditional remedies	Conservation of medicinal plants	No protection in health policy (Albert and Porter 2015; Sohtun 2016; Dutta 2022; Shullai et al. 2023)
Mental health stress	Postpartum depression, anxiety	Informal peer support via women's groups	Overlooked in public health frameworks (Keeni et al. 2018; Dutta 2022; Marbaniang and Singh 2022; Shullai et al. 2023)
Water scarcity	Infection risk, poor hygiene during pregnancy	Community-led water management	Inadequate infrastructure for women (Sohtun 2016; Soreide 2022; Shullai et al. 2023; Nengnong et al. 2024; Nongkynrih et al. 2024)
Gendered access barriers	Reduced care and resources for women	Women-led resilience efforts (e.g. Seed banks)	GESI missing in adaptation policies (Leonetti et al. 2007; Shangpliang 2012; Mawkhroh 2016; Bordoloi 2021; Dutta 2022)

essential to Indigenous health systems—poses a direct threat to maternal care, especially in remote areas where formal healthcare access is limited (Albert 2014; Patel and Gebreyes 2020; Nengnong et al. 2024; Shangpliang 2020; 2012; Diengdoh and Wahlang 2016). The erosion of these resources undermines both the ecological fabric and cultural continuity of traditional knowledge systems central to Khasi maternal health practices. Climate-related stressors have compounded maternal health challenges across the state. Disasters often result in the destruction of health infrastructure and interruptions in transportation, further marginalizing rural women. Heatwaves and water scarcity have intensified malnutrition and waterborne diseases, elevating maternal and neonatal risk (Nongsiej and Dutta 2024; Nongkynrih et al. 2024). Moreover, mental health stressors linked to climate disruption remain under-acknowledged in current health frameworks despite evidence of significant psychological burdens on affected women (Contractor et al. 2018). Gendered dimensions of climate vulnerability are particularly pronounced in Meghalaya's rural communities. Women are responsible for water and food security but face structural barriers such as limited land ownership and reduced access to credit and extension services (Pratiwi et al. 2016; CIFOR 2015; Ravera et al. 2016). Case studies from Bihar and Uttarakhand, as well as international contexts such as Kenya and South Africa, demonstrate that when women are equipped with knowledge, resources, and agency, they lead transformative climate responses—managing seed banks, practicing agro-biodiversity conservation, and advocating for sustainable land use (Puzyreva and Roy 2018; Caswell and Jang 2024).

In the Khasi Hills districts, traditional health systems have adapted to environmental stressors through diverse, community-driven strategies. Local herbal remedies, rooted in generational knowledge, continue to play a vital role in maternal health. At the same time, community-based water management ensures clean water access in the face of climate-induced scarcity (Ravera et al. 2016). These Nature-Based Solutions (NbS), particularly when informed by Gender Equality and Social Inclusion (GESI) frameworks, offer context-specific, sustainable responses to health challenges. However, their scalability remains limited without institutional support, scientific validation, and integration into broader policy frameworks. Traditional healers, or *nongai dawai kynbat*, are pivotal in these adaptive responses. Their expertise in local flora and community-based health enables them to act as critical intermediaries in climate-resilient healthcare. Collaborations between *nongai dawai* and scientific researchers could yield innovative health protocols while also fostering respect for traditional epistemologies (Albert 2014; Albert and Porter 2015; Albert, Porter and Green 2023). Likewise, organizations such as the *Seng Kynthei*, embedded in local community life, are uniquely positioned to bridge gaps between traditional knowledge, climate adaptation strategies, and biomedical health systems. Their role in advocating for women's health and resilience is increasingly vital as climate disruptions escalate. Table 12.2 presents a comparative overview of maternal health outcomes across three community health delivery models—traditional-only, biomedical-only, and integrated approaches—highlighting variations in postnatal care coverage, referral rates, and community satisfaction.

Globally, successful models of integrated health systems—such as China's Traditional Chinese Medicine (TCM) policies and Kerala's fusion of Ayurveda with allopathy—illustrate how Indigenous systems can complement biomedical frameworks when supported by policy and mutual respect (Xu and Jang 2009; Cheung et al. 2020; Joseph et al. 2021; Nedungadi et al. 2023). Meghalaya could draw from these examples to build inclusive, pluralistic healthcare strategies that acknowledge the lived realities and agency of Indigenous women. Ultimately, building resilience to climate change demands that adaptation strategies move beyond technocratic solutions. They must engage with the lived experiences of women, incorporate Indigenous knowledge, and embrace rights-based frameworks that promote equity and sustainability. Programs such as the LDC Initiative for Effective Adaptation and Resilience (LIFE-AR) illustrate how such integration can center local voices in climate governance (Patel and Gebreyes 2020). Incorporating GESI principles in both policy and

Table 12.2 Comparative outcomes—traditional versus integrated practices

Community type	Postnatal coverage	Referral rate	Community satisfaction	Notes
Traditional-only (*nongai dawai* only)	Low to medium	Low	High cultural trust	Lack emergency care (Albert 2014; Albert and Porter 2015; Chaand et al. 2019)
Biomedical-only (PHC/ASHAs)	Medium	High	Moderate trust	Cultural mismatch (Albert and Porter 2015; Albert and Green 2023)
Integrated (*Seng Kynthei* + ASHA)	High	High	High satisfaction	Shows adaptive potential (Govt. of Meghalaya 2021; Albert, Porter and Green 2023)

practice is essential for fostering communities capable of withstanding and thriving amidst climate uncertainty.

12.4.3 Policy Review and Analysis

Existing health policies in Meghalaya, particularly under the National Health Mission (NHM), have made commendable efforts to address maternal and child health challenges. Flagship schemes such as the Janani Suraksha Yojana (JSY) and the Janani Shishu Suraksha Karyakram (JSSK) promote institutional deliveries ensuring free healthcare services targeted to pregnant women and newborns to reduce mortality rate (Ministry of Health and Family Welfare 2025). While these initiatives have made significant strides, however, these programs were not designed with climate-sensitive healthcare in mind, and they often fall short of addressing the growing vulnerabilities of mothers in the face of climate change particularly in ecologically fragile regions. Building bridges between biomedical and traditional systems—rather than replacing one with the other—is therefore increasingly crucial for improving maternal outcomes in culturally and ecologically sensitive ways, especially in the context of climate change. Environmental degradations are already affecting maternal health in Meghalaya increasing the risks of malnutrition, waterborne diseases, and limited access to care during emergencies. In this context, the exclusion of Indigenous knowledge systems further undermines policy efficacy. Traditional Khasi healers (*nongai dawai kynbat*), who offer vital maternal and neonatal care—particularly in rural and climate-affected areas—remain outside formal policy frameworks. Their contributions of climate-adaptive practices such as the use of locally sourced medicinal plants and context-specific dietary recommendations, are seldom acknowledged in health planning. Additionally, ASHA and Anganwadi workers who are central to service delivery are not trained to engage with these Indigenous frameworks or the emerging health challenges posed by environmental changes (Albert and Porter 2015). Promising models from other parts of the world have demonstrated that collaborative systems involving traditional healers and biomedical professionals can improve health outcomes, especially when climate vulnerabilities intersect with structural inequities (Aborigo et al. 2015; Contractor et al. 2018). Internationally, several regions offer valuable lessons. China's Traditional Chinese Medicine (TCM) is integrated into public healthcare policies, creating synergies between indigenous and biomedical systems (Xu and Jang 2009; Cheung et al. 2020). Across Africa, collaborations between traditional and biomedical practitioners have improved care in areas such as HIV/AIDS management (Mills et al. 2005; Zuma et al. 2017; Mutola et al. 2021), while closer to home in Kerala, Ayurvedic and allopathic systems are offered in parallel through state support, demonstrating the feasibility of policy-led integration (Joseph et al. 2021; Nedungadi et al. 2023). These models are premised on valuing structured collaboration and local capacity building.

Meghalaya's current policy architecture appears to remain largely oriented toward biomedical frameworks which may limit its capacity to adequately respond to the layered and evolving impacts of climate change on maternal health. The absence of more integrated and climate-sensitive approach in existing policies may not be able to fully address the complex and context-specific challenges facing maternal well-being in the state's changing environmental landscape. The predominantly top-down orientation could be a critical limitation of Meghalaya's public health strategy. Indigenous knowledge systems—such as those upheld by *nongai dawai kynbat*—are not fully recognized or integrated into state healthcare frameworks. Despite progressive legislation such as the Khasi Hills Autonomous District Council's Protection and Promotion of Khasi Traditional Medicine Act (2011), integration remains superficial (Albert 2014). Traditional healers often refer patients to biomedical facilities when needed but this reciprocity is rarely returned (Albert and Porter 2015). This asymmetry reflects broader structural issues in India's medical pluralism where codified systems like AYUSH are privileged over non-codified and community-based practices like Khasi medicine.

The persistence of policy silos has deepened the divide between modern and Indigenous systems; Khasi traditional medicine, rooted in local biodiversity and cultural practice, continues to be underutilized due to lack of institutional support (Shangpliang 2020). *Nongai dawai*, who are often respected community members, remain marginalized despite their significant role in maternal care (Albert and Porter 2015). Policies must move beyond tokenism to include traditional healers in training, planning, and service delivery. Environmental degradation driven by deforestation and mining has also jeopardized access to medicinal plants that form the backbone of Khasi traditional medicine. This has direct implications for maternal and neonatal health particularly for women who depend on these resources for prenatal and postpartum care (Dutta 2022; Shangpliang 2012). Gender-responsive climate adaptation strategies—such as community-led reforestation—can help safeguard these ecosystems while enhancing healthcare resilience. All of these developments underscore the need for holistic policies that combine economic resilience with accessible and community-based healthcare.

12.4.4 Community Insights: Role of the Seng Kynthei

Within the Khasi community, the *Seng Kynthei* have emerged as central actors in promoting maternal health and early childcare awareness. Rooted in their close alignment with local customs and social norms, these collectives of women function as trusted connectors between households and formal healthcare systems, building essential bridges of trust and encouraging wider community engagement (Albert 2014). Their activities span from hosting health camps to circulating information on nutrition and maternal practices, ensuring even women in the most remote or marginalized settings receive timely, culturally sensitive health information. Just as importantly, *Seng Kynthei* initiatives offer safe and supportive environments where women can openly discuss topics often considered taboo—such as reproductive health, family planning, and maternal nutrition (Shangpliang 2012). The scope of their work extends beyond service delivery into advocacy, challenging entrenched gender norms that constrain women's autonomy over their own health choices (Dutta 2022). In areas where public health infrastructure is minimal or lacking, these groups have filled critical gaps. Some Khasi villages, for instance, have seen measurable health improvements through community-driven nutrition schemes that combine traditional dietary knowledge—often shaped by seasonal and climate-dependent food systems—with modern nutritional guidelines leading to reduced anemia among pregnant women and better child health outcomes (Albert 2014). Community health volunteers bolster these gains—often women who are members of the *Seng Kynthei*—may serve as a vital link between medical personnel and local families. Their involvement helps ensure consistent maternal care and timely post-natal follow-up (Shangpliang 2012). At the same time, Khasi women have responded proactively to environmental challenges, adopting resilient agricultural practices such as intercropping, seed preservation, and sustainable harvesting. These adaptations not only enhance local food security but also mitigate the effects of climate volatility (Patel and Gebreyes 2020). Incorporating these grassroots innovations into formal climate resilience policies would facilitate a synergy between Indigenous expertise and scientific approaches—especially valuable in contexts where rural women face layered vulnerabilities (Liswanti et al. 2020). Designing effective maternal and climate health interventions demands a grounded understanding of the everyday realities faced by Khasi women and frontline health workers. However, despite their critical roles in caregiving, food production, and community health, women's insights are frequently left out of policy dialogues (CIFOR 2015; Gumucio et al. 2022). Local accounts consistently point to deteriorating access to clean water and medical care—issues aggravated by ecological stress. Health professionals working in the Khasi Hills districts also highlight the need for integrated approaches, combining Indigenous healing systems with biomedical care to respond more flexibly to community needs (Albert 2014). Strengthening partnerships among traditional healers, women's groups, and public health providers can not only expand the reach of services but also foster deeper trust and cooperation. More importantly, incorporating women into decision-making spaces ensures that policies reflect the realities and values of those most directly affected.

As discussed earlier, Indigenous systems of knowledge offer critical tools for improving maternal health, particularly in settings where biomedical resources are scarce. Grassroots organizations like the *Seng Kynthei* exemplify how culturally grounded strategies can build health but also climate resilience from the ground up (Puzyreva and Roy 2018). Policymakers could collaborate with Autonomous District Councils (ADCs) to initiate village consultations aimed at identifying and safeguarding traditional health knowledge. These spaces would allow women to shape health initiatives that respect and reflect local cultural frameworks (Borkotoky 2016; CIFOR 2015). For such integration to be meaningful, traditional practices must not only be recognized but also validated and incorporated into healthcare systems. This includes developing training content for medical professionals that covers Indigenous birthing practices, medicinal plant use, and sustainable, climate-resilient nutrition strategies rooted in local ecologies (Albert and Porter 2015). At the same time, policymakers must establish formal protections to safeguard the intellectual property rights associated with traditional knowledge, prevent misappropriation, and encourage knowledge-sharing between communities and researchers (Dutta 2022). Despite their central roles in both public health and environmental management, Khasi women continue to encounter systemic hurdles—from unequal access to education and financial services to limited healthcare options. Efforts to empower them through access to microfinance, adaptive health training, and climate-resilient technologies can significantly enhance their capacity to lead change at the community level (Puzyreva and Roy 2018). Promoting women's involvement in governance structures is equally important. Their lived experiences as caregivers and keepers of traditional wisdom position them to offer comprehensive, community-oriented perspectives on maternal and climate health (CIFOR 2015). Women's participation in developing solutions—such as climate-resilient cropping methods or water conservation systems—can simultaneously address nutrition, sanitation, and maternal care gaps. Gender-sensitive policies must also confront the cultural barriers that suppress women's agency and innovation. Educational campaigns that challenge restrictive gender norms and elevate women's roles in both health and environmental decision-making are critical. Joint efforts by government agencies, NGOs, and local organizations can help create environments where women's contributions are recognized and amplified (Patel and Gebreyes 2020). In this context, networks such as those supported by and enriched by the *Seng Kynthei* become essential to building robust maternal health systems but also to fostering community resilience in the face of climate change. These groups play a key role in translating medical advice into culturally relevant practices, mobilizing resources, and creating pathways for community-driven healthcare delivery (Albert 2014) in climate-vulnerable regions. To maximize their impact, policy frameworks should include sustained funding, skill development programs, and structured collaboration with public health authorities. Equipping these groups to act as first responders in maternal emergencies, for example, could significantly improve outcomes for women in remote areas. Likewise, health governance bodies should integrate their voices into discussions, enabling them to represent local realities and propose grounded solutions (Shangpliang 2012). A participatory, locally anchored approach—combining traditional knowledge and modern medicine—ultimately offers the most sustainable path forward. Supporting women's leadership and leveraging community networks not only enhances maternal health outcomes but also reinforces broader community resilience in the face of ecological and structural challenges (Dutta 2022). A healthcare system rooted in equity and cultural integrity is not only possible—it is already taking shape through the work of women like those in the *Seng Kynthei.*

The *Seng Kynthei* as grassroots women's collectives embedded in Khasi village life are not only maternal health advocates but also vital agents of community-based climate adaptation. Their proximity to local environmental realities—ranging from declining water sources to food insecurity—places them at the frontline of both public health and climate resilience. While their roles in promoting maternal and reproductive health are well-recognized, what is less acknowledged is how these same interventions help mitigate the compounding effects of climate change on women's health outcomes. In fact, by supporting traditional food systems, sustainable harvesting, and community-led health surveillance, the *Seng Kynthei* contribute to a form of ecosocial resilience—one that bridges health, ecological knowledge, and gendered leadership in climate-vulnerable areas. Their initiatives such as promoting local nutrition practices in response to climate-induced food scarcity or facilitating care during extreme weather events reflect an adaptive capacity deeply rooted in Indigenous knowledge and collective caregiving, so the integration of their roles into formal climate and health policy frameworks is critical. These women's networks exemplify how climate change mitigation and maternal health strategies can be synergized at the community level, especially in remote areas where public services are fragile or absent. As a response to intersecting crises of health and climate there is a need to recognize and strengthen their position as mediators of both cultural continuity and environmental adaptation to play a pivotal role in ensuring inclusivity and sustainability.

12.4.5 Policy Pathways and Strategic Recommendations for Integrated Maternal Healthcare

Drawing from the empirical and contextual findings of the study, several targeted strategies emerge to enhance maternal health outcomes in the Khasi Hills districts, particularly through an integration of Indigenous and formal health systems. These recommendations directly address the gaps in health service delivery, cultural mismatches, and structural exclusions that weaken both biomedical and traditional healthcare systems. Given these findings, the following strategies are proposed to strengthen gender-sensitive, culturally responsive, and community-grounded maternal healthcare in the Khasi Hills districts. First, there is a need to revise ASHA and Anganwadi training protocols to incorporate Indigenous Khasi health practices, particularly those related to maternal care. Doing so would improve cultural competency and facilitate more meaningful engagement with rural communities. Equally important is the formalization of referral and collaboration pathways between *nongai dawai* (traditional healers) and biomedical practitioners, which would enhance continuity of care while fostering mutual recognition and respect across knowledge systems. In addition, recognizing and resourcing *Seng Kynthei* groups as formal health mediators could strengthen maternal health outreach and education. These women's collectives can bridge the gap between policy frameworks and local realities with an explicit mandate for maternal health. To evaluate the feasibility of integrated health delivery, pilot programs should be launched that align traditional knowledge with formal maternal healthcare interventions. Supporting these initiatives, microfinance and capacity-building programs must be introduced for women engaged in traditional healthcare as well as climate-resilient health entrepreneurship; this economic empowerment is critical for the adaptive capacities of communities facing climate-induced health vulnerabilities. It is also essential to establish community-led health archives that document Indigenous medical knowledge through participatory and cooperative research methodologies, and these archives can serve as repositories of context-specific healing practices and reinforce the intergenerational transmission of health knowledge. But most crucial is embedding Gender Equality and Social Inclusion (GESI) frameworks into maternal health policy would ensure that Indigenous knowledge and gender equity remain central to both public health and climate adaptation efforts in Meghalaya. Collectively, these strategies not only offer context-sensitive improvements to healthcare delivery but also advocate for systemic equity and sustainability through grassroots participation. They offer a pathway toward building the trust between parallel systems of knowledge and improving maternal health resilience in a climate-stressed region.

12.5 Conclusion

The recommendations above offer a framework for developing policy innovation at the intersection of public health, Indigenous knowledge, and gender-responsive climate adaptation. To conclude, this study affirms the urgent need to bridge traditional and formal healthcare systems by centering Indigenous women not merely as beneficiaries, but as active agents of change within their communities. Through the contextual analysis of maternal health challenges in Meghalaya's Khasi Hills districts, the study underscores how gender, culture, and environment intersect to shape health vulnerabilities. Despite the matrilineal system, Khasi women continue to face significant structural and institutional barriers to accessing quality healthcare—barriers that are further intensified by climate-induced risks. Yet, the presence of trusted community institutions such as the *Seng Kynthei* offers an opportunity to recalibrate healthcare delivery toward models that are more culturally grounded, locally embedded, and resilient. The work advances current scholarship by highlighting the potential of medical pluralism as a pragmatic framework for integrated care in Indigenous contexts. It contributes to ongoing debates on how to make health systems more inclusive particularly under conditions of environmental precarity and gender disparity. Crucially, it foregrounds the role of Indigenous knowledge not as supplementary, but as central to achieving sustainable and equitable health outcomes.

Going forward, future research should examine how local governance structures—such as the *Dorbar Shnong* and Autonomous District Councils—can be more effectively aligned with both national policy frameworks and global sustainable development goals. Particular emphasis should be placed on documenting Indigenous health practices, evaluating integrative pilot models, and expanding leadership and participation opportunities for Indigenous women in healthcare governance. This study calls for a reimagining of maternal health systems that do not merely adapt to climate change but are transformed through inclusive, community-led, and culturally resonant practices. If Meghalaya's approach can be deepened and refined through such frameworks with the potential to serve as a replicable model for other Indigenous and climate-vulnerable regions across the Global South. To build resilient futures, we must not only listen to Indigenous women—we must let them lead!

Authors Contributions The authors are responsible for the conceptualization, literature review, analysis, and writing of the manuscript in its entirety.

Funding None. This research did not receive any specific grant from funding agencies in the public, commercial, or not-for-profit sectors.

Potential Conflicts of Interest The authors declare no potential conflicts of interest with respect to the research, authorship, and/or publication of this article.

Informed Consent This study is based on qualitative analysis of secondary sources and theoretical frameworks.

Data Availability Statement This study is based on publicly available secondary sources, including government reports, academic literature, and policy documents. All references are cited within the manuscript and are accessible through standard academic or institutional channels

Abbreviations

ASHA	Accredited Social Health Activist
ADCs	Autonomous District Councils
AYUSH	Ayurvedic, Yoga and Naturopathy, Unani, Siddha and Homeopathy
CAG	Comptroller and Auditor General (of India)
GESI	Gender Equality and Social Inclusion
JSSK	Janani Shishu Suraksha Karyakram
JSY	Janani Suraksha Yojana
KHADC	Khasi Hills Autonomous District Council
LIFE-AR	Least Developed Countries (LDC) Initiative for Effective Adaptation and Resilience

MoHFW	Ministry of Health and Family Welfare
NFHS	National Family Health Survey
NRHM	National Rural Health Mission
POCSO Act	Protection of Children from Sexual Offences
TCM	Traditional Chinese Medicine

References

Aborigo RA, Allotey P, Reidpath DD (2015) The traditional healer in obstetric care: a persistent wasted opportunity in maternal health. Soc Sci Med 133:59–66. https://doi.org/10.1016/j.socscimed.2015.03.042

Albert SM (2014) Medical pluralism among the indigenous peoples of Meghalaya, Northeast India—Implications for health policy. DrPH thesis. London School of Hygiene & Tropical Medicine. https://researchonline.lshtm.ac.uk/id/eprint/1856013

Albert S, Porter J (2015) Is 'mainstreaming AYUSH' the right policy for Meghalaya, northeast India? BMC Complement Altern Med 15:288. https://doi.org/10.1186/s12906-015-0818-x

Albert S, Porter J, Green J (2023) The Nongai Dawai Khasi healers of Meghalaya—A tribal understanding of the human potential. In: Reddy S, Guite N, Subedi B (eds) Ethnomedicine and tribal healing practices in India. People, Cultures and Societies: Exploring and Documenting Diversities. Springer, pp 183–200. https://doi.org/10.1007/978-981-19-4286-0_10

Awiti AO (2022) Climate change and gender in Africa: a review of impact and gender-responsive solutions. Front Clim 4:895950. https://doi.org/10.3389/fclim.2022.895950

Bonney PG (2016) Syiemship in Khasi society: The Syiemship of Nongstoin. In: Lyngdoh CR (ed) Revisiting traditional institutions in the Khasi-Jaintia Hills. Cambridge Scholars Publishing, pp 80–92

Bordoloi P (2021) Women in agriculture: an ecological and socio-economic study of Meghalaya, North East India. Biol Forum—An Int J 13(4):532–536. https://www.researchtrend.net

Caswell C, Jang N (2024) Mainstreaming gender equality and social inclusion in nature-based solutions for climate change adaptation. International Institute for Sustainable Development (IISD). https://www.jstor.org/stable/resrep57244

Borkotoky D (2016) The constitutional-traditional interface: the autonomous district council and traditional institutions in the Khasi Hills. In: Lyngdoh CR (ed) Revisiting traditional institutions in the Khasi-Jaintia Hills. Cambridge Scholars Publishing, pp 163–178

Chaand I, Horo M, Nair M, Harshana A, Mahajan R, Kashyap V, Falero F, Escruela M, Burza S, Dasgupta R (2019) Malnutrition in Chakradharpur, Jharkhand: an anthropological study of perceptions and care practices from India. BMC Nutrition 5:35. https://doi.org/10.1186/s40795-019-0299-2

Cheung H, Doughty H, Hinsley A, Hsu E, Lee TM, Milner-Gulland EJ, Possingham HP, Biggs D (2020) Understanding traditional Chinese medicine to strengthen conservation outcomes. People and Nature 2(4):1016–1030. https://doi.org/10.1002/pan3.10166

CIFOR (2015) Gender, climate change and forests in Southeast Asia. https://www.cifor.org/publications/pdf_files/Brief/GenderClimateBrief5.pdf

Contractor SQ, Das A, Dasgupta J, Van Belle S (2018) Beyond the template: the needs of tribal women and their experiences with maternity services in Odisha, India. Int J Equity Health 17(1):134. https://doi.org/10.1186/s12939-018-0850-9

Comptroller and Auditor General (2019) Executive Summary. https://cag.gov.in/uploads/download_audit_report/2019/04%20EXECUTIVE%20SUMMARY_DH-060657520347147.64692106.pdf

Department of Health & Family Welfare, Government of Meghalaya (n.d.) Hospitals in Meghalaya. https://meghealth.gov.in/hospitals.html

Das M (n.d.) Changing family system among a matrilineal tribe in India. http://archive.iussp.org/Brazil2001/s10/S12_04_Das.pdf

Diengdoh BH, Wahlang BB (2016) Climate change mitigation and traditional institutions: can REDD+ revitalise age-old Khasi regimes of community-based forest management? In: Lyngdoh CR (ed) Revisiting traditional institutions in the Khasi-Jaintia Hills. Cambridge Scholars Publishing, pp 200–226

Dutta B (2022) Mining, displacement, and matriliny in Meghalaya: gendered transitions. Routledge

Government of Meghalaya. (2021). *Mobilising collaborative actions to improve nutrition status of SAM and MAM children: A case study of Meghalaya.* Social Welfare Department (Women & Child Development). https://meghealth.gov.in/docs/Meghalaya%20Health%20Policy%202021.pdf

Gumucio T, Hansen J, Carr ER, Huyer S, Chiputwa B, Simelton E, Partey S, Schwager S (2022) Enhancing climate services design and implementation through gender-responsive evaluation. Frontiers in Climate 4:908602. https://doi.org/10.3389/fclim.2022.908602

International Institute for Population Sciences (IIPS) and ICF. 2021. National family health survey (NFHS-5), India, 2019–21: Meghalaya. IIPS, Mumbai.https://www.nfhsiips.in/nfhsuser/publication.php

Institute of Bioresources and Sustainable Development (IBSD). (2023). Traditional healers meet at IBSD Meghalaya Node, Upper Shillong. https://ibsd.gov.in/ibsd/home/index.php/gallery/view?id=42. Accessed 14 Jan 2025

Joseph SM, Iyer DS, Pillai RV (2021) Ayurvedic response to COVID-19 Pandemic in Kerala, India and its impact on quarantined individuals—a community case study. Front Public Health 9:732523. https://doi.org/10.3389/fpubh.2021.732523

Keeni M, Takashino N, Fuyuki K (2018) Women's issues in Meghalaya: role of the government, NGOs, and the village leadership. Int J Gend Women's Stud 6(2):98–107. https://doi.org/10.15640/ijgws.v6n2p10

Kharbangar S (2017) A study of Khasi indigenous religion: a comparative study of some ritualistic practices. Doctoral dissertation. North Eastern Hill University.

Khyriemmujat S, Basaiawmoit AB (2016) Ka Pomblang: a depiction of Khasi polity. In: Lyngdoh CR (ed) Revisiting traditional institutions in the Khasi-Jaintia Hills. Cambridge Scholars Publishing, pp 38–51

Laloo MN (2016) Relations of traditional institutions with modern constitutional institutions. In: Lyngdoh CR (ed) Revisiting traditional institutions in the Khasi-Jaintia Hills. Cambridge Scholars Publishing, pp 138–149

Leonetti DL, Nath DC, Hemam NS (2007) Women's reproductive lives and the roles of their mothers and husbands among the matrilineal Khasi. Curr Anthropol 48(6):861–890

Liswanti N, Tamara A, Djoudi H (2020) Climate finance and gender on the ground: Insights from mitigation and adaptation interventions in Indonesia. Center for International Forestry Research. https://www.jstor.org/stable/resrep29214

Lyngdoh A (2017) A study of the matrilineal social system and its implications on masculinity among the Khasis. Doctoral thesis, Martin Luther Christian University. Shillong, India

Lyngdoh CR (2020) Traditional political institutions of the Khasis. In: Marak Q, Chaudhuri SK (eds) The cultural heritage of Meghalaya. Routledge, pp 555–567

Lyngdoh F (2016) The Kur and Dorbar in the Khasi traditional polity. In: Lyngdoh CR (ed) Revisiting traditional institutions in the Khasi-Jaintia Hills. Cambridge Scholars Publishing, pp 21–37

Marbaniang D, Singh NK (2022) Women empowerment in Khasi society of Meghalaya. Int J Acad Res Dev 7(3):17–24. https://www.academicjournal.in

Mawkhroh I (2016) Women's participation in the Dorbar: a case study in West Khasi Hills. In: Lyngdoh CR (ed) Revisiting traditional institutions in the Khasi-Jaintia Hills. Cambridge Scholars Publishing, pp 150–162

Mawlong BL (2016) Khasi matrilineal culture: tradition, continuity, and change. In: Lyngdoh CR (ed) Revisiting traditional institutions in the Khasi-Jaintia Hills. Cambridge Scholars Publishing, pp 188–199

Meherali S, Nisa S, Aynalem YA, Kennedy M, Salami B, Adjorlolo S, Ali P, Silva KL, Aziato L, Richter S, Lassi ZS (2024) Impact of climate change on maternal health outcomes: an evidence gap map review. PLOS Global Public Health. https://doi.org/10.1371/journal.pgph.0003540

Mills E, Cooper C, Anis AH, Tyndall M (2005) Traditional African medicine in the treatment of HIV. Lancet Infect Dis 5(8):465–467. https://doi.org/10.1016/S1473-3099(05)70195-7

Ministry of Health and Family Welfare. (2025). Update on maternal and child health indicators under NHM. Government of India. https://www.mohfw.gov.in/?q=pressrelease/update-maternal-and-child-health-indicators-under-nhm

Miri S (2022) Khasi religion. In: Long JD, Sherma RD, Jain P, Khanna M (eds) Hinduism and tribal religions. Springer Nature, pp 775–777

Mutola S, Pemunta NV, Ngo NV (2021) Utilization of traditional medicine and its integration into the healthcare system in Qokolweni, South Africa: prospects for enhanced universal health coverage. Complement Ther Clin Pract 43:101386. https://doi.org/10.1016/j.ctcp.2021.101386

National Rural Health Mission (2009) Report of the 3rd common review mission: Meghalaya, 4th to 11th November 2009. https://nhm.gov.in/images/pdf/monitoring/crm/3rd-crm/meg_3rd_crm_report.pdf

Nedungadi P, Salethoor SN, Puthiyedath R, Nair VK, Kessler C, Raman R (2023) Ayurveda research: emerging trends and mapping to sustainable development goals. J Ayurveda Integr Med 14(6):100809. https://doi.org/10.1016/j.jaim.2023.100809

Nelson V, Meadows K, Cannon T, Morton J, Martin A (2002) Uncertain predictions, invisible impacts, and the need to mainstream gender in climate change adaptations. Gend Dev 10(2):48–55. https://doi.org/10.1080/13552070215913

Nengnong CB, Passah M, Wilson ML, Bellotti E, Kessler A, Marak BR, Carlton JM, Sarkar R, Albert S (2024) Community and health worker perspectives on malaria in Meghalaya, India: covering the last mile of elimination by 2030. Malar J 23:83. https://doi.org/10.1186/s12936-024-04905-2

Nongkynrih J, Haobijam J, Devi LK (2024) A comparative study to assess the placental weight, maternal, and neonatal outcomes among anemic and non-anemic pregnant mothers in selected healthcare settings of Meghalaya. Int J Obstet Gynaecol Nurs 6(2):130–138. https://doi.org/10.33545/26642298.2024.v6.i2b.171

Nongsiej S, Dutta I (2024) Economic and social consequences of the pandemic: with special reference to Khasi women of rural belts. In: Women in the changing world. Palki for Women's Cell, Teachers Unit, Lakhimpur Girls College and Publication Cell. pp 42—49

Patel S, Gebreyes BY (2020) What is effective climate adaptation? Case studies from the least developed countries. International Institute for Environment and Development. http://pubs.iied.org/10209IIED

Pde S (2020) Folklore and the Khasi identity. Doctoral thesis, Jadavpur University. Jadavpur University

Pereira FW, Khatso V (2007) Customary Laws in north east India: Impact on women. National Commission on Women, New Delhi

Pratiwi NAH, Rahmawati YD, Setiono I (2016) Gender and climate change adaptation. In: Mainstreaming gender in climate change adaptation: a case study from Cirebon, Indonesia. International Institute for Environment and Development, London

Puzyreva M, Roy D (2018) Adaptive and inclusive watershed management: assessing policy and institutional support in Kenya. International Institute for Sustainable Development. http://www.jstor.com/stable/resrep21909.9

Ravera F, Martín-López B, Pascual U, Drucker A (2016) The diversity of gendered adaptation strategies to climate change of Indian farmers: a feminist intersectional approach. Ambio 45(Suppl. 3):S335–S351. https://doi.org/10.1007/s13280-016-0833-2

Ray R (2023) 'Tribes of trees' and 'economical geology': Narrating space and people in *Placing the frontier in British North-East India*. Oxford University Press. https://doi.org/10.1093/oso/9780192887085.003.0005

Shangpliang RM (2012) Forest legislations and livelihood strategies: Khasi women in rural Meghalaya. Sociol Bull 61(3):479–493

Shangpliang RM (2020) Forest and tribal life with special reference to the Khasis. In: Marak Q, Chaudhuri SK (eds) The cultural heritage of Meghalaya. Routledge, pp 537–554

Shullai WK, Singh AS, Agarwal M (2023) The matrilineal society of Meghalaya: Trends of family planning and its implications on maternal health. Int J Reprod, Contracept, Obstet Gynecol 12(7):2062–2069. https://www.ijrcog.org

Singh A, Sundaram SP, Ningombam JD (2024) Undernutrition and its determinants among under-five children in a tribal community of Meghalaya. J Fam Med Prim Care 13:340–347

Sohtun EO (2016) NGOs and development: a study of three NGOs in Meghalaya. IOSR J Hum Soc Sci 21(3):40–46. https://doi.org/10.9790/0837-2103044046

Soreide KN (2022) Changing land rights in the Khasi Hills of Meghalaya: a gendered approach. In: Linkenbach A, Verma V (eds) State, law and Adivasi: shifting terrains of exclusion. Sage, pp 247–269. https://perspectivia.net/publikationen/psigs/linkenbach_adivasi/soreide_meghalaya

Xu J, Yang Y (2009) Traditional Chinese medicine in the Chinese health care system. Health Policy (Amst, Neth) 90(2–3):133–139. https://doi.org/10.1016/j.healthpol.2008.09.003

Zuma T, Wight D, Rochat T, Moshabela M (2017) Traditional health practitioners' management of HIV/AIDS in rural South Africa in the era of widespread antiretroviral therapy. Glob Health Action 10(1):1352210. https://doi.org/10.1080/16549716.2017.1352210

Chapter 13
Legal, Technological and Regulatory Challenges to the Integration of Telemedicine into Climate Change Mitigation-Empirical Insights from Bengaluru City, India

V. Chaitra, Renuka Joseph, and K. Hema

Abstract The role of telemedicine in mitigating climate change looks promising as there is growing evidence of it contributing to reducing carbon footprints, saving energy consumption, and reducing waste accumulation, but the existing research is not exhaustive. Telemedicine provides multiple benefits, including remote healthcare access and expert consultation; however, constraints in terms of insufficient information about the patient's medical history, possibility of misdiagnosis from technical limitations increases the risk of medical negligence litigation. Only by addressing these challenges, telemedicine can effectively contribute to climate change mitigation. *Firstly,* this paper discusses and reviews the literature on the impact of telemedicine on the environment. *Secondly,* it critically examines the guidelines and reported judgements to explore the challenges in establishing standard of care in telemedicine negligence claims, and the technological and regulatory challenges of its integration into climate mitigation. The methodology also includes questionnaires to collect data from the key stakeholders of the telemedicine system: doctors and patients in Bengaluru Urban, a growing hub for telehealth services in India. Despite the Telemedicine Guidelines 2020, certain issues persist. In conclusion, this would contribute to some practical suggestions for more reforms in the telemedicine governance of India and guiding policy-making in the integration of telemedicine, environment protection, and public health.

Keywords Telemedicine · Climate change · Healthcare consumer · Standard of care · Telemedicine Guidelines · Sustainable healthcare

V. Chaitra (✉) · R. Joseph · K. Hema
School of Law, Christ University, Bengaluru, Karnataka, India
e-mail: chaitra.v@christuniversity.in

R. Joseph
e-mail: renuka.joseph@christuniversity.in

K. Hema
e-mail: hema.k@christuniversity.in

P. Ghosh et al. (eds.), *Climate Crises and Public Health Challenges*, Disaster Risk Reduction, https://doi.org/10.1007/978-981-95-9709-3_13

13.1 Introduction

13.1.1 Background of the Study

Climate change is an unfortunate reality which is affecting every dimension of human life globally. The IPCC defines it *"as a long-term alteration of temperature and typical weather patterns of a region, primarily driven by anthropogenic greenhouse gas emissions resulting from industrial activity, deforestation, and fossil fuel combustion"* (Intergovernmental Panel on Climate Change 2023). Healthcare is a significant contributor to the climate crisis accounting for 4.4% of global emissions (Bosurgi 2019) and thus telemedicine can be an environmentally sustainable alternative. In India, the health care sector contributes 2% of its greenhouse gas emissions (Health Care Without Harm 2019) and Bengaluru is one the substantial contributors for India's greenhouse gas emissions, particularly from transportation (DH Contributor 2020).

Telemedicine (remote diagnosis and treatment through video conferencing and virtual technologies) grew in popularity when the Covid 19 pandemic spread across the globe and people isolated themselves at their homes to prevent the risk of infection. Despite the end of the pandemic, the utilization of telemedicine is rising (d'Arville et al. n.d.; Weiner et al. 2021) as it provides multiple benefits such as increased access to healthcare for patients in remote areas, expert consultation, reasonable medical costs, and decreased risk of exposure to contagious disease for patients and hospital staff. India has kept pace with this development evidenced by the rising number of private telemedicine platforms and in particular, the launching of *"eSanjeevani—the National Telemedicine Service of India"* by the Government of India in the year 2020, known to be the world's largest telemedicine implementation in the primary healthcare that recorded providing services to 10 crore beneficiaries with 57% being women and around 12% senior citizens (Ministry of Health and Family Welfare 2023).The EY-IPA study indicates that the Indian telemedicine market will reach USD 5.5 billion by 2025 (PTI 2020). In the context of increasing awareness and concerns of mitigating climate change, the relevance of telemedicine virtual visits has also increased as there is favorable evidence of it contributing to reducing carbon footprints, saving energy consumption, and reducing waste accumulation (Theil et al. 2023). However the existing research is not exhaustive (Haleem et al. 2021). A systematic review study which analyzed empirical findings from the UK, USA, Canada, Australia, and few parts of Europe indicates that the amount of carbon saved per consultation ranged between 0.70 and 372 kg CO2 but there is a lack of comprehensive study regarding the medical specialties examined (Purohit et al. 2021).

The application of telemedicine also helps in theattainment of various sustainable development goals namely, SDG 3(Good Health and Well-being), SDG 11(Sustainable Cities and Communities), and 13(Climate Action) by world nations. Telemedicine enhances access to healthcare, especially in remote regions, thereby improving health outcomes and reducing health inequalities, aligning directly with

SDG 3. It facilitates early diagnosis, remote consultations, and continuity of care, even during emergencies or pandemics (Anwar et al. 2019). In relation to SDG 11, telemedicine supports the development of sustainable urban systems by reducing the need for physical travel to healthcare facilities, thereby lowering traffic congestion and minimizing strain on urban infrastructure (Li et al. 2025). It also nurtures inclusivity by making healthcare accessible for people with mobility issues or those living in geographically isolated areas. Moreover, telemedicine plays a vital role in combating climate change (SDG 13) by significantly reducing carbon emissions associated with patient and provider travel (Tuckson et al 2017). Leveraging the digital platforms can cut down on the environmental footprint of healthcare delivery, promoting greener health systems. Thus, telemedicine is not merely a healthcare tool, but a catalyst for holistic sustainable development.

Despite the plethora of existing international literature on telemedicine's impact on mitigating climate change, there are no empirical findings in the Indian context. In the current situation, though there are several compelling reasons to adopt telemedicine, there are unique challenges associated with it. Researchers and regulators have pointed out constraints in terms of insufficient information about the patient's identity and medical history, virtual diagnosis, and treatment replacing physical examination and misdiagnosis resulting due to technical limitations, all of which have the possibility of resulting in medical negligence litigation. The question also arises with respect to the liability of telemedicine platforms which are e-commerce entities. Only by addressing these challenges, telemedicine can be effectively integrated into the healthcare system and contribute to climate change mitigation. The evidence-based literature from other jurisdictions are significant, but they lack longitudinal studies and are limited to developed nations. In addition, there is scarce research on other important dimensions of telemedicine, such as its regulatory and institutional challenges, which are equally vital for its inclusion in the climate change mitigation framework and public health system.

13.1.2 Objectives

The study is presented in two parts. *Firstly,* the paper aims to evaluate the potential of telemedicine as a tool for climate change mitigation by reducing different greenhouse gases emissions. It begins by synthesizing existing literature to understand how telemedicine contributes to building climate-resilient healthcare systems. *Secondly,* the study highlights the current challenges faced by telemedicine practices in India—particularly in their capacity to effectively respond to the growing climate crisis. It critically examines the telemedicine guidelines in India and judicial precedents to understand the risks and accountability concerns from a legal perspective. The study explores the challenges relating to the fulfilment of lack of standard of care requirements in medical negligence claims in telemedicine and the technological, infrastructure, and regulatory challenges of its integration into climate mitigation.

Furthermore, it examines how the existing telemedicine system can be improved to respond to the growing public health challenges due to the climate crises.

13.1.3 Significance of the Study

The legal, infrastructure, and institutional challenges associated with telemedicine also constitute a public health concern for the rural and natural disaster prone area population who cannot access both in-person healthcare as well as telehealth services. Therefore, it is critical to strengthen the telemedicine system in India not only to mitigate climate change but also to reduce the healthcare inequities among the vulnerable population. Though the Telemedicine Guidelines of India, 2020 attempts to validate telemedicine consultations in India, various issues affecting healthcare consumers remain unaddressed. The study suggests that in this technology era, the traditional definitions of standard of care in the context of telemedicine require reconsideration. In conclusion, this would contribute to some practical suggestions for more reforms in the telemedicine governance of India and guiding policy-making in the integration of telemedicine, environment protection, and public health delivery.

13.2 Data and Methods

The chapter adopts an integrated method approach including doctrinal research and qualitative empirical analysis to explore the intersection of telemedicine and climate change. The doctrinal study involves a critical evaluation of the domestic legal frameworks related to climate change mitigation, telemedicine governance, and public health including statutes, judicial decisions and official reports, and policies of concerned ministries. Secondary sources include interdisciplinary academic literature and legal commentaries.

The qualitative empirical study was conducted through questionnaire and semi-structured interviews of the key stakeholders of the telemedicine system: doctors and patients in Bengaluru Urban, a growing hub for telehealth services in India. The qualitative data collection and analysis specifically relates to understanding the operation of telemedicine and the challenges faced by the patients and healthcare providers, its impact on healthcare and climate change. The participants included the doctors from various specialties who use telemedicine and patients were selected based on their prior use and with minimum two years' experience in availing the telemedicine services. Purposive sampling was employed to contact the study respondents and collect data to include relevant insights and their experiences. It was chosen to select information-rich participants for the study. Thematic analysis was used to analyze the qualitative data collected. The questionnaire and interview responses were thoroughly read and were categorized into major themes aligning with the research objectives. These qualitative responses were also quantified and the findings presented as percentages to understand the prevalence of specific issues relating to telemedicine and its potential to mitigate climate change.

13.3 Results and Discussion

13.3.1 Telemedicine and Climate Change Mitigation: The Intersection

The increasing popularity of the impact of telemedicine on the environment is timely and relevant and should be dealt with meticulous focus and prudence. Traditional healthcare has been contributing to environmental contamination and the situation will worsen if not acted promptly (Zeynep 2024). The statistics related to the same, from the pre-COVID-19 pandemic itself, show that in developed nations, the healthcare sector contributes an estimated 3–8% of total emissions, while the percentages are anticipated to be lower in the developing regions (Holmner et al. 2012) The U.S. healthcare sector ranks as the second-highest energy consumer among all industrial sectors in the country, with inpatient facilities being the second most energy-intensive commercial buildings nationwide (Holmner et al. 2012). Similarly, the National Health Service (NHS), one of Europe's largest employers, produces a carbon footprint of roughly 20 million tons of CO2 annually, accounting for over 4% of the UK's yearly emissions (Climate and Health Alliance Organisation 2012). NHS facilities also incur energy costs exceeding £410 million per year (NHS Carbon Reduction Strategy for England 2009). The Central Pollution Control Board of India, in 2020, reported that the total biomedical waste generated everyday is around 800 tons (Jaiswal et al. 2024).

Telemedicine helps in reducing healthcare's environmental footprint amid the rising concerns with respect to climate change. The existing literature in this field shows that there is an inextricable nexus between telemedicine and climate change. Like other climate change mitigation strategies, telemedicine is regarded as a highly effective technology for reducing greenhouse gas emissions by reducing the need for travel and transportation. One such study of review of 20 teledermatology studies has found that synchronous and asynchronous telemedicine systems can lessen patients' travelling 70% and 43% respectively (Woottoon et al. 2011). The advantages of telemedicine can be significantly correlated to home care programs and outpatient consultations as it helps in reducing travel. For instance, the telemedicine initiative at UC Davis, California, facilitated 13,000 outpatient consultations over five years, saving 4.7 million travel miles and reducing CO2 emissions by 1,700 tons (Yellowlees et al. 2010). Comparable benefits have been observed in Scotland (Wootton & Tait 2010) and Wales (Lewis et al. 2009). A study showed that the telemedicine services in Catalonia, Spain, between January 2018 and June 2019, resulted in savings of 192,682 km in travel distance, 3,779 h of travel time, and 15,664 Euros (Alaball et al. 2019). In Canada, a study indicated that telemedicine could replace over 11 million annual home visits by nurses, cutting approximately 120 million kilometers of travel and 33,220 tons of GHG emissions (Scott et al. 2009). In a region of Portugal, teleconsultations in specialties such as neurology, dermatology, rehabilitation, and general surgery contributed to a 455-ton reduction in vehicle greenhouse gas emissions (Oliviera et al. 2013). The rehabilitation department at the University of Sweden demonstrated that using telemedicine instead of in-person consultations reduced carbon dioxide emissions by a factor of 40 to 70 (Holmner et al. 2014). These examples demonstrate the substantial potential of telemedicine as a large-scale climate mitigation strategy, especially when considering its cumulative effects.

The use of telehealth technologies, including e-health records, e-prescriptions, and surveillance systems, only contribute to gains in terms of climate benefits but also most notably accepted as a means for improving the efficiency of healthcare services. The positive effect of electronic documentation on greenhouse gas emissions is significant. Although it can decrease traveling, for instance, when patients can access prescriptions/ medical reports online, the main environmental benefit stems from reducing paper waste. This, in turn, lowers the ecological footprint of the paper industry and helps curb deforestation. However, before the COVID-19 pandemic, the telemedicine system was fraught with many barriers, including limited experience with information technology, inadequate infrastructure, user apprehension toward telemedicine services, and a lack of organizational planning and preparedness.

The COVID-19 pandemic and the subsequent period have seen a significant surge in patients utilizing telemedicine healthcare services, driven by necessity. Telemedicine proved to be a crucial ally during the COVID-19 pandemic, facilitating a transition from predominantly in- person consultation and care to an effective and widely accepted digital mode. This shift was exemplified by virtual consultations, primarily video calls conducted by doctors through secure platforms integrated with patients' medical records, enabling the prescription of follow-up tests, scans, and medications. (Croymans et al. 2020). A study in 2020 indicated that a healthcare company conducted 3,015,530 medical appointments, of which 640,122 were digital, comprising 495,913 video consultations and 144,209 telephone appointments, averaging 3,700 digital appointments daily (Serra et al. 2022).These digital solutions were estimated to reduce CO2 emissions by 6,655 net tons, with 1,957 tons saved from digital appointments and 4,698 tons from 3,064,646 medical reports downloaded to digital health folders and in particular, the patient satisfaction was very strong, with a net promoter score (NPS) of 62.1% and 48.7% of these users opting for repeat digital appointments (Serra et al. 2022).

Thus, there are a plethora of studies which indicate that telemedicine is making a positive impact on climate change by minimizing environmental contamination. The benefits of digital transformation outweigh not only in the field of healthcare but also within human activities, subsequently influencing energy consumption and CO2 emissions. However, the unexplored areas in this context include a comprehensive carbon-cost benefit analysis of telemedicine usage which can be difficult because the various factors contributing to its carbon footprint have not been thoroughly investigated. In specific, the environmental impact and costs associated with the production, distribution of data, constant and everyday usage, and disposal of ICT remain unexamined.

13.3.2 Challenges of Telemedicine in Addressing Climate Crisis in India

Telemedicine relies heavily upon Information and Communication Technology (ICT), where the patient examination may take place online in real-time, or their data is saved and examined later by the doctors without real-time interaction. In both cases, the significant barrier is the digital divide across the country in terms of digital literacy, affordability, and accessibility of the technology. Further, technical issues like slow internet connectivity, poor video/audio quality, delay may add to inadequate communication between the doctor and the patient. In comparison to physical consultations, virtual consultations lack in-person consultation and physical examination of the patient. It places a significant responsibility on the doctors to promptly determine whether the patient needs in-person consultation which is connected to their professional judgement. However, this is coupled with restricted access to the overall patient's condition due to the absence of physical examination, incomplete health history and high dependence on patients' description of the symptoms. The liability of telemedicine platforms in case of medical negligence remains unclear. Though telemedicine has clear merits, these limitations can result in medical error, misdiagnosis and delays, consequently leading to medical negligence litigation. Studies across the USA show that in 2021, 70% of telemedicine claims were related to diagnosis, with 25% accounting for cancer misdiagnosis, while 20% were related to stroke misdiagnosis (How to Mitigate the Risks of Telemedicine, 2022; Saada and Deoli n.d.). Studies from the National Library of Medicine, published in 2022, show an alarming annual number of 5.2 million cases filed related to medical malpractice across various healthcare settings in India (Singh 2024). Thus it is imperative to address the legal, regulatory, and technological challenges, and mitigate medical negligence risks in telemedicine before effectively integrating it into climate change mitigation.

13.3.2.1 Legal and Regulatory Framework—An Overview and Gaps

Promoting telemedicine not only revolutionizes health and medical services accessibility but also addresses climate change concerns. Distance is a critical factor in telemedicine; it reduces the need for travel for appointments, decreases the emission of greenhouse gases, and decreases operational footprints, thus aligning telemedicine with sustainability goals, health care promotion, and environmental protection. Though telemedicine is a potential and promising solution for addressing climate change and transforming the medical and health sectors, it presents a wide range of legal and regulatory challenges. The leading legal issues include patient's sensitive health data and privacy, informed consent, standard of care, prescription of medicines, cross-border liability and malpractices, misdiagnosis or erroneous treatments. The list of problems should not stop a country from allowing telemedicine. Prohibiting telemedicine is no solution in this technology-driven world.

The Telemedicine Practice Guidelines, 2020 (hereinafter referred to as TPG) recognizes and regulates the practice of telemedicine in India. The interpretation of these guidelines should be used in conjunction with other national clinical standards and protocols—the Indian Medical Council Act, 1956, the Indian Medical Council (Professional Conduct, Etiquette and Ethics Regulation 2002), and others (Venkatesh et al. 2022). The TPG mandates that the identity of the patient and the Registered Medical Practitioner (RMP) should be disclosed explicitly (Para 3.2 of the TPG, 2020). The TPG restricts telemedicine consultation to first aid, life-saving measures, counselling and advice on referral and most importantly in emergency cases the patient must be advised for an in-person visit at the earliest (Para 1.4.2 of the TPG, 2020). There is the presumption of informed consent in cases where the patient initiates the telemedicine services, but in other cases, consent must be explicit (Para 3.4 of the TPG, 2020). This presumption of informed consent seems questionable. Informed consent can be obtained when the patients are given important information, including possible health risks and benefits of medical treatment. Upon providing such information, the patient can decide if they want to be treated or tested (Para 3.4 of the TPG, 2020). When a patient initiates a telemedicine procedure, only free consent can be presumed but not informed consent. Free consent cannot be equated with informed consent as the patient is unaware of the potential health risks and benefits of the telemedicine consultation. Informed consent should always be obtained and can never be presumed; such presumptions would defeat the very purpose of the TPG. When the RMP is consulting a minor, it has to be done only in the presence of an adult (Para 3.2.3 of the TGP, 2020). However, this would cause great inhibition to the minors who want to discuss their mental health/reproductive health in the presence of their natural/legal guardians. To overcome this, informed consent must be obtained by the guardian, but minors should determine if the parent/guardian should be present during the telemedicine treatment process to discuss such health concerns.

The TPG also imposes limitations on prescribing certain medicines to prevent negligence from incorrect medicine prescriptions. List "O" medicines indicate the "over the counter" medicines that can be prescribed in any type of telemedicine consultations which are used for common health conditions and may be deemed necessary during public health emergencies (Para 3.7.4 of TPG, 2020). List "A" medicines can be prescribed by the RMP in the first consultation if it is a video consultation and in the follow-up consultations for refills (Para 3.7.4 of TPG, 2020). "List B" drugs can be prescribed in the follow-up consultations provided there was prior in-patient consultation for the same medical condition (Para 3.7.4 of TPG, 2020). The TPG also provides a list of "Prohibited Drugs" that cannot be prescribed in any telemedicine consultation (Para 3.7.4 of TPG, 2020). Though breach of confidentiality by the RMP amounts to misconduct, if there is reasonable evidence to believe that such breach is a consequence of a technology breach or by a person other than the RMP, the RMPs will not be held liable (Para 3.7.1.3 of TGP, 2020). However, the guidelines stipulate that the RMPs should exercise a reasonable degree of care in hiring such services (Para 3.7.1.3 of TGP, 2020). Concerning misconduct and professional ethics, the Indian Medical Council Act, 1956 shall be the regulating legislation for telemedicine (Para 3.7.1.4 of TGP, 2020). In addition, The Digital Personal Data Protection Act 2023 read with the Draft Digital Personal Data Protection Rules, 2025 classifies healthcare providers and entities as data fiduciaries and they are obligated to obtain explicit consent from data principals before collecting, using, and sharing their data as well as taking appropriate security measures for the protection of such data.

The TPG is binding only on the RMPs, but not on other medical practitioners like dentists, nurses or the practitioners of traditional medicine like Ayurveda, Unani, Siddha, and Homeopathy (Nishith Desai Associates, Telemedicine in India: Legal, regulatory, and tax overview, 2024). There are separate sets of guidelines regulating these practices of medicines. Furthermore, the TPG excludes telemedicine consultations outside the jurisdiction of India. Internet being borderless, and TPG being territorial specific, the conundrum arises when there is cross-border telemedicine practice. Having fragmented and half-baked guidelines is not effective. A unified and comprehensive legislation should be enacted to regulate telemedicine across all the forms of medical practice.

Currently, India has the IT Act, 2000 and the rules enforced thereunder and the Digital Personal Data Protection Act 2023 (DPDPA) to regulate data. Neither of the legislation defines "health data". It should be noted that health-related and medical records and other personal health-related data are considered as "Sensitive Personal Data or Information" under the Data Protection Rules, framed under the IT Act, 2000. The IT Act of 2000 and the Data Protection Rules require individuals to be informed about how their data will be used, but they do not specifically apply to healthcare. The European Union's General Data Protection Regulation (GDPR), 2018, under Recital 35 provides an inclusive list of what can be considered as "Health Data". Similarly, the United States has the Health Insurance Portability and Accountability Act (HIPAA), which focuses exclusively on healthcare data protection. Once DPDPA is implemented, it will replace the Information Technology (Reasonable Security Practices and Procedures and Sensitive Personal Data or Information) Rules of 2011, introducing an updated framework for data protection. However, not defining "health data" under the DPDPA calls for immediate attention if health and medical-related data is to be protected adequately.

The pandemic COVID-19 accelerated the drafting and enforcement of the TPG as social distancing was crucial, yet access to basic healthcare and medical facilities remained a fundamental necessity. As a consequence, TPG was adopted to address the immediate situation of the pandemic, which resulted in the framework better suited for the short term, lacking an overarching framework. Though the TPG provides robust guidelines to regulate telemedicine in India, nevertheless, there are certain shortcomings and gaps that may hinder the growth of the telemedicine sector. The provisions for teleconsultation regarding minors, the question of informed consent and cross-border consultations need a relook. With the experiences faced during the pandemic, there is a dire need to draft and enact uniform and robust legislation to regulate telemedicine that can encompass all forms of medical practice. Tackling these legal gaps will not only ensure the doctor-patient relationship in telemedicine but also establish telemedicine as a way forward to address climate change.

13.3.2.2 Legal Accountability for Medical Negligence in the Context of Telemedicine

The standard of care that is expected from doctors is explicit and validated in traditional/in person healthcare. It is universally established that medical practitioners have a duty of care to their patients. The Supreme Court of India has validated distinct duties of doctors: "*a duty of care in deciding whether to undertake the case, a duty of care in deciding what treatment to give and a duty of care in the administration of that treatment*" (L.B. Joshi v. T.B. Godbole AIR 1969 SC 128). Any breach of these duties results in the civil wrong/tort of medical negligence. The standard of care expected is that of an ordinary competent medical professional exercising an ordinary degree of professional skill (Jacob Mathew v. State of Punjab [(2005) 6 SSC 1). A doctor is not liable for medical negligence if he has followed a practice that is accepted as proper by a responsible body of medical experts skilled in that particular art (Bolam v. Friern Hospital Management Committee (1957) 1 WLR 582, Jacob Mathew v. State of Punjab (2005) 6 SSC 1). In addition, if the doctor chose to follow one procedure and not another which resulted in the failure of medical/surgical treatment, he cannot be held liable if the procedure which was in fact adopted was one which was acceptable to medical science as on that date (Jacob Mathew v. State of Punjab (2005) 6 SSC 1). However, the application of the same standard of care within the setting of telemedicine can be difficult as it is relatively new and still evolving. The complex questions include, How is the history and examination in the initial consultation documented? In what cases is it safe to offer virtual consultation for the treatment of a disease? If there is a medical error, should the error be attributed only to the doctor or the telemedicine platform as well? (Rowland et al. 2022).

For the first two questions, the TPG has addressed it partially. It mandates the recording of patient interaction through phone logs, email, video meetings etc. and maintaining of the patient records, reports, documents, images, diagnostics, data etc. (Telemedicine Practice Guidelines 2020, Para 3.7.2). It restricts telemedicine consultation to first aid, life-saving measures, counselling and advice on referral; however, it is still unclear in which cases it is safe to rely upon telemedicine services. The question that needs a thorough investigation is whether telemedicine is suitable for all branches of medicine and all types of health concerns.

As far as the third question regarding liability is concerned, it has to be dealt with in two parts separately i.e., the liability of the doctors and telemedicine platform in cases of negligence. Telemedicine negligence does not change the application of established legal doctrines of medical negligence applicable to doctors, such as doctor-patient relationship and standard of care. Remote consultations via telephone have been in practice since the invention of the telephone and it is established that in the cases of diagnostic or treatment advice to the patient via the telephone by the doctor constitutes a doctor-patient relationship (Bienz v. Central Suffolk Hosp., 557 N.Y.S.2d 139, 139–40 (App. Div. 1990). The TPG explicitly recognized that registered doctors can provide telemedicine consultations within any part of India which also partially negated the Supreme Court's observation in the case of Martin F. D'Souza v. Mohd. Ishfaq *"No prescription should ordinarily be given without actual examination. The tendency to give prescriptions over the telephone, except in an acute emergency, should be avoided"* (2009) 3 SCC 1). In India, there are no explicit guidelines on the standard of care expected in telemedicine as on date no medical negligence cases have been decided by the apex court on the subject matter. Though the Bolam test is the standard rule, there is still a need for specific guidance on good professional practice regarding the practice of telemedicine in India (Stanberry 2006) as it is in the transformative stage. The American Telemedicine Association, have formulated specific guidelines for different branches of medicine regarding telemedicine services to ensure quality medical care (Pantanowitz et al. 2014; American Telemedicine Association n.d.).

Telemedicine platforms provide diverse services, with some scheduling appointments only, others offering online consultations with prescription and some offering online consultations accompanied by doorstep delivery of medicines (Debayan Bhattacharya 2022) and the potential for medical negligence is present when they provide online consultation and medical services. In such cases, the question of the liability of the telemedicine platforms for the medical negligence by doctors registered on their platforms is very nuanced. In law, a person can be vicariously held liable for the tort (civil wrongs) committed by others in case of certain relationships i.e., the principal is liable for the agent's acts, the master for the servant's actions, and the employer for the employee's conduct within the course of their employment. In addition, in an employer-employee relationship it also depends upon the manner of execution of the work; an independent contractor, though employed to do the work, the manner of his work is not controlled by the person who has employed him. On the other hand, an employee acts under the direct authority i.e., control and supervision of his employer. Thus, under tort law the employer is not vicariously held liable for the wrong committed by independent contractors. The discussion that follows here is whether the telemedicine platform qualifies as an employer and has significant control over the manner of the work carried out by doctors. In this regard the platforms explicitly provide in their terms and conditions that they are merely e-commerce marketplaces serving only as a link between the doctors and patients without providing any clinical advice on their own, ruling out the possibility of the platform engaging in a doctor-patient relationship with the patient (Practo Terms and Conditions, n.d.). Another modern test to determine the employer-employee relationship in cases where the employer is not able to control the manner of work due to its skilled nature is the hire and fire test i.e., the person who employs him, is his paymaster and has the power to discharge him. Applying this test, the hospitals can be brought within the purview of employers. In the case of *Smt. Savita Garg vs. The Director, National Heart Institute, V. Krishnakumar v. State of Tamil Nadu & Ors. and then in Maharaja Agrasen Hospital v. Rishabh Sharma,* the Supreme Court of India observed that *"when a patient goes to a private hospital/clinic, he goes by the reputation of the hospital and with the hope that proper care will be taken by the hospital authorities. Since these hospitals charge a fee for the services rendered by them, they are duty bound to bestow the best care, and if the hospital fails to discharge their duties through their doctors being employed on job basis or employed on contract basis, it is the hospital which has to justify"* (AIR 2004 SC 5088; [2015] 8 S.C.R. 100; 2020 (6) SCC 501). In *Smt. Rekha Gupta v. Bombay Hospital Trust & Anr.,* the hospital denied its liability by stating that it only provided infrastructural and other support facilities for the treatment and did not have any role in recommending any diagnosis to the aggrieved patient. The National Consumer Disputes Redressal Commission rejected this argument and held that a hospital cannot escape liability as they have employed the doctors, and nursing staff and further pointed out that the fees for doctor's consultations are generated by the hospital and they also charged a commission of 20% on each doctor's consultation (2003)CPJ160(NC). Though these existing tort principles and apex court judgements are convincing regarding the liability of hospitals, they are flawed in relation to telemedicine. Unlike hospitals, telemedicine does not provide any infrastructure and supporting facilities and only aids the patients in locating appropriate medical care from certified doctors and thus cannot be held liable in the same way.

As far as the liability of the telemedicine platforms in the capacity of intermediary is concerned, the Information Technology Act 2000 exempts the intermediaries from the liability for the acts of the third parties when they are merely hosting the third party information or data on their platforms with the stipulation that the intermediaries observe due diligence, do not commence the transmission, select the receiver of the transmission and select or modify the data included in that transmission, and observe due diligence (Sect. 79 of the Information Technology Act, 2000). With respect to the telemedicine platform, it merely acts as a facilitator by providing a list of registered medical practitioners to be selected by a patient, resulting in the transmission of data initiated/selected by the patients (Treelife 2020). Consumer Protection (E-Commerce) Rules, 2020 mandates the e-commerce service providers to provide accurate descriptions and other content on their platforms i.e., the details of the seller of the goods/services, whether registered or not, their geographic address, customer care number, their customer ratings, and all the relevant information to empower the consumers to make informed decisions (Rule 5 of the Consumer Protection (E-Commerce) Rules, 2020). Therefore, telemedicine platforms can avoid legal liability with necessary diligence on their part by verifying the credentials of the doctors registered on their platform, ensuring that the doctors are conversant with the operation of the telemedicine platform including its legal and technical aspects through regular training and periodic inspections, adhering to proper documentation procedures, and complying with the platform's policies (Arogya Legal 2021).

It is unrealistic to impose the same liability standards on telemedicine platforms as that of traditional hospitals because the former do not provide any physical infrastructure and involve in-person patient examination. The applicability of stringent rules and the fear of litigation can deter innovation and development of telemedicine.

Therefore, it is recommended that these platforms are held liable when there is a direct breach of duties committed by them, namely when they mishandle patient data, provide misleading information on their platform, or fail to verify the credentials of the doctors registered on their platforms.

13.3.2.3 Technological Challenges

The success of telemedicine and its integration into climate change mitigation largely depends upon the infrastructure and internet penetration in a country. Studies indicate that disparity in access to high-speed internet, digital literacy, and socioeconomic circumstances restrict entry and lead to underutilization of the benefits these modern advancements in telemedicine can offer (Mallipeddi et al. 2017). As of March 2024, out of total internet subscribers of 954.40 million in India, 398.35 million were rural internet subscribers and as of April 2024, out of 6,44,131 villages in the country (villages data as per Registrar General of India), 6,12,952 villages are having 3G/4G mobile connectivity (Press Release 2024). Though the number of internet users has significantly increased, the internet speed in India is comparatively slower evidenced by the data published as of November 2024, the country is ranked 25th in mobile internet speed and 91st for broadband speed (TOI Tech Desk, 2024) (India's Mobile and Broadband Internet Speeds—Speedtest Global Index 2024). The figures clearly indicate that initiatives to improve internet connectivity in rural areas look optimistic, but there is a need to create awareness about the benefits of telemedicine as well as its positive impact on the environment. About 70% of healthcare facilities in India are in cities, which cater to about 30% of the country's population (Technology Information, Forecasting and Assessment Council, 2022). The long-distance travel for medical purposes is from the large proportion of the population residing in the rural part of India as the specialized health-care centers are mostly situated in the cities. Tapping into this population could significantly reduce the environmental pollution arising due to the healthcare industry. However, there is currently no recorded authentic information available in India on how much the rural population rely upon telemedicine or continue to prefer in-person visits.

13.3.3 Empirical Insights from Bengaluru Urban

Bengaluru Urban is a growing hub for telehealth services in India. The empirical qualitative study was carried out to explore and understand the benefits and challenges of telemedicine services in Bengaluru Urban and its potential to mitigate climate change. The methodology included questionnaires and semi-structured interviews to collect data from key stakeholders of the telemedicine system: doctors and patients. The questionnaire and semi-structured interviews contained both closed and open-ended questions. The qualitative data collection and analysis identified the themes relating to understanding the operation of telemedicine, technological challenges and other barriers faced by the patients and healthcare providers, its impact on healthcare and climate change.

13.3.3.1 Medical Practitioner/Doctors Responses and Perspectives

The prevailing opinion among the doctors in the study population is that telemedicine has reduced the number of in-person visits for primary care and minor ailments. It is not suitable for all types of health concerns and is prioritized in minor ailments and follow-up treatments. While telemedicine technology has been a boon for the healthcare sector in reducing health disparities, the absence of the physical examination of the patient is a common concern in accurately assessing the health condition of the patient. This is followed by the lack of patient trust in telemedicine, technical issues, and the lack of personal interaction with the patient (Fig. 13.1). In comparison to the urban population, the rural population tends to prefer telemedicine services less (Fig. 13.2). Around 80 per cent of the doctors suggested that telemedicine has the potential to minimize environmental pollution. The remaining per cent of the study population was unexpectedly not aware of the impact of telemedicine on the environment (Fig. 13.3). The major technological challenge faced by telemedicine patients is digital literacy/ less familiarity with technology followed by poor internet connectivity, privacy concerns, and technical problems within the telemedicine platforms (Fig. 13.4). Few medical practitioners emphasized that accessing telemedicine is better than not having any healthcare access, particularly in underserved rural areas (Fig. 13.5).

The empirical data reveals that telemedicine is considered as a prospective alternative and is widely acknowledged by the medical practitioners. It's valued for its

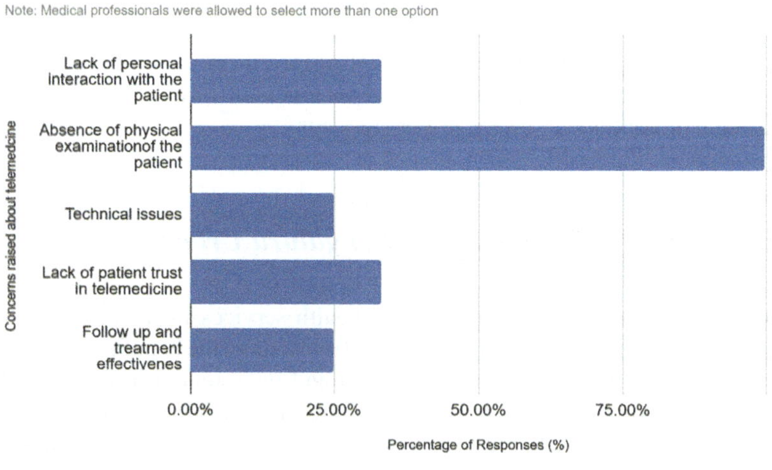

Fig. 13.1 Perceived concerns of telemedicine in comparison with in-person visits among medical practitioners

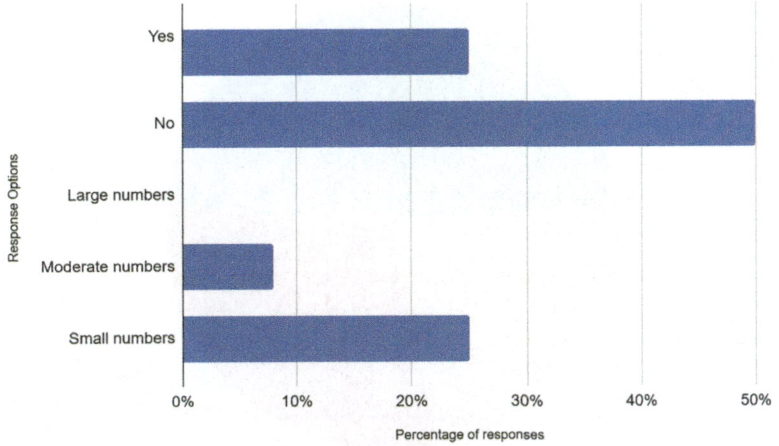

Fig. 13.2 Perception of rural patient engagement with telemedicine services

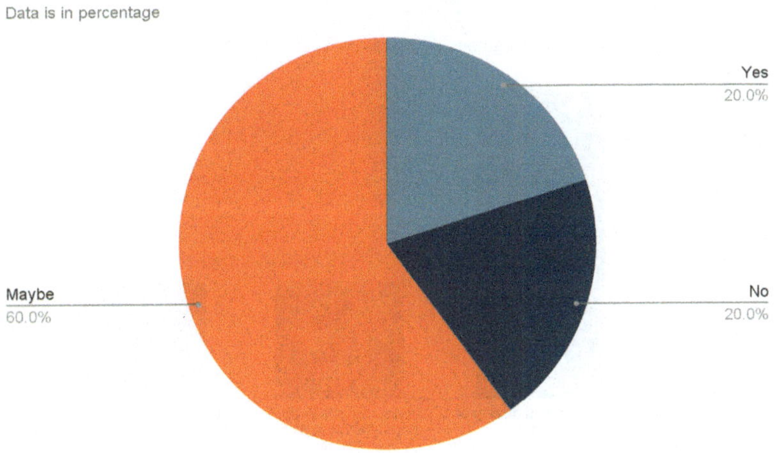

Fig. 13.3 Perceived environmental benefits of telemedicine in clinical practice by the medical professionals

potential to bridge health accessibility and reduce climate change effects—especially by limiting unnecessary travel and usage of infrastructure. However, key limitations according to the medical practitioners include, inability to conduct physical examinations, diagnostic accuracy, and patient trust. They believe that telemedicine is primarily useful for non-emergency services and minor health issues, especially during the pandemic and affirm that it is not suitable and a viable option for all types of health concerns. Having effective guidelines can ensure quality standards, protect patient rights, practitioners' accountability, and bridge the accessibility gaps. It can

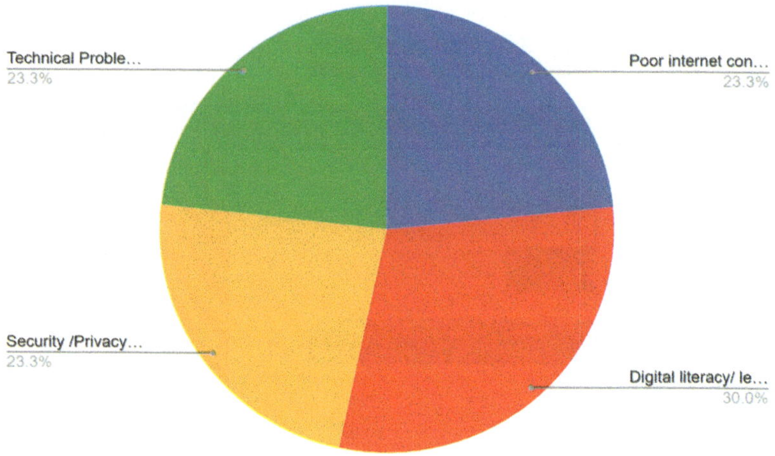

Fig. 13.4 Perceived technological problems by the medical practitioners while using telemedicine platforms

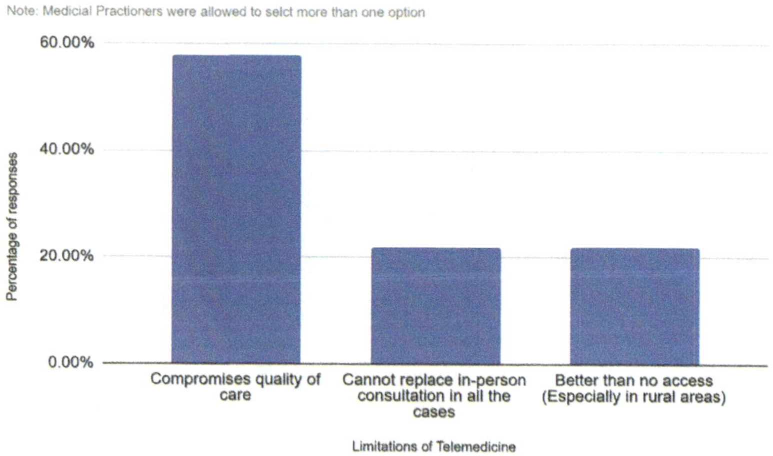

Fig. 13.5 Perceived limitations of telemedicine

address challenges such as diagnostic accuracy, patient privacy, while promoting equitable access to health care.

13.3.3.2 Patient's Responses and Perspectives

Most of the patients in the study population who used telemedicine services find it useful for minor ailments, are cost-effective, less commuting, and save time. Many stated that unstable internet, lack of clarity in audio/video communication, privacy and security concerns were some of the main issues they faced while others believe that they are not good with the use of technology (Fig. 13.6). They do prefer in-person visits for urgent/ serious health needs as they find receiving medical attention in face-to-face settings with physical examination is more satisfactory and reliable than in virtual settings (Fig. 13..7). Around 41 per cent of the patients in the study respondents suggested that they would use telemedicine more instead of visiting hospitals if it helps to minimize the environmental pollution caused due to the health-care industry (Fig. 13.8). The remaining per cent of the study population was not very much aware of the impact of telemedicine on the environment. Among the patients, the majority of them preferred telemedicine for non-emergency situations when there were minor ailments or non-serious health conditions, while the rest of the patients were comfortable to use it in both emergency and non-emergency situations (Fig. 13.9). It is imperative to note that 54% of the patients expressed their reluctance to access telemedicine based on their previous experience, whereas 45% of them viewed it positively, indicating a divergent perception (Fig. 13.10).

The empirical data reveals that patients find telemedicine useful for minor health concerns and viable when there are no other alternatives available. Poor internet connectivity and access, technical and digital skills and privacy concerns are the inhibitions expressed by the patients in using telemedicine. Having clear legal protocols for virtual consultations, establishing grounds of malpractice in the cases of negligence and providing an accessible mechanism for grievance redressal could shift patients towards using telemedicine.

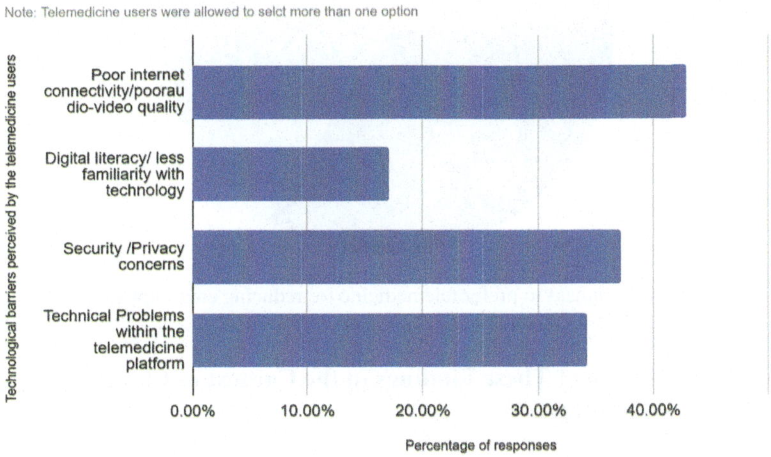

Fig. 13.6 Technological barriers experienced in telemedicine practice

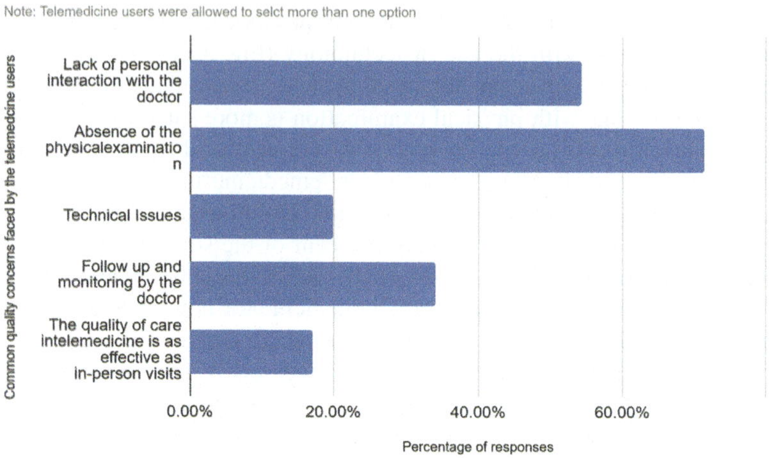

Fig. 13.7 Patients views on quality of care in telemedicine compared to in-person visits

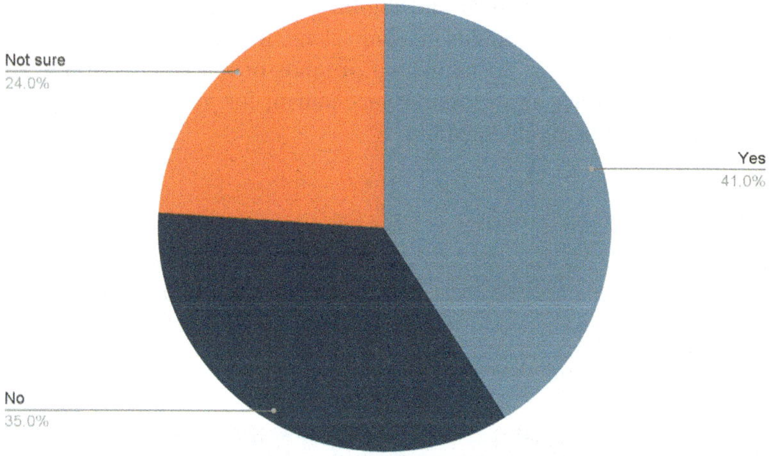

Fig. 13.8 Patients' willingness to prefer telemedicine for reducing environmental pollution

13.3.3.3 Implication of These Findings in the Context of Climate Change Mitigation

The findings indicate that there is a need to create awareness among the public as well as the medical community about the benefits of telemedicine and its positive impact on the environment. It also suggests that, though not advisable in cases of serious health needs, patients can be encouraged to consider telemedicine as a mode of treatment, especially in minor ailments, cases of follow-up visits to minimize the patient's travel and protect other resources from overexploitation. The fundamental

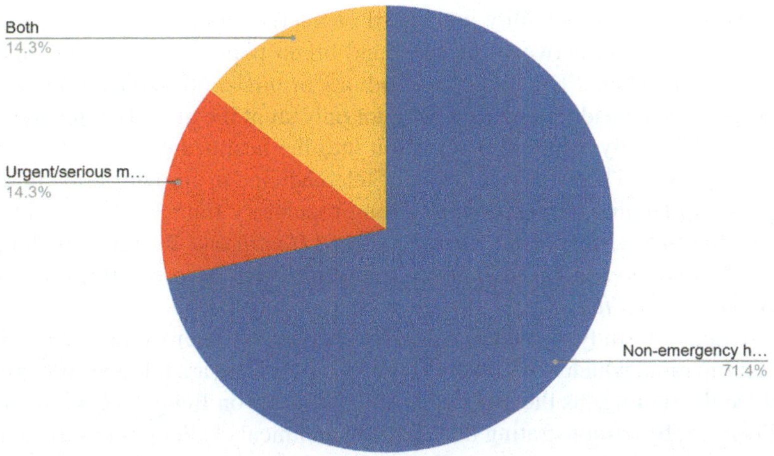

Fig. 13.9 Perceived usage of telemedicine by the patients during emergency and non-emergency health conditions

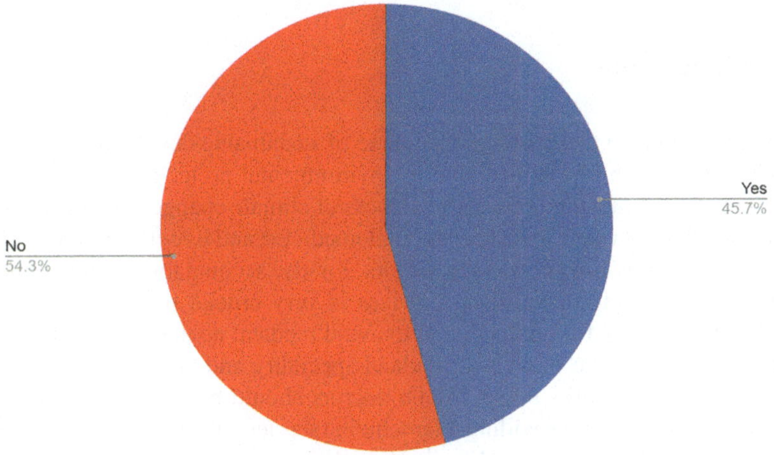

Fig. 13.10 Perceived notion on telemedicine by the patients based on their previous experience

proposition of this research is that telemedicine has potential to reduce carbon emissions and improve the Air Quality Index (AQI) which is also supported by previous studies. A study conducted at Virginia with respect to pediatric telemedicine showed that 69,454 gallons of fuel equivalent to 618 metric tons of $CO2$ was saved (Grabski et al. 2024). In the Netherlands, telemedicine nursing care generated around 60% lower greenhouse gas emissions and fossil fuel usage (Van Bree et al. 2025). In addition, it holds the potential to provide access to medical services to rural areas thus achieving equitable public health outcomes. Investment in the digital infrastructure to support and improve stability and speed of internet connectivity is very important to bridge the gap between the rural and urban population allowing equitable access to telemedicine. The empirical findings in this study indicate that if these challenges are left unaddressed, it would not only stunt the growth of telemedicine, but also consequently affect its integration into the healthcare system and climate change mitigation. Therefore, inclusion of telemedicine in climate action plans and public health planning strategies with strong regulatory framework can support in achieving the SDGs, in specific, *Target 3 (Good Health and Well-being), Target 13 (Climate Action), and Target 9 (Industry, Innovation, and Infrastructure), and Target 10 (Reduced Inequalities).*

The empirical study provides valuable insights on telemedicine and climate change mitigation, which is relatively an under-researched area. It does not intend to generalize the findings as the qualitative study focuses on Bengaluru urban population. However, by demonstrating the legal and technical challenges of telemedicine, its integration with mitigation of climate change in the Bengaluru urban area, a thriving hub for telemedicine as well as a city grappling with climate change-related consequences, these findings can inform better policy making.

13.4 Conclusion

This research adds to the evolving discourse of sustainable development by formulating the integration of telemedicine in environmental protection, which has been largely ignored despite the growing awareness of climate change. However, the findings suggest that the usage and success of telemedicine and its contribution to climate change mitigation is linked to many factors. Firstly, accountability and gaining the trust of the public in telemedicine services is very crucial by laying down good professional practices for telemedicine to avoid medical negligence. More research is needed to explore the branch of medical speciality and types of health concern where telemedicine works better. While doctors should be held liable for medical negligence committed in providing telemedicine services, the telemedicine platforms should not be held liable unless they are directly involved as stringent application of rules and the fear of litigation can deter innovation and development of telemedicine.

Awareness about telemedicine must be created by tapping into rural populations which could significantly contribute to reducing environmental pollution as they travel long distances for medical purposes. It also largely depends upon the telehealth infrastructure and internet penetration in a country.

More evidence-based research in India focusing upon the role of telemedicine in climate change mitigation can provide in-depth insights on the topic. Quantitative study comparisons of carbon emission from telemedicine and traditional healthcare systems would lead to creating more awareness among the general public and medical community. Assessment of policy frameworks that can promote telemedicine as a sustainable healthcare paradigm is the need of the hour. The evaluative studies on the financial impacts of telemedicine usage for both patients and healthcare providers, emphasizing its cost effectiveness can be useful. Comparative research of regional or international telemedicine operations to explore best practices can provide valuable insights to the Indian virtual healthcare system.

13.5 Potential Conflicts of Interest

On behalf of all authors, the corresponding author states that there is no conflict of interest.

13.6 Informed Consent

Informed consent was obtained from all individual participants included in the study.

13.7 Data Availability Statement

The datasets during and/or analyzed during the current study available from the corresponding author on reasonable request.

Acknowledgements We are grateful to all the study respondents i.e., doctors and patients who participated in empirical data collection on telemedicine use in Bengaluru Urban, India.

Authors Contribution All authors have contributed equally to this work.

Funding The research was not supported by any funding.

Appendix

Questionnaire for Doctors/Medical Practitioners

1. What are the common concerns associated with telemedicine virtual visits in comparison to in-person visits?
2. What types of health issues do doctors handle through telemedicine, and do you think telemedicine is suitable for all kinds of health concerns?
3. Do you see rural patients opting for telemedicine for consultation and treatment?
4. Do you view telemedicine as a tool for reducing the negative impact on the environment due to healthcare, including lesser travel, reduced use of medical resources and less wastage?

5. What are the technological problems generally faced while using telemedicine platforms?
6. In your experience, does telemedicine compromise with the quality of care in comparison to in person visits as the former do not include physical examination of patients and there is heavy reliance on their description of symptoms? If yes, please elaborate your experience.

Questionnaire for Patients

1. Did you face any technological problems like audio or video issues/ less familiarity with technology/security concerns/ glitches within the platforms while using telemedicine?
2. In your experience, is the quality of medical care in telemedicine not as effective as in- person visits? If yes, what is/are the common concern/s?
3. Would you prefer using telemedicine more instead of visiting hospitals if it helps to minimise the environmental pollution caused due to the healthcare industry?
4. Have you used telemedicine services for non-emergency health concerns or even in case of urgent and serious medical needs?
5. Based on your previous experiences would you prefer to use telemedicine instead of in-person hospital visits?

References

Alaball JV, Parella JF, Segui FL, Pena J (2019) Impact of a telemedicine program on the reduction in the emission of atmospheric pollutants and journeys by road. Int J Environ Res Public Health 16(22). https://doi.org/10.3390/ijerph16224366

American Telemedicine Association (n.d.) Practice Guidelines. ATA. https://www.americantelemed.org/resource_categories/practice-guidelines/

Anwar S, Prasad R, Chowdhary BS et al (2019) A telemedicine platform for disaster management and emergency care. Wireless Pers Commun 106:191–204. https://doi.org/10.1007/s11277-019-06273-6

Arogya Legal (2021) The liability conundrum for telemedicine platforms in India: striking a balance between vicarious and intermediary liability. Arogya Legal. https://arogyalegal.com/2021/article/the-liability-conundrum-for-telemedicine-platforms-in-india-striking-a-balance-between-vicarious-and-intermediary-liability/. Retrieved 8 Jan 2025

Bosurgi R (2019) Climate crisis: healthcare is a major contributor, global report finds. BMJ. https://doi.org/10.1136/bmj.l5560

Van Bree EM, Snijder LE, Ossebaard HC, Brakema EA (2025) Environmental impact of physical visits and telemedicine in nursing care at home: comparative life cycle assessment. J Med Internet Res. https://doi.org/10.2196/67538

Climate and Health Alliance Organisation (2012) Sustainability in the NHS Health check 2012

Croymans D, Hurst I, Han M (2020) Telehealth: the right care, at the right time, via the right medium. NEJM Catalyst. https://doi.org/10.1056/CAT.20.0564

d'Arville L, Mehra S, Boulton A, Kapur V (n.d.) Heeding the Call to Reinvent Healthcare Delivery. Bain and Company. https://www.bain.com/insights/heeding-the-call-to-reinvent-healthcare-delivery/

Debayan Bhattacharya D (2022) Analysing the liability of digital medical platforms for medical negligence by doctors. NUJS Law Rev 15(2)

DH Contributor (2020) Namma CO_2 emission is high, and it can make one ill. Deccan Herald. https://www.deccanherald.com/india/karnataka/bengaluru/namma-co2-emission-is-high-and-it-can-make-one-ill-807423.html

Grabski DF, Meyer MJ, Gander JW (2024) Pediatric telemedicine visits reduce greenhouse gas emissions. J Clim Chang Health 17:100309. https://doi.org/10.1016/j.joclim.2024.100309

Haleem A, Javaid M, Singh R, Suman R (2021) Telemedicine for healthcare: capabilities, features, barriers, and applications. Sens Int 2. https://doi.org/10.1016/j.sintl.2021.100117

Health Care Without Harm (2019) Health care's climate footprint: How the health sector contributes to the global climate crisis and opportunities for action. Report No. 4. https://global.noharm.org/sites/default/files/documents-files/5961/HealthCaresClimateFootprint_092319.pdf

Holmner A, Rocklov J, Ng N, Nilsson M (2012) Climate change and eHealth: a promising strategy for health sector mitigation and adaptation. Glob Health Action 5(1). https://doi.org/10.3402/gha.v5i0.18428

Holmner A, Ebi K, Lazuardi L, Nilsson M (2014) Carbon footprint of telemedicine solutions—unexplored opportunity for reducing carbon emissions in the health sector. Plos one. https://doi.org/10.1371/journal.pone.0105040

How to Mitigate the Risks of Telemedicine (2022) Sciarabba Walker & Co., LLP. https://swcllp.com/how-to-mitigate-the-risks-of-telemedicine/. Retrieved 7 Jan 2025

India's Mobile and Broadband Internet Speeds—Speedtest Global Index (2024) Speed test. https://www.speedtest.net/global-index/india. Retrieved 8 Jan 2025

Intergovernmental Panel on Climate Change (2023) Climate change 2023: Synthesis report. https://www.ipcc.ch/report/ar6/syr/

Jaiswal S, Shesadri D, Boina N (2024) Biomedical waste management and its impact on climate change in India. Express Healthcare. https://www.expresshealthcare.in/news/biomedical-waste-management-and-its-impact-on-climate-change-in-india/445813/#:~:text=According%20to%20the%20Central%20Pollution,infectious%20organisms%2C%20and%20expired%20medicines

Lewis D, Tranter G, Axford A (2009) Use of videoconferencing in Wales to reduce carbon dioxide emissions, travel costs and time. J Telemed Telecare 15(3). https://doi.org/10.1258/jtt.2009.003010

Li Z, Xiang L, Ning J, Li W, Huang Y, Xiao X (2025) Pathways to sustainable health care development: study on the carbon reduction potential of telemedicine in China. J Med Internet Res 27:e63927. https://doi.org/10.2196/63927

Mallipeddi S, Goda D, Yerram S, Varghese A, Ande JR (2017) Telemed Beyond: Navig Front Med Technol 2(1). https://hal.science/hal-04495006v1

Ministry of Health and Family Welfare (2023) Press Release: Press Information Bureau. https://pib.gov.in/PressReleasePage.aspx?PRID=1899855. Retrieved 7 Jan 2025

NHS Carbon Reduction Strategy for England (2009) Saving Carbon, improving health. Lean Business Ireland. https://www.leanbusinessireland.ie/includes/documents/1232983700_dGKJ_saving_carbon,_improving_health_nhs_carbon_reducti%5B1%5D.pdf

Nishith Desai Associates (2024) Telemedicine in India- The future of medical practice. Nishith Desai Associates. https://www.nishithdesai.com/fileadmin/user_upload/pdfs/research_Papers/Telemedicine-in-India.pdf

Oliviera T, Barlow J, Bayer S (2013) Teleconsultations reduce greenhouse gas emissions. J Health Serv Res Policy 18(4). https://doi.org/10.1177/1355819613492717

Pantanowitz L, Dickinson K, Evans A, Hassel L, Henricks W, Henricks W, Lennerz J, Lowe A, Parwani A, Riben M, Smith D, Tuthill M, Weinstein R, Wilbur D, Krupinski E, Bernard J (2014) American telemedicine association clinical guidelines for telepathology. J Pathol Inform 5(1). https://pubmed.ncbi.nlm.nih.gov/25379345/

Practo Terms and Conditions (n.d.) Terms and Conditions. Practo. https://www.practo.com/company/terms

Press Release (2024) Universal connectivity and Digital India initiatives reaching to all areas, including tier-2/3 cities and villages. Ministry of Communication. https://pib.gov.in/PressReleasePage.aspx?PRID=2040566

PTI (2020) Telemedicine Market in India to Reach USD 5.5 billion by 2025: EY-IPA Study. Financial Express. https://www.financialexpress.com/life/telemedicine-market-in-india-to-reach-usd-5-5-%20billion-by-2025-ey-ipa-study-2078029/

Purohit A, Smith J, Hibble A (2021) Does telemedicine reduce the carbon footprint of healthcare? A systematic review. Futur Healthc J 8(1):e85–e91. https://doi.org/10.7861/fhj.2020-0080

Rowland SP, Fitzgerald E, Lungren M, Lee E, Harned Z, McGregor A (2022) Digital health technology-specific risks for medical malpractice liability. NPJ Digit Med 5(1). https://doi.org/10.1038/s41746-022-00698-3

Saada N, Deoli A (n.d.) Malpractice Risks with Telehealth: The Do's and Don'ts. Rutgers University, New Jersey State Policy Lab. https://policylab.rutgers.edu/malpractice-risks-with-telehealth-the-dos-and-donts/

Scott R, Perverseff T, Lefebre N (2009) Reducing environmental impact: an example of how e-health can reduce environmental impact and concomitantly improve health. In: Proceedings of the 1st annual conference on e-health: the virtual dimensions of health and environment empower, Enhance, Enforce, vol 10, no 10

Serra C, Tanarro A, Cummings C (2022) Impact on the reduction of CO2 emissions due to the use of telemedicine. Sci Rep 12. https://doi.org/10.1038/s41598-022-16864-2

Singh SG (2024) Over 5.2 MN medical malpractice cases filed in India annually, shows data. Business Standard. https://www.business-standard.com/finance/personal-finance/over-5-2-mn-medical-malpractice-cases-filed-in-india-annually-shows-data-124042900073_1.html

Stanberry B (2006) Legal and ethical aspects of telemedicine. J Telemed Telecare 12(4). https://pubmed.ncbi.nlm.nih.gov/16774696/

Technology Information, Forecasting and Assessment Council (2022) Current Trends in Telemedicine in India. Department of Science and Technology India. https://www.tifac.org.in/images/latestreports/Telemedicine_Study_Report(Final%20version).pdf

Theil C, Mehta N, Sejo CS, Qureshi L, Moyer M, Valentino V, Saleh J (2023) Telemedicine and the environment: life cycle environmental emissions from in-person and virtual clinic visits. NPJ Digit Med 87. https://www.nature.com/articles/s41746-023-00818-7

TOI Tech Desk (2024) Top 10 countries with fastest mobile internet 2024; check where India ranks. Times of India. https://timesofindia.indiatimes.com/technology/tech-news/top-10-countries-with-fastest-mobile-internet-2024-check-where-india-ranks/articleshow/116678539.cms

Treelife (2020) Data privacy for telemedicine patforms. Treelife. https://treelife.in/startups/data-privacy-for-telemedicine-platforms/

Tuckson J, Edmunds M, Hodgkins M (2017) Telehealth. N Engl J Med 377(16):1585. https://doi.org/10.1056/NEJMsr1503323

Venkatesh U, Aravind GP, Velmurugan A (2022) Telemedicine practice guidelines in India: global implications in the wake of the COVID-19 pandemic. World Med Health Policy 14(3). https://doi.org/10.1002/wmh3.497

Weiner J, Bandeian S, Hatef E, Lans D, KW L, Liu A (2021) In-person and telehealth ambulatory contacts and costs in a large US insured cohort before and during the COVID-19 pandemic. AMA Netw Open 1(4(3)). https://pubmed.ncbi.nlm.nih.gov/33755167/ Retrieved 7 Jan 2025

Wootton R, Tait A (2010) Environmental aspects of health care in the Grampian NHS region and the place of telehealth. J Telemed Telecare 16(4). https://doi.org/10.1258/jtt.2010.004015

Woottoon R, Bahaadinbeigy K, Hailey D (2011) Estimating travel reduction associated with the use of telemedicine by patients and healthcare professionals: proposal for quantitative synthesis in a systematic review. BMC Health Serv Res 11

Yellowlees PM, Chorba K, Parish M, Wynn Jones H, Nafiz N (2010) Telemedicine can make healthcare greener. Telemed J E Health 16(2). https://doi.org/10.1089/tmj.2009.0105

Zeynep (2024) The role of the health sector in tackling climate change: a narrative review, Health Policy. Health policy 143. https://doi.org/10.1016/j.healthpol.2024.105053

Part III
Navigating Climate Crises with Health Policies: Adaptation, Resilience Building, and Future Outlook

Chapter 14
Evaluating the Impacts of Climate Change on Public Health Surveillance Systems across Southeast Asia: A Systematic Review

Rumbidzai Matindike

Abstract Southeast Asia experiences a very high rate of vulnerability to climate-related health challenges. High temperatures, extreme weather events, and shifting disease patterns are common in this region. The emergence of new disease variants is threatening the region, thus intensifying existing health challenges. These new variants include waterborne diseases, infectious diseases, and heat-related illnesses. This requires the implementation of public health surveillance systems as an effective control measure in the fight against these health risks; however, it is also critical to understanding the nature of the new variants and their relation to climate change. This study seeks to investigate how climate change affects public health surveillance systems in Southeast Asia, paying attention to their ability to cope with the intensity of climate-induced health challenges. This study utilizes the Systematic Review and Meta-Analyses (PRISMA), a structured approach that synthesizes findings from scientific studies and existing policy documents. The results show alarming evidence of a lack of effective policies that directly focuses on climate change impacts on public health, a shortage of skilled and qualified health professionals to effectively assist in mitigating and responding to climate-induced health risks. This study, therefore, will alert policymakers to the need for improving policies aiming at a more focused action-oriented climate-health approach. Additionally, communities in Southeast Asia will be equipped with how to fight and cope with climate change health challenges, which is a crucial measure that will be highlighted by this review paper.

Keywords Climate change · Health challenges · Public health · Surveillance · Vulnerability

---

R. Matindike (✉)
Department of Geography Environmental Management Energy Studies, University of Johannesburg, Johannesburg, Gauteng, South Africa
e-mail: matindikerumbidzai@gmail.com

P. Ghosh et al. (eds.), *Climate Crises and Public Health Challenges*, Disaster Risk Reduction, https://doi.org/10.1007/978-981-95-9709-3_14

14.1 Introduction

Climate change is a phenomenon which has ripped the world apart particularly affecting public health worldwide. Scholars noted that these challenges include the intensification of vector-borne diseases such as malaria and dengue, an increased frequency of extreme weather events, worsening air pollution, and rising waterborne illnesses (Garzon et al. 2024). Additionally in Southeast Asia climate change is anticipated to increase existing health and security threats, emphasizing the urgent need for improved surveillance systems and regional cooperation to address the consequences of severe weather events and related challenges (Brennan 2021).

It is therefore important to design and implement policies which are shaped to adapt and mitigate existing and emerging climate-related health risks. At the core of these systems are public health surveillance mechanisms, which are instrumental in identifying disease patterns, tracking outbreaks, and enabling timely interventions, giving credence to the importance of public health surveillance systems and their role in the identification and mitigation of climate-related risks (Lawson et al. 2020). Public health surveillance involves the continuous systematic collection, analysis, and interpretation of health-related data to monitor disease trends and health outcomes (Asala 2023). Due to the intensity of climate change, health challenges are also increasing, leading to these surveillance systems becoming unsustainable, thus failing to cope with the emergence of new disease variants and erratic weather conditions. However, it is critical that these systems are improved to mitigate these challenges.

Southeast Asia is vulnerable to the effects of climate change due to its typical combination of environmental, socioeconomic, and demographic factors (Amul 2022). The region frequently experiences climate change-related events such as typhoons, floods, and droughts, which disrupt livelihoods and increase public health risks. High population density, rapid urbanization, and dependence on climate-sensitive industries, including agriculture and fisheries, further worsen these vulnerabilities (Li et al. 2023). For example, the rising incidence of dengue fever in the region has been closely linked to increasing temperatures and shifting rainfall patterns, emphasizing the critical need for strong public health responses. Additionally, despite recent socioeconomic progress, typhoid fever remains a significant public health concern, worsened by antimicrobial resistance and inadequate surveillance infrastructure (Bhutta et al. 2018).

Many countries in Southeast Asia face significant challenges in adapting their public health surveillance systems to the evolving impacts of climate change (Garzon et al. 2024). Climate change impacts are particularly evident in Third World Countries, where infrastructure is inadequate, unreliable Internet access and insufficient data storage capabilities, which hinders effective disease surveillance (Alhassan and Wills 2024). Absence of finance, scarcity of new technological resources, and inability to afford hospitalization often create a bias in tracking the progress of mitigating climate-related illnesses. Building resilient public health surveillance systems in the region requires addressing these limitations (Saleem et al. 2014).

There is an urgent need for a well-trained workforce equipped to manage and utilize modern surveillance technologies effectively. Training programmes should focus on integrating new technologies and methodologies, enabling health professionals to enhance data collection, analysis, and response capacities. Strengthening such systems will be essential for effectively addressing the health challenges posed by climate change in Southeast Asia.

Public health surveillance systems are critical for addressing the health impacts of climate change in Southeast Asia (Angkurawaranon et al. 2014). These systems facilitate the continuous collection, analysis, and circulation of health data, enabling the early detection of disease outbreaks and the implementation of interventions that are based on evidence. However, challenges such as inadequate resources, limited regional collaboration, and the insufficient integration of climate-related health data often undermine their effectiveness (Nisha et al. 2022). The inclusion of climate-sensitive health indicators is essential for enhancing disease early warning systems and informing public health policy and planning (Thomson et al. 2019). As the impacts of climate change increase, it is important to evaluate the capacity of these systems and their ability to address emerging health threats and to identify opportunities for enhancing their resilience.

In addition to improving system functionality, understanding the intersection of climate change and public health surveillance encourages regional collaboration and knowledge sharing. Southeast Asia is a region characterized by immense diversity in culture, language, religion, and political systems (Gin 2019). This diversity comes from the region's historical interactions with major cultures, such as India and China, and its colonial past (Koldunova and Yefimova 2017). Such diversity presents challenges and opportunities for addressing several health issues, such as spreading infectious diseases. Due to similarities in the experiences shared by countries in this region, this presents an opportunity for shared viewpoints and collaboration between countries, coupled with comparisons of surveillance systems.

A fundamental aspect of resilience is adaptability. By understanding how climate change impacts the functioning of public health surveillance systems, healthcare institutions can identify vulnerabilities and implement strategies to enhance system adaptability. Climate change can potentially disrupt disease monitoring by altering transmission patterns, necessitating adjustments to surveillance protocols. This includes integrating climate-sensitive indicators in data collection and analysis, such as those that monitor changes in environmental conditions, including temperature, precipitation patterns, and extreme weather events (Trewin et al. 2021).

Furthermore, a comprehensive understanding of these impacts facilitates the incorporation of predictive modelling and early warning systems into public health surveillance. Early warning systems utilize data from diverse sources, including environmental sensors, weather forecasts, and even social media, to improve the accuracy of disease predictions (Garzon et al. 2024). Analysing the climate through epidemiological data can help improve the ability to cope with disease outbreaks as people can effectively plan; for example, dengue outbreaks can be tracked by forecasting the rainfall patterns and temperature variations. This approach is very important in the prevention of disease outbreaks and the mitigation of climate-related diseases.

In Southeast Asia, public health surveillance systems are of great importance across the region; however, literature is scarce on the relationship between climate change and public health surveillance systems in this region. Although there has been notable literature on the health impacts of climate change and the role of surveillance systems, these two have been appropriated as different areas of study. There is, however, a scarcity in investigating the relationship between these two and how they both intersect in the emergence of climate-induced health risk. This study, therefore, bridges this gap, necessitating the implementation of policies tailored to prepare and respond to climate-induced health risks.

This study is a systematic review and seeks to evaluate or review existing literature to address the objective of evaluating the effectiveness of existing public health surveillance systems in Southeast Asia in addressing emerging health risks associated with climate change. Specifically, it seeks to investigate the effectiveness of surveillance systems in forecasting and responding to climate-induced health risks, coupled with the prevention of outbreaks. Finally, it will alert policymakers to the need to improve existing surveillance systems with stronger and efficient systems, which will effectively mitigate climate-induced risks and adapt to the emerging health challenges.

14.2 Methodology

14.2.1 Inclusion and Exclusion Criteria

This study synthesizes past data from multiple studies; systematic reviews can provide more accurate and reliable results through meta-analysis, which combines data statistically to increase the power and precision of the findings (Sgarbossa et al. 2022). Systematic reviews are considered very reliable, and they often produce factual evidence. They support policymakers and practitioners by providing a combined, summary of the available evidence on a particular issue without bias, enabling informed decision-making (Luchini et al. 2021). The researcher followed a structured methodology that included the formulation of a research question, an exhaustive literature search, the selection of relevant studies, critical appraisal, and the synthesis of results. Advanced tools and frameworks, such as the PRISMA checklist and flow diagram, were also used by the researcher to manage and visualize data, thereby improving the clarity and transparency, and making it easier to interpret findings.

Databases help with the sorting out of large volumes of data in a structured way, which is essential for many applications in scientific research. The researcher used the Scopus database in this study because it is widely recognized for its usefulness in bibliometric analyses, research evaluations, and science policy assessments. Scopus is frequently used for systematic literature reviews due to its strong query engine and advanced visualization features (Valente et al. 2022).

Scopus is commonly selected because of its comprehensive quality assurance procedures and enriched metadata records, even though both WoS and Scopus provide a comprehensive citation report. The study made use of Boolean operators, which are used to retrieve information making it easier to manage and obtain search results quicker. Primary Boolean operators AND, OR, and NOT are used to retrieve targeted and relevant search results by combining or excluding keywords. Boolean operators are essential for optimizing the efficiency of search engines and information retrieval systems because they allow users to formulate highly targeted search queries (Picalho et al. 2022). Figure 1 shows the steps involved in the inclusion and exclusion process of the bibliometric analysis.

On 4 November 2024, 112 articles were retrieved through a data search in the Scopus database. Keywords closely related to the topic of study were used to locate the articles. The selection process was guided by a set search criterion, which ensured that the articles found aligned with the study's objectives. As shown below, the

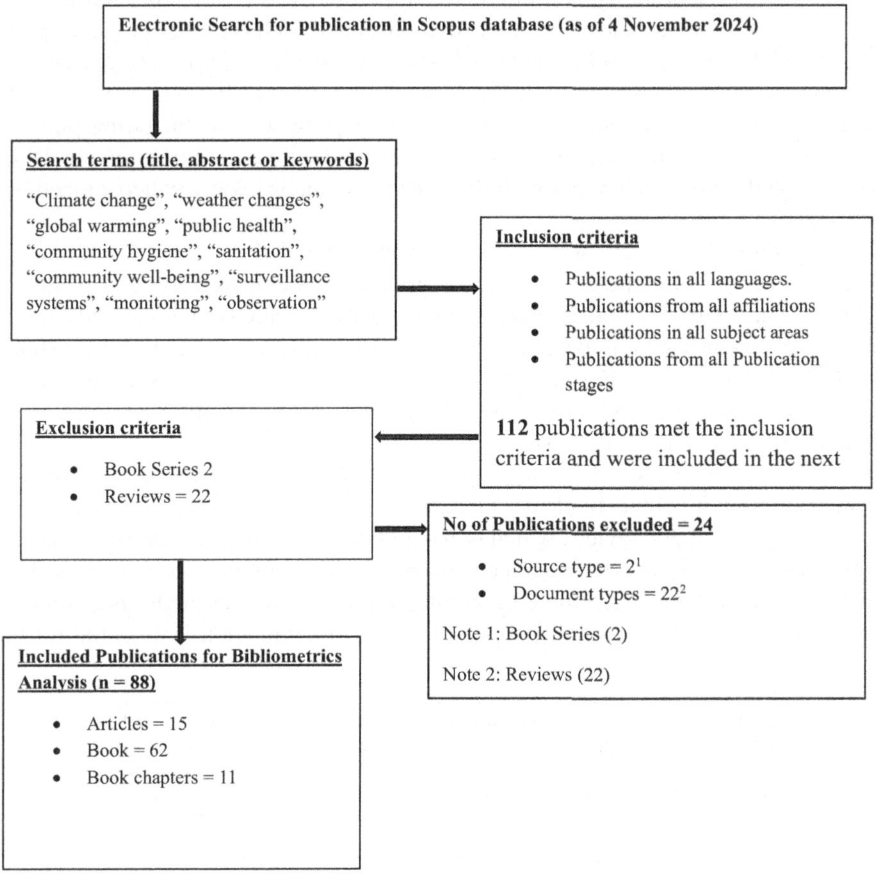

Fig. 1 Steps followed in the inclusion and exclusion of documents for bibliometric analysis

keywords were separated using Boolean operators after being added to the search database (Fig. 2).

The researcher found a total of 112 articles by applying the criterion of selecting only English-language articles. It was found that all the available documents met this language requirement. Many non-native English speakers feel inferior because of the language barrier, which can lower their self-esteem and result in lower quality work (Yakhontova 2021). Publishing in English significantly raises the visibility and impact of research because it is the primary language of science and is widely understood by people worldwide (Dos Santos et al. 2021). To progress their careers and achieve worldwide recognition, researchers, particularly those in non-English speaking countries, usually need to publish in English (Bozbıyık and Uysal 2022).

Publications from all affiliations were included in the search. Research on complex topics such as climate change and public health surveillance spans multiple disciplines, including environmental science, public health, epidemiology, and policy studies. Academic affiliations in research may include university research centres, institutional collaborations, and international partnerships (Hottenrott and Lawson 2022). By including articles from all affiliations, the search is more likely to capture a range of diverse approaches and methodologies, resulting in more accurate and comprehensive findings. Limiting the search to specific affiliations or institutions could inadvertently exclude important studies or perspectives. Including publications from all affiliations ensures a more neutral and unbiased representation of the research landscape, which is particularly vital in systematic reviews where objectivity is essential.

Publications from every subject area were also included in the inclusion criteria. These subject areas included biological, agricultural, medical, earth and planetary sciences, and environmental sciences, among others. Research outcomes and citation impact have been shown to be enhanced by interdisciplinary research (IDR) (Zhang et al. 2021). By integrating research from various disciplines, this approach can capture studies that approach the topic from multiple perspectives, yielding more comprehensive and practical insights.

Finally, publications at the every stage of publication were included in the review. A more thorough synthesis of the available evidence is necessary for a thorough and reliable systematic review, which is ensured by including works at every stage of publication. This expands the search results as well. To communicate new ideas, data, and discoveries to the scientific community and the public, research dissemination is essential at various stages of publication. This dissemination is essential for the

(ALL (climate AND change) OR ALL (weather AND changes) AND ALL (global AND warming) AND ALL (public AND health) AND ALL (community AND hygiene) AND ALL (sanitation) OR ALL (community AND wellbeing) AND ALL (surveillance) AND ALL (surveillance AND systems) OR ALL (monitoring) OR ALL (observation) AND ALL (south AND east AND Asia) AND ALL (southern AND Asia))

Fig. 2 Keywords used in the literature search

progress of research and development (Madden et al. 2024). After the inclusion stage, the articles advanced to the exclusion stage.

In total, 112 publications met the inclusion criteria and advanced to the next stage of analysis. However, because two book series often provide theoretical insights or general discussions rather than the empirical or applied data that systematic reviews typically require, they were excluded. Such content lacks the precise, fact-based conclusions needed to effectively address the research questions. Furthermore, 22 review articles were removed based on the type of document. The emphasis was maintained on primary studies in accordance with the Preferred Reporting Items for Systematic Reviews and Meta-Analyses (PRISMA) guidelines to ensure that the systematic review contains direct evidence from original research sources. This approach enhances the reliability of the review by prioritizing first-hand empirical data.

A total of 88 publications comprising 62 books, 11 book chapters, and 15 journal articles were ultimately selected for bibliometric analysis. In fields like public health surveillance and climate change, where theoretical knowledge and policy discussions are essential, books and book chapters are included because they can cover more complex or in-depth aspects of the subject. By including a range of publication types, a more comprehensive perspective on the research topic is guaranteed. According to (Aleixandre-Benavent et al. 2017), books often give readers a more in-depth and comprehensive analysis of subjects than journal articles, giving them a solid understanding of the theoretical foundations and historical development of a field. Although journal articles primarily concentrate on empirical data, books and book chapters provide case studies, interdisciplinary approaches, and in-depth theoretical frameworks to enhance the analysis. Journals are also a vital source of citation data for assessing the significance and scope of research outputs (Surekha et al. 2024). By combining these publication types, the bibliometric analysis can capture the topic's theoretical and empirical facets, providing a comprehensive and nuanced understanding of the research landscape.

Eighty-eight data files were reassessed after extraction. A standardized Microsoft Excel template was created to make the procedure easier. According to (Schwabish 2023), Microsoft Excel is still a widely used and adaptable tool for data analysis, automation, and visualization in a variety of domains. Its ability to integrate data from multiple sources and compare sizable protein datasets makes it especially useful in bioinformatics research (Schmidt and Frishman 2006). Important details like the study design, methodology, population results, authorship, and publication year were all included in the template.

The number of articles that were found, eliminated, and included was documented in compliance with the Preferred Reporting Items for Systematic Reviews and Meta-Analyses (PRISMA) guidelines. The reviewer developed the template, extracted and analysed the data, and resolved any discrepancies through synthesis. A graphic representation of this procedure is shown in Fig. 3.

88 files in the Scopus database qualified for preliminary examination. 11 articles were removed as duplicate records during the identification phase. Redundant entries can compromise data integrity, resulting in inaccurate analysis and erroneous

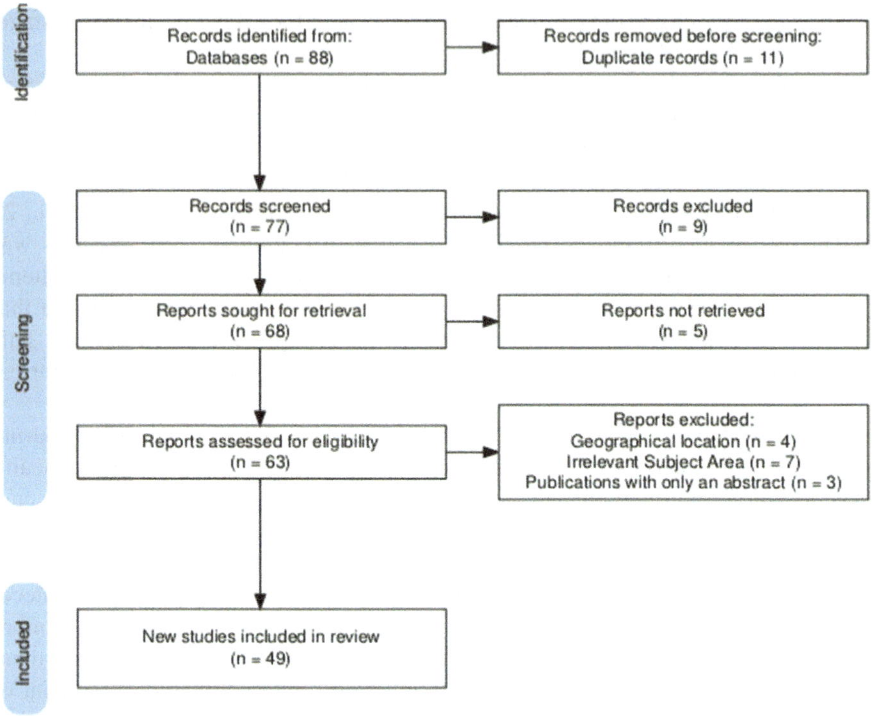

Fig. 3 PRISMA flow diagram detailing steps in identifying documents for analysis

conclusions. It is essential to eliminating duplicates from your data to ensure its accuracy and consistency, hence facilitating reliable, data-driven insights (Azeroual et al. 2023). Certain authors, papers, or publications may appear more significant than they are due to duplication. This may result in policymakers or practitioners making wrong decisions.

Following the elimination of duplicates, 77 articles advanced to the further phase of the screening process. At this time, it was not feasible for the researcher to retrieve 5 articles. Errors or faults in the filesystem can result in issues that affect files inaccessible, potentially leading to data loss (Fryer et al. 2014). Researchers may also be unable to access certain files due to restricted entry to secure databases or insufficient authorization. Additionally, 4 files were excluded as they did not originate from South or Southeast Asia, which was the primary focus of the study. The researcher also eliminated files pertaining to regions that were not the primary focus, such as Africa or Europe, to maintain relevance. Eliminating files containing erroneous or irrelevant geographical data facilitates confidence and utilization of a dataset (Singh and Rafiei 2018).

Three articles were eliminated because they only had abstracts. Assessing the rigor and validity of research through abstracts may prove challenging. Access to full-text publications is necessary to understand critical aspects such as the study's design, the random assignment of participants to groups, and the implementation of blinding (Dundar et al. 2006). In the absence of this information, it is challenging to assess the quality and reliability of the research. Ultimately, seven articles were excluded as they were not part of the geographical area that was under research by the researcher and qualified as irrelevent. Incorporating irrelevant files can obscure the analysis and hinder the objectives of the investigation. Based on these criteria, 49 publications were retained for further analysis and assessment.

14.3 Results and Discussion

49 articles were retrieved and transferred to a Microsoft Excel spreadsheet to facilitate assessment of the literature data that was available. Digital object identifier (DOI) was the major factor that was used to identify articles, and the data also included the year of publication, the name of the article, and bibliographic information such as author names. The reason behind using DOIs is that they provide a unique and permanent link to digital objects, which include articles from journals, thereby ensuring that these resources can be reliably located and accessed over time (Gorodzha 2018).

Articles retrieved were qualitatively analysed together with the extracted data that corresponded. The study used VOSviewer software (version 1.6.20) (van Eck and Waltman 2014), a specialized tool for constructing and visualizing bibliometric networks. VOSviewer was used to create bibliographic maps and visualize the occurrences and co-occurrence networks of key terms. The advantage of using VOSviewer for bibliometric analysis is that it can create maps that are easy to understand while providing all the information. These maps illustrate relationships among various bibliometric elements, such as authors, keywords, institutions, and countries, based on metrics like co-authorship, citations, and co-occurrence. VOSviewer has very strong visualization capabilities, such as the identification of patterns and key trends in data. By leveraging VOSviewer, this study was able to provide a comprehensive and visually accessible representation of the key themes and relationships within the analysed literature.

Bibliographic analysis is the most used method of analysing data because of its ability to find the relationship between keywords from research articles presented as linked clusters (Masenyama et al. 2022). According to van Eck and Waltman (2014), constructing a bibliometric map involves four key steps, which are selecting the counting method (binary or full), determining the minimum number of term occurrences, calculating the relevance score for co-occurring terms, and using the selected terms to generate the map.

In this study, the titles, abstracts, and key terms of the retrieved articles were used as input data in VOSviewer, enabling a graphical representation based on the co-occurrence of key terms. This visualization facilitated the identification of relationships and clusters, providing insights into the thematic structure of the analysed literature.

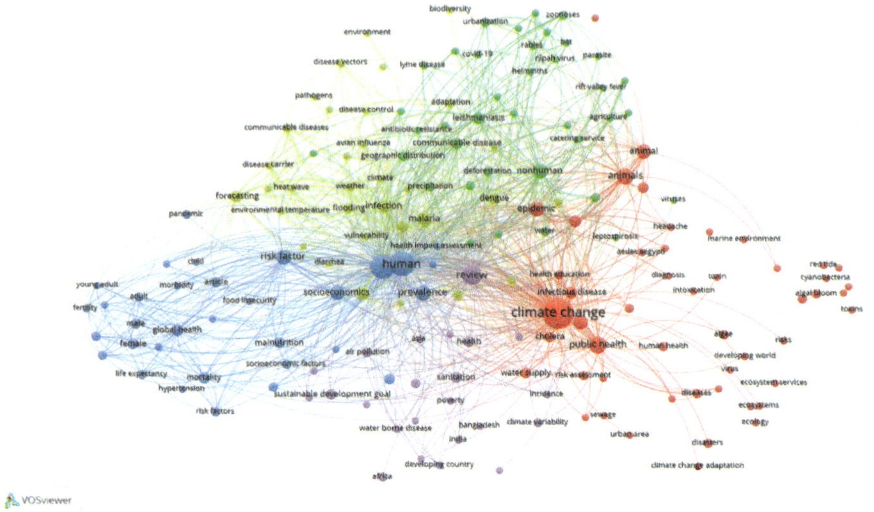

Fig. 4 A keyword co-occurrence map for climate change impacts. The bigger the node, the larger the weight, and the closer the node, the stronger the connection

Co-occurrence analysis is a technique employed to uncover relationships and patterns within datasets by analysing the frequency with which entities, such as terms, genes, and authors, appear together (Moresi and Pinho 2021). This method highlights the terms or keywords that appear the most and how they relate to each other, making it easier to understand the themes and trends in a research article. By employing co-occurrence analysis, this study was able to map critical connections within the research field, providing valuable insights into its conceptual and methodological landscape. Figure 4 depicts a network map categorizing the identified literature into three distinct conceptual clusters, represented by blue, red, and green colours. These clusters are further elaborated in Table 1, which provides a detailed visualization of the keywords, link strengths, and the associated cluster themes.

In the red cluster, the theme "Integrating Climate Change into Public Health Strategies and Health Education Initiatives" emerges as a focal point of this research. The rising impacts of climate change in Southeast Asia present substantial challenges to public health systems, necessitating an urgent and coordinated response. Climate change demonstrates the highest total link strength within the research network, underscoring its centrality to this study. Evaluating link strength is critical for analysing network connectivity, resilience, and overall structural dynamics. This approach facilitates the identification of critical connections that sustain the integrity and performance of the network (Qi et al. 2019).

In the purple cluster, the theme "The role of sanitation in mitigating waterborne diseases", supports that waterborne diseases have always been a public health problem in Southeast Asia. Sanitation and general hygiene play a vital role in the fight against these diseases, with the region being prone to unhealthy practices such

Table 1 Cluster colour and proposed themes

Colour recognition	Selected keyword	Occurrence	Total link strength	Cluster theme
Red	Climate change	31	422	Integrating climate change into public health strategies and health education initiatives
	Public health	10	184	
	Health education	3	58	
Purple	Review	13	288	The role of sanitation in mitigating waterborne diseases: A review of current challenges and solutions
	Sanitation	4	80	
	Waterborne diseases	2	46	
Blue	Humans	19	392	Socioeconomic risk factors influencing human vulnerability
	Risk factor	8	184	
	Socio economics	7	172	
Yellow	Environmental temperature	2	71	Predicting vulnerability to environmental temperature variations
	Forecasting	5	126	
	Vulnerability	2	60	
Green	Geographic distribution	3	72	Geographic distribution of communicable diseases linked to deforestation
	Communicable diseases	3	88	
	Deforestation	3	50	

as the consumption of contaminated foods and inadequate sanitation. Many Southeast Asian countries rely on onsite sanitation systems (OSS), such as septic tanks and cesspools, which are frequently poorly designed, maintained, and managed. These poorly designed OSS contribute to water pollution and facilitate the spread of waterborne diseases (Koottatep et al. 2018). The overpopulation in rural areas in this region also contributes to severe sanitation challenges as a result of rapid population growth and insufficient infrastructure development (Nguyen et al. 2014). Diseases such as diarrhoea and helminth infections are usually due to poor sanitation facilities, and this has proven to be a major public health concern in Southeast Asia (Han et al. 2019).

The blue cluster highlights the theme, "Socioeconomic risk factors that exacerbate human vulnerability in Southeast Asia". This region experiences severe economic challenges, there is a high rate of poverty, and the economy is highly unstable, making it very difficult for this region to fight natural disasters. According to (Li et al. 2021), these difficulties make the development of strategies to fight and adapt to climate change unsustainable. Economic problems often lead to poor infrastructure in communities, which are marginalized, with these communities often facing problems such as lack of proper housing and poor sanitation systems, especially in rural communities, increasing their vulnerability to climate change and hazards.

The yellow cluster focuses on the theme "Predicting vulnerability to environmental temperature variations". Southeast Asia is experiencing very warm weather conditions, which are projected to increase in the future. Southeast Asia has always experienced very harsh weather conditions, and it is prone to heavy rainfall, cyclones, and severe heatwaves. The projected increase in temperature is expected to affect the rainfall patterns in this region, exposing it to severe droughts and floods (Sentian et al. 2022). Conversely, high poverty rates in this region increase the pressure on resources, increasing the region's vulnerability to climate change. Lack of financial support is another factor that further disables them from potentially fighting and adapting to climate change. High temperatures and harsh weather conditions greatly affect the people in Southeast Asia (Huynh et al. 2014).

The green cluster highlights the theme "Role of deforestation in the emergence and spread of communicable diseases in Southeast Asia". Deforestation changes the environment severely, disturbing the ecosystem and the behaviour of certain inhabitants, directly affecting human beings. The disturbance of habitats of certain species promotes their contact with human beings, enabling the transmission of diseases (Davidson et al. 2019). Additionally, (MacKenzie and Williams 2009) noted that cutting down trees and land degradation create favourable conditions for species like mosquitoes, increasing vulnerability to vector-borne diseases.

Density Visualization of Co-authorship.

VOSviewer offers an interactive platform for visualizing bibliometric data, enabling users to zoom into specific areas and explore detailed information about individual authors and their connections (Li and Wei 2022). This functionality efficiently identifies key authors, research clusters, and prominent collaborative focal points within a given field. In density visualizations, regions with higher densities of co-authorship links are highlighted more prominently, facilitating the recognition of influential researchers, institutions, and countries that occupy central positions within the research network (Ramezani et al. 2021). Additionally, VOSviewer provides filtering options, such as by the number of publications or citations, allowing users to focus on the most impactful authors and collaborations (Ye et al. 2007).

The density visualization underscores the implications of climate change on surveillance systems, with particular emphasis on co-authorship dynamics. Co-authorship networks often exhibit structures, reflecting the presence of central authors or institutions that engage in extensive collaboration (Liu and Ge 2022). Analysing co-authorship networks provides insights into trans-regional and international collaboration patterns, which are vital for understanding the global research landscape. Institutions with high publication outputs frequently emerge as leaders in academic influence and collaborative networks. The Table 2 illustrates the number of authors affiliated with selected institutions. For this study, the researcher randomly selected ten institutions, presented as follows.

Co-authorship refers to the collaborative partnership between two or more researchers who jointly contribute to the creation and publication of scholarly works, such as journal articles, conference proceedings, and books. It serves as a key indicator of academic collaboration and plays a crucial role in advancing knowledge, fostering interdisciplinary research, and strengthening professional networks. Over the twentieth and twenty-first centuries, co-authorship has grown significantly, with a marked increase in collaborative publishing across numerous disciplines (Hermanowicz and Lei 2023).

Additionally, the data presented highlights a notable rise in co-authorship, including an increasing number of collaborations involving universities in Africa, such as the University of Cape Town and Stellenbosch University. Furthermore, there is growing scholarly attention on surveillance research in regions such as Southeast Asia, reflecting the global expansion of collaborative efforts in this domain.

Figure 5 illustrates the density of authors who have conducted research on the impacts of climate change on surveillance systems in Southeast Asia. Each cluster represents the total number of authors, with many likely affiliated with the same institution to facilitate more effective communication during the research process. Collaborative efforts are known to enhance research productivity and impact, as demonstrated by the higher citation rates often observed in co-authored publications across various fields (Nisha et al. 2022).

Recommendations

Strengthening Regional Collaboration: Enhancing partnerships among Southeast Asian nations through the establishment of regional centres for climate-health monitoring can facilitate the sharing of data, expertise and resources while improving cross border coordination for more effective public health surveillance.

Table 2 Affiliation and number of authors within institutions

Affiliation	Number of authors
Russian Academy of Sciences	6
University of Exeter	5
Stellenbosch university	5
The University of Manchester	4
Klinikum der Universität München	4
Universitatea din Bucuresti	4
Chinese university of Hong Kong	4
University of Pennsylvania	3
National university of Singapore	3
University of cape town	3

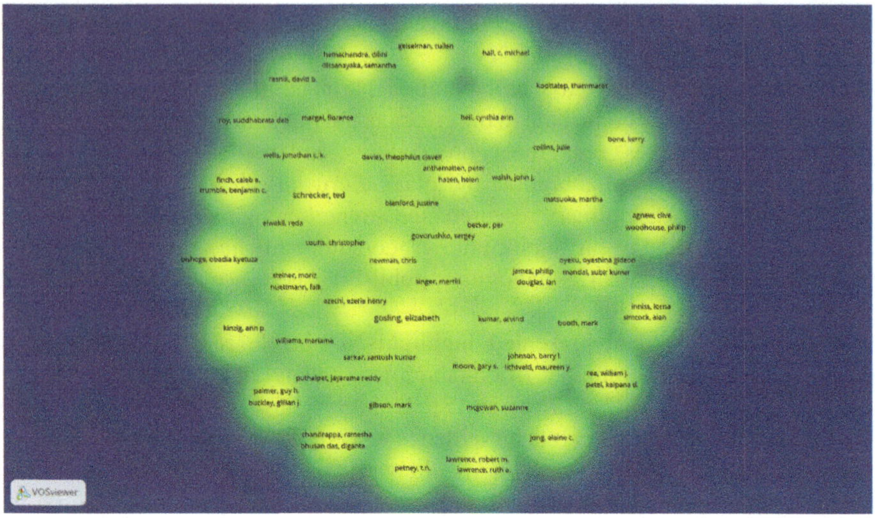

Fig. 5 Co-authorship map for climate change impacts on surveillance in Southeast Asia

1. Governments should allocate resources to update surveillance systems, incorporating advanced technologies such as artificial intelligence, machine learning, and geospatial tools to improve the detection and prediction of climate-related health risks.
2. Public health strategies should be introduced that address the impacts of climate change by combining adaptive measures, such as climate-sensitive disease modelling and early warning systems, into existing health frameworks.
3. Emphasis should be placed on engaging local communities in solving public health issues, particularly in rural and remote areas, to strengthen data collection and response mechanisms.
4. Provide targeted training programmes for public health professionals to equip them with the skills needed to address climate-related health challenges. This includes expertise in environmental health, data analytics, and disaster preparedness.
5. Encouraging collaborations among public health researchers, climate scientists, and policymakers to better understand and address the complex interactions between climate change and public health.

By implementing these recommendations, Southeast Asian nations can build resilient public health surveillance systems that can mitigate the impacts of climate change and safeguarding the health of their populations.

14.4 Conclusion

Climate change poses significant challenges to public health surveillance systems across Southeast Asia, increasing existing vulnerabilities and introducing new risks. Increased incidences of vector-borne diseases, extreme weather events, and climate-induced displacement are placing exceptional pressure on these systems. The capacity of surveillance systems to detect, monitor, and respond to public health threats is often constrained by limited resources, fragmented data systems, and insufficient regional collaboration. This study highlights the importance of integrating climate change considerations into public health strategies to ensure the resilience and effectiveness of surveillance systems. Collaborative research, robust infrastructure, and advanced technologies are essential to addressing the intersection of climate change and public health in the region.

Disclosure Statement The author(s) declare that there are no conflicts of interest related to this research. This research has been conducted in compliance with ethical guidelines, and no personal or professional relationships have influenced the outcomes or interpretations presented in this work. For transparency, the author(s) affirm their commitment to academic integrity and objectivity in the pursuit of this research.

References

Al Saleem N, Househ M, El Metwally A (2014) Challenges in building health surveillance systems in Saudi Arabia. Stud Health Technol Inf

Aleixandre-Benavent R, González De Dios J, Castelló Cogollos L, Navarro Molina C, Alonso-Arroyo A, Vidal-Lnfer A, Lucas-Domínguez R (2017) Bibliometrics and indicators of scientific activity (1). The evaluation of research and scientific activity in pediatrics through bibliometrics. Acta Pediatr Espanola 75(1–2):18–25. https://www.scopus.com/inward/record.uri?eid=2-s2.0-85018553703&partnerID=40&md5=3102d2e7d8723935a584ba0679ac20a3

Alhassan JAK, Wills O (2024) Public health surveillance through community health workers: a scoping review of evidence from 25 low-income and middle-income countries. BMJ Open 14(4):Article e079776. https://doi.org/10.1136/bmjopen-2023-079776

Amul GG, Ang M, Kraybill D, Ong SE, Yoong J (2022) Responses to COVID-19 in Southeast Asia: diverse paths and ongoing challenges. Asian Econ Policy Rev 17(1):90–110. https://doi.org/10.1111/aepr.12362

Angkurawaranon C, Jiraporncharoen W, Chenthanakij B, Doyle P, Nitsch D (2014) Urbanization and non-communicable disease in Southeast Asia: a review of current evidence. Public Health 128(10):886–895. https://doi.org/10.1016/j.puhe.2014.08.003

Asala AF (2023) Surveillance: the role of observation in epidemiological studies. In: Statistical approaches for epidemiology: from concept to application, pp 129–145. https://doi.org/10.1007/978-3-031-41784-9_8

Azeroual O, Nikiforova A, Sha K (2023) Overlooked aspects of data governance: workflow framework for enterprise data deduplication. In: 2023 International conference on intelligent computing. Communication, Networking and Services, ICCNS, 2023

Bhutta ZA, Gaffey MF, Crump JA, Steele D, Breiman RF, Mintz ED, Black RE, Luby SP, Levine MM (2018) Typhoid fever: way forward. Am J Trop Med Hyg 99(3):89–96. https://doi.org/10.4269/ajtmh.18-0111

Bozbıyık M, Uysal HH (2022) A closer look at the doctoral writing practices in an English-medium instruction University in Turkey. In: Multilingual education, vol 40, pp 175–196. https://doi.org/10.1007/978-3-030-88597-7_9
Brennan E (2021) Climate change and security threats in Southeast Asia. In: Handbook of security and the environment, pp 132–148. https://doi.org/10.4337/9781789900668.00018
Davidson G, Chua TH, Cook A, Speldewinde P, Weinstein P (2019) The role of ecological linkage mechanisms in plasmodium knowlesi transmission and spread. EcoHealth 16(4):594–610. https://doi.org/10.1007/s10393-019-01395-6
Dos Santos SM, Fraumann G, Belli S, Mugnaini R (2021) The relationship between the language of scientific publication and its impact in the field of public and collective health. J Sci Res 10(1):S78–S87. https://doi.org/10.5530/JSCIRES.10.1S.24
Dundar Y, Dodd S, Williamson P, Dickson R, Walley T (2006) Case study of the comparison of data from conference abstracts and full-text articles in health technology assessment of rapidly evolving technologies: does it make a difference. Int J Technol Assess Health Care 22(3):288–294. https://doi.org/10.1017/S0266462306051166
Fryer D, Qin M, Sun J, Lee KW, Brown AD, Goel A (2014) Checking the integrity of transactional mechanisms. In: Proceedings of the 12th USENIX conference on file and storage technologies, FAST 2014
Garzon JL, Ferreira O, Plomaritis TA, Zózimo AC, Fortes CJEM, Pinheiro LV (2024) Development of a Bayesian network-based early warning system for storm-driven coastal erosion. Coast Eng 189:Article 104460. https://doi.org/10.1016/j.coastaleng.2024.104460
Gin OK (2019) Bonds that bind: shared historical and sociocultural characteristics of Southeast Asia. Suvannabhumi: Multidiscip J Southeast Asian Stud 11(1):71–100. https://www.scopus.com/inward/record.uri?eid=2-s2.0-85169083576&partnerID=40&md5=f4b61b165b171914ce7ec8ae023d6a04
Gorodzha LV (2018) DOI—solution of the problem of determination placement of the electronic publication in the internet. Tech Electrodyn 2018(2):95–96. https://www.scopus.com/inward/record.uri?eid=2-s2.0-85042474212&partnerID=40&md5=59a479fa9d474312f943cd61c7e20393
Han KT, Wai KT, Aye KH, Kyaw KW, Maung WP, Oo T (2019) Emerging neglected helminthiasis and determinants of multiple helminth infections in flood-prone township in Myanmar. Trop Med Health 47(1):Article 1. https://doi.org/10.1186/s41182-018-0133-6
Hermanowicz JC, Lei MK (2023) Organizational location and propensity for coauthorship in sociology. Stud High Educ 48(4):630–642. https://doi.org/10.1080/03075079.2022.2159362
Hottenrott H, Lawson C (2022) What is behind multiple institutional affiliations in academia. Sci Public Policy 49(3):382–402. https://doi.org/10.1093/scipol/scab086
Huynh NT, Lin W, Ness LR, Occeña-Gutierrez D, Trân XD (2014) Climate change and its impact on cultural shifts in east and Southeast Asia. In: A world after climate change and culture-shift, pp 245–302. https://doi.org/10.1007/978-94-007-7353-0_12
Koldunova E, Yefimova L (2017) Political systems and political culture of the southeast Asian states. In: Is non-western democracy possible? A Russian perspective, pp 417–462. https://doi.org/10.1142/9789813147386_0012
Koottatep T, Chapagain SK, Polprasert C, Panuvatvanich A, Ahn KH (2018) Sanitation situations in selected southeast Asian countries and application of innovative technologies. Environ Dev Sustain 20(1):495–506. https://doi.org/10.1007/s10668-016-9892-6
Lawson ET, Alare RS, Salifu ARZ, Thompson-Hall M (2020) Dealing with climate change in semi-arid Ghana: understanding intersectional perceptions and adaptation strategies of women farmers. GeoJournal 85(2):439–452. https://doi.org/10.1007/s10708-019-09974-4
Li J, Wei R (2022) VOSviewer application status and its Knowledge Base. J Libr Inf Sci Agric 34(6):61–71. https://doi.org/10.13998/j.cnki.issn1002-1248.21-0843
Li L, Jin J, He R, Kuang F, Zhang C, Qiu X (2023) Effects of social capital on farmers' choices of climate change adaptation behavior in Dazu District, China. Clim Dev 15(2):110–121. https://doi.org/10.1080/17565529.2022.2061403

Li Z, Deng X, Zhang Y (2021) Evaluation and convergence analysis of socio-economic vulnerability to natural hazards of belt and road initiative countries. J Clean Prod 282:Article 125406. https://doi.org/10.1016/j.jclepro.2020.125406

Liu J, Ge H (2022) Collaboration mechanisms and community detection of statisticians based on ERGMs and kNN-walktrap. Comput Stat Data Anal 168:Article 107372. https://doi.org/10.1016/j.csda.2021.107372

Luchini C, Veronese N, Nottegar A, Shin JI, Gentile G, Granziol U, Soysal P, Alexinschi O, Smith L, Solmi M (2021) Assessing the quality of studies in meta-research: review/guidelines on the most important quality assessment tools [review]. Pharm Stat 20(1):185–195. https://doi.org/10.1002/pst.2068

MacKenzie JS, Williams DT (2009) The zoonotic flaviviruses of southern, south-eastern and eastern Asia, and Australasia: the potential for emergent viruses. Zoonoses and Public Health

Madden JC, Kandarova H, Neuhaus W, Osborne N, Paini A, Seabra R, Trigwell S (2024) Journeying through journals: the publishing process and how to maximise research impact. Altern Lab Anim 52(6):334–351. https://doi.org/10.1177/02611929241286624

Masenyama A, Mutanga O, Dube T, Bangira T, Sibanda M, Mabhaudhi T (2022) A systematic review on the use of remote sensing technologies in quantifying grasslands ecosystem services. GIScience Remote Sens 59:1000–1025

Moresi EAD, Pinho I (2021) How to identify the emerging topics of a research topic. New Trends Qual Res 9:46–55. https://doi.org/10.36367/ntqr.9.2021.46-55

Nguyen V, Nguyen-Viet H, Pham-Duc P, Wiese M (2014) Scenario planning for community development in Vietnam: a new tool for integrated health approaches? Glob Health Action 7(1):Article 24482. https://doi.org/10.3402/gha.v7.24482

Picalho AC, Lucas ERO, Amorim IS (2022) Boolean logic applied to the construction of search expressions. AtoZ 11. https://doi.org/10.5380/atoz.v11i0.81838

Qi M, Tan S, Deng H, Wu J (2019) Evaluating link significance in maintaining network connectivity based on link prediction. Chaos 29(8):Article 083120. https://doi.org/10.1063/1.5091608

Ramezani A, Shahmoradi L, Azadeh F, Sheikhshoaei F, Atlasi R, Namazi N, Larijani B (2021) Visualization of knowledge flow in interpersonal scientific collaboration network endocrinology and metabolism research institute. Iran J Diabetes Metab 20(1):71–82. https://www.scopus.com/inward/record.uri?eid=2-s2.0-85147877660&partnerID=40&md5=cbaadfa2d310cbfaf1543ee7082ac79c

Schmidt T, Frishman D (2006) PROMPT: a protein mapping and comparison tool. BMC Bioinform 7:Article 331. https://doi.org/10.1186/1471-2105-7-331

Schwabish J (2023) Data visualization in excel: a guide for beginners, intermediates, and wonks. https://doi.org/10.1201/9781003321552

Sentian J, Payus CM, Herman F, Kong VWY (2022) Climate change scenarios over Southeast Asia. APN Sci Bull 12(1):102–122. https://doi.org/10.30852/sb.2022.1927

Sgarbossa N, Cobaisse MI, Cianciulli GG, Bracchiglione J, Franco JVA (2022) Systematic reviews: Key concepts for health professionals. Medwave 22(9):Article e2622. https://doi.org/10.5867/medwave.2022.09.2622

Singh SK, Rafiei D (2018) Strategies for geographical scoping and improving a gazetteer. In: The web conference 2018—proceedings of the world wide web conference, WWW 2018

Surekha S, Sindhu S, Veerappan S, Arvinth N (2024) Bibliometric study: natural and engineering sciences. Nat Eng Sci 9(2):376–385. https://doi.org/10.28978/nesciences.1574466

Thomson M, Lyon B, Ceccato P (2019) Climate matters in health decision-making. In: The Palgrave handbook of global health data methods for policy and practice, pp 263–281. https://doi.org/10.1057/978-1-137-54984-6_14

Trewin B, Cazenave A, Howell S, Huss M, Isensee K, Palmer MD, Tarasova O, Vermeulen A (2021) Headline indicators for global climate monitoring. Bull Am Meteorol Soc 102(1):E20–E37. https://doi.org/10.1175/BAMS-D-19-0196.1

Valente A, Holanda M, Mariano AM, Furuta R, Da Silva D (2022) Analysis of academic databases for literature review in the computer science education field. In: Proceedings—Frontiers in education conference, FIE

van Eck NJ, Waltman L (2014) Visualizing bibliometric networks. In: Ding Y, Rousseau R, Wolfram D (eds) Measuring scholarly impact: methods and practice. Springer International Publishing, pp 285–320. https://doi.org/10.1007/978-3-319-10377-8_13

Yakhontova T (2021) What nonnative authors should know when writing research articles in English. J Korean Med Sci 36(35):1–7. https://doi.org/10.3346/JKMS.2021.36.E237

Ye Q, Wu B, Yang X, Pei X (2007) Visualization the backbone of social networks by given vertices and link reduction. IET Conference Publications

Zhang L, Sun B, Jiang L, Huang Y (2021) On the relationship between interdisciplinarity and impact: distinct effects on academic and broader impact. Res Eval 30(3):256–268. https://doi.org/10.1093/reseval/rvab007

Chapter 15
Public Governance in the Face of Health Crisis and Climate Change: Impact on Socio-Ecological Resilience in Two South Asian Countries

Johanna R. Ñaupari

Abstract Public management is transcendental to face the crises that a nation is going through; the ability of governments to anticipate and coordinate effective responses based on the national reality, guaranteeing the distribution of resources and essential services, is decisive to face disturbances. This capacity, translated as resilience, is the ability to manage unexpected changes or disturbances, in a socio-ecological approach that understands the interaction of society and nature as a synergistic whole. This research aims to analyze how public management is dealing with the climate and health crisis in some South Asian countries (Nepal and Pakistan) and how it impacts socio-ecological resilience; this analysis will be done using the seven principles of resilience proposed by researchers from the Stockholm Resilience Centre, investigating how government decisions increased the state of vulnerability given its importance in transforming adversity into an opportunity for reconstruction, strengthening social ties and contributing to a sustainable future. Finally, it is concluded that adaptive policies, investments in resilient infrastructure, and social support mechanisms would make the difference between extreme vulnerability and the ability of communities to overcome the impacts of these global crises.

Keywords Resilience · Climate crisis · Climate change · Vulnerability · Impacts

15.1 Introduction

Climate change and health crises represent two of the greatest contemporary global challenges, with profound interrelated effects on social and ecological systems. These dual crises not only strain public health infrastructures but also destabilize economic and environmental conditions, particularly in regions already marked by systemic

J. R. Ñaupari (✉)
Climate Change and Sustainable Development, San Antonio Abad del Cusco University, Cusco, Peru
e-mail: 232372@unsaac.edu.pe

P. Ghosh et al. (eds.), *Climate Crises and Public Health Challenges*, Disaster Risk Reduction, https://doi.org/10.1007/978-981-95-9709-3_15

vulnerabilities. South Asia is one such region, characterized by recurrent extreme climatic events such as floods, heatwaves, and cyclones, which exacerbate existing fragilities in public health and local economies. As these environmental impacts increase, governments' capacities to respond effectively become critically important in building socio-ecological resilience.

While climate adaptation and public health preparedness have been addressed independently in an expanding body of the literature, a significant gap persists in understanding how these domains intersect in high-risk regions. This study seeks to address that gap by examining the convergence of climate and health governance in two South Asian countries—Nepal and Pakistan—selected for their shared vulnerability yet contrasting institutional capacities. In particular, it investigates how adaptive public management strategies, including policy innovation and emergency response mechanisms, shape national capacities to respond to complex socio-ecological risks. What distinguishes this study from previous work is its integrated and comparative approach. Rather than analyzing climate or health governance in isolation, it focuses on their intersection, situating national strategies within a broader regional context. This allows for the identification of shared challenges, as well as institutional and policy divergences that influence varying levels of resilience. The analytical framework draws on recent high-impact studies such as Wild et al. (2025), who examine the integration of climate and health policies; Arvin et al. (2023), who explore institutional resilience in South Asia; and Brett et al. (2024), who analyze governance under complex and cascading risks. These works provide the conceptual foundation for this study by highlighting the importance of cross-sectoral coordination and adaptive governance in building sustainable and equitable resilience pathways.

By engaging with these theoretical perspectives and applying them to empirically grounded case studies, this article contributes to a more nuanced understanding of how governance systems can be leveraged to enhance resilience in highly vulnerable socio-ecological contexts. In doing so, it offers evidence-based insights to inform coherent and integrated policy frameworks in regions confronting the combined effects of climate and health emergencies.

15.2 Methodology

This research was conducted using a qualitative approach, focused on the analysis of public policies, governmental decisions, and their impacts on the adaptive capacity of communities in the face of unforeseen changes. This approach enables an in-depth understanding of the institutional and sociopolitical processes that shape responses to climate and health crises in contexts of high vulnerability. Two countries in South Asia were chosen: Nepal and Pakistan, due to their high vulnerability to climate and health crises, as well as their dependence on the government's capacity to anticipate and manage these crises. The selection of these countries allows for a comparative and regional analysis, considering contextual differences in terms of infrastructure, resources, and health systems (Sterk et al. 2017).

Data collection was carried out through a systematic documentary review of official reports, regulatory frameworks, strategic plans, and technical documents published between 2015 and 2024 by governmental entities, multilateral organizations, and non-governmental organizations. A thematic coding matrix, based on categories derived from the theoretical framework (socio-ecological resilience), was employed as the main analytical tool, allowing for a transversal and comparative reading of the documentary content. The analytical process was structured into three stages: (a) identification and selection of relevant documents; (b) thematic categorization based on resilience principles; and (c) comparative analysis between the two countries.

From an ethical standpoint, although the study did not involve fieldwork or the collection of personal data, principles of transparency and traceability in source usage were strictly observed, ensuring the responsible and well-documented use of public information.

The theoretical framework guiding this research is based on the seven principles of resilience proposed by researchers at the Stockholm Resilience Centre (Stockholm Resilience Centre, n.d.) which provide a lens to analyze the capacity of socio-ecological systems to absorb disturbances, adapt to change, and maintain their functionality. Particular attention is given to elements such as actor diversity, institutional learning, multi-level governance, and the interdependence between social and ecological systems (Table 15.1).

15.3 Results

The relationship between climate change and health crises from a governmental perspective lies in the need for governments to implement preventive and adaptation policies to mitigate the effects of climate change on public health. This includes strengthening health systems, managing the risks of natural disasters, and ensuring climate justice, focusing on protecting the most vulnerable communities.

Using the seven Stockholm principles of resilience allows us to identify and analyze how shocks such as health and climate crises disrupt the resilience (Stockholm Resilience Centre, n.d.) of Nepal and Pakistan, revealing the structural and operational weaknesses that limit their ability to face and recover from these challenges as follows.

Table 15.1 Methodology description (constructed by the author)

Component	Description
Research approach	Qualitative research focused on the analysis of public policies and governmental decisions. Objective: to understand their impact on communities' adaptive capacity in the face of crises
Case selection	Countries: Nepal and Pakistan. Selection criteria: high vulnerability and critical dependence on governmental action. Sampling method: theoretical and purposive
Document sample	Approx. 18 national policy documents. Selection criteria: relevance, applicability, and direct connection to climate and health crisis management
Data collection	Systematic document review. Sources: official reports, legal frameworks, strategic plans (2015–2024). Issued by governments, multilateral agencies, and NGOs
Analytical tool	Thematic categorization matrix, developed based on the socio-ecological resilience theoretical framework
Phases of analysis	a. Identification and selection of relevant documents; b. Thematic categorization based on resilience principles; c. Comparative analysis between Nepal and Pakistan
Ethical considerations	No fieldwork or personal data collection involved. Ensured transparency and traceability in the use of public information sources
Theoretical framework	Based on the seven principles of resilience (Stockholm Resilience Centre). Focus on actor diversity, institutional learning, multilevel governance, and social-ecological interdependence

15.3.1 Nepal

Principle 1. Maintain Diversity and Redundancy of Species, Landscape Types, Actors, and Institutions

Nepalese face constant vulnerability to natural disasters, as evidenced by the recent 6.4 magnitude earthquake that killed 128 people, and the nearly 5,700 disasters that have occurred in the last year, from earthquakes to landslides and floods. These events reflect how climate change exacerbates natural phenomena, increasing their frequency and intensity (United In The Face Of Disaster: Nepal Navigates Climate Risks 2023). In this context, this principle that promotes maintaining diversity and redundancy is fundamental in the diversification of infrastructure, resources, and response strategies that can help communities recover faster and mitigate the impact of future disasters.

Extending beyond the immediate physical impacts, these disasters expose critical vulnerabilities within Nepal's health infrastructure, that combines public and private services under a decentralized model faces challenges in effective implementation due to the lack of institutional and policy reforms at the provincial and federal levels. Although there are various levels of care, from basic services managed by local governments to specialized care at the federal level, the lack of structural reform according to constitutional mandates has limited its effectiveness (Khatri et al. 2023). In this context, this principle that advocates maintaining the diversity and redundancy of actors and institutions is crucial. Diversity in the management of health services, both at the public and private levels, can offer redundancy in care, which strengthens the system in the face of crises and improves the capacity to respond to diverse health needs in the country.

Principle 2. Manage Connectivity of Resources, Species, and People

Nepal's economy faces significant challenges due to its vulnerability to climate change, which affects key sectors such as agriculture, tourism, and infrastructure. These climate impacts not only threaten the well-being of communities, but also hinder sustainable economic growth (World Bank 2024). In this context, this principle becomes relevant as it involves cooperation and synergy between the government, international organizations, and local communities. By strengthening collaborative networks and sharing resources and knowledge, their capacity to adapt to the effects of climate change can be improved, promoting more inclusive and sustainable economic development. Nepal is highly vulnerable to natural disasters, such as the earthquakes of 2015 and 2022, landslides, and environmental degradation, all of which pose significant threats to its infrastructure and ecosystem (BTI Nepal Country Report 2024).

Consequently, the intersection between environmental degradation and public health becomes increasingly evident; disruptions in air and land transport, along with problems in supply chains, significantly aggravate health crises, especially in rural or underserved areas. These disruptions hamper access to health services, amplifying morbidity and mortality as medical resources cannot reach the most vulnerable communities in time. The reduction in access to health services due to these crises further exacerbates the situation, putting the lives of many people at risk in an already fragile context (Ebi et al. 2020).

Principle 3. Manage Slow Variables and Feedbacks

Water resource management is influenced by slow variables and feedbacks related to climate change, especially due to its dependence on transboundary rivers, such as the Ganges and Brahmaputra, which they share with China and India. These rivers are critical for irrigation, drinking water, and hydroelectric power generation in Nepal, but any alteration in their flow, driven by climate change, dam construction, or deforestation, can have a profound impact on agriculture and water security (Karki 2024). Differences in water management and competition for these resources between countries can lead to disputes and conflicts, exacerbated by variations in flow that occur gradually but significantly over time.

Correspondingly, the health sector is also shaped by slow feedbacks that accumulate over time and affect long-term outcomes, such as the ongoing burden of communicable, maternal, neonatal and nutritional diseases and the increasing prevalence of non-communicable diseases, are influenced by slow feedbacks, which reflect social, economic, and environmental factors that affect public health in the long term. Despite progress in health indicators, slow feedbacks, such as deficiencies in health infrastructure, nutrition, and access to health care, exacerbate the situation. High mortality rates from these diseases (71.1% of all deaths) indicate that improvements in health are not evenly or rapidly distributed, requiring a long-term approach to address these inequalities and improve health systems to reduce the burden of these diseases (Dumka et al. 2024).

Principle 4. Promote Complex Adaptive Systems Thinking

These farmers face multiple challenges, such as limited access to modern agricultural technology, poor infrastructure, and adverse weather conditions, which exacerbate their food insecurity. Their high dependence on rain-fed agriculture and lack of adaptation to climate change reinforce a cycle of poverty and low yields (Kandel et al. 2024). An adaptive systemic approach would make it possible to understand and address the interlinkages between these factors, promoting integrated and adaptive solutions that improve farmers' resilience to climate change and food insecurity.

A similar systems-based perspective is needed in the health sector, where structural and contextual complexities also demand adaptive responses, where known gaps and barriers, such as geography, affect service delivery. However, there is a lack of systematic analysis of how internal and external factors influence outcomes (Adhikari et al. 2022). The complexity of the system requires a flexible and continuous approach that allows for identifying and adjusting the factors that affect its functioning and overcoming challenges in comparing evidence.

Principle 5. Stimulate Learning

In Nepal, around 770,000 children between the ages of 5 and 12 are still not in school, and only half of students in grades 3, 5, and 8 reach the expected performance levels in Nepali and mathematics. In addition, attendance in early childhood education remains low, with only 51% coverage (United Nations International Children's Emergency Fund [UNICEF] 2017). These educational challenges are exacerbated by the country's vulnerability to climate change, which disrupts access to education due to natural disasters, forced displacement, and lack of adequate infrastructure, affecting the academic and social future of new generations.

In this context, community-based educational initiatives represent a valuable opportunity to bridge existing gaps and strengthen local capacities. Community learning centers can be key to rural transformation, offering educational programs that foster local development. To be successful, it is crucial that these centers receive the necessary support, including resources, a solid legal base, political backing, and active civic participation (Sharma 2015). If these requirements are met, learning centers can contribute to improving the living conditions of communities, allowing people to develop their capacities, make informed decisions, and participate fully in development, while continuing to learn and work with dignity. Nepal continues to face considerable structural challenges, including a low degree of economic and social development, limited educational attainment among its population, geographic disadvantages associated with its landlocked status, and a nascent state apparatus that remains in the process of consolidation (BTI Nepal Country Report 2024).

Principle 6. Broaden Participation

The principle of widening participation is key to integrating climate change mitigation into Nepal's socio-economic development, especially considering its status as a least developed country. Although education is recognized as a requirement for the implementation of mitigation measures, it is rarely considered as a tool to improve decision-making by authorities. Rather than being seen as a tool for mitigation, education is more associated with adaptation, risk management, and health threats (Havukainen et al. 2022). To address this gap, it is crucial to widen participation and involve communities, businesses, and authorities in the educational process, ensuring that climate change mitigation is more effectively integrated into policies and programs, not only at the grassroots level, but also at higher decision-making levels.

However, translating the principle of broad participation into practice remains a complex challenge, especially in contexts of crisis response and institutional fragmentation; achieving broad participation is difficult in contexts such as the response to the Ebola epidemic and the Nepal earthquake, where diverse organizations were involved, from small volunteer groups to large international non-governmental organizations and government responses. The variability in the size, infrastructure, and resources of these organizations complicates effective coordination and inclusive participation in planning and implementation processes. Despite the diversity of actors involved, the limited administrative or logistical infrastructure of some groups and differences in their capacities make it difficult to integrate all actors into a coordinated and equitable response (Penta 2023).

Principle 7. Promote Polycentric Governance Systems

Promoting polycentric governance in Nepal faces challenges due to the complex combination of sociopolitical volatility and biophysical impacts of climate change, which increase the country's vulnerability to disasters. The overlapping of multiple governance paradigms, such as response and recovery, disaster risk management, climate change adaptation, and federalized risk reduction, hinders effective coordination between different levels of government and stakeholders (Vij 2023). Implementing a polycentric approach requires overcoming these difficulties, ensuring smoother and more coherent collaboration between different actors, which would allow for more efficient and adaptive disaster management.

This difficulty in achieving effective multi-level coordination is also reflected in Nepal's decentralized health sector, where federal reforms have introduced parallel governance challenges; under-federalism has faced multiple challenges, such as weaknesses in infrastructure, shortage of qualified personnel, and lack of coordination between different levels of governance. In addition, absenteeism of health workers and confusion over roles and responsibilities have complicated the effective implementation of the new model (Wasti et al. 2023). In this context, promoting polycentric governance systems becomes relevant, as this approach could improve coordination, accountability, and clear allocation of responsibilities, allowing for a more efficient and equitable response in the health system. However, there have been no systematic studies exploring the perspectives of key actors on the opportunities and challenges arising from federalization of the health system. Nepal's political environment has been marked by persistent instability, with key political parties frequently unable to reach consensus on critical issues, and when agreements are made, they often require prolonged periods of negotiation. The bureaucratic system, essential for the effective implementation of governance, is plagued by significant deficiencies, including a lack of professionalism, entrenched bureaucracy, and widespread nepotism and favoritism (Shrestha 2021).

15.3.2 Pakistan

Principle 1. Maintain Diversity and Redundancy of Species, Landscape Types, Actors, and Institutions

In terms of the response to climate change, this country has opted to reduce fossil fuels, turning its efforts in recent decades to nuclear energy (Iftikhar and Abid 2023).

Pakistan has more than 50 years of experience in nuclear power generation, and the IAEA has praised its safety record and the quality of human resources in this sector (IAEA 2021).

Currently, the country operates six nuclear plants that generate approximately 3,500 MW of electricity with third-generation plants, equipped with advanced multilayer safety systems (IAEA 2022) seeking to promote affordability, focus on the consumer, reduce dependence on imported fuels, and encourage economic growth and public–private collaboration in the energy sector. Its extensive experience in nuclear generation, together with favorable policies, creates an enabling environment for new investments in nuclear plants.

Simultaneously, the health sector has emerged as a strategic priority, particularly in strengthening emergency response capacity during times of crisis, the response to the shortage of medical products and equipment sought to create a health system that would respond more quickly to the emergency that was being experienced through investment in equipment and the improvement of hospital capacities, thus reflecting an effort for a rapid and efficient response attitude in the health system, ensuring that there are sufficient resources and options to face future crises (Tavakkoli et al. 2022). Enhancing Pakistan's long-term growth potential necessitates far-reaching governance reforms, particularly in the areas of fiscal capacity, institutional integrity, regulatory efficiency, and accountability mechanisms (Muhula 2024).

Principle 2. Manage Connectivity of Resources, Species, and People

Poverty and lack of economic opportunities in farming communities increase pressure on resources, as in the case study of Qurumber National Park in Gilgit, depleting their ecosystem services and making it difficult to conserve their resources (Ali et al. 2023).

This pattern of structural strain is also evident in the healthcare sector, where institutional capacity was severely tested during the COVID-19 crisis, government policies failed due to public indifference, and the easing of lockdown worsened the situation, causing stress in health workers, overload in laboratories, and lack of space in intensive care units, while costs of care increased and collapsed (Khalid and Ali 2020). Furthermore, the COVID-19 pandemic served as a catalyst to accelerate the adoption of e-services and advance digital transformation efforts within Pakistan's public sector (Muhammad et al. 2024).

Principle 3. Manage Slow Variables and Feedbacks

Social inequality, war conflicts, and development challenges increase vulnerability to climate change (Martine 2022). A serious problem is the poor facilities for irrigating crops, since around 70% of the planted area is irrigated and where the average annual drought rate is 40%. Added to this is migration due to lack of opportunities and the conflict that exists (Ali et al. 2021).

These intersecting challenges not only affect rural livelihoods and agricultural sustainability, but also drive population displacement and broader socio-economic disruptions; due to war conflicts in countries such as Pakistan, people have been displaced to other countries or safer places, leaving behind an abandonment of activities that allowed them to have economic income from various formal activities or gave them the economic possibility of accessing a decent health system (Martine 2022). Despite notable progress, Pakistan continues to face major challenges, including inadequate infrastructure, limited financial resources, weak institutional capacity, and restricted access to advanced technologies (Muhammad et al. 2024).

Principle 4. Promote Complex Adaptive Systems Thinking

Pakistan faces water scarcity, and although water and climate change policies exist, their implementation is constrained by financial, political, and institutional challenges. The Indus River Basin water system faces problems of water storage, transportation, and distribution, and integrated approaches are not implemented due to conflicts between national and provincial mandates (Ali et al. 2021).

This lack of integrated coordination is not limited to the environmental sector, but was also evident in the initial response to the COVID-19 pandemic, financing measures were introduced to the emergency, and emergency mechanisms to manage the acquisition of medicines and health technologies began as soon as the first case of contagion was reported. The response of medical staff to this situation was to reorient themselves in a little over a month, however, multisectoral coordination did take a while to function efficiently (Tavakkoli et al. 2022).

Principle 5. Stimulate Learning

The knowledge gap on climate change in developing countries is still large, which is why it is important to improve awareness about global warming and solutions such as climate engineering, obtaining feedback to adapt and improve strategies. Differences in the perception of climate change show the need to promote education and access to information on climate risks, thus strengthening resilience and local participation in the face of climate change (Hussain et al. 2023).

This gap in knowledge and perception is not limited to environmental issues, but has also been evident in the public response to health crises such as the COVID-19 pandemic; in a study conducted to understand the attitude of the population toward the vaccine, most participants expressed interest in knowing the safety and efficacy of the vaccine before receiving it, as they feared being used as test subjects or suffering adverse effects related to the vaccine. Highlighting the need to provide sufficient knowledge and awareness through evidence-based studies, some participants believed that measures such as handwashing, wearing masks, and social distancing would be enough to protect against virus. In addition, there was a predominant anti-vaccine attitude, as many thought that the vaccine would not be useful, considering that virus is similar to the flu, a rapidly changing virus. Therefore, they claimed that the vaccine would not reduce the spread or severity of the infection. These beliefs reflect insufficient knowledge and the spread of myths, exacerbated by the changing statements made by the World Health Organization (WHO) in response to the evolving COVID-19 situation. All of this underscores the urgent need to provide robust evidence-based education to the public so that people can more easily accept the acceptance of the vaccine (Z. Khan et al. 2021).

Principle 6. Broaden Participation

Climate change has intensified care work for women, affecting their health and well-being, especially in a context of gender inequality. Women must travel further, work more, and take on greater care responsibilities, a task generally considered feminine, which also falls on girls, limiting their opportunities. Despite these challenges, women are beginning to organize and raise their voices on key issues related to climate change (Butt et al. 2020).

Addressing such complex and intersectional challenges requires robust national research frameworks capable of informing inclusive and equitable policy responses; the creation of the Pakistan Medical Research Council (PMRC) and its focus on expanding national and international engagement reflects its effort to strengthen medical research in Pakistan, establishing key collaborations for the advancement of science and its linkage to national development (Rasheed et al. 2021).

Principle 7. Promote Polycentric Governance Systems

At this time, there is no adequate way to effectively manage the multiple problems facing the country, such as climate change and economic crises (Ali et al. 2021), and to ensure the sustainable development of the agricultural sector, science-based strategies adapted to local contexts are crucial to solve issues such as water scarcity and how to improve technology that allows for sustainable forms of irrigation for cotton crops that improve the economy of farmers.

This broader institutional fragility not only affects environmental and agricultural management, but also undermines the country's capacity to respond to public health emergencies, highlighting the urgent need for more coherent and inclusive governance. Given the major challenges, this country faced in responding to COVID-19, polycentric governance could enhance coordination among different levels of government (national, provincial and local) and other key actors, such as the private sector and international organizations.

Pakistan, with a large population and limited financial capacity compared to countries such as China or the United States, needs to improve its health facilities and infrastructure. During the COVID-19 pandemic crisis, the number of hospitals was not sufficient. Cooperation among local, provincial, and national authorities would be key to implement adequate measures and ensure the availability of sufficient resources to control the transmission of the virus (Waris et al. 2020). Governance structure is marked by instability and weakness, with the country lagging behind its regional peers in key institutional and governance indicators. This underscores the necessity for a comprehensive evaluation and reform, particularly in areas such as accountability, transparency, and the provision of timely justice (Khan 2022).

Local governments have enacted an average of seventy laws, enhancing development and service delivery. Although their budgets are smaller compared to higher levels of government, they have implemented a participatory legislative model. Challenges such as limited resources and experience have impeded the process, yet they have succeeded in creating essential and specific laws. This demonstrates gradual progress in legislative capacity and local self-governance, in line with the 2015 Constitution, providing a solid foundation for future legislative endeavors (Bhusal 2023).

15.4 Discussion

Grounded in the theoretical framework of socio-ecological resilience, this study investigates the role of public governance in addressing the compounded effects of climate change and health crises in Nepal and Pakistan. Through a qualitative, document-based analysis of 18 official policies issued between 2015 and 2024, and guided by the seven resilience principles of the Stockholm Resilience Centre, the research identifies critical institutional and infrastructural vulnerabilities. Findings emphasize the relevance of governance diversity, policy integration, and inclusive participation as key components of resilience. The study concludes that adaptive, cross-sectoral strategies are vital to enhancing resilience in both environmental and health systems. Compared to indicator-based or quantitative assessments, this method enabled a deeper understanding of how resilience is framed and operationalized in national strategies. Similar studies in Bangladesh and Sri Lanka have also revealed implementation gaps within strong legal frameworks, while the inclusion of health governance in this analysis adds a distinct, cross-sectoral contribution.

One key finding of this study is the limited preparedness of health systems in Nepal and Pakistan to address the combined impacts of climate change and health crises, which undermines their socio-ecological resilience. This situation is comparable to that of several African countries, where low institutional capacity to manage climate-related health risks has also been identified (Leal Filho et al. 2021). Similarly, there is a lack of effective integration between mitigation and adaptation policies, as observed in Ghana, where fragmented strategies hinder overall effectiveness (Owusu et al. 2023). In contrast, in the Philippines, previous disaster experience and strong community engagement enhanced emergency preparedness (Bollettino et al. 2020). Moreover, Ñaupari and Ferrer (2024) found that in Peru, public management decisions during COVID-19 further weakened social resilience due to their failure to leverage institutional diversity and learning—echoing the patterns identified here. Recent studies also emphasize the urgent need for governance frameworks prioritizing ecosystemic and human rights-based approaches to address both climate and health crises (McAneney et al. 2016; Kaslow 2022). These contrasts highlight the importance of articulating cross-sectoral policies linking health, environment, and social equity as a foundation for strengthening resilience.

The findings of this study also underscore the need for more integrated and adaptive policymaking to address overlapping vulnerabilities. In Nepal, clarifying institutional roles and strengthening decentralized capacities are crucial to improve crisis response. In Pakistan, enhancing regulatory efficiency and fostering coordination among sectors and governance levels would support more effective risk management. Policymakers in both countries should incorporate resilience principles—particularly diversity, connectivity, and participation—into the design and modification of public policies. Doing so can help bridge current gaps in implementation, ensuring that governance frameworks are not only comprehensive on paper but also operational in practice. These insights are relevant beyond South Asia and may inform policy reform in other developing countries facing similar governance and climate-health challenges.

While the study offers valuable insights, it is also constrained by its reliance on secondary data, specifically the analysis of policy documents, which may not fully reflect informal practices or outcomes at the local level. The lack of fieldwork and stakeholder input restricted the triangulation of results with community realities. However, this also opens avenues for future research to build on these findings through participatory approaches and empirical studies. Despite this limitation, the qualitative method provided a robust foundation for understanding institutional narratives and comparing resilience frameworks across two diverse national contexts. The study's strength lies in its cross-sectoral lens, which contributes novel insights to the international literature on socio-ecological resilience and governance under compound crises.

Beyond its regional relevance, this study contributes to a broader understanding of how governance systems in developing countries manage overlapping climate and health crises. Several findings—particularly those concerning coordination challenges, institutional fragmentation, and the need for inclusive and adaptive frameworks—may be applicable to other countries facing similar structural constraints. By integrating environmental and health governance under the lens of socio-ecological resilience, the study provides a replicable model for analyzing governance capacity in fragile contexts. This reinforces its value not only for South Asia but also for global scholarship on resilience, especially in the Global South, where systemic vulnerabilities require context-sensitive yet transferable approaches.

15.5 Conclusions

Effective public management is key to building resilience to climate and health crises. In Nepal and Pakistan, the government's ability to anticipate risks, coordinate responses, and distribute resources significantly impacts socio-ecological resilience. Using the seven resilience principles as an analytical framework reveals the importance of adaptive governance, infrastructure investment, and inclusive decision-making.

In Nepal, decentralization and sociopolitical instability weaken coordination, particularly in health and climate policies. Fragmented governance has limited the coherence of actions, making polycentric governance and greater stakeholder engagement essential. Deficiencies in infrastructure and resource management underscore the need for integrated and flexible approaches that involve both public and private actors.

On the other hand, Pakistan is showing some progress through diversification and redundancy, investing in health infrastructure, and investing in nuclear energy to ensure stable supply. However, the COVID-19 pandemic and climate change-related challenges highlight persistent coordination challenges across sectors and levels of government. Strengthening polycentric governance and local participation is crucial for an effective response and long-term adaptation.

Both countries emphasize the shared need for systemic, inclusive, and forward-looking strategies that reduce vulnerability and promote resilience. Applying resilience principles supports the development of public policies capable of addressing complex crises and building a sustainable future.

Moreover, the findings from this comparative study may serve as a reference for other countries in the Global South facing similar structural and institutional challenges. By highlighting the value of integrated governance frameworks and cross-sectoral policy alignment, this research contributes to broader discussions on climate-health resilience in fragile contexts. Future studies could expand this approach through participatory and empirical research, further enriching the understanding of how resilience can be operationalized in real-world settings.

15.6 Informed Consent

Informed consent was obtained from all the individual participants included in the study.

15.7 Data Availability Statement

The datasets during and/or analyzed during the current study are available from the corresponding author on reasonable request.

Funding The research was not supported by any funding.

References

Adhikari B, Mishra SR, Schwarz R (2022) Transforming Nepal's primary health care delivery system in global health era: addressing historical and current implementation challenges. Glob Health 18(1). https://doi.org/10.1186/s12992-022-00798-5

Ali U, Jing W, Zhu J, Omarkhanova Z, Fahad S, Nurgazina Z, Khan ZA (2021) Climate change impacts on agriculture sector: a case study of Pakistan. Ciência Rural 51(8). https://doi.org/10.1590/0103-8478cr20200110

Ali A, Ali Shedayi A, Raza H, Hussain E, Hasnain S (2023) Economic valuation of ecosystem services of qurumber national Park in Gilgitbaltistan, Pakistan. Karakoram Int Univ 29(1). https://www.researchgate.net/publication/366976480

Bhusal T (2023) Participatory local governance in rural Nepal: the primacy of informality. Dev Policy Rev 41(6):e12724. https://doi.org/10.1111/dpr.12724

Bollettino V, Alcayna-Stevens T, Sharma M, Dy P, Vinck P (2020) Public perception of climate change and disaster preparedness: evidence from the Philippines. Clim Risk Manag 30. Article 100250. https://doi.org/10.1016/j.crm.2020.100250

BTI Nepal Country Report (2024) BTI 2024. https://bti-project.org/en/reports/country-report/NPL

Butt MN, Shah SK, Yahya FA (2020) Caregivers at the frontline of addressing the climate crisis. Gend Dev 28(3):479–498. https://doi.org/10.1080/13552074.2020.1833482

Dumka N, Gurung A, Hannah E, Goel S, Kotwal A (2024) Understanding key factors for strengthening Nepal's healthcare needs: health systems perspectives. J Glob Health Rep 8. https://doi.org/10.29392/001c.94931

Ebi KL, Harris F, Sioen GB, Wannous C, Anyamba A, Bi P, Boeckmann M, Bowen K, Cissé G, Dasgupta P, Dida GO, Gasparatos A, Gatzweiler F, Javadi F, Kanbara S, Kone B, Maycock B, Morse A, Murakami T, Capon A (2020) Transdisciplinary research priorities for human and planetary health in the context of the 2030 agenda for sustainable development. Int J Environ Res Public Health 17(23):8890. https://doi.org/10.3390/ijerph17238890

Havukainen M, Mikkilä M, Kahiluoto H (2022) Climate policy reform in Nepal through the lenses of the institutional analysis and development framework. Sustainability 14(12):7391. https://doi.org/10.3390/su14127391

Hussain A, Sipra HFK, Waheed A, Ukhurebor KE (2023) Exploring the academic perceptions of climate engineering in developing countries. Atmósfera. https://doi.org/10.20937/atm.53264

Iftikhar M, Abid I (2023) Nuclear energy in the global and Pakistan contexts: trends and perspectives. South Sustain 4(1):e077. https://doi.org/10.21142/ss-0401-2023-e077

International Atomic Energy Agency, IAEA (2021) IAEA projections for nuclear power growth increase for second year amid climate, energy security concerns. https://www.iaea.org/newscenter/pressreleases/iaea-projections-for-nuclear-power-growth-increase-for-second-year-amid-climate-energy-security-concerns

International Atomic Energy Agency, IAEA (2022) IAEA projections for nuclear power growth increase for second year amid climate, energy security concerns. https://www.iaea.org/newscenter/pressreleases/iaea-projections-for-nuclear-power-growth-increase-for-second-year-amid-climate-energy-security-concerns

Kandel GP, Bavorova M, Ullah A, Pradhan P (2024) Food security and sustainability through adaptation to climate change: lessons learned from Nepal. Int J Disaster Risk Reduct 101:104279. https://doi.org/10.1016/j.ijdrr.2024.104279
Karki B (2024) The threats of climate change in Nepal: natural catastrophes and global conflict frontiers. Unity Journal 5(1):141–156. https://doi.org/10.3126/unityj.v5i1.63182
Kaslow F (2022) Climate change, global health, and human rights: governance and accountability frameworks. Glob Environ Polit 22(1):88–103. https://doi.org/10.1162/glep_a_00614
Khalid A, Ali S (2020) COVID-19 and its challenges for the healthcare system in Pakistan. Asian Bioeth Rev 12(4):551–564. https://doi.org/10.1007/s41649-020-00139-x
Khan S (2019) Policy diagnostics for Pakistan's water security challenge. World Water Policy 5(1):8–14. https://doi.org/10.1002/wwp2.12006
Khan Z, Allana R, Afzal I, Ali A, Mariam O, Aslam R, Shah I, Allana A, Haider M, Jandani R, Khan Z, Siddiqui A, Shah J, Butt U (2021) Assessment of attitude and hesitancy toward vaccine against COVID-19 in a Pakistani population: a mix methods survey. Vacunas 23:S26–S32. https://doi.org/10.1016/j.vacun.2021.08.002
Khan F (2022) Good governance: prosperous Pakistan. ISSN-L: 2958-5732. https://ndu.edu.pk/issra/insights.php
Khatri RB, Assefa Y, Durham J (2023) Multidomain and multilevel strategies to improve equity in maternal and newborn health services in Nepal: perspectives of health managers and policymakers. Int J Equity Health 22(1). https://doi.org/10.1186/s12939-023-01905-7
Leal Filho W, Nagy GJ, Bortolini R, de Oliveira LB, Minhas A (2021) Climate change, health, and cities: perspectives from low-income countries in Africa and Asia. Int J Environ Res Public Health 18(6):2907. https://doi.org/10.3390/ijerph18062907
Martine G (2022) A perfect storm of war, pandemic, climate crises and global dis-governance: unfolding implications for population dynamics and policies. Rev Latinoam Población 16:e202210. https://doi.org/10.31406/relap2022.v16.e202210
McAneney J, McAneney D, Musulin R, Walker G, Crompton R (2016) Government-sponsored natural disaster insurance pools: a view from down-under. Int J Disaster Risk Reduct 15:1–9. https://doi.org/10.1016/j.ijdrr.2015.11.004
Muhammad A, Su H, Muhammad M, Naqib U, Ali A (2024) An analysis of E-governance in Pakistan from the lens of the Chinese governance model. Heliyon 10(5):e27003. ISSN 2405-8440. https://doi.org/10.1016/j.heliyon.2024.e27003
Muhula R (2024) Transforming governance for a stronger Pakistan. World Bank Blogs. https://blogs.worldbank.org/en/endpovertyinsouthasia/transforming-governance-stronger-pakistan
Ñaupari JR, Ferrer A (2024) The effect of public management on social resilience in Peru in the face of COVID-19: lessons for the Global South. Int J Sustain Policy Pract 20(2):49–84. https://doi.org/10.18848/2325-1166/CGP/v20i02/49-84
Owusu MF, Adu J, Dortey BA, Gyamfi S, Martin-Yeboah E (2023) Exploring health promotion efforts for non-communicable disease prevention and control in Ghana. PLOS Global Publ Health 3(9):e0002408. https://doi.org/10.1371/journal.pgph.0002408
Penta S (2023) Interdisciplinary collaboration in international crisis medical relief: a look at the Nepal Earthquake and Ebola relief efforts. Disaster Med Public Health Prep 17. https://doi.org/10.1017/dmp.2022.270
Rasheed R, Rizwan A, Javed H, Sharif F, Zaidi A (2021) Socio-economic and environmental impacts of COVID-19 pandemic in Pakistan—an integrated analysis. Environ Sci Pollut Res 28(16):19926–19943. https://doi.org/10.1007/s11356-020-12070-7
Sharma TN (2015) Education for rural transformation: the role of community learning centers in Nepal. J Educ Res 4(2):87–101. https://doi.org/10.3126/jer.v4i2.12391
Shrestha S (2021) Practices of good governance in Nepal. Int J Res-GRANTHAALAYAH 9(6):59–66. https://www.granthaalayahpublication.org/journals/granthaalayah/article/view/5687/5510
Sterk M, Van De Leemput IA, Peeters ET (2017) How to conceptualize and operationalize resilience in socio-ecological systems? Curr Opin Environ Sustain 28:108–113. https://doi.org/10.1016/j.cosust.2017.09.003

Stockholm Resilience Centre (n.d.) Principles for building resilience sustaining ecosystem services in social-ecological systems. Stockholm: University of Stockholm. https://www.stockholmresilience.org/research/research-news/2015-02-19-applying-resilience-thinking.html

Tavakkoli M, Karim A, Fischer FB, Llamas LM, Raoofi A, Zafar S, Fruchtman CS, De Savigny D, Takian A, Antillon M, Muñoz DC (2022) From public health policy to impact for COVID-19: a multi-country case study in Switzerland, Spain, Iran and Pakistan. Int J Public Health 67. https://doi.org/10.3389/ijph.2022.1604969

Ting L, Watts N (2024) Human rights-based approaches to public health emergencies: climate, COVID-19, and the path forward. Health Hum Rights J 26(1):65–78

United In the Face of Disaster: Nepal Navigates Climate Risks (2023) https://unsdg.un.org/latest/stories/united-face-disaster-nepal-navigates-climate-risks

United Nations International Children's Emergency Fund (2017) Educational program. https://www.unicef.org/nepal/education

Vij S (2023) Polycentric disaster governance in a federalising Nepal: interplay between people, bureaucracy and political leadership. Policy Sci 56(4):755–776. https://doi.org/10.1007/s11077-023-09510-9

Waris A, Atta U, Ali M, Asmat A, Baset A (2020) COVID-19 outbreak: current scenario of Pakistan. New Microbes New Infect 35:100681. https://doi.org/10.1016/j.nmni.2020.100681

Wasti SP, Van Teijlingen E, Rushton S, Subedi M, Simkhada P, Balen J, Adhikari SR, Adhikary P, Balen J, Bajracharya B, Bhandari S, Bhattarai S, Gautam S, Karki A, Karki JK, Koirala B, Lee ACK, Marahatta SB, Neupane R (2023) Overcoming the challenges facing Nepal's health system during federalisation: an analysis of health system building blocks. Health Res Policy Sys 21(1):117. https://doi.org/10.1186/s12961-023-01033-2

World Bank (2024)The world bank in Nepal. https://www.worldbank.org/en/country/nepal/overview

Chapter 16
Navigating Sexual and Reproductive Health Challenges of Adolescents and Women: Health System Adaptations against a Rising Sea and Changing Climate in Coastal Regions of Bangladesh

Mahin Al Nahian, Tanvir Ahmad, Mithun Dutta, Prianka Barman, Ashraful Alam Siddique, and Peter Kim Streatfield

Abstract The changing climate and a rising sea have accelerated salinity intrusion in coastal surface and groundwater and had detrimental impacts on women's sexual and reproductive health and rights (SRHR) in Bangladesh. We investigate how coastal adolescents and women are challenged in managing their SRH needs due to salinity intrusion and how the local health systems can be strengthened to meet such SRH demands. We implemented a cross-sectional survey in 135 clusters from 6 high-salinity and 3 low-salinity coastal unions and interviewed 6092 ever-married 15–49-year-old women. About 94% of female respondents experienced any health complications in the last 6 months. The child marriage situation is quite critical in coastal areas; about 75% of girls marry before the legal age of 18 and 33% marry even

M. Al Nahian
Initiative for Climate Change and Environment, International Centre for Diarrhoeal Disease Research, (icddr,b), Mohakhali, Dhaka, Bangladesh

M. Al Nahian (✉) · T. Ahmad · P. Barman · A. A. Siddique · P. K. Streatfield
International Centre for Diarrhoeal Disease Research, Mohakhali, Dhaka, Bangladesh
e-mail: mahin.nahian@icddrb.org

T. Ahmad
e-mail: tanvir.ahmad@icddrb.org

P. Barman
e-mail: prianka.barman@icddrb.org

A. A. Siddique
e-mail: ashraful.siddique@icddrb.org

P. K. Streatfield
e-mail: pkstreatfield2011@gmail.com

M. Dutta
Save the Children, Gulshan—2, Dhaka, Bangladesh
e-mail: mithun.dutta@savethechildren.org

P. Ghosh et al. (eds.), *Climate Crises and Public Health Challenges*, Disaster Risk Reduction, https://doi.org/10.1007/978-981-95-9709-3_16

before 15. Cesarean delivery takes place in 61% of cases in high-salinity areas with 15% of babies reported smaller than normal at birth. In both high- and low-salinity areas, free or low-cost medicines, pediatric doctors, nutritional supplements, and mental health support are widely advocated by respondents for children. For women, the need for female doctors is crucial, with antenatal and postnatal care provision. Diagnostic tests at the sub-district level, specialized facilities for pregnant mothers in cyclone shelters, and low-cost medicine support for women are also demanded considering SRH-specific health conditions. Local SRH demand-driven policy and programs, health workforce strengthening, resource allocation for climate-resilient health systems, community participation in the health decision making, and climate change and health research are essential for achieving 'climate justice' for coastal adolescents and women in Bangladesh.

Keywords Climate change · Salinity intrusion · Sexual and reproductive health (SRH) · Menstrual hygiene · Maternal health · Coastal Bangladesh

16.1 Introduction

The coastal regions of Bangladesh are at the forefront of the global climate crisis, where geographic vulnerabilities intersect with socioeconomic challenges to create a perfect storm of environmental and human insecurity (Islam et al. 2025; Abir et al. 2024; Roy et al. 2022). With over 710 kilometers of coastline, these areas are exposed to the full brunt of climate change impacts, including intensified cyclones, sea level rise (SLR), and salinity intrusion (Talukder et al. 2023; Alam et al. 2018). Bangladesh is experiencing SLR faster than the global average about 7.4 mm/year in the southeast and 5.5 mm/year in the southwest region (Nishat and Mukherjee 2013). In Bangladesh, about 50 million people live below 5 m above the mean sea level (Wong et al. 2014). These phenomena are not just environmental concerns; they are deeply intertwined with the daily lives, socioeconomic activities, health, and well-being of coastal populations, particularly women and adolescents (Akter et al. 2024; Talukder et al. 2023; Sultana and Luetz 2022; Brammer 2014).

Salinity intrusion, due to anthropogenic drivers, climatic factors, and SLR has far-reaching implications for water quality, agricultural productivity, and public health. Coastal residents, already facing precarious livelihoods, now confront an escalating crisis of water insecurity that exacerbates existing inequalities (Bayabil et al. 2020; Uddin et al. 2019; Brammer 2014).

Gender disparities are particularly pronounced in coastal Bangladesh, with women bearing a disproportionate burden due to their roles in household water management, caregiving, and managing menstrual and reproductive health (Reimann et al. 2023; Assefa et al. 2021). From the point of sexual and reproductive health and rights (SRHR), women and adolescents are faced with critical challenges, especially in meeting their menstruation, pregnancy, and other reproductive health needs within an acute freshwater crisis in the saline coastal environment (Kayser et al.

2019; Baker et al. 2018; Caruso et al. 2015). Furthermore, in the absence of such critical resources, in saline-prone areas, risks of infections and long-term health complications are compounded by existing systemic barriers, including limited healthcare infrastructure, entrenched gender norms, and the recurrent devastation caused by natural disasters (Zakar and Iqbal 2024; Chakraborty et al. 2019; Brammer 2014).

Within this context, the study explores the lived experiences of women and adolescents in coastal Bangladesh as they navigate the intersecting challenges of salinity intrusion and SRHR. By examining both high and low-salinity areas, the research aims to uncover the specific barriers these populations face and identify opportunities for health system adaptation. Through this lens, the study not only highlights urgent policy needs but also explores strategies to foster resilience and equity in a rapidly changing climatic and social landscape.

16.2 Methods and Materials

16.2.1 Study Areas

As groundwater salinity is the main environmental/CCSLR-induced driver in the study, we used the groundwater salinity monitoring station data from the Bangladesh Water Development Board (BWDB) to identify the study areas exposed to salinity intrusions. Using a cut-off value of electrical conductivity, EC > 3000 μS/cm, we delineated the 'high-salinity' and 'low-salinity' study areas (Nahian et al. 2018). The study covered 9 upazilas in 7 coastal districts across 3 divisions, representing the southeast, central, and southwest coastal regimes of Bangladesh (Map 16.1).

16.2.2 Sample Design and Sample Size

We used the probability proportional to size (PPS) sampling condition to derive 15 unique mouzas/study clusters from the selected upazilas (sub-districts). Considering a 'case to control' match ratio of 2:1, the study sample was:

a. High-salinity areas = 6 Upazilas x 15 Mouzas/study clusters x 40 HHs = 3600 HHs

b. Low-salinity areas = 3 Upazilas x 15 Mouzas/study clusters x 40 HHs = 1800 HHs

In case of nonresponse and withdrawal, an additional 20% of HHs were kept in the sampling framework. So, the total sample size was (5400/0.8) 6750 HHs and 50 HHs from each cluster/mouza. In each selected cluster, about 150 HHs were listed after identifying certain landmarks.

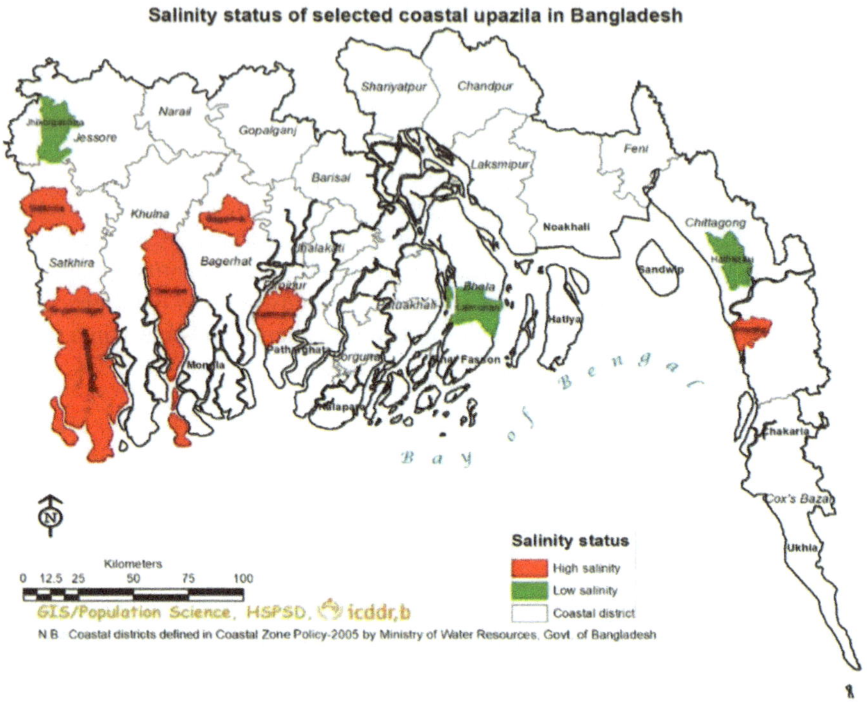

Map 16.1 Study area map showing the 'high-salinity' and 'low-salinity' areas/study clusters

16.2.3 Questionnaires

Two questionnaires were used: Household and Women's. The Household Questionnaire collected demographic, health, and socioeconomic data, along with information on climate change impacts and natural disaster experiences. The Women's Questionnaire focused on health, water use, salinity-related issues, blood pressure, and menstrual health. The study protocol was approved by the Ethical Review Committee of the International Centre for Diarrhoeal Disease Research, Bangladesh (icddr,b).

16.2.4 Data Collection

Data collection was done in two phases: the first phase occurred from October 16 to November 1, 2023, and the second from November 6 to December 8, 2023. In total, data were collected from 6396 households and 6092 eligible 15–49 years of age ever-married women within the households. Monitoring and debriefing sessions were held during and after each phase to ensure quality and consistency.

16.2.5 Response Rates

Out of 18,931 listed households, 6750 were selected, with 6396 interviews completed. A total of 6698 eligible women were identified, and 6092 were successfully interviewed. The response rates were 95% for households and 91% for women aged 15–49.

16.2.6 Data Processing and Analysis

All the data analysis was done with STATA 15 software (StataCorp 2017). Sampling weights were calculated based on household and women's nonresponses. Final data analysis was carried out with weighted samples.

16.2.7 Limitations

We feel timing of the survey was a very critical factor for the data collection and hence, our initial research design prioritized data collection in the summer season. However, due to unforeseen circumstances, data collection moved in the postmonsoon season, due to which we failed to capture some of the essence of CCSLR-induced health vulnerability in our data.

16.3 Results

16.3.1 Climate Change and Disaster Landscape in Coastal Bangladesh

Approximately 93% of households in coastal areas experienced at least one disaster in the past decade. Cyclones and storm surges were the most frequent, with high-salinity areas experiencing these events 4 times more often than low-salinity areas. While asked about climate change, the oldest members/head in a household informed notable climate changes over the past decade, including longer summers with increased heat, shorter monsoons, erratic rainfall patterns, heightened cyclone intensity, and salinity intrusion (Fig. 16.1).

Respondents in both high- and low-salinity regions exhibit a strong consensus on long-duration summer seasons and increasing heat. In low-salinity areas, a larger proportion of respondents (83%) report shorter monsoon seasons compared to high-salinity areas (69%). Both areas recognize the erratic nature of rainfall, with slightly more awareness in low-salinity areas (87%) than in high-salinity areas (84%).

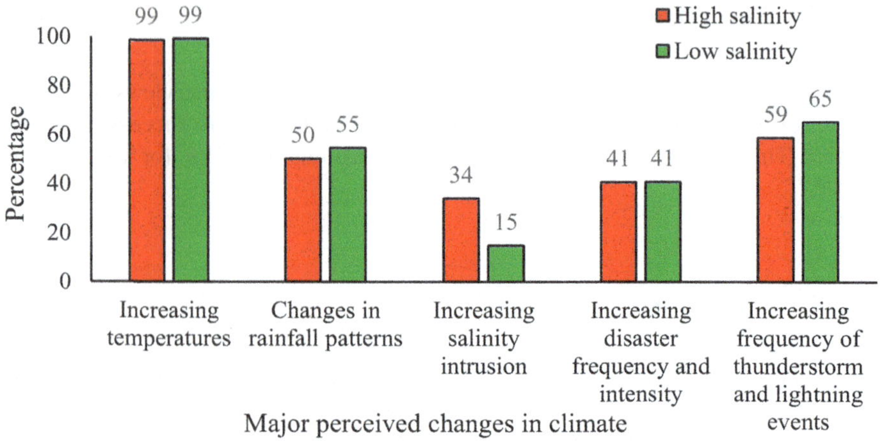

Fig. 16.1 Major indicators of climate change in coastal Bangladesh

Respondents in low-salinity areas report a slightly higher perception of increased intensity and frequency of natural disasters (i.e., cyclone and storm surge, flood, and river erosions) and summer windstorms and lightning events. Respondents in high-salinity areas are more concerned with the accelerated salinity intrusion issue. The respondents further elaborated on the various impacts of climate change, sea level rise, and salinity intrusion on their households, including increased water insecurity and soil salinity in high-salinity areas, income loss, and higher medical expenses. In contrast, low-salinity areas faced greater medical costs.

16.3.2 Women Respondents' Background Information and Socioeconomic Characteristics

The study was focused on adolescents and women 15–49 years of age, ever-married, and living in the coastal areas. Considering such inclusion criteria, most female respondents were of 20–34 age group, with the largest group being 25–29 (17%), indicating a young population. The age groups 20–24 and 25–29 have slightly higher percentages in high-salinity areas. Age groups 30–34, 35–39, 40–44, and 45–49 have almost equal distribution. About 18% were born in their current location, and 27% have lived there for 10 years or less. Religiously, 80% of the population is Muslim, with more diversity in high-salinity areas (77% Muslim, 23% Hindu, Buddhist, or Christian) than in low-salinity areas (93% Muslim). Most women (95%) were married, with slight variations between regions. Nearly 49% of women completed secondary education, and education levels were slightly higher in high-salinity areas. The majority of the women respondents (85%) were housewives.

16.3.3 Health Conditions among the Women Respondents

Nearly 94% of women surveyed reported experiencing some form of illness or negative health condition in the 6 months preceding the survey. Specifically, 99% of women in low-salinity areas and 92% in high-salinity areas reported health-related issues, reflecting their heightened vulnerability to climate-related health problems (Fig. 16.2). Interestingly, and contrary to our initial belief, women from low-salinity areas seemed to be suffering more from different negative health complications in comparison to high-salinity areas, like stomach aches (31 versus 43%), gastric ulcers (37 versus 77%), etc. Fever, colds, coughs, and headaches were more common in low-salinity areas (91%) than in high-salinity areas (76%). Even issues like skin diseases and gynecological problems were more frequently reported in low-salinity areas (39 and 29%) compared to high-salinity areas (24 and 8%), warranting further investigation. Only high blood pressure is slightly more common in high-salinity areas (14%) than in low-salinity areas (12%). Conditions like typhoid and hyperbilirubinemia (jaundice) are relatively rare in both regions.

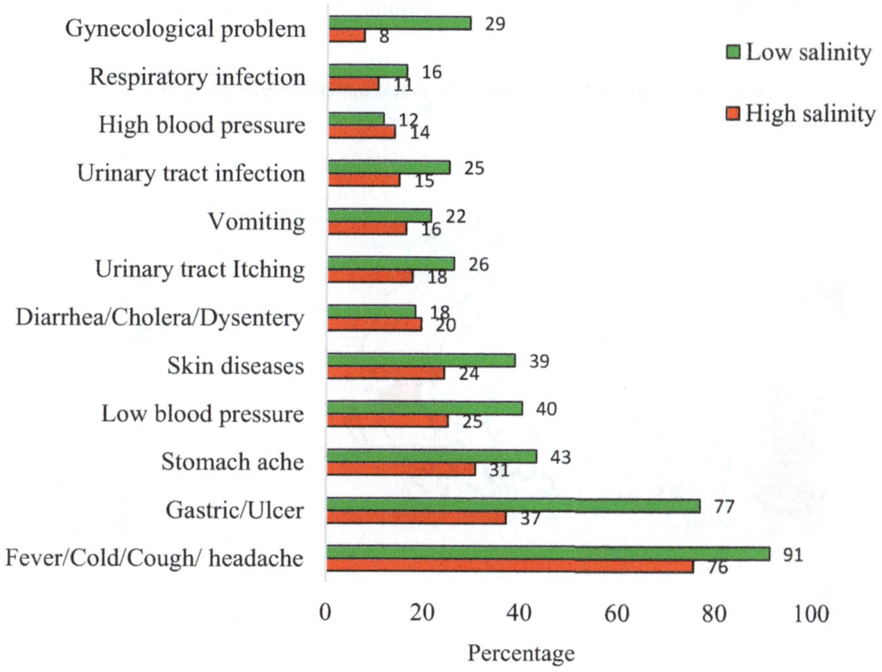

Fig. 16.2 Percentage of 15–49 aged women reporting major health complications, suffered in the last 6-months in coastal Bangladesh

16.3.4 *Age at First Marriage and Prevalence of Child Marriage among Coastal Women*

Women who married at ages 11–14 are more common in high-salinity areas, accounting for 33%, whereas in low-salinity areas, it is 29%. Marriages in the age group 15–17 years were very similar, with 42% in high-salinity areas and 44% in low-salinity areas.

Child marriage seems to be alarmingly high among coastal women (Map 16.2). It was found that about 75% of young girls got married before the legal minimum age of 18 years of age, and the child marriage prevalence is just marginally higher in high-salinity areas (75%) than in low-salinity areas (74%). Interestingly, even though Lalmohon from Bhola is a low-salinity island in the Meghna estuary, it has very high child-marriage prevalence; on the other hand, both high- and low-salinity upazillas in Chittagong district have the lowest child marriage statistics, even below the national average.

Alarmingly, it was found that about 33% of girls underwent marriage before even reaching their 15th birthday. In high-salinity areas, about 33% of girls got married within 15 years of age, whereas in low-salinity areas, the prevalence was 30%.

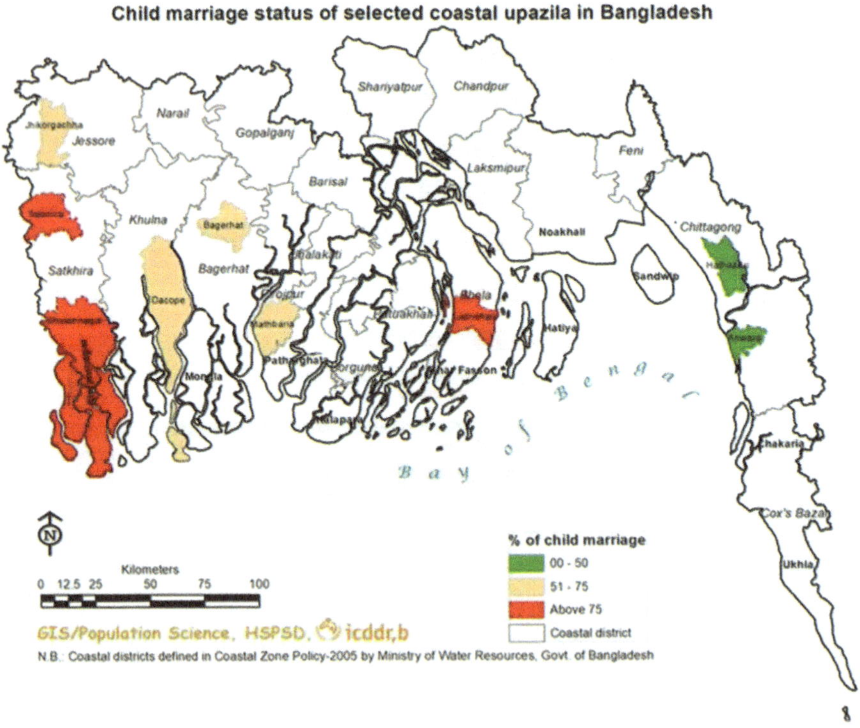

Map 16.2 Child marriage status of selected coastal Upazila in Bangladesh

Furthermore, child marriage is more prevalent among Muslim communities (78%) than Hindu or other religions' (63%).

16.3.5 Menstrual Regularity, Materials Used, and Problems from Poor Menstrual Hygiene

Regular menstruation is reported by 63% of women, while 15% experience irregular periods. A notable proportion of women also report cessation of menstruation due to family planning methods (6%). A higher percentage of women (61%) use cloths during menstruation, followed by sanitary napkins (45%). The use of cloths is higher in low-salinity areas (69%) compared to high-salinity areas (60%). We found napkins were more widely used by the adolescent girls of 15–19 years of age (87%), whereas women aged 20–49 were more dependent on cloths (65%) during their menstruation.

Women in high-salinity areas experienced more severe challenges due to poor menstrual hygiene management compared to low-salinity areas. Vaginal infection or itching was reported by 19% of women in high-salinity areas, significantly higher than the 11% reported in low-salinity areas. Abnormal or irregular menstruation was also more common in high-salinity areas (19%) compared to low-salinity areas (11%). Other problems, such as vaginal irritation (10 versus 9%), stress (13 versus 11%), and domestic violence (1 versus 0.2%) were also reported. Gynecological problems are marginally higher among the 15–19 years group than the 20–49 years women.

It is quite positive that women sought consultation and treatment from medical professionals in case of menstrual challenges. In high-salinity areas, 49% consult MBBS doctors, while 58% do so in low-salinity areas. Village doctors are also commonly consulted, with 36% of women in high-salinity areas and 28% in low-salinity areas seeking their care.

In both high- and low-salinity areas, surface water (river, pond, ditch, canal, etc.) was the primary source for menstrual hygiene management, with 52% usage in high-salinity areas and 42% in low-salinity areas. Coastal shallow tube wells, with lower salinity, were the second most preferred option in both areas, while deep tube wells were minimally used. Almost 89% of women from both high- and low-salinity areas reported their concern about the poor water quality/perceiving the water as unsafe for use due to salinity during menstruation.

Another important concern is the use of birth control pills to stop or delay menstruation. Such practice was found to be higher in high-salinity areas (19%) compared to low-salinity areas (13%).

16.3.6 Family Planning Practices

Overall, 88% of the women in the total study population have used family planning methods.

Overall, 64% of women in the study used any family planning (FP) method, with 66% in high-salinity and 57% in low-salinity areas. Modern FP method use was 55% in high-salinity and 49% in low-salinity areas. Oral contraceptives were preferable in high-salinity (30%) areas and injectables in low-salinity (20%) areas. Pharmacies were the most common source of modern family planning, used by 38% overall, with 48% in low-salinity and 36% in high-salinity areas.

16.3.7 Status of Antenatal Care Check-Ups during Pregnancies

About 91.2% of women had received at least 1 ANC. Among those who received ANC, about 36.5% of women had 4 or more ANC. In low-salinity areas, 4 or more ANC was received by 40% women, just about the national statistics, whereas in the high-salinity areas, the status is 36% (see Map 16.3).

16.3.8 Mode of the Last Delivery

We found significant differences in the mode of delivery between high- and low-salinity areas. In high-salinity areas, the majority of deliveries were cesarean sections, accounting for 61% of births. In contrast, only 40% of deliveries in low-salinity areas were cesarean sections. Overall, 56% of all births are delivered via cesarean sections, which is quite high compared to the reported 45% of all births delivered via cesarean section in the BDHS 2022 survey (National Institute of Population Research and Training (NIPORT), & ICF 2023).

Normal deliveries were more common in low-salinity areas (60%) compared to high-salinity areas (37%). This reflects comparatively better overall health conditions, more accessible prenatal care, and fewer complications during pregnancy and childbirth due to negative climatic and environmental conditions between the high-salinity and low-salinity areas. Assisted vaginal deliveries were relatively rare in both regions, but slightly more prevalent in high-salinity areas (2%) compared to low-salinity areas (1%). A notable concern is the proportion of babies born in the coastal areas reported as smaller than normal in both high-salinity (15%) and low-salinity (10%) areas (see Fig. 16.3).

Interestingly, C-section found to be more prevalent among the Hindu and non-Muslim communities (61%) compared to Muslim communities (55%).

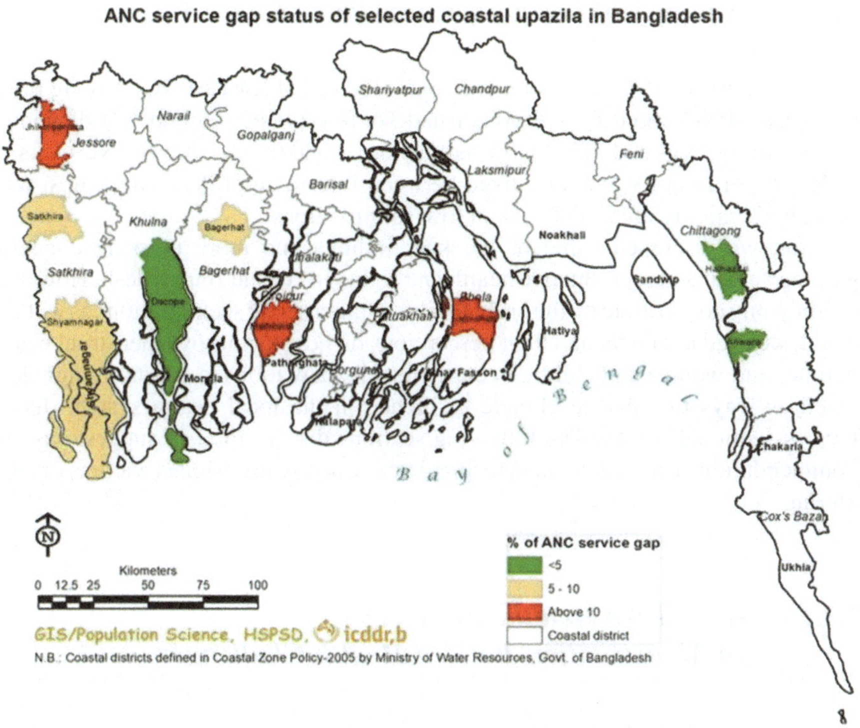

Map 16.3 ANC service gap status of selected coastal Upazila in Bangladesh

Fig. 16.3 Cesarean (C-section) delivery status in coastal areas

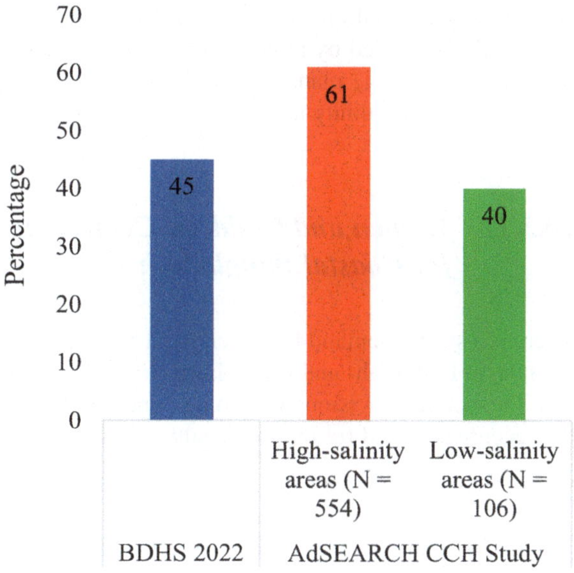

16.3.9 Frequency of Postnatal Care (PNC) Check-Ups

It was found, about 76% of women or babies received at least 1 PNC. Among those who received PNC, about 79% of women had 3 or fewer PNC. Around 80% of women in high-salinity areas and 77% in low-salinity areas had three or fewer PNC visits. A smaller proportion of women received between four to eight PNC check-ups, with 14% in high-salinity areas and 22% in low-salinity areas.

So far, we discussed some of the key findings and more pressing concerns regarding sexual and reproductive health and rights in coastal Bangladesh. However, we didn't only concentrate on the problems but rather tried to seek solutions as well. In general, we tried to understand their perceptions on how to improve menstrual health outcomes and women and children-centric health adaptations in coastal Bangladesh as well as ways to enhance climate resilience in the local health system during the cross-sectional survey. The following sections discuss the key suggestions and recommendations from the women respondents, who are the primary victims of such suffering.

16.3.10 Specialized Healthcare Needs for Menstruation-Related Health Challenges

Arranging a female doctor was a key requirement, with 94% of women in low-salinity areas and 87% in high-salinity areas indicating this need. Additionally, providing free or low-cost medicines was more commonly requested in low-salinity areas (84%) compared to high-salinity areas (69%). The provision of free or low-cost sanitary pads was similarly needed by almost half of the women in both areas. Awareness meetings on safe hygiene management during menstrual periods were more frequently requested in low-salinity areas (37%) than in high-salinity areas (26%).

16.3.11 Women and Children-Centric Health Adaptations for Coastal Bangladesh

There is a strong demand for specialized healthcare provisions for both children and women, in both high- and low-salinity areas. Most of the women respondents advocate for special provisions for women and children in health facilities, highlighting the recognized need for tailored healthcare services for these groups.

- For women, the need for female doctors is significant, with over 80% in both areas supporting this. Special care for pregnant mothers and access to antenatal/postnatal care is crucial for 29–32% of respondents, along with a smaller percentage advocating for transportation support for mothers.

- For children, 51% in high-salinity areas and 66% in low-salinity areas sought free or low-cost medicines for children. Provision of pediatric doctors in local health facilities was highly demanded, supported by 64% in high-salinity areas and 71% in low-salinity areas. Nutritional supplements and mental health support for children and adolescents receive moderate backing, with about 32% and 16%, respectively.
- Emergency services like diagnostic tests at the Upazila/Sub-district level were prioritized by 38% in high-salinity areas and 48% in low-salinity areas. A smaller number of women emphasize special provisions for pregnant mothers in cyclone shelters.
- Cost and support services show 38% to 47% of women supporting free or low-cost medicines for women, while mental health support garners limited support.

16.3.12 Measures for Healthcare Resilience in Disaster-Prone Climatic-Vulnerable Areas

The suggestions from women respondents emphasize several critical points for improving healthcare resilience in disaster-prone climate-vulnerable areas.

- **Affordability of Medication**: The most common suggestion in both high- and low-salinity areas was providing medicine freely or at low cost, supported by 69% of women in high-salinity areas and 81% in low-salinity areas. Additionally, 56% of women in high-salinity areas and 66% in low-salinity areas emphasize the need for reliable access to medicines.
- **Specialized Health Professional**: The provision of specialized doctors considering local disease patterns was another key measure highlighted by women in both high and low-salinity areas. In high-salinity areas, 57% of women recommend this measure, while 68% of women in low-salinity areas also supported it.
- **Health Facility Infrastructure**: Building health facilities on elevated ground was a widely supported measure to mitigate healthcare facility access problem during waterlogging and flooding events. In high-salinity areas, 24% of women suggest this approach, whereas 28% of women in low-salinity areas advocated for it. Additionally, establishing makeshift temporary health facilities in cyclone shelters was supported by 15% of women in high-salinity areas and 26% in low-salinity areas.
- **Emergency Health Services**: Emergency transport was recommended by 7% in high-salinity and 10% in low-salinity areas. Health worker-led awareness sessions were also suggested by 13% women in high and low-salinity areas.
- **Support for Vulnerable Groups**: 30% of women in high-salinity and 40% in low-salinity areas stressed the importance of specialized health support for vulnerable groups, including children, the elderly, and pregnant women.

- **Diagnostic and Laboratory Services**: Affordable diagnostic services were advocated by 14% in high-salinity and 28% in low-salinity areas, indicating a greater need in low-salinity regions.
- **Other Measures**: Ensuring safe food and water was suggested by 5% in high-salinity and 3% in low-salinity areas, though it was a lower priority compared to other measures.

16.4 Discussion

This novel study tried to establish a sexual and reproductive health (SRH) baseline scenario for CCSLR-exposed coastal women and adolescents in Bangladesh. Coastal communities in Bangladesh have already identified salinity intrusion as the major CCSLR-induced gradual climatic/environmental threat with the most severe consequences on public health outcomes (Roy et al. 2022). Our study highlights the challenges faced by coastal women and adolescents in addressing their sexual and reproductive health (SRH) needs due to the effects of salinity intrusion. It also explores context-specific health adaptations and climate resilience in local health systems to meet the SRH demands.

Increased surface and groundwater salinity in coastal areas is gradually impacting access to safe water, vital for domestic consumption, hygiene management, cooking, cleaning, homestead production, and irrigation. Though, access to drinking water seems relatively better in both high and low-salinity areas, however, there are underlying disparities to explore. In high-salinity regions, only 40% use deep tube wells, compared to 70% in low-salinity areas as deep tube wells have relatively higher salt concentrations in water (Nahian et al. 2018). Interestingly, alternative water sources and rainwater are only utilized in high-salinity areas, probably due to acute water insecurity. This has critical policy implications for water infrastructure-related future project implementations.

Respondents from both regions are aware of climate change, with high-salinity areas focusing on salinity intrusion and water issues, while low-salinity areas stress agricultural losses and irrigation water shortages. Erratic rainfall, shorter monsoons, and increased disasters are recognized, with low-salinity areas facing higher agricultural losses (47%) due to water shortages, compared to high-salinity areas (26%). This may be due to, limited agricultural works ongoing in low-salinity areas, which had already become obsolete in high-salinity areas, due to higher salt content in water and soil. High-salinity areas are more concerned with salinity affecting water and soil (38% facing soil salinity). These differences highlight the need for region-specific adaptation strategies, including water management, irrigation improvements, and disaster preparedness.

Approximately 93% of households in coastal regions have experienced a disaster in the last decade, highlighting their vulnerability to extreme events like cyclones and storm surges. The impact is similar in both high-salinity (93%) and low-salinity (97%) areas, though high-salinity areas face these events four times more often. This difference may stem from factors like proximity to the sea, poor management of embankments, and shortage of disaster risk reduction facilities.

The findings highlight a significant health burden on women, with about 94% of female respondents experiencing some kind of health complications in the last 6 months. Gastric ulcers and gynecological problems are more common in low-salinity areas, while high-salinity areas see higher rates of high blood pressure, diarrhea, and stomach ailments. Low-salinity areas report more fever, colds, coughs, headaches, and skin diseases. High blood pressure is slightly higher in high-salinity areas, while low blood pressure is less common. Addressing these issues requires targeted healthcare services tailored to the unique challenges of each region.

Child marriage remains alarmingly high, with 75% marrying before 18 and 33% before 15 years old, particularly in high-salinity areas. Despite high awareness (84%) of legal marriage-age laws, sociocultural norms and economic pressures perpetuate early marriages, exacerbating the vulnerability of young women. This trend demands urgent policy interventions to enforce marriage-age laws.

High-salinity areas report higher use of cloths for menstrual hygiene (60%) and associated health issues such as vaginal infections (19%). These findings highlight the urgent need for affordable and safe menstrual hygiene products, as well as awareness campaigns, to address hygiene management challenges. Improved access to sanitary products and community-based educational initiatives could significantly enhance menstrual health management and reduce related health risks.

The reliance on surface water for hygiene and menstrual management, particularly in high-salinity areas (52%), underscores the impact of salinity intrusion on water quality. The perception of water as unsafe (89% in both areas) for hygiene and menstrual management further emphasizes the urgent need for improved water infrastructure and management.

Women in high-salinity areas experience more severe menstrual hygiene-related issues compared to those in low-salinity areas. Notably, 19% of women in high-salinity regions reported vaginal infections or itching, and irregular menstruation, compared to 11% in low-salinity areas for both issues. Such issues demand heightened attention.

While menstruation is a natural part of life, there can be reasons why women want and need advice from health providers. Only about one in nine sought such advice, and in half of the cases, it was from qualified doctors and another one-third from unqualified village doctors. Only a very small proportion of women respondents sought suggestions from community-based female health workers. However, this has some practical complications; without the actual expertise and knowledge of the specific circumstances, such advice may result in undesirable health complications. This also emphasizes the capacity enhancement need of the community health workers to strengthen the local health system.

While 88% of women have used family planning methods, modern methods are less prevalent in low-salinity areas (49% vs. 55% in high-salinity areas). Pharmacies and community clinics are the primary sources, with high-salinity areas relying more on community health workers, reflecting accessibility challenges. Enhancing access to family planning services and promoting modern contraceptive methods in low-salinity areas could address these gaps and improve reproductive health outcomes.

A higher proportion of women in both high-salinity (64.3%) and low-salinity (60.4%) areas have 3 or fewer antenatal care (ANC) visits, indicating limited prenatal care access. Low-salinity areas have a slightly higher percentage (36.1%) of women receiving 4–8 ANC visits compared to high-salinity areas (32.3%), but overall ANC visits remain low, with only around 3.4% attending 9 or more visits. Such issues need further exploration and targeted intervention design considering the contextual barriers and enablers.

Another very concerning finding is the high level of deliveries performed by cesarean section (C-sections). This level is almost half of all deliveries in these study areas, particularly high in the high-salinity areas. In low-salinity areas, 60% of deliveries were normal vaginal births, compared to only 37% in high-salinity areas, where the majority (61%) were cesarean sections. Overall, 56% of deliveries across both areas were C-sections. These findings suggest that higher salinity might be associated with an increased likelihood of cesarean deliveries, possibly through some 'environmental-physiological' pathway to impact maternal or fetal conditions. However, much more research is needed to enhance our understanding of such issues.

In coastal Bangladesh, there is a pressing need for women- and children-centric healthcare. Women respondents emphasized the importance of female doctors, with over 80% supporting their presence, and advocated for specialized services such as antenatal and postnatal care. For children, access to affordable medicines is a priority, with 51% in high-salinity and 66% in low-salinity areas supporting free or low-cost options. Pediatric care is also highly valued, with 64% in high-salinity and 71% in low-salinity areas calling for more pediatricians. Nutritional and mental health support for children receive moderate support. Additionally, many women highlight the need for emergency services and diagnostic tests at the Upazila/Sub-district level, as well as low-cost or free healthcare options for both women and children, emphasizing the importance of cost-effective, accessible health services.

Women respondents also highlighted several key measures for enhancing health-care resilience in disaster-prone, climate-vulnerable areas of coastal Bangladesh. The most common suggestion is making medications affordable, with 69% of women in high-salinity areas and 81% in low-salinity areas supporting free or low-cost access. Reliable access to medicines is also emphasized, particularly in low-salinity areas (66%). Specialized medical care tailored to local disease patterns is advocated by over half of women in both areas. To address inundation risks during waterlogging and flooding, building health facilities on the elevated ground was a widely supported measure, as was establishing temporary makeshift healthcare/emergency treatment facilities in cyclone shelters.

16.5 Conclusion

The coastal regions of Bangladesh are grappling with a dual burden of climate vulnerability and systemic challenges, disproportionately impacting the sexual and reproductive health (SRH) of women and adolescents. This study underscores the multifaceted barriers posed by salinity intrusion, erratic climatic patterns, and socioeconomic inequities that exacerbate health risks and limit access to essential healthcare services.

Findings reveal alarming rates of child marriage, inadequate menstrual hygiene management, and insufficient antenatal and postnatal care, particularly in high-salinity areas. The overreliance on surface water and high rates of cesarean deliveries further highlight the pressing need for comprehensive health interventions. Despite widespread awareness of climate change and its impacts, the resilience of local health systems remains constrained by infrastructural inadequacies, cultural norms, and economic barriers.

To address these challenges, the study advocates for targeted, region-specific strategies that prioritize the needs of women and adolescents. Key recommendations include:

1. **Strengthening Water Infrastructure**: Investing in desalination plants, rainwater harvesting, and safe water distribution networks to mitigate the impacts of salinity on hygiene and health.
2. **Enhancing Healthcare Access**: Establishing climate-resilient health facilities, ensuring the availability of female healthcare providers, and expanding access to affordable and quality SRH services.
3. **Promoting Health Education**: Implementing community-based educational campaigns to raise awareness about menstrual hygiene, family planning, and the importance of regular antenatal care.
4. **Policy and Advocacy**: Enforcing laws against child marriage and promoting socioeconomic initiatives that empower women and delay early marriage.
5. **Integrated Health and Climate Research**: Advancing research on the intersection of climate change and health to inform evidence-based policies and interventions.

Achieving 'climate justice' for women and adolescents in Bangladesh's coastal regions requires a collaborative effort involving policymakers, health professionals, and community stakeholders. By fostering inclusive and adaptive health systems, we can not only address the immediate health needs of vulnerable populations but also build long-term resilience in the face of a changing climate.

Acknowledgments This study was funded by the Department of Foreign Affairs, Trade, and Development (DFATD), through Advancing Sexual and Reproduction Health and Rights (AdSEARCH), Grant number: SGDE-EDRMS-#9926532, Purchase Order 7428855, Project P007358. The cross-sectional survey was implemented under the authority of the National Institute of Population Research and Training (NIPORT), Medical Education and Family Welfare Division, Ministry of Health and Family Welfare, the Government of Bangladesh. Associates for Community and Population Research (ACPR), a private research agency was engaged to collect data. The funding for data collection was provided by the Government of the People's Republic of Bangladesh through the Ministry of Health and Family Welfare. icddr,b is also grateful to the Governments of Bangladesh and Canada for providing core/unrestricted support. We also thank the survey respondents for their valuable time and input in the study. We acknowledge the contribution of Muhammad Zahirul Haq, Data Management Supervisor, Health Systems and Population Studies Division of icddr,b for preparing the informative maps.

References

Abir TM, Das T, Ferdous F (2024) Building community resilience and adaptation to climate change: integrating disaster risk reduction in the coastal regions of Bangladesh. J Coast Res 40(4):792–802. https://doi.org/10.2112/JCOASTRES-D-23-00043.1

Akter T, Hoque MA-A, Mukul SA, Pradhan B (2024) Coastal flood induced salinity intrusion risk assessment using a spatial multi-criteria approach in the South-Western Bangladesh. Earth Syst Environ. https://doi.org/10.1007/s41748-024-00399-9

Alam E, Momtaz S, Bhuiyan HU, Baby SN (2018) Climate change impacts on the coastal zones of Bangladesh: perspectives on tropical cyclones, sea level rise, and social vulnerability. In: Islam MN, van Amstel A (eds) Bangladesh I: climate change impacts, mitigation and adaptation in developing countries. Springer, pp 145–166

Assefa GM et al (2021) Gender equality and social inclusion in relation to water, sanitation and hygiene in the Oromia region of Ethiopia. Int J Environ Res Public Health 18(8):4281. https://www.mdpi.com/1660-4601/18/8/4281

Baker KK, Story WT, Walser-Kuntz E, Zimmerman MB (2018) Impact of social capital, harassment of women and girls, and water and sanitation access on premature birth and low infant birth weight in India. PLoS One 13(10):e0205345. https://doi.org/10.1371/journal.pone.0205345

Bayabil HK, Li Y, Tong Z, Gao B (2020) Potential management practices of saltwater intrusion impacts on soil health and water quality: a review. J Water Clim Chang 12(5):1327–1343. https://doi.org/10.2166/wcc.2020.013

Brammer H (2014) Bangladesh's dynamic coastal regions and sea-level rise. Clim Risk Manag 1:51–62. https://doi.org/10.1016/j.crm.2013.10.001

Caruso BA, Sevilimedu V, Fung IC-H, Patkar A, Baker KK (2015) Gender disparities in water, sanitation, and global health. Lancet 386(9994):650–651. https://doi.org/10.1016/S0140-6736(15)61497-0

Chakraborty R, Khan KM, Dibaba DT, Khan MA, Ahmed A, Islam MZ (2019) Health implications of drinking water salinity in coastal areas of Bangladesh. Int J Environ Res Public Health 16(19):3746. https://www.mdpi.com/1660-4601/16/19/3746

Islam MT, Hossain MM, Ha-Mim NM, Hossain MZ, Sikder SK (2025) Climate change impacts on shoreline migration and community livelihood resilience: evidence from coastal Bangladesh. Front Sustainability 5. https://doi.org/10.3389/frsus.2024.1525483

Kayser GL, Rao N, Jose R, Raj A (2019) Water, sanitation and hygiene: measuring gender equality and empowerment. Bull World Health Organ 97(6):438–440. https://doi.org/10.2471/BLT.18.223305

Luetz J (2018) Climate change and migration in Bangladesh: empirically derived lessons and opportunities for policymakers and practitioners. In: Leal Filho W, Nalau J (eds) Limits to climate change adaptation. Springer, pp 59–105

Nahian MA, Ahmed A, Lázár AN, Hutton CW, Salehin M, Streatfield PK (2018) Drinking water salinity associated health crisis in coastal Bangladesh. Elementa: Sci Anthropocene 6:2. https://doi.org/10.1525/elementa.263

National Institute of population research and training (NIPORT), & ICF (2023) Bangladesh demographic and health survey 2022: key indicators report. NIPORT and ICF

Nishat A, Mukherjee N (2013) Sea level rise and its impacts in coastal areas of Bangladesh. In: Shaw R, Mallick F, Islam A (eds) Climate change adaptation actions in Bangladesh. Springer, pp 43–50

Reimann L, Vafeidis AT, Honsel LE (2023) Population development as a driver of coastal risk: current trends and future pathways. In: Cambridge prisms: coastal futures, vol 1, p e14. https://doi.org/10.1017/cft.2023.3

Roy B, Penha-Lopes GP, Uddin MS, Kabir MH, Lourenço TC, Torrejano A (2022) Sea level rise induced impacts on coastal areas of Bangladesh and local-led community-based adaptation. Int J Disaster Risk Reduct 73:102905. https://doi.org/10.1016/j.ijdrr.2022.102905

StataCorp (2017) Stata 15 software, StataCorp LLC

Sultana N, Luetz JM (2022) Adopting the local knowledge of coastal communities for climate change adaptation: a case study from Bangladesh. In: Frontiers in climate, vol 4. https://doi.org/10.3389/fclim.2022.823296

Talukder B et al (2023) Collective intelligence for addressing community planetary health resulting from salinity prompted by sea level rise. J Clim Chang Health 10:100203. https://doi.org/10.1016/j.joclim.2023.100203

Uddin MN et al (2019) Mapping of climate vulnerability of the coastal region of Bangladesh using principal component analysis. Appl Geogr 102:47–57. https://doi.org/10.1016/j.apgeog.2018.12.011

Wong PP, Losada IJ, Gattus J-P, Hinkel J, Khattabi A, KL MI, Saito Y, Sallenger A (2014) Coastal systems and low-lying areas. In: Field CB, Barros VR, Dokken DJ, Mach KJ, Mastrandrea MD, Bilir TE, Chatterjee M, Ebi KL, Estrada YO, Genova RC, Girma B, Kissel ES, Levy AN, MacCracken S, Mastrandrea PR, White LL (eds) Climate change 2014: impacts, adaptation, and vulnerability. Part a: global and sectoral aspects. Contribution of working group II to the fifth assessment report of the intergovernmental panel on climate change. Cambridge University Press, pp 361–409

Zakar R, Iqbal S (2024) Editorial: social determinants of women's health in low and middle income countries. In: Frontiers in global women's health, vol 5. https://doi.org/10.3389/fgwh.2024.1482047

Chapter 17
Breaking the Cycle of Vulnerability: Climate Crises and Public Health Challenges in Bangladesh

MD Shiyan Sadik, Bushra Noor, and Sarah Afsari

Abstract Bangladesh, one of the nation's most vulnerable to climate change, faces escalating public health challenges as environmental crises intensify. This study explores the multifaceted health impacts of climate change, including direct effects such as heatwaves, air pollution, and the rising prevalence of vector-borne diseases, as well as indirect consequences like mental health burdens and disruptions to food security. Vulnerable groups, including children, the elderly, and low-income populations, are disproportionately affected, exacerbating existing health inequities. While Bangladesh has developed policies such as the Health National Adaptation Plan (HNAP) and the Climate Change Strategy and Action Plan, these frameworks primarily focus on disaster response and lack comprehensive preventive measures. Key gaps include limited integration of mental health, insufficient cross-sectoral coordination, and underrepresentation of marginalized communities in adaptation strategies. Additionally, challenges such as inadequate funding, corruption, and weak infrastructure further hinder effective implementation. This research highlights the urgent need for a robust, climate-resilient healthcare system that aligns with international frameworks like the Paris Agreement. Proposed strategies include developing a comprehensive HNAP with a focus on preventive measures, strengthening institutional capacity, and investing in sustainable healthcare infrastructure. Enhanced disease surveillance, integration of mental health services, workforce training, and community engagement are emphasized to ensure equitable healthcare delivery. Addressing governance issues, the study calls for transparent funding mechanisms, robust monitoring, and data-driven decision-making to bridge existing gaps. By adopting a holistic, inclusive approach that prioritizes vulnerable populations and leverages global best practices, Bangladesh can mitigate the health impacts of climate change and build resilience for future challenges. This study provides actionable insights to guide policymakers in achieving sustainable public health outcomes.

MD S. Sadik (✉) · B. Noor · S. Afsari
North South University, Dhaka, Bangladesh
e-mail: Shiyan.sadik@northsouth.edu

S. Afsari
e-mail: Sarah.afsari@northsouth.edu

P. Ghosh et al. (eds.), *Climate Crises and Public Health Challenges*, Disaster Risk Reduction, https://doi.org/10.1007/978-981-95-9709-3_17

Keywords Climate change · Public health · Bangladesh · Climate resilience · Health policy

17.1 Introduction

17.1.1 Background and Context

South Asia and Southeast Asia is a dynamic region, holding over one-fifth of the world's population (Sivakumar and Stefanski 2010). With densely populated countries, it also encompasses fast developing economies. The fast-growing demography has resulted in relentless building of infrastructures and rapid urbanization. This socio-economic adverse impact of climate change on the physical and mental health of the critically vulnerable communities of Bangladesh. While there are adequate policies on the rehabilitation and resilience against environmental disaster, there are shortcomings in the policy when it comes to addressing health issues due to the crisis. Understanding the impact of climate change on the public health will foster a clearer picture on the need for hospitals, clinics and health services in those regions, allowing the government to plan properly on the expansion of healthcare systems in the country, in specific to the need of the people in the region, thus addressing health disparities in the country. An in-depth understanding of the impact depending on the specific demographic and geological characteristics would allow us to elaborate the consensus of knowledge development and policy development for Bangladesh on the consequences of the climate crisis. It will allow us to address the literature gap in the consequences of the climate crisis on human health in particular. Thus, allowing us to make necessary changes to our policies, ensuring healthier lifestyles for the communities. By closing the gap in literature, the research can be used as a policy making tool for targeted interventions, resource allocation and long-term healthcare infrastructure planning.

17.1.2 Impact of Climate Crises on Physical and Mental Health in Bangladesh

17.1.2.1 Direct Health Impacts

The greenhouse gasses being released into the atmosphere is responsible for global warming, which has resulted in extreme weather events, such as heatwaves, heavy rainfall, and drought (Arsad et al. 2022; Abbass et al. 2022). Among such adverse impacts the intensity of heatwaves has become more frequent and extreme over the past few years, making it very difficult to move and work outside. The latest Intergovernmental Panel on Climate Change (IPCC) report further strengthens the statement, showing that heat waves have become more intense and common in the past 50 years (Arsad et al. 2022; IPCC 2014). The adverse impact has resulted in the highest record setting temperature in Bangladesh. With each year a new record setting high temperature. During summer seasons, the country experiences the hottest days, with the average temperature of 30 °C. In 2023, certain regions experienced temperatures soaring to 40 °C, with some areas reaching as high as 41 °C, marking the highest recorded temperatures in the past six decades. In the consecutive year, the temperature was even higher, with an average temperature of 40–42 °C, making it the hottest year experienced by all the districts of Bangladesh (Mustajib 2024). The extreme weather posed great vulnerability to heat strokes, vomiting, diarrhea, heat exhaustion, headache, pneumonia, shortness of breath, dehydration, and more (Mustajib 2024). A research done on the impact of heatwaves on mortality and morbidity by Arsad et al. (2022), shows that patients of cardiopulmonary diseases, renal disease, diabetes, mental disease and people living in urban areas were particularly vulnerable to the extreme heatwaves. Due to the extreme heat waves, the death of four people was reported (Mustajib 2024). Although various measures and reports have addressed rising temperatures, the government's actions toward adaptation and mitigation remain unclear and lack transparency in their implementation and strategic direction. The country needs to implement sustainable adaptive measures to tackle the series of heatwaves, however for an economically struggling country like Bangladesh, transitioning to cleaner energy and reducing carbon emissions is not an easy pursuit. In the year 2022, the country had produced a total of 109,075,300 tons of fossil CO_2 emissions with a yearly change of +3.57%, contributing to only 0.29% of the total world's CO_2 emission (Bangladesh CO_2 Emissions - Worldometer n.d.). The adverse impacts that the country faces are a result of the global contribution.

Bangladesh's air pollution has become a rather great environmental constraint for all of its residents. It has been claimed by Dr. Nazmul Hossain that breathing air in Dhaka city is equivalent to smoking a pack of cigarettes every two days. However, there are only articles to support the statement, indicating lacking's in research analysis of such a pressing environmental issue. On December 10th of 2024, an article on "Dhaka's Air Quality Plunges to Hazardous Levels: Equivalent to Smoking 22 Cigarettes a Day" was published Greentime news reporter, stating that Dhaka's Air Quality Index has reached a staggering 394, where the healthy scale for AQI is from 0 to 50, moderate is 51–100, unhealthy is 151–200, very unhealthy is 200–300, and highly unhealthy is 301–500 (Khandker et al. 2022). Research on Bangladesh's air pollution and its consequences by Khandker et al. (2022), has highlighted various sources to air pollution, that include anthropogenic sources such as burning fossil fuel, coal, wood, open burning of waste and agricultural residues, emissions from motor vehicles and more. Natural sources include windblown dust, sea spray, forest fires, and methane gas emitted by animals. The research also states that 85% of the local air pollution in Dhaka is contributed by brick kilns, surface dust, and vehicles. The short- and long-term exposure to such high level pollution has made the citizens highly vulnerable to respiratory infections, heart disease, and lung cancer (Khandker et al. 2022). Further analysis has shown that 80% of the PM2.5-related deaths could be reduced if the country followed the guidelines set for 10 $\mu g/m^3$ by WHO (Khandker et al. 2022; WHO 2021).

Climate change severely affects the rate of vector-borne diseases (VBDs), well established in many researches done (Githeko et al. 2000). According to The World Health Organization (WHO), malaria, dengue, and the Zika virus pose significant global health threats as the climate change alters their transmission patterns (Jibon et al. 2024; Item 2023). IPCC (2014) has recorded the increase in the VBDs in recent years, and has asserted that if proper measures are not taken to mitigate the impacts, further increase in the level of VBDs may be experienced. The increased temperature may be linked with the frequent outbreaks of dengue outbreaks, however to establish the former statement, further research and study is required to fill up the gap (Jibon et al. 2024). However, it can be stated that the rising temperature and rise in sea level has made suitable conditions for the spread of VBDs such as dengue, malaria, and chikungunya (Jibon et al. 2024). It is important for policymakers to prioritize yearlong surveillance and early warning systems, proper resource allocation, further research, and easy access health services (Jibon et al. 2024). The most vulnerable to such VBDs are the children, the sick, old people, and people living in slums and urban areas with poor hygiene measures.

Bangladesh being a tropical country experiences six different seasons; summer, rainy season, autumn, late autumn, winter and spring. Each season plays a crucial role in the culture, heritage, and economy of his country. Agriculture in this country has been a crucial part, with most of its food grown and supplied locally, which is heavily influenced by the seasonal characteristics such as temperature, rainfall, humidity, day length, etc. (Sikder and Xiaoying 2014). As the climate crisis extends, these parameters are significantly altered resulting in inconsistent and poor harvest. The output is further hindered by natural disasters, such as cyclones, salinity intrusion, extreme temperature, drought, and floods (Sikder and Xiaoying 2014). Among the crops, the staple food of Bangladesh, rice cultivation, is expected to be heavily affected. The overall rice production is expected to decrease by an average of 7.4% within the period 2005–2050 (Sikder and Xiaoying 2014). As the agricultural output gets constrained, the food supply falls short, with the existing supply being accessible only to a certain group of people in the economy. Pushing a mass of people toward malnutrition followed by other health complications.

17.1.2.2 Indirect Health Impacts

As the communities face climatic crises, they are met with both physical health and mental health hurdles. The mental ill-health driven due to climate change can be classified into two ways: (1) Immediate post-traumatic stress and depression (2) Longer term development and well-being interruption (Hayward and Ayeb-Karlsson 2021). Immediate post-traumatic stress and depression often occurs right after a disaster has struck. While experiencing loss of life and loved ones and widespread destruction is in itself a traumatic experience (Hayward and Ayeb-Karlsson 2021; Akhter et al. 2015; Kabi and Khan 2017; Hasan et al. 2020), there are other factors that contribute to greater discomfort during disasters such as cyclones. Hurdles such as losing homes, agricultural land, and livestock may result in intense anxiety and stress (Hayward and Ayeb-Karlsson 2021). Such experiences create trauma and the victims often suffer from PTSD (Hayward and Ayeb-Karlsson 2021; Akhter et al. 2015). Such disasters can also amplify previous traumas. The trauma mixed with financial instability may result in loss of hope for security and social acceptance in the future (Hayward and Ayeb-Karlsson 2021; Akhter et al. 2015).

Long term development and well-being is affected and expressed long after the disaster has struck. With poor infrastructure and security, communities suffer both financially, and to ensure security of their families. Children often struggle to continue schooling due to closing down of schools (Hayward and Ayeb-Karlsson 2021). Broken bridges are often mended using banana rafts, making it harder to commute, other factors such as fear of being bitten by snakes and drowning contribute further to the inaccessibility of proper education (Hayward and Ayeb-Karlsson 2021). The near-death experiences shape the way the children think, as their play revolves around "death and burial", leading to long-term changes to their cognition and personality (Hayward and Ayeb-Karlsson 2021; Hasan et al. 2020).

17.1.3 Case Studies: Vulnerable Communities

Dhaka is a densely populated city. Majority of the people migrate to Dhaka from the rural areas. Factors such as low economic growth, imbalance in resource management, unemployment and underemployment has led to a big proportion of the people of this city to be poor. Thus, making it difficult to have proper facilities and planned accommodation of homes in this city. As a result, open spaces, wetlands, and more are encroached, causing harm to the environment (Sinthia 2020). A case study done on the Korail slum of Dhaka by Sinthia (2020), has found that the settlements are temporary habitats, found in vulnerable conditions due to low income and insecure tenure system. The tenants live there by paying a comparatively high prices and with a few civic services such as water, electricity, etc. owners of the settlements are often under the shadow of political leaders and musclemen, hence there is no scope for the tenants to complain about the poor services. Such dynamics of the political power and increasing population results in the expansion of encroached areas, and slums. Settlements as such cause environmental degradation, with poor sanitation and hygiene, posing threat to public health. Land encroachment may result in cutting down trees, further decreasing the city's ability to capture carbon.

A study done on Linking Coastal Disasters and Migration: A Case Study of Kutubdia Island, Bangladesh by Rahman et al. (2014) has shown that only 13% of the families residing in the island have lived there for at least three generations. After the southeastern part of Kutubdia Island experienced coastal erosion and flooding in August 2012, residents had left one part of the island for the mainland, simultaneously residents from mainland had settled in other parts of the island, either not knowing of the environmental problems, or not caring of the problems associated with living in the islands. As the global temperature rises, causing the sea levels to rise, the population around the coastal areas will eventually have to relocate and migrate. Bangladesh will have to plan ahead of time in order to foster a systematic migration in the future, as 17% of the land will be submerged by 2050 (Islam n.d.). Otherwise, mass movement may result in chaotic, unmanageable movement within the already populated country, further degrading the environment and may lead to increased deforestation rates.

Research performed on the different parameters affecting the rural agriculture of Bangladesh has shown that the productivity of agriculture heavily depends on the productivity of land (Rahman and Anik 2020). An increase in annual-rainfall and long-term-temperature significantly enhance production (Rahman and Anik 2020). Such factors combined with a various leguminous plant being grown in the field in rotation will allow enhanced production. However, the research also implies how different archeological areas may alter the efficiency of production. Floodplain being the most suitable, followed by low-laying floodplain, and coastal-plain agro ecologies (Rahman and Anik 2020). However, with the increasing variability in rainfall and temperature due to climate change, agricultural productivity is severely hindered, hence it is important to ensure other factors are kept in close control to maintain adequate food supply in the market.

17.2 Current Policy Landscape

Bangladesh follows a disaster management protocol that is more human right based, addressing equity and equality, inclusivity, and good governance (Chisty et al. 2022). There are four national instruments: the Disaster Management Act, the National Disaster Management Policy, the National Plan for Disaster Management (NDPMs), and the Standing orders on Disasters (SOD) (Chisty et al. 2022). The Disaster Management act provides the legal framework for the country's disaster events, formulates rules and provides jurisdiction on penalty and more (Chisty et al. 2022). The National Disaster Management Policy reflects upon the preparedness of the community and acts as a guide for the development projects to take place (Chisty et al. 2022). The National Plan for Disaster Management has a specific time frame and works on mitigating risks involving any disaster, every five years, changes are made to the policy, learning from the mistakes and lacking from the previous strategy (Chisty et al. 2022). Finally, the SOD provides a guideline for all the stakeholders in every level of the government, along with the role of the society in every step of the disaster management steps (Chisty et al. 2022).

17.2.1 Policy and Practice: Existing Gaps

Putting health-related concerns on the backburner when it comes to addressing climate-related problems is not only an issue the south east Asian countries face. According to reports from WHO less than 0.5% of international climate finance is currently allocated to health-related projects (World Health Organization 2018). This is a pressing concern, as the efforts to develop a nation would hold little to no value if its workforce remains at significant risk of decline, particularly for a country like Bangladesh, which is projected to lose 17% of its land over the next 25 years (BIPR n.d.).

If one were to assess the state of the country's healthcare sector, independent of the climate crisis, it is important to note that earlier this year, over 700 hospitals and clinics were closed for operating without the requisite accreditation (Foyez 2024). This action followed a series of troubling reports concerning malpractice, underscoring the deep-rooted issues within the healthcare system that hinder its capacity to deliver quality and safe care to the population. There are reports of over 1000 Private hospitals running without license (Foyez 2024), 223 health facilities are left unused, machineries are left unused (Moral and Moral 2022), ambulances received from India left untouched (Prothom Alo English Desk 2022a, b) and many ICU facilities left unused or left unrepaired for years (Prothom Alo English Desk 2022a, b). And most of these unused facilities are located in rural regions that are vulnerable to natural disasters (Roy 2024). In order to effectively identify gaps in the country's healthcare sector in addressing climate crisis-related issues, it is essential to gain a comprehensive understanding of existing healthcare policies and assess why they have yet to yield tangible and functional results (Table 17.1).

A thorough analysis of the table exhibits notable gaps and inefficiencies in Bangladesh's public health policies, particularly in their approach to addressing climate change-related health challenges. Most policies prioritize disaster response over preventive, long-term health resilience measures, leading to a reactive approach rather than addressing root causes. There is a lack of integration between climate adaptation, public health, and urban planning, which undermines a unified response to health risks like age or gender specific vulnerabilities, vector-borne diseases and waterborne illnesses etc. Marginalized and vulnerable groups, including those in rural and coastal regions are insufficiently prioritized, further exacerbating health inequities. Mental health impacts of climate crises remain largely unaddressed, reflecting a narrow view of climate-related health risks. Moreover, public awareness and community engagement efforts are minimal, failing to build grassroots capacity for adaptation. Weak enforcement mechanisms and fragmented governance hinder effective implementation, while the absence of robust data collection systems prevents evidence-based decision-making. And most importantly, funding and resource allocation lack clarity, leaving critical initiatives underfunded and disconnected from broader climate resilience goals.

Table 17.1 Public health policies in Bangladesh related to the climate crisis

Policy	Focus areas	Key gaps and limitations
National Water Policy (1999)	Flood-proofing, pollution control, early warning, wastewater management, industrial regulation, arsenic mitigation, water safety	Poor climate–health linkage, inadequate disease focus, low integration of marginalized communities, limited adaptation strategies, enforcement gaps
National Adaptation Programme of Action (NAPA) (2005)	Disaster preparedness, climate-smart agriculture, stress-tolerant crops, climate-adapted WASH infrastructure, social protection, gender inclusion	Weak linkage between public health and climate, minimal urban strategies, poor focus on vector-borne disease, funding and research gaps, low public engagement
Bangladesh Climate Change Strategy and Action Plan (BCCSAP) (2009)	Disaster preparedness, pollution mitigation, water and sanitation, vector control, healthcare capacity building, awareness campaigns, adaptation integration	No actionable timeline, poor coordination, insufficient funding and research, overlooked mental health, weak inclusion of vulnerable groups and communities
Health Policy (2011)	Pollution control, disaster response, water and sanitation, air quality, disease prevention, urban health awareness, coastal health resilience	Lacks sectoral integration, insufficient proactive measures, underrepresentation of mental health, urban planning gaps, no clear funding roadmap, weak enforcement
Bangladesh Health National Adaptation Plan (HNAP) (2016–2030)	Leadership, climate-resilient systems, disease surveillance, training, adaptation research, disaster preparedness, cross-sector collaboration, sanitation, green hospitals	Limited focus on vulnerable groups, weak implementation, unclear funding strategies, poor coordination, minimal urban health focus, lack of gender
National Plan for Disaster Management (NPDM) 2021–2025	Cyclone, flood, earthquake, tsunami, and landslide risk reduction, arsenic mitigation, biological hazard planning, food and health resilience	Prioritizes disaster response over prevention, vague health provisions, insufficient investment in health infrastructure, poor focus on marginalized groups
Bangladesh Delta Plan 2100	Urban drainage, salinity control, health and food security, infrastructure, ecosystem restoration, water quality, transboundary cooperation	Weak health integration, overlooked vector disease risk, low investment in health systems, poor sanitation and grassroots engagement, no climate–health surveillance

17.2.2 Comparative Analysis Between Foreign Countries and Bangladesh's HNAPs

Bangladesh's Health National Adaptation Plan (HNAP) provides a comprehensive framework to build climate resilience within the health sector, recognizing the country's high vulnerability to extreme weather events and climate-sensitive diseases. Aligned with the WHO's operational framework, it outlines strategic objectives across ten components, including leadership, health workforce, risk monitoring, and climate-informed programs. However, despite its structured design, the plan faces significant shortcomings. It lacks specific, measurable adaptation actions and gives limited attention to broader environmental health determinants, mental health, and climate-resilient infrastructure. The monitoring and evaluation framework is weak, with no clear indicators or community engagement metrics. Additionally, the plan fails to translate its recognition of vulnerable populations, geographically tailored approaches, and community involvement into actionable strategies. Workforce training lacks detail, coordination mechanisms are underdeveloped, and there is a critical absence of localized research on climate–health linkages, ultimately weakening the plan's effectiveness in fostering a truly resilient and inclusive health system. To develop a functional and implementable climate-resilient healthcare program for Bangladesh, it is essential to thoroughly examine and analyze the Health National Adaptation Plans (HNAPs) from other regions. By doing so, we can identify diverse, proven solutions that have successfully contributed to the establishment of accessible and effective healthcare systems (Table 17.2).

A comparative analysis of Bangladesh's Health National Adaptation Plan (HNAP) reveals significant gaps in structure, coordination, and execution when measured against the HNAPs of countries like Ethiopia, Nepal, New Zealand, Fiji, and Uganda. Bangladesh's plan lacks strong cross-sectoral collaboration, particularly in inter-ministerial coordination and community engagement—areas where Ethiopia and Nepal demonstrate robust mechanisms for unified action. Its financing framework is also notably weak, with no clear strategy for sustainable resource allocation, unlike Uganda, where financial support has been secured through coordinated efforts among government ministries, international donors, and technical partners (The Rockefeller Foundation 2025). Bangladesh's HNAP leans heavily on reactive measures rather than adopting forward-looking strategies such as heatwave management and air quality control, which are integral components of New Zealand's proactive approach. Moreover, the plan offers limited attention to vulnerable populations and mental health—both of which are central to the HNAPs of Fiji and Nepal. Perhaps most critically, Bangladesh's plan lacks concrete monitoring indicators, timelines, and a structured research agenda, further limiting its ability to measure progress or adapt to emerging climate–health challenges. At first glance, Bangladesh's HNAP may appear comprehensive; however, considering the dynamic nature of climate-related challenges and examining how other countries have developed diverse, context-specific policies—and continuously revised them in response to emerging obstacles—it becomes clear that Bangladesh falls short in several critical areas. Unlike

Table 17.2 Key focus areas of HNAPs from different countries

HNAPs	Focuses
Uganda: National Health Adaptation Plan	• Aimed at safeguarding public health from the effects of climate change • Climate training for healthcare workers • Climate-proofing infrastructure
Ethiopia Health National Adaptation Plan 2024–2024	• Emphasizes strong intersectoral coordination to address climate and health challenges effectively • Training health professionals and community health workers to mitigate climate-related health risks • Supports building climate-resilient healthcare facilities and adopting sustainable technologies • Enhances early warning systems and disease surveillance for improved outbreak forecasting and response • Promotes health and climate research and integrates climate change into education curricula
Nepal: Climate Change Health Adaptation Strategies and Action Plans of Nepal (2017–2021)	• Create a climate change unit in the Ministry of Health • Develop district-level early warning systems for health risks • Train health workers and researchers on climate–health risks • Add climate–health topics to school and university curricula • Strengthen surveillance for vector, water, and foodborne diseases • Build 50 eco-friendly, climate-resilient health facilities

(continued)

Table 17.2 (continued)

HNAPs	Focuses
New Zealand: Health National Adaptation Plan 2024–2027	• Create Climate Health Action Plans (CHAPs) for regional needs • Partner with locals and indigenous communities for equitable adaptation • Embed climate–health adaptation into all health policies • Form a scientific advisory group with mātauranga Māori experts • Develop a national heat warning system and heatwave definitions • Help local governments create regional climate–health emergency plans • Assess climate-sensitive diseases and health risks nationwide • Collaborate on health, climate, and equity in housing and urban decisions • Monitor HNAP action progress across the health system • Research climate impacts on mental health, diseases, and healthcare
Australia: National Health and Climate Strategy Implementation Plan 2024–2028	• Build a climate-resilient health system to protect population health from climate impacts • Decarbonize the health system through a comprehensive plan to achieve net-zero emissions • Develop a national heat warning system with localized heatwave definitions for better response • Establish a Climate and Health Expert Advisory Group to guide adaptation policies • Enhance collaboration with First Nations stakeholders to co-design climate–health policies • Implement air quality monitoring systems to address climate-induced respiratory risks • Strengthen primary care, aged care, and preventive health services to adapt to climate challenges • Foster international collaboration to build climate-resilient health systems globally • Support health system adaptation through investments in research and workforce training • Integrate health considerations into climate policy across sectors via a health in all policies approach

(continued)

Table 17.2 (continued)

HNAPs	Focuses
Samoa: Climate Health Strategy	• Integrate climate considerations into health policies • Enhance surveillance for climate-sensitive diseases • Build capacity in climate–health data and risk mapping • Promote community education on hygiene and climate risks • Upgrade water, sanitation, and waste systems • Address malnutrition through nutrition education and food security • Include mental health in disaster responses • Improve labs for diagnosing climate-related diseases • Strengthen cross-sector collaboration on health risks • Boost maternal and child health in vulnerable areas
Fiji: National Climate and Health Adaptation Plan	• Strengthen the Climate Change and Health Unit (CCHU) with focal points • Develop a Health Sector Decarbonization Strategy for net zero • Conduct health-specific vulnerability and capacity assessments • Upgrade the disease surveillance system with real-time GIS monitoring • Deliver climate-resilient upgrades to 20 key healthcare facilities • Mandate climate impact assessments in all health policies • Promote sustainable healthcare waste management nationwide • Support nature-based solutions like tree planting and green spaces • Launch small grants for local climate–health research • Train health professionals on climate resilience and sustainability

countries that demonstrate dedicated efforts in adapting their plans through regular updates and targeted strategies, Bangladesh's HNAP lacks strong cross-sectoral coordination, sustainable resource mobilization, proactive planning, equity-focused interventions, and clear accountability mechanisms. These gaps significantly undermine the plan's potential to build a truly climate-resilient health system.

17.3 Proposed Strategies for Building Climate-Resilient Healthcare Systems

Choosing strategies to build a Climate-resilient Healthcare system for Bangladesh is not as straightforward as it might seem as it requires an in-depth understanding of the problems the country is facing, a thorough evaluation of the existing HNAPs (Health National Adaption Policies) and going through the suggestions made by international organizations like WHO and UNFCC to structure an efficient and faceable Health Adaptation Policy plan. Based on observing the current state of the country's climate and healthcare system on the primary stage the problems identified are: In order to bring impactful results, it's not enough to just address the issues and suggest a solution, the solution needs to be a long-term sustainable solution.

According to various reports from the World Health Organization (WHO), specific guidelines have been recommended for structuring an effective Health National Adaptation Plan (HNAP). The 2015 WHO Operational Framework for Building Climate-Resilient Health Systems outlines six core building blocks essential for health systems: leadership, workforce, information systems, technology, service delivery, and finance (World Health Organization 2015). In another report from 2021, WHO identified six quality criteria necessary for a well-structured HNAP:

(a) Leadership and enabling environment
(b) Cross-sectoral coordination and policy coherence
(c) Comprehensive coverage of climate-sensitive health risks
(d) Comprehensive coverage of adaptation options and actions
(e) Resourcing
(f) Monitoring, evaluation, and reporting.

The Ministry of Health must lead the development of the HNAP with clearly defined roles, adequate resources, and formal government endorsement. The plan should prioritize addressing health impacts through active engagement with various sectors and evidence-based decision-making, which should be informed by vulnerability assessments. It must comprehensively address climate-sensitive health risks, ensure cross-sectoral coordination, and include targeted adaptation measures. Resource planning, monitoring, and evaluation are critical components. Furthermore, the plan must incorporate the inclusive participation of local communities and vulnerable groups to ensure effective implementation (World Health Organization 2021).

Even all that is not enough as we also have to consider factors like, external factors such as political, economic, social, technological, environmental, and legal dynamics significantly influence public health outcomes. Contemporary healthcare organizations, in particular, operate in a challenging and dynamic environment characterized by these complex factors (Mosadeghrad et al. 2023).

The proposed strategies aim to address the gaps and inconsistencies in Bangladesh's current public health and climate adaptation policies, particularly concerning the health challenges exacerbated by the climate crisis. While it is evident that formulating a comprehensive and flawless Health National Adaptation Plan (HNAP) for Bangladesh requires extensive observation and years of data collection, the strategies focus on identifying and addressing the key components that such a plan must include. Drawing on well-established HNAPs from other countries, these strategies also integrate recommendations from various international institutions and relevant research literature.

17.3.1 Enhancing Policy Frameworks

To build a healthcare system truly resilient to climate disasters, the first critical step is the development of a robust policy framework, starting with a comprehensive Health National Adaptation Plan (HNAP) that aligns with global frameworks like the Paris Agreement, ensuring long-term sustainability and preparedness. In order to tackle issues like asynchronization between different ministries and slow paced plan progression a key strategy would involve integrating health consideration into climate policy through a "Health in All Policies" approach that would tie different ministries (like health, environment and finance) with a separate regulatory body, which can ensure more effective tracking of climate-related health issues, quick funding allocation and expedites solutions by fostering better communication and coordination across sectors. This approach will address the delays and information gaps that often arise due to lack of intersectoral collaboration, necessitating the creation of a multi-sectoral Climate and Health coordination mechanism. Additionally, it is essential to focus on the diverse health risks exacerbated by climate change, including extreme weather-related injuries and fatalities, heat-related illnesses, respiratory diseases, waterborne diseases, zoonoses, vector-borne diseases, malnutrition, foodborne illnesses, noncommunicable diseases, mental health issues, and the impact of damaged healthcare facilities (World Health Organization 2021). A policy framework must be inclusive of vulnerable populations, particularly children, pregnant women, the elderly, and the injured. The most critical health issue that Bangladesh has yet to adequately address, exacerbated by climate change, is mental health. A recent article in the American Journal of Nursing highlights that individuals affected by climate-related disasters often develop symptoms of acute stress disorder, PTSD, depression, and anxiety (Azim 2017). This underscores the urgent need to integrate mental health into the national policy framework, as evidenced by tragic cases of farmers who have taken their own lives following devastating losses. A study conducted in the United States found a 15% increase in suicides among rural male farmers between 2000 and 2013 during periods of severe drought (Mosadeghrad et al. 2023). To effectively address these mental health challenges, the policies must be developed in collaboration with environmental disaster specialists, various stakeholders, and international scientific experts, working closely with local communities to ensure comprehensive, data-driven interventions.

17.3.2 Strengthening Institutional Capacity

Establishing regardless of how detailed the design is, it would all go in vain if there are no regulatory bodies within the institutions to foresee the plans and monitor its progression. Strengthening institutional capacity is essential to ensure the effective implementation of climate–health policies. This can be achieved by establishing a dedicated climate change unit within the Ministry of Health to oversee and coordinate efforts. Task forces comprising experts from health, climate, and infrastructure sectors should be formed to streamline planning and execution, ensuring an integrated approach to addressing climate-related health challenges. Additionally, the development of subnational health adaptation plans tailored to regional and city-specific needs is vital for localized resilience. Regular review and monitoring of climate–health policies are crucial to adapt strategies to evolving challenges, and the mandatory inclusion of climate impact assessments in all health policies will ensure comprehensive and forward-looking decision-making.

17.3.3 Investing in Infrastructure and Resources

Building climate-resilient healthcare infrastructure is imperative to ensure continuity of care during extreme weather events and other climate-induced disruptions. This involves constructing flood-proof and eco-friendly hospitals and healthcare centers designed to withstand natural disasters. Mobile healthcare units should be deployed to reach populations in remote or disaster-prone areas, ensuring equitable access to care. To reduce dependency on nonrenewable resources, healthcare facilities should adopt renewable energy solutions such as solar power. Establishing sentinel sites for monitoring climate-sensitive diseases (CSDs) and noncommunicable health issues linked to climate change is essential for proactive response measures. Furthermore, equipping healthcare centers and hospitals in high-risk areas with climate-resilient technologies and infrastructure is critical. Ensuring uninterrupted functionality of all national and regional public health laboratories, including those for referral sample testing, is a cornerstone of maintaining robust health systems under climate stress.

17.3.4 Leveraging Technology for Transparency and Efficiency

The integration of digital platforms can significantly enhance the efficiency and accountability of climate–health initiatives. Strengthening health and climate research is a foundational step, which can be supported by launching small grants for local climate–health studies. Establishing an online database for climate-sensitive health data and creating integrated systems to facilitate data sharing and analysis between health and climate sectors will improve decision-making processes. Real-time tracking of healthcare resources and climate impacts through robust data platforms is essential for timely responses. Additionally, mobile applications should be developed to provide district-level early warning systems, notify communities about disease outbreaks, and offer detailed guidance on safety measures. Surveillance systems for climate-sensitive diseases must be enhanced with real-time GIS monitoring and risk mapping to better predict and manage health risks. Upgrading laboratories to effectively diagnose climate-related diseases is another critical component to ensure preparedness and resilience in the healthcare system.

17.3.5 Optimizing Resource Utilization

Efficient resource management is vital to prevent wastage and ensure the sustainability of healthcare systems, particularly in the face of climate challenges. Transparency and efficiency in the procurement and distribution of medical supplies can be achieved through a comprehensive, publicly accessible database that tracks unused resources, enabling their allocation to areas of need. Establishing an online forum for citizens to report issues with healthcare facilities, coupled with a defined response timeline, can enhance accountability and service delivery. Notably, the healthcare sector contributes approximately 5% of global carbon emissions (Campbell-Lendrum et al. 2023), making the adoption of environmentally friendly technologies imperative to reduce its carbon footprint. In climatically vulnerable regions where high-tech medical facilities may be impractical, eco-friendly, renewable energy-powered healthcare centers with resilient designs offer a sustainable solution. Additionally, recycling and repurposing healthcare resources wherever possible will further minimize waste and support a more sustainable healthcare system.

17.3.6 Developing Climate Change Informed Health Workforce

Healthcare professionals are central to building resilient healthcare systems, particularly in the face of climate-related challenges. It is essential to equip healthcare providers with the necessary skills to address the health impacts of climate change through targeted training programs. Expanding the capacity of health professionals in climate and health-related subjects, alongside developing a network of disaster health workers, community representatives, and volunteers, will enhance preparedness and response capabilities. Financial and nonfinancial incentives should be offered to encourage healthcare workers to serve in disaster-prone regions, addressing the persistent challenge of workforce shortages in vulnerable areas. Additionally, expanding telemedicine infrastructure will ensure the continuity of healthcare delivery during and after climate-induced disruptions, bridging gaps in access and maintaining essential health services for affected communities.

17.3.7 Population Outreach and Awareness

Incorporating climate–health topics into school and university curricula is essential for fostering awareness and preparing future generations to address the intersection of climate change and public health. Equitable adaptation strategies can be strengthened by partnering with local and indigenous communities to identify vulnerable groups and design accessible evacuation plans tailored to their needs during disasters. Government-supported collaboration with local political leaders can enhance communication with communities and facilitate unified efforts to mitigate health risks. Promoting community education on critical issues such as hygiene, climate risks, maternal and menstrual health, aged care, and primary medical aid will empower individuals with the knowledge to improve resilience and reduce vulnerabilities in the face of climate challenges. Incorporating the participation of stakeholders like politicians, locals, trained medical professionals, educators and disaster specialists will help us communicate with people in a deeper level and ensure the safety of people at all levels.

17.3.8 Ensuring Effective Implementation

The effective implementation of climate-resilient healthcare strategies is crucial for their success and requires robust mechanisms to regularly assess progress and evaluate the impact of climate–health initiatives. Establishing a stringent anti-corruption framework within the multisectoral system is essential to monitor funding, resource allocation, and utilization effectively. All national and international contracts related to climate–health resilience projects must undergo multiple revisions by diverse oversight bodies and stakeholders to identify potential discrepancies or loopholes that could lead to exploitation. Collaboration with global organizations and local stakeholders is vital for mobilizing the technical and financial resources needed for sustained efforts. Environmental Impact Assessments (EIAs) must be conducted for all projects before progression, with the resulting reports made publicly accessible. To ensure smooth operation and transparency, annual audits by an impartial third party—ideally comprising international specialists—should be mandated to eliminate opportunities for report fabrication. Additionally, enforcing strict adherence to established climate–health policies and frameworks through formal mandates will strengthen accountability and drive consistent progress toward developing resilient healthcare systems.

17.4 Moving Forward: Financing and Governance

Bangladesh's Health National Adaptation Plan (HNAP) outlines several potential funding sources to support climate-resilient healthcare initiatives. These include allocating specific funds within the health sector for climate adaptation projects, utilizing the Bangladesh Climate Change Trust Fund (BCCTF) to finance health adaptation measures, and leveraging donor funding and international agencies like WHO and UNDP to address resource gaps. However, the primary challenge in Bangladesh lies not in sourcing funds but in ensuring their effective management, equitable distribution, and appropriate allocation.

Bangladesh has garnered international recognition for its initiatives to utilize global funds, such as the Green Climate Fund (GCF), and domestic frameworks like the Bangladesh Climate Change Resilience Fund (BCCRF) and the Climate Change Trust Fund to combat climate challenges. Yet, the implementation has often been marred by mismanagement and corruption, undermining the impact of these financial resources. As a result, vulnerable populations, particularly those most affected by climate disasters, frequently lack adequate public health support.

The scale of this issue is significant. A 2020 report from Transparency International Dhaka revealed that approximately US $8 million allocated for climate mitigation projects between 2018 and 2020 was embezzled. Poor fund management led to incomplete projects and unmet objectives. Similarly, a Transparency International Bangladesh (TIB) report highlighted that over 54% of climate mitigation project funds were misused, with numerous solar panels becoming nonfunctional shortly after installation. Instead of addressing pressing needs such as freshwater shortages in regions like Shyamnagar and Assasuni upazilas in Satkhira, funds were frequently diverted to less impactful initiatives (BonikBarta n.d.) For example, Prothom Alo reported that between 2019 and 2023, the Bangladesh Climate Change Trust (BCCT) approved 322 projects, of which 209 focused on installing solar streetlights at a cost of Tk 3.38 billion (Rahman and Rahman 2024), an investment with limited relevance to climate adaptation. In contrast, only 90 projects targeted critical areas such as biodiversity conservation, eco-tourism, afforestation, and clean water supply. Very little attention has been given to leveraging climate funds for long-term, sustainable healthcare programs, with most initiatives limited to temporary health camps in safer zones of vulnerable areas.

To effectively address these systemic issues and establish a robust, climate-resilient healthcare framework, a multidisciplinary approach is essential. The following strategies are recommended.

17.4.1 Establish a Climate-Resilient Healthcare Fund

Create a dedicated fund within the broader climate finance framework to ensure targeted investments in healthcare services for climate-affected communities.

This fund should prioritize developing healthcare infrastructure, training medical personnel, and ensuring access to essential services during and after climate disasters.

17.4.2 Policy Formation and Regulation

Develop specific policies to allocate a fixed percentage of climate funds to public health initiatives. These policies should emphasize equitable resource distribution, the establishment of emergency funds for climatically sensitive regions, preventative measures, vaccination programs for diseases like dengue and cholera exacerbated by climate change, and strict penalties for corruption and exploitation of disaster victims.

17.4.3 Transparency Mechanisms

Establish anti-corruption committees to oversee all projects and ensure accountability. Additionally, create a digital platform to track climate fund allocations for public health, providing real-time data on fund distribution, project progress, and outcomes. This platform would empower communities to monitor the impact and raise concerns when necessary.

17.4.4 Capacity Building

Allocate adequate funding for training programs led by experienced international professionals to prepare healthcare workers for high-stress environments in disaster-prone areas. Training should include mental health support for disaster survivors and treatment for diseases linked to environmental changes. Disaster healthcare providers should also receive incentives in recognition of the risks they undertake.

17.4.5 Research Grants

Provide national and international research grants to ensure access to up-to-date information on environmental changes and their health impacts. Such research would guide regular updates to policy plans, safeguarding the well-being of all communities.

17.4.6 Independent Auditing

Conduct regular audits of health-related climate fund projects by independent bodies to mitigate corruption and ensure efficient use of resources. This approach would build donor confidence and strengthen public trust.

17.4.7 Community-Based Healthcare

Incorporate community-driven approaches to bridge the gap between policy and implementation. Local health workers and community leaders should play pivotal roles in identifying health needs, communicating available resources, and ensuring effective service delivery.

17.5 Conclusion

In conclusion, although Bangladesh being one of the most vulnerable countries to climate crisis and its impacts, it is only fair to say it is far behind in formulating effective and efficient policies and framework to tackle the impacts of climate change, especially on the physical and mental health. In order to effectively implement the policies and formulate a framework that is effective in tackling the impacts of climate change, the country will have to strengthen its institutional capacity, leverage technology for further transparency and efficiency, optimize resource utilization, develop climate change informed health workforce, outreach population and raise awareness and lastly ensure effective implementation of the policy. The country will also have to ensure adequate funding to implement the policies. Establishing a climate-resilient healthcare fund, ensuring complete transparency, building capacity, aiming for research grants, and community-based healthcare will all contribute to a sustainable financial strategy.

Such multifaceted approach will allow Bangladesh to integrate strong policy framework. The development of a comprehensive Health National Adaptation Plan (HNAP), aligned with global frameworks like the Paris Agreement, is essential for addressing the diverse health risks posed by climate change.

However, the success of such multifaceted steps will depend on collaborative effort of the community, executives and policy makers. By adopting a holistic, collaborative approach, Bangladesh can create a healthcare system that not only responds to the immediate health impacts of climate change but also prepares for future challenges, ensuring the health and well-being of all its citizens.

Authors Contribution All authors have contributed equally to this work.

Funding The research was not supported by any funding.

Data Availability Statement The datasets during and/or analyzed during the current study available from the corresponding author on reasonable request.

Potential conflicts of interest On behalf of all authors, the corresponding author states that there is no conflict of interest.

Informed Consent Informed consent was obtained from all individual participants included in the study.

References

Abbass K, Qasim MZ, Song H, Murshed M, Mahmood H, Younis I (2022) A review of the global climate change impacts, adaptation, and sustainable mitigation measures. Environ Sci Pollut Res 29(28):42539–42559. https://doi.org/10.1007/s11356-022-19718-6

Akhter SR, Sarkar RK, Dutta M, Khanom R, Akter N, Chowdhury MR, Sultan M (2015) Issues with families and children in a disaster context: A qualitative perspective from rural Bangladesh. Int J Disaster Risk Reduct 13:313–323. https://doi.org/10.1016/j.ijdrr.2015.07.011

Arsad FS, Hod R, Ahmad N, Ismail R, Mohamed N, Baharom M, Osman Y, Radi MFM, Tangang F (2022) The impact of heatwaves on mortality and morbidity and the associated vulnerability factors: a systematic review. Int J Environ Res Public Health 19(23):16356. https://doi.org/10.3390/ijerph192316356

Azim MD (2017, November 27) Impacts of climate change on mental health of farmers. The Financial Express. https://thefinancialexpress.com.bd/views/opinions/impacts-of-climate-change-on-mental-health-of-farmers-1511707679

Bangladesh Climate Change Strategy and Action Plan 2009 (2009) In Asia Pacific Energy Organization. Bangladesh Ministry of Environment and Forests. https://policy.asiapacificenergy.org/sites/default/files/Bangladesh%20Climate%20Change%20Strategy%20and%20Action%20Plan%202009.pdf

Bangladesh CO_2 Emissions - Worldometer (n.d.) https://www.worldometers.info/co2-emissions/bangladesh-co2-emissions/

Bangladesh Delta Plan 2100 Bangladesh in the 21st Century (2018) In Local Government Engineering Department. Bangladesh Ministry of Planning. https://oldweb.lged.gov.bd/UploadedDocument/UnitPublication/1/756/BDP%202100%20Abridged%20Version%20English.pdf

Bangladesh Health National Adaptation Plan (HNAP) (2018) In Ministry of Health and Family Welfare. Government of the People's Republic of Bangladesh. https://iedcr.portal.gov.bd/sites/default/files/files/iedcr.portal.gov.bd/page/b1dfa518_356c_4acc_8cb5_074929c46748/2024-04-17-08-17-bbc5c98051 6bc29ec86232155531e76e.pdf

Bangladesh Health Policy 2011 (2012) In Bangladesh National Information Board. Bangladesh Ministry of Health and Family Welfare. https://bangladesh.gov.bd/sites/default/files/files/bangladesh.gov.bd/policy/b58fb9a0_1984_4491_a980_0c23bd476b21/66.%20Health%20Policy%202011.pdf

Bangladesh shuts down hundreds of unlicensed hospitals, facilities (2024, January 2) Benar News. https://www.benarnews.org/english/news/bengali/hospitals-closed-02012024134822.html

BIPR (n.d.) BIPR. https://bipr.jhu.edu/BlogArticles/31-Climate-Change-Exposes-Bangladesh-to-Greater-Risk.cfm#:~:text=BIPR%20%7C%20Climate%20Change%20Exposes%20Bangladesh%20to%20Greater%20Risk&text=The%20World%20Risk%20Index%202023,of%20the%20country's%20agricultural%20land

BonikBarta (n.d.) Severe fresh water crisis in southwest coastal region. Severe Fresh Water Crisis in Southwest Coastal Region. https://bonikbarta.net/home/news_description/383414/Severe-fresh-water-crisis-in-southwest-coastal-region

Campbell-Lendrum D, Neville T, Schweizer C, Neira M (2023) Climate change and health: three grand challenges. Nat Med 29(7):1631–1638. https://doi.org/10.1038/s41591-023-02438-w

Change NIPOC (2023) Climate change 2022—impacts, adaptation and vulnerability. https://doi.org/10.1017/9781009325844

Chisty MA, Muhtasim M, Biva FJ, Dola SEA, Khan NA (2022) Sendai framework for disaster risk reduction (SFDRR) and disaster management policies in Bangladesh: how far we have come to make communities resilient? Int J Disaster Risk Reduct 76:103039. https://doi.org/10.1016/j.ijdrr.2022.103039

Clean Air Act-2019 Bangladesh, World Bank - Google Search (n.d.) https://www.google.com/search?q=Clean+air+act-2019+bangladesh%2C+World+Bank&oq=Clean+air+act019+bangladesh%2C+World+Bank&aqs=chrome..69i57.21841j0j7&sourceid=chrome&ie=UTF-8

Climate Adaptation Strategy for Health (2013) Climate adaptation strategy for health, Samoa. https://www.health.gov.ws/wp-content/uploads/2022/07/2Climate-Adapation-Strategy-for-Health-.pdf

Climate Change Health Adaptation Strategies and Action Plans of Nepal (2017–2021) (2017) UNFCC. Government of Nepal Ministry of Health and Population Ramshahpath, Kathmandu. https://www4.unfccc.int/sites/NAPC/Documents/Parties/Nepal%20HNAP.pdf

Climate funds and corruption in Bangladesh (2017, November 6) [Video]. dw.com. https://www.dw.com/en/climate-funds-and-corruption-in-bangladesh/video-41073503

Correspondent S (2024, February 11) Over 1,000 private medical centres run without licence. The Daily Star. https://www.thedailystar.net/health/healthcare/news/over-1000-private-medical-centres-run-without-licence-3541471

Dietz T, Shwom RL, Whitley CT (2020) Climate change and society. Ann Rev Sociol 46(1):135–158. https://doi.org/10.1146/annurev-soc-121919-054614

Foyez A (2024, February 1) Bangladesh shuts down hundreds of unlicensed hospitals, facilities. Benar News. https://www.benarnews.org/english/news/bengali/hospitals-closed-02012024134822.html#:~:text=We%20have%20listed%20more%20than,Moinul%20Ahsan%20told%20BenarNews

Githeko AK, Lindsay SW, Confalonieri UE, Patz JA (2000) Climate change and vector-borne diseases: a regional analysis. PubMed 78(9):1136–1147. https://pubmed.ncbi.nlm.nih.gov/11019462

Hasan MT, Adhikary G, Mahmood S, Papri N, Shihab HM, Kasujja R, Ahmed HU, Azad AK, Nasreen M (2020) Exploring mental health needs and services among affected population in a cyclone affected area in costal Bangladesh: a qualitative case study. Int J Ment Health Syst 14(1). https://doi.org/10.1186/s13033-020-00351-0

Hayward G, Ayeb-Karlsson S (2021) 'Seeing with Empty Eyes': a systems approach to understand climate change and mental health in Bangladesh. Clim Chang 165(1–2). https://doi.org/10.1007/s10584-021-03053-9

Health Adaptation Plan (2024) Ministry of Health and Medical Services, Fiji. https://www.health.gov.fj/wp-content/uploads/2024/11/Fiji-HNAP.pdf

Health National Adaptation Plan 2024–2027 (2024) Ministry of Health, New Zealand. https://www.health.govt.nz/system/files/2024-10/Health%20National%20Adaptation%20Plan%202024-2027_0.pdf

Health National Adaptation Plan-II (2024–2028) (2024) In Wateraid. Ministry of Health, Ethiopia. https://www.wateraid.org/et/sites/g/files/jkxoof401/files/2024-08/HNAP%20II.pdf

IPCC (2014) Food security and food production systems. In: Field CB (ed) Impacts, adaptation, and vulnerability. Part A: global and sectoral aspects. Contribution of working group II to the fifth assessment report of the intergovernmental panel on climate change. Cambridge University Press, Cambridge (UK)

Islam MA (n.d.) BCAS: article. https://www.bcas.net/article-full-desc.php?article_id=11

Isunju BJB (2024b, November 12) John Bosco Isunju. TIME. https://time.com/7172450/dr-john-bosco-isunju/

Item (2023, August 11) https://www.who.int/emergencies/disease-outbreak-news/item/2023-DON481

Jibon MJN, Ruku SRP, Islam ARMT, Khan MN, Mallick J, Bari AM, Senapathi V (2024) Impact of climate change on vector-borne diseases: exploring hotspots, recent trends and future outlooks in Bangladesh. Acta Trop 259:107373. https://doi.org/10.1016/j.actatropica.2024.107373

Kabi R, Khan HTA (2017) Study on the health status of coastal people in Bangladesh after Cyclone SIDr and Aila. Eur Sci J (ESJ) 13(15):10. https://doi.org/10.19044/esj.2017.v13n15p10

Khandker S, Mohiuddin A, Ahmad SA, McGushin A, Abelsohn A (2022) Air pollution in Bangladesh and its consequences. Research Square (Research Square). https://doi.org/10.21203/rs.3.rs-1184779/v1

Koons E (2024, May 28) Heat wave in Bangladesh: record-setting temperatures. Climate Impacts Tracker Asia. https://www.climateimpactstracker.com/heat-wave-in-bangladesh-record-setting-temperatures/

Moral S, Moral S (2022, November 8) Wastage in health sector: 233 establishments lie idle. Prothomalo. https://en.prothomalo.com/bangladesh/73edcejemx

Moral S, Moral S (2024, June 21) Half the ICUs at Bangabandhu Medical are out of order. Prothomalo. https://en.prothomalo.com/bangladesh/l1zxlu8yli

Mosadeghrad AM, Isfahani P, Eslambolchi L, Zahmatkesh M, Afshari M (2023) Strategies to strengthen a climate-resilient health system: a scoping review. Glob Health 19(1). https://doi.org/10.1186/s12992-023-00965-2

Mustajib S (2024, April 26) Extreme heatwaves in Bangladesh: the environmental governance perspectives. The Diplomat. https://thediplomat.com/2024/04/extreme-heatwaves-in-bangladesh-the-environmental-governance-perspectives/

National Adaptation Plan of Bangladesh (2022) Ministry of Environment, Forest and Climate Change. Government of the People's Republic of Bangladesh. https://moef.portal.gov.bd/sites/default/files/files/moef.portal.gov.bd/npfblock/903c6d55_3fa3_4d24_a4e1_0611eaa3cb69/National%20Adaptation%20Plan%20of%20Bangladesh%20%282023-2050%29%20%281%29.pdf

National Health and Climate Strategy Implementation Plan, 2024–2028 (2024) Department of Health and Aged Care, Australia. https://ghhin.org/wp-content/uploads/national-health-and-climate-strategy-implementation-plan-2024-2028-1.pdf

National Plan for Disaster Management (2021–2025) (2020) Ministry of Disaster Management and Relief. Government of the People's Republic of Bangladesh. https://modmr.portal.gov.bd/sites/default/files/files/modmr.portal.gov.bd/page/a7c2b9e1_6c9d_4ecf_bb53_ec74653e6d05/NPDM2021-25%20DraftVer5_23032020.pdf

National Water Policy (1999) Ministry of Water Resources. Government of the People's Republic of Bangladesh. https://mowr.gov.bd/sites/default/files/files/mowr.portal.gov.bd/files/32e67290_f24e_4407_9381_166357695653/National%20Water%20Policy%20(English).pdf

Prothom Alo English Desk, and Prothom Alo English Desk (2022, June 10) Ambulance received from India remains unused. Prothomalo. https://en.prothomalo.com/bangladesh/local-news/ambulance-received-from-india-remains-unused

Prothom Alo English Desk, & Prothom Alo English Desk (2022, November 14) X-ray machines at most Feni govt hospitals unused for years. Prothomalo. https://en.prothomalo.com/bangladesh/local-news/h6izde81xy

Rahman S, Anik AR (2020) Productivity and efficiency impact of climate change and agroecology on Bangladesh agriculture. Land Use Policy 94:104507. https://doi.org/10.1016/j.landusepol.2020.104507

Rahman MK, Paul BK, Curtis A, Schmidlin TW (2014) Linking Coastal disasters and migration: a case study of Kutubdia Island, Bangladesh. Prof Geogr 67(2):218–228. https://doi.org/10.1080/00330124.2014.922020

Rahman A, Rahman A (2024, March 29) Climate funds being spent for installation of street lights. Prothomalo. https://en.prothomalo.com/bangladesh/2h1knj11xy
Roy SD (2024, December 19) New building of Lalmonirhat hospital lies unused. The Daily Star. https://www.thedailystar.net/health/healthcare/hospitals/news/new-building-lalmonirhat-hospital-lies-unused-3780236
Sikder R, Xiaoying J (2014) Climate change impact and agriculture of Bangladesh. J Environ Earth Sci 4(1):35–40. https://www.iiste.org/Journals/index.php/JEES/article/download/10247/10453
Singh BB, Singh M, Singh D (2021) An overview of climate change over South Asia: observations, projections, and recent advances. Springer eBooks, pp 263–277. https://doi.org/10.1007/978-981-16-2221-2_12
Sinthia SA (2020) Analysis of urban slum: case study of Korail slum, Dhaka. World Acad Sci, Eng Technol, Int J Econ Manag Eng 14(11):416–430. https://publications.waset.org/10011623/pdf
Sivakumar MVK, Stefanski R (2010) Climate change in South Asia. Springer eBooks, pp 13–30. https://doi.org/10.1007/978-90-481-9516-9_2
The Rockefeller Foundation (2025, March 14) Rockefeller foundation invests in Uganda's climate change health national adaption plan (H-NAP). https://www.rockefellerfoundation.org/news/rockefeller-foundation-invests-in-ugandas-climate-change-health-national-adaptation-plan-h-nap/
Transparency International (2024, July 22) TI Bangladesh report highlighting corruption in climate mitigation projects—climate and corruption case atlas—climate governance integrity programme. Transparency.org. https://www.transparency.org/en/projects/climate-governance-integrity-programme/climate-corruption-atlas/ti-bangladesh-report-highlighting-corruption-in-climate-mitigation-projects
World Health Organization (2015) Operational framework for building climate resilient health systems. https://iris.who.int/handle/10665/189951
World Health Organization (2018) COP24 special report: health and climate change. https://iris.who.int/handle/10665/276405
World Health Organization: WHO (2019, July 30) Air pollution. https://www.who.int/health-topics/air-pollution
World Health Organization (2021) Quality criteria for health national adaptation plans. UNFCC. ISBN: 978-92-4-001898-3. UNFCC. https://unfccc.int/sites/default/files/resource/WHO_Quality_criteria_for_HNAPs.pdf

Chapter 18
The Future of Public Health in a Changing Climate: An Overview on South Asia

Debraj Mukhopadhyay, Tanisha Biswas, Asmita Ghosh, Arunabha Dey, Soham Basu, Aniket Patil, Ayeshik Chakraborty, Kritika Jain, Vibhor Dudhraj, Priyanka Sarkar, Bhupal Bhattacharya, and J. Swaminathan

Abstract Climate change threatens public health across South Asia, hitting vulnerable communities the hardest. To address this issue regional governments and the World Health Organization (WHO) have undertaken several critical projects. MEDLINE, PubMed, EMBASE, Cochrane library, and Web of Science databases

---

D. Mukhopadhyay (✉)
Department of Medical Sciences and Public Health, University of Cagliari (UniCa), Monserrato, Cagliari, Sardinia, Italy
e-mail: debraj.mukhopadhyay@unica.it

T. Biswas
Department of Dental Surgery, Sankalchand Patel University (SPU), Visnagar, Gujarat, India
e-mail: drtanisha17@gmail.com

A. Ghosh
Department of Surgical and Interventional Sciences, McGill University, Montreal, Canada
e-mail: asmita.ghosh@mail.mcgill.ca

A. Dey
Department of Geography, McGill University, Montreal, Canada
e-mail: arunabha.dey@mail.mcgill.ca

S. Basu
Department of Forest Ecology, Mendel University, Brno, Czech Republic
e-mail: sohambasu7@gmail.com

A. Patil
Department of Biostatistics and Demography, International Institute of Population Sciences (IIPS), Deonar, Mumbai, Maharashtra, India
e-mail: aniketpatil1902@gmail.com

A. Chakraborty
Department of Public Health, Institute of Public Health Kalyani, The West Bengal University of Health Sciences, Kolkata, West Bengal, India
e-mail: drayeshik@gmail.com

K. Jain
Department of Public Health, Torrens University, Surry Hills, NSW, Australia

P. Ghosh et al. (eds.), *Climate Crises and Public Health Challenges*, Disaster Risk Reduction, https://doi.org/10.1007/978-981-95-9709-3_18

were searched up to June 2025 to identify relevant articles. A gray literature search was also conducted using the Google Scholar platform. Countries like as Bangladesh, Bhutan, Nepal, Indonesia, and India are pioneering National Adaptation Plans of Action (NAPAs) and health vulnerability assessments, focusing on three main areas: strengthening healthcare systems, fighting diseases spread by mosquitoes and other vectors, and improving water and sanitation infrastructure. While progress has been made, the region faces ongoing hurdles in managing long-term health risks from climate change. Looking ahead, it is critical that healthcare systems evolve to be more robust to the difficulties posed by climate change, while public health measures be smoothly incorporated into larger climate adaptation efforts. Success will depend on governments partnering and firmly incorporating climate change issues into their health programs. This review emphasizes the important need to expand adaption measures, boost community understanding about the complex relationships between climate and health, and improve monitoring of climate's effects on human well-being.

Keywords Climate Change · Public Health · Health Adaptation · Recommendations

18.1 Introduction

Landless farming communities and the urban poor in low- and middle-income countries are only two examples of the people who will be impacted by climate change, which poses a variety of environmental risks. Assessing the future effects of weather and climatic variability on population health begins with an understanding of the current effects. Globally, climate change is a serious problem (Gould and Rudolph 2015). Since the 1950s, the Earth's surface temperature has increased by an average of 1.5 °C (Cooley et al. 2023; Masson-Delmotte et al. 2021; Pachauri et al. 2014). South Asia, which includes Afghanistan, Bangladesh, Bhutan, India, the Maldives, Nepal,

e-mail: drkritikajain6@gmail.com

V. Dudhraj
School of Health and Allied Sciences, Sohna, Sohna Rural, Haryana, India
e-mail: vibhordudhraj23@gmail.com

P. Sarkar
Department of Medical Biotechnology, University of Siena (UniSi), Siena, Tuscany, Italy
e-mail: p.sarkar1@student.unisi.it

B. Bhattacharya
Department of Law, Raiganj University, Raiganj, West Bengal, India
e-mail: bpb@raiganjuniversity.ac.in

J. Swaminathan
School of Allied Health Sciences and Management (SAHSM), Delhi Pharmaceutical Sciences and Research University (DPSRU), Sainik Farm, New Delhi, India
e-mail: swamilingam@gmail.com

Pakistan, and Sri Lanka, is the most densely populated and populous geographical region in the world, with a population of around 1.84 billion people, or one-fourth of the world's total population, according to 2019 World Bank estimates (Sarker et al. 2016). Approximately 8.5% (Zhang et al. 2023) of people worldwide live in extreme poverty, while 65% of South Asians reside in rural regions (Tiwari et al. 2022). The linkages between climate, poverty, and health are clear. The climate issue exacerbates health concerns, with extreme weather events inflicting extensive damage to global health. By 2030, the negative health consequences of climate change might force at least 44 million people into extreme poverty (Jafino et al. 2020). Since 2008, the World Health Organization (WHO) has made climate change and human health a major area of study. Nonetheless, the WHO's First International Conference on Health and Climate, held in Geneva in August 2014 (Patz et al. 2014), highlights the significance of the effects of climate change on human health from the standpoints of global change and health (Neira 2014). There are several ways that climate change may impact health. These include the consequences of more frequent and severe heat waves, fewer fatalities from cold-related causes, more frequent and severe droughts and floods, shifts in the distribution of vector-borne illnesses, and an increased risk of malnourishment and disasters (Haines et al. 2014). The most recent sixth Intergovernmental Panel on Climate Change (IPCC) (Kikstra et al. 2022) Assessment placed a lot of emphasis on the climate change forecast for South Asia. These scenarios show rising temperature extremes and warming trends that are in line with those seen over the previous century throughout the majority of Asia (Naidoo et al. 2022). According to the Fourth IPCC Assessment Report (Solomon et al. 2007) South Asian countries, whose economy mostly depend on agriculture, natural resources, forestry, and fisheries, are predicted to suffer greatly from climate change, particularly through an increase in floods and droughts. Approximately 65% of South Asians (Wijewickrama et al. 2024) reside in rural regions, and 75% of them are impoverished, vulnerable, and at danger of climate change-related suffering (Parsons et al. 2024). Climate-related risks (Van Valkengoed et al. 2024) include variations in rainfall intensity and the potential for El Niño (Ariska et al. 2024) impacts to modify the South Asian monsoon pattern. Increased temperatures can make people, including workers, crops, and livestock more vulnerable (Touch et al. 2024). The fear that catastrophic occurrences would be made worse by climate change has been heightened by recent, historic rains and hailstorms in northern India in late March and early April 2015 (Fressoz and Locher 2024). Devastating natural catastrophes struck South Asia between 2020 and 2024, highlighting the area's susceptibility to severe weather conditions. With gusts of up to 185 km/h, Cyclone Amphan, the strongest Bay of Bengal cyclone since 1999 (Singh et al. 2024), made landfall in Bangladesh and India in May 2020 (Bandyopadhyay et al. 2021). More than 128 people were killed, millions were evacuated, and $13 billion was damaged, mostly in West Bengal and Odisha (Nirupama and Mammen 2024). During the same year, from May to October, South Asia experienced catastrophic floods due to relentless monsoon rains, affecting India, Bangladesh, Nepal, and Pakistan (Mirza and Dixit 2024). Over 1200 lives were lost, and millions were displaced as homes, farmland,

and infrastructure were submerged. Bangladesh saw one-third of its territory underwater, highlighting the need for resilient disaster management (Dewan and Extremes 2015; Mirza and Dixit 2024). In 2024, India faced another severe monsoon-driven flood crisis from May to September, with states like Assam, Uttar Pradesh, and Kerala bearing the brunt (Sahu). These events caused immense economic losses and disrupted livelihoods, exacerbating food insecurity and health crises. Together, these disasters emphasize the urgency of climate-resilient infrastructure, effective early warning systems, and sustainable environmental practices to mitigate the escalating impacts of climate change in South Asia. Due to the severe agricultural devastation caused by these storms, several farmers took their own lives. Significant price hikes for food were also seen. South Asia's densely populated low-lying mega-deltas have also been recognized as a high-risk region (Ford et al. 2016). Due to a few variables, South Asians are particularly vulnerable as the effects of climate change intensify. High population densities, vast regions near sea level (especially in the Maldives and coastal Bangladesh, including the Sundarbans, the delta that stretches into India), and agricultural reliance on major rivers fed by Himalayan glaciers that are susceptible to melting are some examples of the physical and human geography of the region. High rates of poverty and income inequality, as well as dysfunctional national institutions and governance frameworks, all contribute to this vulnerability. It is crucial to acknowledge that effective governance and widespread collaboration may lower the human cost of the rising disasters predicted by climate change. Responses to some calamities have significantly improved in South Asia. For instance, Cyclone Bhola, a category 3 storm (the greatest category, which is a measure of maximum wind speed, is 5) killed at least 300,000 people in Bangladesh in 1970 (Cash et al. 2013), however, less than 3500 people died in 2007 as a result of category 4 Cyclone Sidr (Paul 2010). However, as seen by Pakistan's tardy and unfair reaction to the 2005 earthquake, other governing systems are still lacking (Yasir 2009). Therefore, unless prompt corrective action is implemented, climate change and climate-induced severe events will continue to increase the burden of illnesses and the loss of life and property across South Asia. This review emphasizes the crucial significance of increasing adaption strategies to better meet the escalating problems posed by climate change. It emphasizes the need of raising public knowledge and community understanding about the complex and interwoven linkages between climate and health. Communities that promote knowledge and involvement can better plan for and respond to climate-related health threats. Furthermore, the evaluation recommends considerable improvements in the monitoring and tracking of climate change's impact on human well-being. Improved data collection and analysis are critical for formulating informed policies and responses to protect public health in the face of accelerating climate impacts.

18.2 Methods

Free text terms for "Climate Change," "Public Health", "Health Adaptation", and "Future Directions," "South Asia" were used to search the literature in electronic databases: Medical Literature Analysis and Retrieval System Online (MEDLINE) via OVID platform, Excerpta Medica Database (EMBASE) via Ovid platform, PubMed, and Cochrane Library. Search period was from January 1990 to February 2025. In addition, a gray literature search was conducted using the Google search engine platform to identify relevant articles not captured through database searches to perform this narrative literature review.

18.3 Results and Discussion

18.3.1 Readiness of the Health Systems

The provision of healthcare services is disrupted by healthcare facilities' susceptibility to harsh weather occurrences, particularly during emergencies (Capon and Hanna 2009; Kais and Islam 2016). Most of the region's nations are ill-equipped to handle the health threats posed by climate change due to their overworked public health systems. Most nations have early warning systems for disease and vector surveillance, as well as emergency preparedness. However, the catastrophe planning process has not sufficiently addressed climate change. A toolbox for climate-resilient and sustainable healthcare facilities has been created by the US Department of Health and Human Services (HHS) (Gardiner et al. 2019). According to the 2016 Paris Agreement, nations must adopt mitigation and adaptation plans (Table 1), with international collaboration to increase funding and capability and to transfer environmentally friendly technology to underdeveloped nations so that sustainable development may occur (Association and Policy 2023). The United Nations Framework Convention on Climate Change (UNFCCC) mandates that nations create and carry out national adaptation plans (NAPs) in compliance with its requirements (Dawson and Spannagle 2008). Identifying medium- and long-term adaptation needs and creating and implementing suitable efforts are the goals of these programs. Without a clear connection between national climate change objectives and current disaster risk reduction, agricultural, health, and other pertinent policies, adaptation efforts in South Asia have been dispersed. Identifying medium- and long-term adaptation needs and creating and implementing suitable efforts are the goals of these programs. WHO-SEARO, the WHO Regional Office for South East Asia (Vélez et al. 2022) evaluated the national health policies of the member states and discovered that, despite the fact that health is a component of their climate action plans, the integration of climate change into health policy is either minimal or nonexistent (Table 2) (Sen et al. 2017). This results from a conceptual ignorance about the harmful impacts of climate change on human health. Countries are required by the Paris Agreement to track and assess the results of adaptation efforts (Dhimal et al. 2017). However, there is no monitoring or assessment mechanism in place in any of the South Asian nations. Limiting issues have been recognized as capability, resource limitations, and information deficiencies (Nassera 2017). To address the root causes of vulnerability, the UN framework convention mandates that national adaptation plans involve all facets of society. The degree to which nations in the area address social inequality in their action plans varies. Issues including forest degradation, the impact of dams, the loss of natural resources, and urban planning and transportation that directly impact poor and vulnerable communities are not given enough attention (Byravan and Rajan 2012). The fundamental principle of India's strategy is to safeguard the weaker and impoverished segments of society (Ghosh 2009). Since the health impacts of climate change are not gender neutral and most South Asian nations have patriarchal social systems, incorporating a gender-sensitive lens into adaptation planning would help address current disparities. Building community resilience to address these issues has received more attention in recent years.

Note: Extreme weather events and health problems brought on by climate change are threats to South Asia. Priority must be given to adaptation tactics that have a strong health focus. It is necessary to promote community involvement and decentralized efforts to mitigate the consequences of climate change

Planning for and recovering from climate catastrophes will need an understanding of how communities use their resources and recover from the harm caused by climate hazards. For instance, the Maldives' geographic features make it naturally susceptible

Table 1 Strategies for addressing climate change include adaptation and mitigation measures

Adaptation and mitigation strategies	1	The goal of mitigation measures is to reduce greenhouse gas emissions, which can be achieved by using fewer fossil fuels
	2	Examples include avoiding personal automobiles by choosing to walk, bike, or take public transportation
	3	Public health measures to address present and projected climate change concerns are part of adaptation strategies
	4	These include short-term solutions like enhanced infectious disease surveillance and early warning systems for weather changes
	5	Redesigning metropolitan areas to resist catastrophic occurrences and using novel farming techniques are examples of longer-term measures

Source (Sen et al. 2017), (VijayaVenkataRaman et al. 2012)

Note: Extreme weather events and health problems brought on by climate change are threats to South Asia. Priority must be given to adaptation tactics that have a strong health focus. It is necessary to promote community involvement and decentralized efforts to mitigate the consequences of climate change

Table 2 Incorporating health into national climate change adaptation plans in South Asia

Country	Steps taken–action
Bangladesh (Mani and Wang 2014)	The country's climate change action plan contains both adaptation and mitigation methods, although they are rarely integrated into health programs. At the intersection of climate and health, efforts include research, education, and community-based activities.
India (Dogra et al. 2019; Tewary et al. 2021)	The Ministry of Health and Family Welfare's national action plan on climate change and health aims to increase awareness of climate change's health impacts, strengthen the health system's capacity to manage climate-sensitive diseases, assist states in assessing and addressing health vulnerabilities, foster partnerships to integrate health into the climate agenda, strengthen monitoring and research, and close gaps in evidence-based health policies.
Sri Lanka (Kelkar and Bhadwal 2007)	The national plan lists health as one of the nine sectors most susceptible to climate change. Climate change and vector control research is now undertaken, and climate change issues have been included into nutrition policy. Public health professionals also take part in training and awareness activities.
Maldives (Kelkar and Bhadwal 2007)	Efforts are underway to strengthen institutional and technical capacity for addressing climate change and health, but a comprehensive action plan has yet to be produced. A pilot initiative for low-emission, climate-resilient development is being launched to enhance local planning and service delivery, with a focus on community ownership and long-term sustainability.
Bhutan (Climate change adaptation to protect human health. Bhutan. Geneva: World Health Organization 2015; Tandlick et al. 2018)	Health is a major emphasis of the national adaptation strategy, both as a standalone priority and as a cross-sectoral subject. The health sector's adaptation aims include guaranteeing access to safe drinking water, controlling vector-borne infections, strengthening disaster readiness, increasing surveillance, expanding climate and health research capacity, and addressing nutritional consequences. However, prices and timetables remain unclear.
Nepal (Bowen and Ebi 2017; Organization 2017)	The national climate change policy acknowledges public health as a particularly vulnerable sector, encouraging cross-sector collaboration and incorporating climate change concerns into all relevant sector strategies.

(continued)

Table 2 (continued)

Country	Steps taken–action
Pakistan (Kelkar and Bhadwal 2007)	Climate change strategies will assess health vulnerabilities and build community capacity. National health plans will include measures to address the effects of climate change on health. To protect public health, stakeholders will be educated and trained on climate-health concerns, as well as increased readiness for severe events, disease monitoring, and forecasting systems.
Afghanistan (Kelkar and Bhadwal 2007)	The national action plan highlights health as a vulnerable sector and underlines the importance of incorporating climate change issues into the health system. This involves educating medical personnel and hospitals to deal with rising dangers and diseases, as well as promoting public knowledge of climate-related health concerns so that people may take precautions.

to climate change and other natural disasters. Indigenous local knowledge and practices are tried-and-true and in line with local needs and priorities, and its communities have been coping with climate unpredictability for generations (Boon et al. 2012). Prioritizing climate change adaptation measures in the health agenda is essential. Public health and welfare are now not given enough attention in the regional climate conversation. Informing the public about the negative effects of climate change on health is a major responsibility of the health sector. Countries should first concentrate on creating national adaptation strategies and setting up processes for monitoring and evaluation that are in line with international standards. For cross-sectoral initiatives to be successful and long-lasting, strong political commitment is required. Furthermore, there should be a concerted effort to advance scientific and public health studies on climate change and associated illnesses. Strategies for adapting to climate change must be given top priority in the health agenda. Currently, population health and welfare are not given enough attention in the region's climate narrative. Raising public awareness of the negative health implications of climate change is a major responsibility of the health sector. Establishing national adaptation strategies and monitoring and evaluation mechanisms in accordance with international standards should be the top priority for nations. For cross-sector effort to be sustained and coordinated, political will is necessary. It is necessary to support fundamental scientific and public health studies on climate change and related illnesses. Programmatic approaches by themselves are not enough to increase climate change resilience. The social vulnerabilities of local communities must be considered in any efforts to bridge the gaps in readiness to handle the health hazards associated with climate change. Local, national, and regional responses to climate change should be informed by the lived experiences and knowledge of these groups, and representatives of these communities must participate in the decision-making process when developing the region's adaptation strategy. The following are important suggestions for adaptation planning to address vulnerabilities: (i) Evaluate the vulnerabilities, needs, and capacities of stakeholders; (ii) Focus on the most vulnerable and impoverished groups, particularly women and marginalized communities; (iii) Involve diverse stakeholders to bridge the gap between traditional and local knowledge and scientific knowledge; (v) Address both the immediate and long-term effects of climate change; and (vi) Locally inclusive bottom-up approaches should complement the top-down mobilization of resources.

18.3.2 Climate Change and Health-Related Initiatives

For over a decade, the World Health Organization (WHO) has played a central role in supporting and guiding Member states in responding to the health impacts of climate change. Within the WHO South-East Asia Region, numerous regional-, national-, and high-level meetings have been convened and are expected to continue. Because the health effects of climate change are highly context-specific, considerable attention has been given to country-level factors and priorities. Several countries, including Bangladesh (Mani and Wang 2014) Bhutan ("Climate change adaptation to protect human health. Bhutan. Geneva: World Health Organization" 2015; Tandlick et al. 2018) Nepal (Dhimal et al. 2017), Indonesia ("Indonesia Climate Change Sectoral Roadmap–ICCSR") and India (Dogra et al. 2019; Tewary et al. 2021) have undertaken health vulnerability and adaptation assessments. Regional and national training programs on this subject have been implemented. In Bhutan, the health adaption to climate change initiative, carried out from 2010 to 2015, resulted in a sustained focus on the health impacts of climate change ("Climate change adaptation to protect human health. Bhutan. Geneva: World Health Organization" 2015; Tandlick et al. 2018). In 2013, Bangladesh and Nepal initiated climate change adaptation programs with a particular focus on strengthening the resilience of water and sanitation systems. In addition, Bangladesh, Bhutan, Nepal, the Maldives, and Timor-Leste have developed National Adaptation Plans of Action (NAPAs) and submitted those to the secretariat of the United Nations Framework Convention on Climate Change (Tu'anga and Islands 2011; "United Nations Framework Convention on Climate Change. NAPAs received by the secretariat,"). NAPAs provide least developed countries, as designated by the United Nations, with a framework to identify and prioritize actions that respond to their most urgent and immediate climate change adaptation needs, particularly where delays would increase future vulnerability or costs. In parallel, all countries have developed national plans or strategic frameworks to address the health impacts associated with climate change.

Significant progress has been made throughout the South Asian region because of the WHO's persistent efforts over the past ten years to advise and assist member nations in addressing the health effects of climate change. The increased commitment to addressing these issues is shown in country-specific programs including health vulnerability and adaptation evaluations in Bangladesh, Bhutan, Nepal, and India. A proactive approach to identifying and addressing the most pressing health concerns associated with climate change is demonstrated by the creation of National Adaptation Plans of Action (NAPAs) by several countries. It is crucial to keep building on these successes going forward by fostering regional collaboration, bolstering national health adaption plans, and making sure that climate change continues to be a top priority in the medical field.

18.3.3 Understanding Perspectives: Policy Alignment and Structuring

It is critical to improve our understanding of how communities in impacted areas react to the loss of land or livelihoods. Their responses may include migration, either within their nation or across borders, which might put extra demand on already overburdened authorities (Menne and Bertollini 2005). In circumstances of livelihood loss, it is critical to determine whether adequate alternative and lawful revenue sources are accessible, as a lack of such choices may raise vulnerability to serious crime or paid terrorism (Chermak et al. 2025). This knowledge is critical for incorporating climate change risks into national security plans and risk registers, which will guide priorities for adaptation and mitigation efforts. Identifying these behaviors and related hazards is critical for incorporating climate change threats into national security plans and risk assessments (Capon and Hanna 2009). This understanding should help to determine the priority for adaptation and mitigation activities. Climate change inclusion (Asaaga et al. 2024) as a significant concern in national public health strategies is a clear policy lever for promoting appropriate and effective responses to climate change-related health problems. Unless the health risks posed by climate change are clearly articulated in policies, it is difficult to (i) develop appropriate responses; (ii) gain support from partner ministries (such as water, agriculture, and so on) for climate change and health outcomes activities, the majority of which are inherently cross-sectoral; and (iii) secure funding for climate change and health outcomes activities. Climate change issues are frequently addressed indirectly in public health programs, such as through disaster risk management and food security measures; nonetheless, it is critical that these concerns be connected (whenever feasible) to weather and expected climate change. If this is not obvious, many chances to coordinate and work with such initiatives to ensure policies and programmes promote health and well-being will be overlooked, as would efforts to seek cooperative adaptation financing. There is an opportunity here to promote and support nations in exploring alternative and culturally relevant advocacy channels to achieve policy integration. Monitoring, evaluation, and learning indicators (traditionally referred to as M&E) are required to assess and track over time the ability of health systems to plan for and effectively respond to the health hazards associated with climate change. M&E indicators can also help identify best practices for replication and scaling up, major roadblocks to development, and enabling environments. This information can aid improve understanding of the historical success of certain activities and guarantee that future adaption actions are properly developed and conducted. Health systems have a long history of tracking health outcome-specific morbidity and death and determining which individuals, groups, and regions are most vulnerable to an exposure. What distinguishes climate change is that it is expected to alter the location and timing of climate-sensitive health consequences. Surveillance systems may need to be updated to gather data in new areas and at different times of the year. In some nations, surveillance systems may need to start gathering information on new or re-emerging dangers. This preventive approach will assist to avoid epidemics before they begin. In addition to climate-sensitive health outcome measures, indicators are required to evaluate resilience to climate change-related health hazards. It would be beneficial to create a uniform set of indicators for countries in the WHO South Asia Region to track and quantify the efficacy of adaptation policies and initiatives over time. Furthermore, any country may have its own set of indicators for tracking specific concerns.

18.4 Responses and Recommendations

Numerous suggestions have been put up in response to the actual and possible health effects of climate change. These include lowering carbon emissions, doing further research to support the health effects of climate change, educating policymakers, establishing efficient disease surveillance and control systems, and being prepared for disasters, among other things (Costello et al. 2009).

The WHO Regional Framework for Action highlights the following objectives and action points for governments (Bergstrom et al. 2013; Biswas and Rahman 2023; Costello et al. 2009):

18.4.1 Raise Awareness about the Health Impacts of Climate Change

South Asia is increasingly vulnerable to the health impacts of climate change, with rising temperatures and extreme weather events contributing to the spread of diseases and worsening air quality. Millions across the region are already facing heightened risks of heatstroke, respiratory illnesses, and water-borne infections.

(a) Research and share findings on how climate change affects health, and use this knowledge to help businesses and individuals adopt healthier, climate-friendly behaviors.
(b) Build institutional capacity and secure political commitment to achieve climate adaptation and mitigation goals.

(c) National working groups, NGOs, and civil society should collaborate across sectors, regions, and disciplines to develop integrated climate adaptation and mitigation strategies.
(d) Develop training programs and materials to help communities, health workers, and media partners understand and act on climate-related health risks.

18.4.2 Enhance the Ability of Health Systems to Safeguard Against Climate-Related Risks while Significantly Lowering their Greenhouse Gas (GHG) Emissions

As climate change intensifies, it is crucial to strengthen health systems to withstand climate-related threats while also minimizing their own environmental footprint. Building climate-resilient and low-carbon health systems is essential to protect public health now and in the future.

(a) Develop and implement national health action plans that align with existing climate adaptation and mitigation strategies.
(b) Create integrated methods to control climate-sensitive health outcomes, such as integrated vector management and health risk management of catastrophes, by integrating present and predicted climate change threats into current health policies, plans, and programs.
(c) Protect people from climate-related health risks, countries need stronger health systems. This means improving surveillance, monitoring, response capabilities, and risk communication, while building staff capacity. Key challenges vary by country but often include managing air pollution, ensuring food and water security, and controlling disease-carrying insects and other vectors.
(d) Strengthen health systems and emergency response by improving coordination across sectors. This should include mental health support and build on local knowledge and practices that have proven effective. Build early warning systems that use regional and national climate data to respond quickly to current and potential health threats.
(e) Communities should prioritize their health interventions based on timing and local needs—evaluating both benefits and drawbacks—while building their capacity to manage resources effectively.
(f) Establish dedicated climate teams within national health agencies to lead, monitor, and evaluate health adaptation and mitigation efforts. These teams will ensure health remains a central focus in climate action.
(g) Create initiatives that will significantly lower greenhouse gas emissions in the health sector; this will serve as an example of best practices for other industries.

18.4.3 Incorporate Health Safeguards into Climate Risk Reduction Plans across all Key Sectors

Integrating health safeguards into climate risk reduction plans ensures that public health remains a central focus in climate adaptation and mitigation efforts. Sectors such as agriculture, water, urban planning, and disaster management must consider health impacts in their strategies. A cross-sectoral approach is vital to build resilient communities and prevent climate-related health crises.

(a) Reduce health risks from climate change by embedding climate considerations into key development policies. Priority areas should include disaster preparedness, clean energy access, and sustainable transportation like public transit and bike infrastructure.
(b) Encourage the health sector to take an active role in national communications to the UNFCCC and make health concerns a central component of the negotiating process.
(c) Ensure the national climate team actively collaborates with health experts and stakeholders.

It has been noted that catastrophe recovery operations have received more attention from foreign aid than the building of adaptive capability in developing nations (Mirza 2003). Enhancing surveillance, disaster planning, health-seeking behavior during extreme weather events, and overall sustainable resistance to the health risks and impacts of climate change might all be facilitated by capacity development at the local, regional, and national levels. It is also seen to be crucial that the community be involved in group initiatives to increase adaptive ability (Costello et al. 2009). Impacted groups, including those mentioned in the previous section, must be actively involved, especially women, given their known high vulnerability to climate-sensitive health issues. WHO and UNDP launched a project on public health adaptation to climate change in 2010 in response to mounting evidence and awareness of climate-sensitive health concerns (Berry et al. 2018). NAPA focuses on strengthening adaptive capacity of least developing countries (LDCs) to climate change and its impacts and helps them prioritize their action based upon their needs and interests. It is submitted to the secretariat of UNFCCC, which qualifies the country to apply for funding from the LDC Fund (LDCF) of the Global Environment Facility to implement NAPA. Therefore, NAPA serves as the backbone of a country's response for climate change adaptation. Nepal, for example, submitted its NAPA in late 2010 (Dhimal et al. 2017; "Government of Nepal National Adaptation Programme of Action (NAPA) to Climate Change"). It offers a thorough evaluation of the nation's requirements, difficulties, and methods for adapting to climate change in priority sectors. It addresses topics that have an impact on health, such as public health, agriculture and food security, energy and water resources, climate-related catastrophes, forests and biodiversity, and urban infrastructure and settlements. Each of these key categories has suggested actions. With a budget of USD 15 million, the proposed activities for

Nepal's public health adaptation to climate challenges include: (a) reducing the negative effects of climate change on public health through evidence-based research and piloting; (b) empowering communities through public education for responding to these issues; (c) examining disease outbreaks and emergency response; (d) expanding programs on vector-borne, water-borne, and food-borne diseases and disasters; and (e) bolstering forecasting/early warning and surveillance systems on climate change and health. However, it was projected that the cost of treating additional climate-related ailments in Bangladesh, for example, would be USD 2.8 billion, or 3% of its GDP (Kabir 2016). According to the Asian Development Bank (ADB) there will also be huge economic costs from unmitigated climate change. The Bank estimates that in Southeast Asia, the economic cost of climate change could be as high as 6.7% of GDP per year by 2100. This is more than twice the world average (Organization 2013).

Raising awareness, bolstering health systems, and encouraging cross-sector collaboration are all essential components of comprehensive initiatives to address the health effects of climate change. Countries may lessen the health burden brought on by climate change by improving monitoring systems, investing in disaster preparedness, and incorporating climate hazards into national health plans and policies. Furthermore, sustained capacity-building initiatives at the local, regional, and national levels are crucial, as is active community engagement, especially among vulnerable populations. Initiatives like Nepal's NAPA and the WHO's public health adaptation projects offer as examples of how these efforts not only improve resilience but also open doors for climate-conscious, sustainable health practices that other countries may adopt.

18.5 Takeaways and Future Actions

The WHO South Asia region's member governments have made tremendous progress in addressing the health challenges posed by climate change, particularly in terms of forging partnerships. These changes are positive; thus, it is vital to continue fostering them. Numerous health ministries and their personnel around the area are actively and joyfully trying to address climate change and its health impacts. But there are still certain areas that require further study and assistance, such (a) a deeper comprehension of the relationships between health outcomes and country-level efforts connected to the Sustainable Development Goals (SDGs) (Barbier and Burgess 2017); (b) policy integration and framing; and (c) learning, assessment, and monitoring. An outline of these issues is provided below. The 2015 SDGs agreement created a previously unheard-of worldwide policy opportunity to address human development and global sustainability (Barbier and Burgess 2017; Francisco 2008). The SDGs are different from their predecessor, the Millennium Development Goals, in that they aim to combine environmental sustainability and human development goals into a single global agenda, and they apply to all countries, not only the least developed. Planetary health, or human health and the natural systems that sustain it, may relate to and

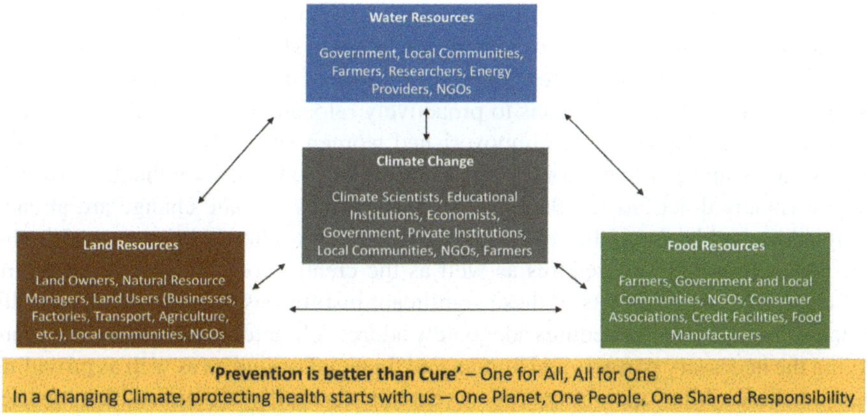

Fig. 1 Building climate resilience through collaborative action

adopt the SDGs' integrated systems approach. Furthermore, if global health initiatives are connected to a process like the SDGs, which have precise goals, metrics, and deadlines, they may be more successful (Bowen and Ebi 2017). Although the WHO South Asian region has made significant strides in addressing the health effects of climate change, more work is required to completely include climate issues into public health initiatives. Building collaborations (Fig. 1), increasing knowledge of the connections between health outcomes, climate change, and the SDGs, as well as strengthening policy integration and monitoring mechanisms, should continue to be the major priorities. Countries may develop more comprehensive and efficient solutions to the health issues brought on by climate change by coordinating health programs with the SDGs' global agenda, which also supports environmental sustainability. The region's resistance to upcoming climate-related health challenges will depend on sustained research, cooperation, and dedication.

18.6 Conclusion

Through a variety of direct and indirect causes, climate change is already negatively impacting people's physical and mental health throughout South Asia. However, as the environmental impact of human activities, especially climate change, increases this century, these consequences are probably going to get much worse. The WHO South Asia Region's member states clearly lack the readiness to handle the health hazards associated with climate change. The significance of sincere cooperation between health and associated ministries, as well as between local and national levels of government and organizational kinds, is at the core of the difficulties. The relationship between climate change and health outcomes necessitates cooperation in order to create the synergies required to provide effective solutions. As usual, the poorest and most vulnerable members of a society suffer the most from the health effects of climate change. These individuals may reside in urban slums with inadequate access to clean water and sanitation, are already at high risk of undernutrition and food insecurity, or lack the financial means to proactively relocate away from disaster-prone areas. The most vulnerable are impoverished women and girls. Because they lack the resources and political clout to act independently, these individuals must rely on governmental actions to adapt. The risks posed by climate change are already being addressed by a number of nonpublic and health-related elements, and inclusive decision-making procedures as well as the creation of policies and programs will capitalize on the efforts of these significant institutions. It is crucial that health policies and planning procedures adequately address climate change, including identifying the necessary partners and human and financial resources as well as providing clear deadlines. Many nations need to increase the capability of their health ministries and other ministries to address the health hazards associated with climate viability and change. Opportunities to increase the health sector's visibility and representation in climate change working groups and committees at the national and regional levels should be encouraged. To help all ministries report on their efforts related to health outcomes and climate change, reporting frameworks are required. In addition to improving population health now, closing these gaps will make people more resilient to climate change. Many South Asian nations are hesitant to limit their own carbon-dependent economic expansion since they are following this example. To provide an alternative to the existing destructive economic paradigm, South Asian nations would be able to follow a path of "green growth" if they were truly world leaders on the problem. While reducing total carbon emissions, several initiatives, such decentralized solar power generation, can promote economic expansion and enhance peoples' quality of life. South Asian countries have a significant chance to take use of their capabilities in technological development research and become global leaders in carbon-neutral development plans.

Acknowledgements On behalf of all authors, the corresponding author want to express gratitude to Samrajnee Sinha Mukherjee, Deb Kumar Mukherjee and Rajdeb Mukhopadhyay, who provided personal as well as moral support and encouragement during this book chapter write-up.

Potential Conflicts of Interest No conflict of interest (None).

Informed Consent As this is a narrative review so, informed consent is not applicable.

Authors Contribution DM and TB designed this work. DM, TB, AG, and AD drafted the manuscript. SB, AP, and AC helped to extract out information from the international policy reports as well as scholarly published works. TB and PS helped to prepare the concise tables and figure. BB, JS, SB, KJ, and VD revised the manuscript. DM and TB accept full responsibility for the conduction of this narrative review and had access to the data, and controlled the decision to publish, the corresponding author DM is the guarantor and attest that all listed authors meet authorship criteria and that no others meeting the criteria have been omitted.

Data Availability Statement The datasets and reports during and/or analyzed during the current review write-up available from the corresponding author on reasonable request.

Funding This work was not supported by any funding.

References

Ariska M, Suhadi S, Irfan M, Iskandar I (2024) Spatio-temporal variations of Indonesian rainfall and their links to indo-Pacific modes. Atmos 15(9):1036

Asaaga FA, Tomude ES, Rickards NJ, Hassall R, Sarkar S, Purse BV (2024) Informing climate-health adaptation options through mapping the needs and potential for integrated climate-driven early warning forecasting systems in South Asia—a scoping review. PLoS One 19(10):e0309757

Association AP (2023) Advancing environmental health and justice: a call for assessment and oversight of health care waste. (APHA Policy Statement Number 20224, Adopted November 2022). A J Environ Occup Health Policy 33(1), 51–59

Bandyopadhyay S, Dasgupta S, Khan ZH, Wheeler D (2021) Spatiotemporal analysis of tropical cyclone landfalls in Northern Bay of Bengal, India and Bangladesh. Asia-Pac J Atmos Sci:1–17

Barbier EB, Burgess JC (2017) The sustainable development goals and the systems approach to sustainability. Econ: Open-Access, Open-Assess E-J 11(1):20170028

Bergstrom R, Caddell R, Chynoweth MW, Ellsworth LM, Henly-Shepard S, Iwashita DK, Lepczyk CA, Miller K (2013) A review of solutions and challenges to addressing human population growth and global climate change. Int J Clim Chang: Impacts Responses 4(3):147

Berry P, Enright PM, Shumake-Guillemot J, Villalobos Prats E, Campbell-Lendrum D (2018) Assessing health vulnerabilities and adaptation to climate change: a review of international progress. Int J Environ Res Public Health 15(12):2626

Biswas RR, Rahman A (2023) Adaptation to climate change: a study on regional climate change adaptation policy and practice framework. J Environ Manag 336:117666

Boon, H., Millar, J., Lake, D., Cottrell, A., & King, D. (2012). Recovery from disaster: resilience, adaptability and perceptions of climate change: national climate change adaptation research facility

Bowen KJ, Ebi KL (2017) Health risks of climate change in the World Health Organization South-East Asia region. WHO South-East Asia J Public health 6(2):3–8

Byravan S, Rajan SJI (2012) An evaluation of India's national actional plan on climate change: Centre for Development Finance (CDF). Humlties Sci

Capon AG, Hanna EG (2009) Climate change: an emerging health issue. New South Wales public health bulletin, vol 20. Csiro Publishing, pp 1–4

Cash RA, Halder SR, Husain M, Islam MS, Mallick FH, May MA, Rahman M, Rahman MA (2013) Reducing the health effect of natural hazards in Bangladesh. Lancet 382(9910):2094–2103

Chermak SM, Freilich JD, Greene-Colozzi E, Klein BR (2025) Open-source research in criminology and criminal justice. Annu Rev Criminol 8

Climate change adaptation to protect human health (2015) Bhutan. World Health Organization, Geneva

Cooley, S., Schoeman, D., Bopp, L., Boyd, P., Donner, S., Kiessling, W., Martinetto P, Ojea E, Racault MF, Rost B, Skern-Mauritzen M. (2023). Oceans and coastal ecosystems and their services

Costello A, Abbas M, Allen A, Ball S, Bell S, Bellamy R, Friel S, Groce N, Johnson A, Kett M, Lee M (2009) Managing the health effects of climate change: lancet and University College London Institute for Global Health Commission. Lancet 373(9676):1693–1733

Dawson B, Spannagle M (2008) United Nations framework convention on climate change (Unfccc). In: The complete guide to climate change. Routledge, pp 392–403

Dewan TH, Extremes C (2015) Societal impacts and vulnerability to floods in Bangladesh and Nepal. Weather Clim Extrem 7:36–42

Dhimal M, Dhimal ML, Pote-Shrestha RR, Groneberg DA, Kuch U (2017) Health-sector responses to address the impacts of climate change in Nepal. WHO South-East Asia J Public Health 6(2):9–14

Dogra N, Kakde V, Taneja P (2019) Decision tool for climate disasters and infectious disease at sub-national level in India: ensuring a paradigm shift in health planning from prevalence to vulnerability. Acta Trop 191:60–68

Ford JD, Cameron L, Rubis J, Maillet M, Nakashima D, Willox AC, Pearce T (2016) Including indigenous knowledge and experience in IPCC assessment reports. Nat Clim Chang 6(4):349–353

Francisco HA (2008) Adaptation to climate change: needs and opportunities in Southeast Asia. ASEAN Econ Bull:7–19

Fressoz, J.-B., & Locher, F. (2024). Chaos in the heavens: the forgotten history of climate change: verso books

Gardiner EP, Herring DD, Fox JF (2019) The US climate resilience toolkit: evidence of progress. Clim Chang 153:477–490

Ghosh P (2009) National action plan on climate change. Prime Minister's Council on Climate Change

Gould S, Rudolph L (2015) Challenges and opportunities for advancing work on climate change and public health. Health 12(12):15649–15672

Government of Nepal National Adaptation Programme of Action (NAPA) to Climate Change. https://lpr.adb.org/sites/default/files/resource/636/nepal-national-adaptation-programme-of-action.pdf.pdf

Haines A, Ebi KL, Smith KR, Woodward A (2014) Health risks of climate change: act now or pay later. Lancet 384(9948):1073–1075

Indonesia Climate Change Sectoral Roadmap - ICCSR

Jafino, B. A., Walsh, B., Rozenberg, J., & Hallegatte, S. (2020). Revised estimates of the impact of climate change on extreme poverty by 2030: the World Bank

Kabir, M. I. (2016). Use of adaptive measures to reduce the impact of climate change on the health sector in Bangladesh

Kais SM, Islam MS (2016) Community capitals as community resilience to climate change: conceptual connections. Int J Environ Res Public Health 13(12):1211

Kelkar U, Bhadwal S (2007) South Asian regional study on climate change impacts and adaptation: implications for human development. Hum Dev Rep 2008:47

Kikstra JS, Nicholls ZR, Smith CJ, Lewis J, Lamboll RD, Byers E, Byers E, Sandstad M, Meinshausen M, Gidden MJ, Rogelj J, Kriegler E (2022) The IPCC Sixth Assessment Report WGIII climate assessment of mitigation pathways: from emissions to global temperatures. Geosci Model Dev 15(24):9075–9109

Mani, M. S., & Wang, L. (2014). Climate change and health impacts: how vulnerable is Bangladesh and what needs to be done?

Masson-Delmotte, V., Zhai, P., Pirani, A., Connors, S. L., Péan, C., Berger, S., Caud N, Chen Y, Goldfarb L, Gomis MI, Huang M (2021). Climate change 2021: the physical science basis. Contribution of working group I to the sixth assessment report of the intergovernmental panel on climate change 2(1), 2391

Menne B, Bertollini R (2005) Health and climate change: a call for action. Br Med J Publ Group 331:1283–1284

Mirza MMQ, Dixit A (2024) Climate change, adaptation, and adaptive governance in water sector in South Asia: the Bangladesh case in the regional context. In: International law, climate change and Bangladesh. Springer, pp 71–92

Mirza MM (2003) Climate change and extreme weather events: can developing countries adapt? Clim Pol 3(3):233–248

Naidoo, I. A., Mc Grenra, D., & Nanthikesan, S. (2022). Thematic evaluation of IFAD'S support for smallholder farmers' adaptation to climate change

Nassera, B. (2017). Insights from national adaptation monitoring and evaluation systems

Neira M (2014) The 2014 WHO conference on health and climate. Bulletin of the World Health Organization, vol 92. SciELO Public Health, pp 546–546

Nirupama A, Mammen JT (2024) Redefining the refugee crisis: examining the status of climate-induced refugees and internally displaced persons with special reference to India. Christ ULJ 13(2):53–88

Organization, W. H. (2013). Post-2015 development agenda

Organization W. H (2017) Status of the development of health national adaptation plan for climate change in South-East Asia

Pachauri, R. K., Gomez-Echeverri, L., & Riahi, K. (2014). Synthesis report: summary for policy makers

Parsons ES, Jowell A, Veidis E, Barry M, Israni ST (2024) Climate change and inequality. Pediatr Res:1–8

Patz JA, Frumkin H, Holloway T, Vimont DJ, Haines AJJ (2014) Climate change: challenges and opportunities for global health. JAMA 312(15):1565–1580

Paul BK (2010) Human injuries caused by Bangladesh's cyclone Sidr: an empirical study. Nat Hazards 54:483–495

Sahu, SK Advanced environmental Science and disaster management. Newredmars Education Pvt Ltd.

Sarker S, Khan A, Mannan MM (2016) Urban population and economic growth: South Asia perspective. Eur J GovMent Econ (EJGE) 5(1):64–75

Sen B, Dhimal M, Latheef AT, Ghosh U (2017) Climate change: health effects and response in South Asia. bmj:359

Singh A, Mishra O, Pandey AP, Bhatla RJE, Science S (2024) Seismic signature of the super cyclone Amphan in bay of Bengal using coastal observatories operating under National Seismological Network of India. Earth and space. Science 11(4):e2023EA003191

Solomon S, Qin D, Manning M, Chen Z, Marquis M, Averyt K, Tignor M, Miller H (2007) IPCC fourth assessment report, vol AR4, p 374

Tandlick, R., Angala, H., Vhiriri, E. P., Ngqwala, N. P., & Srinivas, S. C. (2018). Disaster and health vulnerability assessment for the population of the Kingdom of Bhutan

Tewary A, Mehajan RK, Golechha M, Mavalankar D (2021) Co-developing climate services for resilient public health system in India. Curr Sci 120:1578

Tiwari I, Tilstra M, Campbell SM, Nielsen CC, Hodgins S, Vargas AR, Whitfield K, Sapkota BP, Yamamoto SS (2022) Climate change impacts on the health of south Asian children and women subpopulations-a scoping review. Heliyon 8(10)

Touch V, Tan DK, Cook BR, Li Liu D, Cross R, Tran TA, Utomo A, Yous S, Grunbuhel C, Cowie A (2024) Smallholder farmers' challenges and opportunities: implications for agricultural production, environment and food security. J Environ Manag 370:122536

Tu'anga, T., & Islands, C. (2011). United Nations framework convention for climate change

United Nations framework convention on climate change. NAPAs received by the secretariat

Van Valkengoed AM, Perlaviciute G, Steg LJRA (2024) From believing in climate change to adapting to climate change: the role of risk perception and efficacy beliefs. Risk Anal 44(3):553–565

Vélez C-M, Kapiriri L, Nouvet E, Goold S, Aguilera B, Williams I, Danis M, Essue BM (2022) Examining priority setting in the national COVID-19 pandemic plans: a case study from countries in the WHO-South-East Asia region (WHO-SEARO). Health Policy Open 3:100086

VijayaVenkataRaman S, Iniyan S, Goic R (2012) A review of climate change, mitigation and adaptation. Renew Sust Energ Rev 16(1):878–897

Wijewickrama E, Alam MR, Bajpai D, Divyaveer S, Iyengar A, Kumar V, Qayyum A, Yadav SP, Yadla M, Arruebo S, Bello AK (2024) Capacity for the management of kidney failure in the International Society of Nephrology South Asia region: report from the 2023 ISN global kidney health atlas (ISN-GKHA). Kidney Int Suppl 13(1):123–135

Yasir A (2009) The political economy of disaster vulnerability: a case study of Pakistan earthquake 2005

Zhang B, Wang S, Zscheischler J, Moradkhani H (2023) Higher exposure of poorer people to emerging weather whiplash in a warmer world. Geophys Res Lett 50(21):e2023GL105640

The manufacturer's authorised representative in the EU is Springer Nature Customer Service Centre GmbH, Europaplatz 3, 69115 Heidelberg, Germany. If you have any concerns regarding our products, please contact ProductSafety@springernature.com

Printed and bound by CPI Group (UK) Ltd, Croydon, CR0 4YY
05/10/2026
02228057-0012